T0323355

Sampling Theory

Sampling Theory

For the Ecological and Natural Resource Sciences

DAVID G. HANKIN,
MICHAEL S. MOHR, AND
KEN B. NEWMAN

Great Clarendon Street, Oxford, OX2 6DP,
United Kingdom

Oxford University Press is a department of the University of Oxford.
It furthers the University's objective of excellence in research, scholarship,
and education by publishing worldwide. Oxford is a registered trade mark of
Oxford University Press in the UK and in certain other countries

First Edition published in 2019

Impression: 1

Published in the United States of America by Oxford University Press
198 Madison Avenue, New York, NY 10016, United States of America

British Library Cataloguing in Publication Data

Data available

Library of Congress Control Number: 2019941126

ISBN 978–0–19–881579–2 (Hbk)
ISBN 978–0–19–881580–8 (Pbk)

DOI: 10.1093/oso/9780198815792.001.0001

Printed and bound by
CPI Group (UK) Ltd, Croydon, CR0 4YY

We dedicate our text to the memory and spirit of
Amode Sen (1917–1999) and Scott Overton (1925–2012).

Preface

We three authors have about 120 years of collective experience in research and management in natural resource settings, with a focus on development of survey designs for generating estimates of abundance and other population level metrics, incorporation of such estimates in models of population dynamics, and analysis of population dynamics models, often in the context of harvested fish populations but also for threatened or endangered species of special concern. All of us have also been involved in instruction of statistics classes at the undergraduate and graduate levels. In our collective personal experiences, we have been struck by the obvious and very practical relevance of design-based sampling theory for ecologists and natural resource scientists, but also by the typical absence of formal training in sampling theory by most practicing natural resource and environmental professionals and even by many PhD statisticians. Indeed, one of us (DGH) was not even exposed to the field of sampling theory until Amode Sen (of Sen–Yates–Grundy fame, see Chapter 8) served as a visiting distinguished professor at Humboldt State University for the 1981/82 academic year and taught introductory and advanced sampling theory courses.

Two of us (DGH, MSM) were Amode's "groupies" and we spent many wonderful hours outside the classroom with Amode, discussing sampling theory, famous statisticians who Amode had known personally (especially Gertrude Cox, the first female faculty appointment at North Carolina State and founding editor of Biometrics, among many other achievements), the Indian caste system, enjoying Indian cuisine, and enjoying Amode's sense of delight in people and culture, not just in mathematical formulas. One of us (DGH) found the content of Amode's sampling theory courses to be so interesting and so resource management-relevant, and Amode's frequent presentation of small sample space numerical examples so effective, that he ended up teaching sampling theory at Humboldt State University on a roughly biennial basis from 1983 through 2014. Students originated from a variety of natural resource disciplines (forestry, range, wildlife, fisheries) as well as from biology/ecology and mathematics. It was a pleasure to expose students with such varied backgrounds to the field of sampling theory, to assist them with sampling-related thesis research, and to see what they could do if they were asked to stretch their brains and quantitative skills a bit.

Two of us (DGH, KBN) were also fortunate to have had extensive interactions with Scott Overton, Professor of Statistics at Oregon State University. Scott originally held a joint appointment with Forestry, had keen interest and expertise in theoretical ecology, a special fondness for sampling theory, and had been hoping to develop a sampling theory text of his own. Scott served as advisor for one of us (KBN) in the second year of his Statistics MS Program. Two of us (DGH, KBN) had the pleasure of independently sitting in on Scott's Advanced Sampling Theory course. We both shared an appreciation for Scott's belief that Horvitz–Thompson estimation provided a unifying framework for design-based sampling theory, and we enjoyed Scott's passionate explanations of the important distinctions between "real" finite populations and hypothetical superpopulations, and between design-based and model-based estimation. Scott also helped DGH appreciate that the multi-stage sampling setting allows one to employ an almost bewildering number of sampling strategies and that, perhaps most important, it is completely appropriate to bring everything that one knows about a biological context into the development of an effective sampling strategy.

We have written this sampling theory text with the hope that it may prove of special value to ecologists and natural resource scientists. With a reasonable amount of effort, much of the theory of design-based sampling can be derived and understood by ecologists and natural resource scientists so that when they apply sampling strategies they can "know what they are doing". Call us *old school*—we continue to believe that it is important to understand statistical procedures before they are put into practice. We hope that our text will help scientists to have a clear understanding of what they are doing when they apply simple or sophisticated sampling strategies.

A number of individuals made important contributions to our effective presentation of sampling theory concepts and we wish to give special thanks to these individuals for their assistance. Mohammed Salehi developed the generalized two-stage complete allocation adaptive sampling strategy (Salehi and Seber 2017) at our request and also reviewed our adaptive sampling chapter. Years back, Terry Quinn suggested that it would indeed be possible to provide a worked sample space illustration for adaptive cluster sampling. Tony Olsen reviewed an early draft of our spatially balanced sampling chapter and also responded to our numerous queries concerning GRTS. Trent McDonald reviewed final drafts of the spatially balanced sampling and sampling through time chapters, providing many helpful comments and references. Seth Ricker assisted us in developing an illustrative application of GRTS in a practical, real-life setting. We thank Ian Sherman for encouraging us to develop this text, and Keerthana Sundaramoorthy for her prompt, efficient and helpful advice during the production of the book.

Finally, we wish to acknowledge the generous assistance of the following individuals who provided us with permission to use the fine photographs and artwork presented in our text: Dave Imper (western lily), Mark Allen (Van Duzen River pool:river sequence), Darren Ward (juvenile coho salmon), Thomas Dunklin (mature male coho salmon), Milo Burcham (adult bald eagle on nest with fledgling), Alex Nabaum (number-net illustration), and John Hyde (purse seining for chum salmon). We also extend a special thanks to Shauna Oh for developing the initial cover design concept.

Contents

CHAPTER 1

Introduction

Fig. 1.1 Terrestrial form of the Pacific (coastal) giant salamander, *Dicamptodon tenebrosus*, Redwood National Park, and the original inspiration for this book's cover. Photo credit: D. Hankin.

In this text we attempt to present a rigorous but understandable introduction to the field of sampling theory. Sampling theory concerns itself with development of procedures for random selection of a subset of units, a sample, from a larger finite population, and with how to best use sample data to make scientifically and statistically sound inferences about the population as a whole. The inferences fall into two broad categories: (a) estimation of simple descriptive population parameters, such as means, totals, or proportions, for variables of interest associated with the units of finite populations, and (b) estimation of uncertainty associated with estimated parameter values. Although the targets of estimation are few and simple, estimates of means, totals or proportions see important and often controversial use in management of natural resources and in fundamental ecological research. For example, estimates of total population size and

Sampling Theory: For the Ecological and Natural Resource Sciences. David G. Hankin, Michael S. Mohr, and Ken B. Newman, Oxford University Press (2019). © David G. Hankin, Michael S. Mohr, and Ken B. Newman. DOI: 10.1093/oso/9780198815792.001.0001

associated trends in abundance play key roles in development of harvest policy for exploited species in fisheries and wildlife management settings, and in status reviews and development of recovery strategies for species listed under the Endangered Species Act. Estimates of species abundance or species proportions and associated measures of uncertainty also provide key input values for state variables, such as the true abundances or proportions, in *state-space* models of population dynamics that explicitly separate but also link a *state process* (say, trend in actual abundance of a species) and an *observation process* (process for estimating the value of a state variable) (Newman et al. 2014). Estimates of mean weights of individuals (e.g., of juvenile salmon in a small stream, or of young polar bears foraging off an ice shelf) may provide critically important data for energetic models of growth that may help predict the impacts of climate change on growth and survival of species. A temporal sequence of accurate estimates of mean water quality parameters in a lake may prove key for understanding the seasonal impacts of agricultural runoff and may lead to changes in land use practices. There are thus a wealth of practical and important applications of estimates of simple descriptive finite population parameters. Given the potentially controversial use of such estimates, being able to provide measures of the degree of uncertainty associated with these estimates is critical.

1.1 The design-based paradigm

We stress the classical *design-based* approach to sampling theory in this introductory text. For the design-based approach, a chance randomization scheme is used to select a sample of size n from a finite population consisting of N units. Each unit in the population is recognized to have fixed variable (attribute) values associated with it (e.g., population units may consist of individuals of a given species of tree and each tree, at a given point in time, has fixed diameter at breast height, total height, volume, etc.). Typically, one variable is of primary interest and we refer to this variable as the *target* variable, y. For unit i, we label its unit-specific fixed value as y_i, and we define the population parameters $\mathcal{T}_y = \sum_{i=1}^{N} y_i$ (total) and $\mu_y = \mathcal{T}_y/N$ (mean). We may also take advantage of values of an *auxiliary* variable associated with population units, denoted by x, that can be used to improve accuracy of estimation of target variable population parameters, either through direct incorporation in estimation formulas or indirectly through their influence on the randomization process used to select samples.

In design-based sampling, the randomized procedure(s) used to select sample units determines the *sample space*, the set of all possible samples of size n that can be selected (with or without replacement) from the population of size N. The set of associated sample probabilities that emerge from the randomized selection procedure, not necessarily equal, along with the unit labels that appear in the samples, can be used to calculate first and second order inclusion probabilities—the probabilities that unit i or that units i and j, respectively, appear in a sample of size n selected according to the randomized selection procedure. Associated with each possible sample s is a sample-specific estimate of a population parameter, e.g., $\hat{\mathcal{T}}_y(s)$, calculated by substituting the sample y values (and possibly also x values and/or first order inclusion probabilities) into equations (estimators) used to calculate estimates of target population parameters.

Uncertainty of estimation is measured by the variation in sample estimates over the sample space, variation induced by the interaction of the randomization scheme and estimator with the set of variable values (y, and sometimes also x) associated with population units. If the range of sample estimates is small and the average value of estimates is close to the target population parameter, then an estimator has good accuracy, but if the range of sample estimates is large, or the average value of estimates is far from the target population parameter, then an estimator has poor accuracy. Design-based sampling theorists have developed methods to estimate the variability in sample estimates over the full sample space—the *sampling variance* of an estimator—from just a single selected sample, and they use such estimates to characterize the uncertainty associated with an estimated descriptive population parameter.

We stress this design-based approach in the text for several reasons. First, in many controversial resource management contexts, *objectivity* of estimation is an important virtue of design-based sampling theory (Särndal et al. 1992 p. 21). No model assumptions are made regarding the distribution of the fixed y values over the population units and a chance randomization scheme is used to select sample units. Design-based estimators are therefore *robust* in the same sense that non-parametric statistical methods make no or mild assumptions regarding the distributions of random variables (Brewer and Gregoire 2009) and cannot be faulted by competing researchers or conflicting agencies who might allege deliberate bias resulting from model choice or purposive selection of sample units. Second, from Overton and Stehman (1995): "in finite sampling the populations being sampled and about which inferences are being made are real…and are subject to enumeration and exact description. In contrast, the populations of conventional statistics are hypothetical and represented by mathematical models". Treating the y and x associated with finite population units as fixed values has an undeniable conceptual validity in a finite population. Third, the distribution of x and y over the finite population units will often be poorly represented by any simple probability distribution. For example, asserting that the volume of timber on 1 ha lots follows a normal distribution may be a gross misrepresentation of reality. Fourth, we stress the design-based approach because we believe that it would be difficult to do justice to both design-based estimation and the model-based prediction (see below) approach in a relatively brief text designed for a first course in sampling theory.

We recognize, of course, that the use of models for inference in finite populations is an approach that is consistent with parametric statistics in general, that it has its own logic and validity, and that it may provide an alternative and sound approach for inference. In *model-based prediction* in finite populations, the actual variable values associated with population units are assumed to be realizations of random variables, and the finite population inference problem is to estimate the expected values (i.e., predict the values) of random variables in those $N - n$ units that *did not appear in the sample*, conditioned on the realized values of the random variables that have been observed in the sample of size n. Randomized selection methods used to select sample units are typically not relevant to model-based inference. Predicted values are based on assumed models that may include a dependence of the target variable on one or more auxiliary variables. When the distribution of variable values across units and the relationship between target and auxiliary variables are close to those assumed by models used for prediction, then model-based prediction can generate estimates with errors that may be smaller than those for design-based estimators. However, if the assumed models do not accurately characterize the actual distribution of the target variable and/or the relationship between the target and auxiliary variables, then errors may be large and expected values of model-based predictors may be biased when compared to the fixed population parameters that are the targets of estimation. In model-based prediction, emphasis is put on the value of *balanced* samples, for which the mean values of auxiliary values in the sample are purposely set to equal the mean values in the population (Valliant et al. 2000) to help ensure that model-based prediction is more robust to violations of assumed model structure.

While our emphasis is on design-based inference, we do provide a brief introduction to best linear unbiased (BLU) estimators and model-based prediction of finite population parameters in Chapter 7. Our intention is to provide readers with an appreciation of the fundamentally different perspectives that the design-based and model-based approaches take to estimation of finite population parameters. Both approaches rest on the same basic mathematical foundations of probability and statistics that we summarize in Appendix A. For readers desiring to learn more about model-based prediction, we recommend Valliant et al. (2000), Chambers and Clark (2012), and also Brewer's (2002) text which considers the benefits of combining design-based and model-based approaches.

1.2 Text content and orientation

Material is presented in this text at a level appropriate for and accessible to undergraduate seniors, beginning graduate students (not enrolled in a statistics program), and natural resources or environmental/ecological sciences professionals. Most material presented in the text can be fully understood and appreciated with modest mathematics and statistics training: considerable facility in algebra, including experience with multiple summation notation; at least one semester of calculus, including some experience with partial differentiation; an introductory course in statistical methods, along with an additional unspecified statistics class (including an exposure to analysis of variance); and some prior exposure to, though not necessarily programming proficiency in, R (R Core Team 2018), the statistical/programming/graphics language and environment used to generate the numerical results and figures presented in this text. Depth of appreciation for the subject matter generally would be greatly enhanced by formal background in probability theory, but we do not assume that readers have such background. Our intention is to foster interest in readers to develop or hone the necessary mathematical skills to follow through derivations of important results, especially if these skills are not in their formal backgrounds or are "rusty with lack of use", so that users of this text will be able to develop and successfully apply their own sampling strategies in their own unique application settings.

We believe that the most natural and successful order for working through the text is to directly follow the order of chapters as they appear in the text. Chapter 2 provides a very gentle and largely non-quantitative introduction to important sampling theory definitions and concepts (sampling frame, sampling design, sampling strategy) and properties of estimators (expected value, bias, sampling variance, mean square error) when sampling from finite populations, including justification of randomized selection of units as compared to judgment or purposive selection of units. Important concepts or terms appear in boldface type at first usage to emphasize their importance to readers. Sophistication of presentation and complexity of mathematical notation and sampling strategies gradually increases through the next several chapters. Chapter 3 provides an introduction to equal probability selection methods, with and without replacement, and Chapters 4–6 illustrate application of these equal probability selection methods in the contexts of systematic, stratified, and equal size cluster sampling, respectively. Chapter 7 introduces the explicit use of auxiliary variable information in estimators (ratio and regression estimation), while also providing a brief introduction to BLU estimators and model-based prediction of finite population parameters. Chapter 8 introduces unequal probability selection methods, with and without replacement, and associated estimators (notably the famous Hansen–Hurwitz and Horvitz–Thompson estimators). Selection of units with unequal probabilities, with probabilities based on auxiliary variable values, represents an alternative way to take advantage of auxiliary variable information. Material covered through Chapter 8 is all devoted to sampling from what we call *simple frames*, where the samples are selected directly from the individual units in the populations. (Stratified sampling is a special case where a simple frame is used for independent sample selection from within disjoint sets or strata of population units.) Sampling strategies with complex frames are considered in the following two chapters. Chapter 9 is devoted primarily to two-stage implementation of multi-stage sampling strategies, but we also provide the general framework for many designs with more than two stages of sampling. Chapter 10 is devoted to two-phase implementation of multi-phase sampling. The material presented in the chapters thus far mentioned comprises the "basic stuff" of classical design-based sampling theory and is covered, in various ways, in most sampling theory texts available today.

1.3 What distinguishes this text?

We hope to distinguish our text from previously published sampling theory texts in several different ways. First, we have tried to make the material presented in the text accessible to the reader who may not be "homozygous for the math gene". Toward that end, though we do provide derivations of many important sampling theory results, we also provide small sample space illustrations of the behavior of most of the sampling strategies (selection method and estimator applied to a sampling frame) that we consider. This allows readers to see numerically, rather than imagine in thought, what the sample space actually is. For individuals "heterozygous (or recessive) for the math gene", the small sample space examples often make abstract concepts like sampling variance come alive and have tangible and very real meaning. Sample space examples have been used relatively rarely in existing texts, but we acknowledge their previous effective use in the very small but very useful and repeatedly reissued (1962, 1964, 1968, 1976, 1984) text by Stuart (1984), *The Ideas of Sampling*. Sample space examples also allow us to avoid presentation of numerous fictitious examples for observed data from a single sample; the sample spaces provide an abundance of implicit calculations that can be worked through and confirmed by students and readers if they wish.

Second, we have presented some material that has not frequently or routinely been covered in existing texts. Appendix A provides a review of the mathematical foundations of design-based sampling theory including counting techniques, basic principles of probability theory, properties of discrete random variables, key discrete probability distributions and their clear links to design-based or probability sampling, as well as an overview of the use of Lagrange multipliers and the delta method. Chapter 11 covers adaptive sampling, a topic covered in Thompson (1992, 2002, 2012) and Thompson and Seber (1996) but rarely in other sampling theory texts. More unusual is our coverage of spatially balanced sampling methods [Generalized Random Tesselation Stratified sampling (GRTS) and Balanced Acceptance Sampling (BAS)] in Chapter 12, and our coverage of designs and estimators for sampling through time (including rotating panels and dual frame sampling) in Chapter 13.

Third, we have tried to make extensive use of graphs and figures to convey important sampling theory concepts and especially to help readers understand how the performances (sampling variance or net relative efficiency) of competing sampling strategies may be compared to one another. We also use graphs to portray the shapes of the sampling distributions of design-based estimators, which are often not approximately normal until $n > 20$ or so.

Fourth, we have tried to develop and use a notation that is more consistent than in many previously published sampling theory texts. We use Greek letters to represent population parameters ($\mathcal{T}$ = total, μ = mean, π = proportion, and σ^2 = finite population variance, with a divisor of $N-1$); we always use a caret or *hat* to indicate an estimator or estimate (as in $\hat{\mu}_y$ rather than $\bar{y}$); we use lowercase italic y and x to denote the target and auxiliary population variables, respectively, and uppercase letters to denote random variables; we use uppercase italic S to indicate the population units selected as the random outcome of a sample selection process (*sampling experiment*, see Section A.6.1), and $\hat{\theta} = \sum_{i \in S} f(y_i, x_i)$ as a generic representation of a design-based estimator; we use lowercase italic s to identify the set of units in a particular realized sample of size n and $\sum_{i \in s} y_i$ to indicate the sum of the y values in this sample, thereby avoiding the *re-labeling* of population units implicit in the more commonly used $\sum_{i=1}^{n} y_i$; and we use combinations of uppercase and lowercase letters to define properties relating two finite population variables and properties of estimators (e.g., $\text{Cov}(x, y)$ for covariance, R for the ratio of the means of y and x, $V(\hat{\theta})$ for sampling variance of an estimator $\hat{\theta}$).

Fifth, in the text we provide occasional brief reference to simple R functions or packages that should prove useful for working through problems or for implementing computer-intensive sample selection methods. We do not, however, as a rule present code for what we believe to be relatively simple calculations for which students and other readers should learn to write their own code. Instead, we provide web-based access (www.oup.co.uk/companion/hankin) to either existing R packages or code that we have written ourselves for tasks that we believe to be beyond the level of programming proficiency and/or statistical sophistication that one can reasonably expect from students or professionals using this text as a first course in sampling theory. Examples of such R programs include several methods of selection of unequal probability without replacement samples and for calculation of associated first and second order inclusion probabilities, and programs for implementing the spatially balanced sampling methods described in Chapter 12.

The natural resources/environmental/ecological orientation of the text will be most evident from the problem exercises associated with those chapters for which we believe reasonable "classroom" problems can be developed. Many of these problems have roots in real-world natural resource settings in which the authors have worked. We hope that these problem exercises will help students and readers learn how to answer what is often the most difficult question: "How can we apply these ideas in practice?" We also have frequent references to natural resources/environmental settings/applications within the text of chapters and in the *Chapter comments* section at the conclusion of each chapter, and we begin each chapter with a relevant photograph, usually illustrating a natural world setting where sampling theory can address a topic of interest.

1.4 Recommendations for instructors

If the text were used to teach a one semester sampling theory class to a group consisting of senior undergraduates and beginning graduate students, we recommend using Appendix A, the review of mathematical foundations, as a *reference* and that it only be worked through in detail if students in the course have unusually strong backgrounds and interests in mathematics and probability theory. Instead, we recommend that when concepts or techniques covered in Appendix A are used or noted in other chapters, then the instructor should refer students to Appendix A and perhaps cover a few of the relevant concepts or techniques at that time. It is important that students do not think "this is a math class". Material through Chapter 11 can be easily covered in a one semester course. Content of Chapters 12 and 13 should at least be mentioned and practical importance discussed, but could be covered in some detail if material in other chapters was covered very rapidly.

We highly recommend addition of a weekly (two hour) laboratory/practicum session in addition to lecture/discussion presentation of the material presented in this text. These lab sessions can be used to illustrate computer implementation and evaluation of sampling strategies (using R), especially for unequal probability and spatially balanced selection methods, as well as to illustrate use of R to solve assigned problems (thereby assisting students in developing programming and graphics skills). Lab sessions can also be used as effective class exercises to contrast the performance of simple random sampling and judgment sampling (see agate population exercise described in Section 2.6), and for student presentations of their experiences in development and execution of sampling strategies for estimating simple parameters such as the total number of preserved specimens in a university's ichthyology (fish) collection; the average cost (and average quality rating) of a slice of pizza sold within the local county; the proportion of plants in an herbarium that are in bloom at the time of a survey; the fraction of trees of a given species in a well-defined plot that have diameter at breast height exceeding 1 m; the number of books in a library; the numbers of pieces of trash (carefully defined) per unit area or length in several well-defined areas (neighborhoods) of a small town or city; and numerous other ideas that students will dream up if they are given enough freedom. There is no exercise that is a more effective practical learning experience than identifying a finite population (or study area) of interest, picking a well-identified target of estimation, thinking through how measurements can be made, what sampling frames might make most sense, what selection methods might prove most effective, what estimators might be used, and then executing a small-scale survey, ideally using two alternative sampling strategies, and reporting back to classmates on their experiences.

Finally, after working through this text or taking a class using this text, we hope that all readers or students will develop a deep appreciation for the following joke about statisticians in a wildlife setting.

Three statisticians set out on a bow hunt for deer and hope to make a successful kill of a buck. They are fortunate to locate a fine buck at a decent distance. Statistician 1 lets loose his arrow, but it misses the buck by exactly 6.1 feet to the east of the buck's heart. Statistician 2 lets loose his arrow, but it misses the buck by exactly 6.1 feet to the west of the buck's heart. In excitement, Statistician 3 exclaims "We got him!"[1]

1.5 Sampling theory: A brief history

Sampling theory is a narrowly specialized but important area of the larger body of statistical theory. In Kendall and Stuart's three volume *Advanced theory of statistics* (Kendall and Stuart 1977, 1979, Kendall et al. 1983), the subject of sampling theory occupied just

---

[1] We thank Mathew Krachey, former sampling theory student and now Ph.D. statistician, for passing this joke on to us many years back.

84 of 2,000 pages, about 4%. But work in this small area of statistical theory has remained very active. Volume 6 (Sampling) of the Handbook of Statistics (Krishnaiah and Rao 1988) was 594 pages long, and Volumes 29A and 29B of the Handbook of Statistics (Pfeffermann and Rao 2009a,b) contain 1,340 pages combined devoted to sampling theory.

Numerous published articles have reviewed the roots of sampling theory. Bellhouse (1988), Hansen (1987), and Hansen et al. (1985), in particular, provide excellent reviews of the history and development of the design-based or probability sampling theory that is stressed in this text, along with references for many other previously published histories of sampling theory. Bellhouse (1988) argues persuasively that the development of sampling theory can be seen to have had several paradigm shifts (Kuhn 1970) which were attributable to novel new sampling ideas but also to aggressive promotion of these ideas by their proponents. In the brief description of the historical development of design-based sampling theory that follows, we rely heavily on Bellhouse (1988).

The historical origins of sampling theory can be traced to desires of government agencies for knowledge of population size, production of agricultural crops, average income, and other descriptive population parameters that are important for description and management of a country's economy and social structure. Kiaer (1897, as cited in Bellhouse 1988), Director of the Norwegian Central Bureau of Statistics, was a vigorous early advocate of *representative sampling* which was an alternative to the prohibitively expensive or logistically unfeasible complete enumerations that were judged necessary prior to the acceptance of sampling as an alternative way to obtain (acceptably accurate) estimates of population parameters. At the time, a representative sample was thought to be a sample of units that was an approximate *miniature* of the population, a "correct representation of the whole". (Kruskal and Mosteller (1979a,b,c, 1980) provide historical reviews of the rich, varied, often incorrect and inconsistent meanings that the term *representative sampling* has had in scientific, non-scientific, and statistical settings.) By about 1925, the basic idea of representative sampling (instead of complete enumeration) had been widely accepted, but there had not yet been any agreement that randomized rather than purposive selection was required to achieve a representative sample. From about 1925 to 1935, the British statistician Arthur Bowley became a staunch advocate for equal probability randomized selection of samples. Bellhouse (1988) provides direct quotations from Bowley that suggest that he understood the advantages of systematic sampling as compared to simple random sampling (SRS), but Bellhouse also points out that Bowley's (1926) monograph presented theoretical justifications and suggested methods for calculating uncertainty of estimation for both randomized and purposive selection of units.

All accounts of the history of design-based sampling theory give Neyman's (1934) paper preeminent importance in the paradigm shift that lead to widespread adoption of the randomized selection, design-based approach to generating sample estimates from finite populations. First, Neyman argued (and graphically illustrated) that a then recent purposive large-scale Italian survey had invoked tenuous assumptions of linearity in relations among some variables, had been purposively balanced on some variables, but had generated sample estimates that were inconsistent with available census counts for other variables. Second, Neyman laid out the theory for optimal allocation (ignoring costs) in stratified sampling (*Neyman allocation*, Chapter 5). In optimal allocation, units from different strata typically have different probabilities of inclusion in the overall sample, the first time this idea had been shown to be optimal. Third, Neyman characterized the notion of a *representative sample* as one that would allow estimation of accuracy "irrespectively of the unknown properties of the population studied", based on the behavior of an estimator over all possible randomized sample selections. Fourth, he defined the concept of confidence intervals, based on the notion of repeated sampling within the design-based framework. A few years later, Neyman (1938) developed the theory of double sampling, including the first joint use of cost and variance functions to optimize performance of a given sampling strategy (frame + design + estimator).

The paradigm shift to widespread adoption of randomized design-based sampling strategies for estimation of descriptive population parameters generated an explosive proliferation of important articles and texts. Particularly important were the papers concerning estimation based on unequal probability with and without replacement selection by Hansen and Hurwitz (1943) and Horvitz and Thompson (1952), respectively. Among the many fine sampling theory texts that were produced prior to 1970, we highlight those written by Deming (1950), Cochran (1953), Hansen et al. (1953a,b), Yates (1960), Murthy (1967), and Raj (1968). We stop this listing of texts at 1970 because that year may date another paradigm shift among a minority of statisticians concerned with estimation in finite populations.

Bellhouse (1988) credits Godambe (1955) with generating a paradigm shift that stimulated interest in the theoretical foundations of sampling theory (e.g., Cassel et al. 1977). Godambe's (1955) proof of the non-existence of a unique minimum variance design-based unbiased estimator of the finite population mean stimulated exploration of the use of models in finite population inference. Royall (1970) advanced the *model-based prediction* approach for estimation of finite population parameters. He showed that, if there were a linear model relation between auxiliary and target variables, passing through the origin with variation increasing with x, then the ratio estimator was the best model-based predictor of the finite population total. According to this perspective, randomized selection of units was irrelevant, or even foolish. Indeed, minimum variance of this estimator, given the assumed model, was achieved if the sample were purposively selected to include those n units with largest auxiliary variable values. Given this sample of units, predictions would need to be made only for units with a smaller value of x, for which errors would be smaller because, under the model, variation is less at smaller values of x. (Brewer (1963) had earlier presented similar findings for model-based ratio estimation in finite populations, including the finding that accuracy would be best if the sample contained the units with the largest values of x, but his work was not cited by Royall (1970) and may not have been aggressively promoted.)

Royall's (1970) paper and his subsequent vigorous advocacy for model-based prediction in finite populations generated heated exchanges between practicing proponents of design-based sampling theory and model-based prediction advocates (see, e.g., the paper, comment, rejoinder by Hansen et al. (1983a), Royall (1983), and Hansen et al. (1983b), respectively) for at least the next two to three decades. At least one prominent sampling theorist was first persuaded by the new model-based prediction paradigm but then, on further reflection, proclaimed himself a staunch design-based advocate [contrast Smith (1994) with Smith (1976)], much to the chagrin of Royall (Royall 1994). Publication of Särndal et al.'s (1992) text on model-assisted sampling represents one very important response to the model-based prediction paradigm. This text (and many earlier published papers by Särndal and colleagues) illustrates how models might be used to guide development and adoption of competing sampling strategies, but then estimation remains firmly within the design-based framework. Särndal (2010) provides a recent review of the role of models in sampling theory. Among other things, he notes that the notion of purposive selection of sample units that are *balanced* on a set of auxiliary variables, to achieve *model-robust* estimation within the model-based prediction framework (Valliant et al. 2000), has a new parallel in the design-based setting—the *cube method* (Deville and Tillé 2004, Tillé 2011)—where randomization is used to achieve balancing on auxiliary variables but inference remains within the design-based framework. Thus, one might argue that sampling theory has come *full circle* from the purposively balanced samples that had been recommended prior to Neyman's (1934) landmark paper, but thereafter rejected, to a contemporary recognition in both model-based prediction and design-based estimation that balancing is desirable in both designed-based and model-based settings. Current methods for selecting spatially balanced samples (Chapter 12) are consistent with this notion of balancing used within a design-based framework.

Finally, it is important to distinguish between sample surveys that have a strictly *descriptive* objective—estimation of descriptive population parameters—as compared to those that have an *analytic* or *causal* objective (Hansen 1987). In analytic or causal surveys, primary interest lies not in estimation of the descriptive population parameters themselves, but instead on analysis of the possible *relationships* among survey variables. There is no alternative to conjecture of hypothetical model relations in the context of analytic surveys, but statistical analysis of these relations should acknowledge the probability structure of the sampling design under which the survey data were collected (Korn and Graubard 1999, Lohr 2010 Chapter 10).

CHAPTER 2

Basic concepts

Fig. 2.1 An agate (chalcedony) population used for the judgment sampling exercise described in this chapter. Population also includes agatized petrified wood and jasper collected on the northern California shore. Photo credit: D. Hankin.

In this chapter, we briefly consider the most fundamental terms, concepts, and objectives of sampling theory in the context of a finite population.[1] Presentation is largely at a qualitative level in this chapter, with more sophisticated and better motivated quantitative treatment reserved for subsequent chapters. The intention is to quickly instill a rudimentary but sound conceptual grasp of the basic concepts of sampling theory, including the basic properties of estimators. In subsequent chapters, we will quantify these basic concepts and properties using formal arguments and examples. Important sampling theory terms appear in boldface type at first usage.

---

[1] The term *universe* is sometimes used in place of the term *population* in the sampling theory literature.

Sampling Theory: For the Ecological and Natural Resource Sciences. David G. Hankin, Michael S. Mohr, and Ken B. Newman, Oxford University Press (2019). © David G. Hankin, Michael S. Mohr, and Ken B. Newman. DOI: 10.1093/oso/9780198815792.001.0001

2.1 Terminology

Throughout this text, we will concern ourselves with estimation of a descriptive **population parameter** for a **variable** associated with a **population** of interest containing a finite number of **population units** based on examination of a **sample** (subset) of these units. Illustrative examples of population parameters (mean, proportion, total, respectively) of populations are

- the mean (average) age of male adults living in senior housing facilities throughout the U.S.;
- the proportion of 3rd grade students in North Carolina public schools that are obese;
- the total number of coastal redwood trees exceeding three feet diameter at breast height present on all National Forest lands in California.

We will begin our study of sampling theory with situations in which we can select samples directly from a listing of the units in a population. For example, for the first population, we might select a sample of male adults from a listing of all of the male adults residing in senior housing facilities in the US. For the second population, we might select a sample of students from a listing of all third grade students in North Carolina public schools. These are examples of the simplest kinds of samples that can be selected: samples selected directly from a complete listing of the population units.

A given population may have many different variables and related population parameters associated with it. For example, the population of male adults living in senior housing facilities will also have an average level of education, an average level of savings, an average IQ score, and so on. Typically, we wish to estimate a population parameter for a specific variable that holds primary interest and we refer to this variable as the **target variable**.

An example finite population consisting of $N = 16$ units is displayed in Figure 2.2(a). Each unit $i = 1, 2, \ldots, N$ (in this case, adjoining watersheds) may have multiple attributes or variables associated with it. We use y_i to denote the unit i value for the target variable, y, whereas we use x_i to denote the unit i value for an **auxiliary** variable, x, that may help us to estimate a population parameter of y. Figure 2.2(b) highlights a particular sample of size $n = 4$ units ($i = 3, 6, 10, 16$).

It is not always feasible to select a sample directly from a complete listing of population units. For example, it might be very difficult or impossible to get a listing of the total

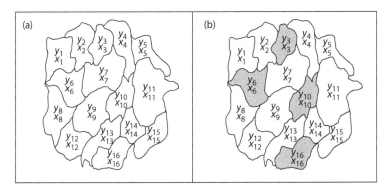

Fig. 2.2 Map (a) depicts an example population of $N = 16$ areal units (watersheds). Map (b) highlights a particular sample of $n = 4$ units. The (x_i, y_i), $i = 1, 2, \ldots, N$, denote the target and auxiliary variable values associated with each population unit.

number of male adults in all senior housing facilities throughout the U.S. Instead, it might be a relatively simple matter to get a listing of all of the senior housing facilities in the U.S. and to select a sample of n senior housing facilities from the N senior housing facilities. Each of the selected facilities would have a certain number of male adult occupants. We could determine the ages of all (or some) of those individuals within the selected facilities, and we could then use that information to estimate the average age of male adults in all senior home facilities. Because it may often not be practical to sample directly from a complete listing of the population units, sampling theorists consider formation of a **sampling frame** to be one of the most critical and challenging tasks. We use the term **simple frame** to describe a setting where the sampling frame consists of a listing of the **population units**, whereas we use the term **complex frame** to describe a setting where the sampling frame consists of a listing of **sampling units** which are well-defined groupings or clusters of the population units. Thus, if we selected a sample of senior home facilities from a listing of all such facilities, we would consider this to constitute selection of sampling units from a complex frame relative to the population units (male adults in all nursing homes).

In many natural resource contexts, the target of estimation may be expressed at a geographic level (as for Figure 2.2). For example, a private timber company may be interested in estimation of total abundance of an endangered species of owl or amphibian present on all N of its forest properties. In this case, sampling frame units consist not of individual owls or amphibians but of specific forest properties which are generally of highly variable size and shape. Here we might associate y_i, unit-specific owl or amphibian abundance, with property i. Similarly, in a stream survey context, where estimation of total fish abundance in a specified reach of stream might be the survey objective, there are many possible alternative sampling frames. One possibility would be to break the entire length of the reach into N 100 m sections, with a unit-specific y_i associated with each section. Another (and often better) sampling frame might consist of a listing of all natural habitat units, of varying size, in the stream reach (Hankin 1984).

2.2 Components of a sampling strategy

We use the term **sampling strategy**[2] to define the complete process whereby (1) a sample of population units is selected from a population (or sampling units from a complex frame), (2) relevant measurements are taken from sampled units, and (3) an estimate of a population parameter is made based on observed measurements. Assuming that the population and target variable of interest have both been carefully defined, development of a sampling strategy requires specification of four interrelated components.

Sampling frame. A listing of the units from which the (initial) sample will be selected. A *simple frame* consists of the basic population units, whereas a *complex frame* consists of sampling units which are well-defined groupings or clusters of the basic population units.

Sample selection methods(s). The method(s) used to select a subset or sample of the units listed in the sampling frame (e.g., selecting a sample of senior housing facilities, or selecting a sample of forest properties), and possibly specification of further methods for selecting subunits in the case of a complex frame (e.g., selecting

---

[2] This terminology follows the usage of, at least, Cassel et al. (1977), Brewer and Hanif (1983), Hedayat and Sinha (1991), Chaudhuri and Stenger (2005), and Brewer and Gregoire (2009).

samples of adult males from within selected housing facilities, or selecting samples of 1 m^2 plot units within selected forest properties).

Observation methods. The explicit protocol(s) that will be used to record or measure y (and sometimes x) of the selected units in a sample. This may be a straightforward measurement of y for the individual units (e.g., a direct response to a survey question), or it may involve use of some model-based approach for estimation of y for the individual units (e.g., use of mark-recapture estimation (Amstrup et al. 2005) to estimate forest property-specific abundance of an owl or amphibian species).

Estimation equations. The explicit formulas that will be used to estimate the population parameter(s) of interest and to calculate an associated error of estimation based on the observations made on the sampled units.

We use the term **sampling design** to refer to a particular choice of sampling frame and an associated sample selection method. These together lead to selection of a particular sample of frame units and/or basic population units. We will learn that the sampling design determines the set of all possible samples that can be selected and plays a key role in determination of errors of estimation.

The vast majority of the material considered in this text will assume that observation methods have been developed which will allow essentially error-free measurement of y (and also x) for all units that appear in a sample. In practice, of course, measurement errors are almost always non-zero and in some cases may be large. We therefore stress at the outset that it is critically important to establish rigorous and repeatable observation protocols that produce measurements of y (and x) that are, to the maximum extent practical, nearly error-free, and reproducible. When it is impossible to establish such protocols, it is critical that a survey has a sound basis for quantifying measurement errors and that these measurement errors are included in the overall error of estimation.

We will defer our consideration of complex sampling frames until later chapters, because it is easier to convey the basic principles of sampling theory for a simple frame. When measurement errors of y are close to zero and the sampling frame consists of the basic population units, then development of a sampling strategy depends only on specification of selection method(s) and specification of estimation equation(s). We therefore stress these two topics—specification of selection method(s) and estimation equation(s)—throughout this text.

2.3 Selection methods

The methods of selecting the units that appear in a sample may be quite simple or quite complex. Unit selection may be random and subject to chance (**probability sampling**), the approach that is the subject of this text, or unit selection may be purposive, based on professional judgment. We may thus refer to a **random sample** or to a **judgment** or **purposive sample**. Although **design-based** sampling theory, the subject that dominates the content of this text, is devoted entirely to randomized schemes for selection of units that appear in samples, some ecologists remain devoted to **representative sampling** in which professional expertise is used to select a particular location for research which is judged to adequately represent, say, a particular vegetation type (Mueller-Dombois and Ellenberg 1974 Chapter 5) or the *average* essential character of a small stream (Williams et al. 2004). A plausible argument may sometimes be made to justify such purposive or judgment sampling when sample sizes are extremely small. But for large sample sizes, it would be irrational to recommend this method of selection for two important reasons. First, judgment sampling lacks *objectivity*; selection of sample units involves subjective, personal choices. Second, it is impossible to describe or predict the statistical properties of judgment selection as a function of sample size or to attach a meaningful measure of uncertainty to estimates based on judgment samples.[3]

The random sample selection methods that we consider in this text all allow for calculation of the **inclusion probability**, π_i, that unit $i = 1, 2, \ldots, N$ in the population is included, by chance, in a random sample S of size n selected from a population of size N following some random selection method. If units are selected instead by purposive methods, then, for a given judge, the inclusion probability is 1 for those units that appear in the judgment sample and 0 for those that do not. Of course, a different judge would likely select a different judgment sample of the same fixed size n. Because inclusion probabilities for units vary among judges, selection of judgment samples lacks the kind of objectivity in sample selection that results when a randomized selection procedure is used instead, and unit selection is subject to chance rather than professional judgment. Randomized selection procedures allow an objective and statistically rigorous basis for inference from samples taken from finite populations.

In Section 2.6 we provide two illustrations of the dangers and limitations of judgment sampling. First, we provide a very brief vignette of a real-life setting where inferences based on judgment selection of representative sample locations appears to have generated seriously erroneous inferences. Second, we review the results from a classroom laboratory exercise that also suggest that judgment sampling can generate poor inferences. To fully appreciate these two illustrations, however, we first need to review the basic properties of estimators.

2.4 Properties of estimators

An **estimator** is a formula or calculation scheme that we use to calculate an **estimate** of a population parameter of interest based on measurements of y taken from units included in a random sample, S. For example, if we wish to estimate the mean of a target variable and we have used equal probability selection methods to select units from a simple sampling frame, then we may estimate the mean of the target variable, $\mu = \sum_{i=1}^{N} y_i / N$, using the following (no doubt familiar) calculation formula (estimator)

$$\hat{\mu} = \frac{\sum\limits_{i=1}^{n} y_i}{n}$$

where the index of the summation is over the $i = 1, 2, \ldots, n$ units that appear in the sample of size n selected from the finite population of size N, y_i denotes the value of the target variable associated with the ith selected unit, and the caret or "hat" over μ indicates an estimator of μ. Note that the sample units in this estimator have been (implicitly) "relabeled" from the population to the sample. For example, if units 5, 7, and 13 had been selected in a sample of size $n = 3$ selected from the $N = 16$ units displayed in Section 2.1,

---

[3] In Chapter 7 we devote some attention to **model-based prediction** of finite population parameters. Although model-based prediction does not generally consider the probabilities associated with selection of sample units, purposive selection of *representative* locations would not generally be recommended for model-based inference, especially if the consequence were that the average values of auxiliary variables in the sample units were greatly different than the average values of these variables in the population (Valliant et al. 2000 Section 3.2).

then the target variable values in the population—y_5, y_7, and y_{13}—would be relabeled as the values—y_1, y_2, and y_3—in the sample.

To avoid the implicit "relabeling" of units implied by the familiar notation used for a sample mean, throughout this text we instead adopt the following equivalent notation for the same estimator

$$\hat{\mu} = \frac{\sum_{i \in S} y_i}{n}$$

where $i \in S$ denotes the fixed unit labels of those n units that have been selected from N and are included in a random sample S. (A formal treatment of this topic is presented in Section A.6.1.) Using this notation, the y_i in a sample remain faithfully attached to their unit labels in the population.

One of the easiest ways to visualize the properties of estimators is to consider patterns of darts thrown at a dartboard. Imagine that the very center of the dartboard, the bullseye, represents the unknown value of a population parameter that we wish to estimate from a sample. There are typically a very large number of distinct samples that could possibly be selected. Each sample would contain a different set of units and associated y values, and would therefore produce a sample-specific estimate—the realized value of an estimator applied to a specific sample of units. (Note, however, that different samples may sometimes yield identical estimates.) Imagine that the different estimates resulting from selection of different samples could be visually represented by the scatter of darts thrown at the dartboard target. Dart locations that are close to the bullseye are *better* (i.e., have higher point values) than dart locations that are far from the bullseye, like an archery target.

Figure 2.3 displays four patterns of dart throws around a bullseye. Together, these patterns convey the most important properties of estimators. Imagine that Patterns (a) and (b) in Figure 2.3 represent the patterns of dart throws made by two highly skilled dart players who are competing at a local tavern, trying to hit the bullseye. They have both just arrived at the tavern; the dart player who produced Pattern (a) was using a set of darts that he had been using for years, whereas the dart player who produced Pattern (b) was using a new set of very expensive darts (never previously used). Pattern (a) is obviously the better pattern of dart throws. On average, the dart locations are centered around the bullseye and the locations are very similar to one another (i.e., the dart throws are highly reliable or precise). Overall, we would characterize this pattern of dart throws as having very high accuracy and very high point value. Pattern (b) is distinguished from Pattern (a) by the average location of the dart throws—on average they are quite far off target. But the reliability or precision of the throws is excellent and comparable to that of Pattern (a). We can recognize that Pattern (b) is less accurate than Pattern (a), because the average location of the darts is so far from the bullseye and the total point value would be less than for Pattern (a). Imagine that Patterns (c) and (d) reflect the locations of dart throws made by the same two individuals, but several hours after they first entered the tavern. (They will be relying on a designated driver to get them home!) On average, the dart throw locations for Pattern (c) remain centered about the bullseye, but the dart throws are now much less precisely located. Pattern (c) is obviously less accurate than Pattern (a) and possibly also less accurate than Pattern (b). For Pattern (d), the average location of dart throws remains far off target, and now the location of dart throws has become highly unreliable or imprecise. Clearly, this is the least accurate pattern of all and would have lowest point value.

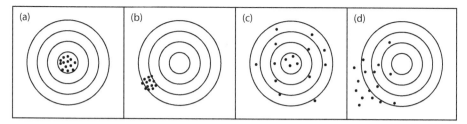

Fig. 2.3 Four patterns of dart throws aimed at the bullseye. Pattern (a) is approximately unbiased, has high precision and high accuracy. Pattern (b) has high precision, but because it is biased it is less accurate than Pattern (a). Pattern (c) is approximately unbiased but, because it has low precision, it has low accuracy. Pattern (d) is biased and has low precision and therefore has the lowest accuracy of the four dart patterns.

The clear conceptual differences among the patterns of dart throws in Figure 2.3, expressed in every day language, have important connections with the fundamental properties of estimators.

Expected value. The average value that an estimator takes on over all possible samples that can be selected according to some selection scheme. (Analogy: the average location of dart throws.)

Bias. The difference between the expected value of an estimator and the target population parameter. If the expected value is exactly equal to the target population parameter, then we have an **unbiased estimator**. (Analogy: the average dart location is exactly equal to location of the bullseye.) If instead the expected value differs from the population parameter (bullseye), then we have a **biased estimator**. Bias may be *positive* or *negative*, i.e., the expected value is greater than or less than the target population parameter.

Sampling variance. Sampling variance measures the variability among the estimates that arise from all possible samples that can be selected. This variability is measured as the average squared difference between estimates and the estimator's expected value. (Analogy: averaged squared distance from individual dart throw locations to the average location of darts.) Sampling variance is inversely related to the reliability or precision of an estimator. Sampling variance will be low when an estimator has high precision or reliability and vice versa.

Mean square error (MSE). Mean square error measures the overall accuracy of an estimator and is equal to the averaged squared difference between estimates and the target population parameter. (Analogy: average squared distance from individual dart locations to the bullseye.)

In Appendix A we show that mean square error equals sampling variance plus the square of the bias. Thus, for an unbiased estimator, mean square error is equivalent to sampling variance. Bias, if large, may make a substantial contribution to mean square error and thereby seriously reduce the accuracy of an estimator. Figure 2.4 uses a triangle analogy (Pythagorean theorem) to visually illustrate the relationships among bias, sampling variance, and mean square error. Note that a very reliable (low sampling variance) estimator with small bias may be a more accurate estimator than an unbiased but unreliable (high sampling variance) estimator.

If a serious attempt is made to "visually memorize" the patterns of dart throw locations displayed in Figure 2.3, the concepts of bias, sampling variance, and mean square error will

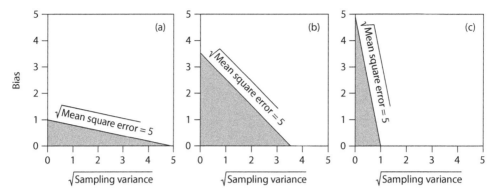

Fig. 2.4 Illustration of the Pythagorean relationship existing between an estimator's bias, sampling variance, and mean square error: bias^2 + sampling variance = mean square error. For each of the three examples, $\sqrt{\text{mean square error}}$ = RMSE = 5 but the bias contributing to that RMSE differs: (a) bias = 1 accounts for a relatively small portion of the RMSE, (b) bias $\approx$ 3.536 and $\sqrt{\text{sampling variance}}$ $\approx$ 3.536 make equal contributions to the RMSE, and (c) bias $\approx$ 4.899 accounts for a relatively large portion of the RMSE.

become solidly imprinted in one's consciousness! The analogy of a pattern of darts thrown at a dartboard with the estimates generated from some sampling strategy is, however, deficient in two important respects. First, the "polar coordinate system" of the dartboard target makes it difficult to convey the notion of bias as being negative (on average, estimates are less than the target value) or positive (on average, estimates are greater than the target value). Second, this same coordinate system also complicates graphical representation of the distribution of possible sample estimates.

2.5 Sampling distribution of an estimator

Consider now a population consisting of $N = 240$ stones, with the variable of interest, y, being stone weight. The frequency distribution of this variable is displayed in Figure 2.5(a). Suppose that we wish to estimate the mean weight of a stone in this population, μ, by a "random selection" of 10 stones. Without getting into the details of the large number of ways that a sample of size $n = 10$ stones might be randomly selected, let us just imagine that we are using a sample selection method for which any particular set of 10 stones that might be selected is equally likely, and that we could in principle draw each of the possible distinct samples of 10 stones that could be selected according to this selection method (the Figure 2.5 caption provides additional details on sample selection). For each such sample, we could estimate μ by the mean weight of the sampled 10 stones, $\hat{\mu}$. We could then construct a frequency distribution of the estimates from all of these possible samples, as in Figure 2.5(b). If this frequency distribution were then scaled so that the total area was equal to one, we would have a graphical representation of the **sampling distribution** of the estimator $\hat{\mu}$. In this form, the histogram of the sampling distribution displays the probability that a sample estimate will fall within a specific bin range (Figure 2.5(b), probability axis). What do you think this distribution would look like if instead the sample size was, e.g., $n = 20$?

The sampling distribution of an estimator can be characterized by a minimum of two parameters: its mean (expected value) and its variance (sampling variance). If the mean

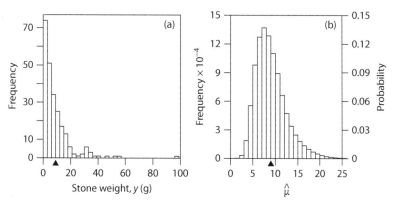

Fig. 2.5 Histograms of (a) population of $N = 240$ stone weights with mean weight $\mu = 9.03$ grams, and (b) sampling distribution of an estimator, $\hat{\mu}$, for the mean stone weight based on random sample selections of $n = 10$ stones, where the estimator was equal to the mean weight of the 10 sampled stones. Individual sample units were drawn at random from the population with equal probability (without replacement), and 10^6 such samples were drawn independently from each other using this same method. The mean of each distribution, 9.03 grams, is pointed to by the upper vertex of the black triangle located below each horizontal axis.

(expected value) of the sampling distribution is equal to the population parameter being estimated, then the estimator is unbiased (as in the example above). Otherwise, the estimator is biased, and the bias may be positive or negative. If the sampling distribution of the estimator has a very small range, then sampling variance will be small (the estimator is precise); if this range is very large, then sampling variance will be very large (the estimator is imprecise). An estimator's overall accuracy is measured by mean square error—sampling variance plus the squared bias.

Note that the sampling distribution of the estimator depicted in Figure 2.5(b) is not normally distributed. Normality is a highly desirable property of the sampling distribution of an estimator, but in Chapter 3 we will learn that the shape of the sampling distribution of an estimator depends on the shape of the distribution of the population variable y (from which the sample is drawn) and on sample size, n.

2.6 Judgment sampling versus random sampling

We begin our illustrations of the dangers of representative (purposive) selection of units with a brief (and greatly oversimplified) review of the history of surveys of spawning escapement (number of adult fish returning to freshwater streams to spawn) for Oregon coastal coho salmon (*Oncorhynchus kisutch*). Oregon coastal coho salmon historically supported extensive commercial and recreational fisheries. For the period 1958–1990, standardized survey protocols were executed in purposively selected *index reaches* ("representative" stream segments) (Jacobs and Cooney 1997). Mean densities (fish-per-mile) in these survey reaches were originally used primarily for qualitative assessment of trends in abundance, for which purpose they were no doubt useful (Chapter 13), but beginning with the passage of the federal Fishery Conservation and Management Act of 1976 and the subsequent introduction of intensive quantitative management of salmon fisheries, index reach densities were "scaled up" to produce quantitative estimates of spawning escapement for the Oregon coast, and these estimates were in turn used to develop harvest management policy. Beginning in 1990, a stratified random survey design was introduced for Oregon coastal coho, replacing the original index reaches by randomly selected reaches, while original index reaches were retained for comparison. Over the period 1990–1997, the mean density in these randomly selected reaches averaged just 27% of the mean density in the index reaches (Jacobs and Nickelson 1998), thereby generating greatly reduced estimates of coho salmon spawning escapement that were at odds with the overly optimistic index-based estimates from previous years that had been used for development of fishery management policy, and likely led to unintentional overfishing of Oregon coastal coho salmon. These much lower, but assumed unbiased, estimates of spawning escapement played an important role in the subsequent determination, in 1998, that Oregon coastal coho salmon were in a seriously depressed state, and they were subsequently listed as "threatened" under the federal Endangered Species Act. Presumably, the original index reaches had either been (a) deliberately selected to represent *excellent* or *good* rather than *average* spawning habitat, or (b) judgment selection of *representative/average* reaches had been quite seriously off the mark.

Our second illustration of the dangers and inadequacy of judgment sampling comes from a repeatable lab experiment that directly contrasts the performance of randomized selection of sample units with judgment selection. One of the simplest randomized selection procedures is **simple random sampling** or **SRS**. With SRS, one selects n sample units from N population units at random with equal probability, without replacement. For example, imagine a box with $N = 100$ balls that are physically identical except for an identifying number, i ($i = 1, 2, \dots, 100$), on the surface of the ball. To draw a simple random sample, we could reach into the box to select one ball at random with equal probability (i.e., all balls in the box being equally likely to be the one selected), record the ball number, and set the ball aside. We could then reach into the box again and select another ball at random with equal probability from the remaining 99 balls, record the ball number, set the ball aside, and so on until we have selected n balls. But, would this sampling approach actually result in a simple random sample? How likely is it that the balls positioned near the top, bottom, and sides of the box have the same probability of being selected as a ball near the middle of the box? This is one reason why casino and lottery games based on a random drawing of numbered ping-pong balls from a container continually mix up the balls (e.g., by rotating the container, injecting pulses of air, etc.). But what if the balls were instead objects—not identical in size and shape—how difficult would it then be to insure that the physical selection of objects was at random with equal probability? For these and other reasons, to insure that a sample is selected in a manner that meets the definition of the random sampling method (here SRS), the sample is usually selected as follows:

(1) the population units are listed by an identifying number, i ($i = 1, 2, \dots N$);
(2) a computer is used to draw a sample of n of these unit numbers according to the definition of the random sampling method using a random number generator; and
(3) these units are then selected for the sample.

Thus, with our box of balls example, we would draw an SRS sample of unit numbers first, and then select the associated balls from the box. In the R statistical/programming/ graphics language and environment (R Core Team 2018), a simple random sample of unit numbers of size $n = 5$ from $N = 100$ can be selected using the following R expression: `sample(x=1:100, size=5, replace=FALSE)`. The result, e.g., {15, 48, 69, 67, 35}, is a listing of the units to include in the sample. If this R expression were executed again, a second SRS sample of unit numbers, selected independently of the first, might result in, e.g., {33, 12, 18, 51, 86}. (In Chapter 3 we present a more formal explanation of SRS and we show that there are about 75.3 million(!) distinct simple random samples of size 5 that can be selected from a population of size 100.)

Jessen (1978) describes an experiment with a population of stones that we have repeated in sampling theory lab sessions with success for many years. We bring to the lab room a large set of agates (a semi-precious gemstone composed of chalcedony) collected from northern California ocean beaches. The entire population of $N = 125$ agates is displayed on a table top (see Figure 2.1), so that students can view all of the population units. The population parameter to be estimated is the mean agate weight, μ. Individual students are asked to select, using their best *judgment*, samples of size $n = 1, 2, 3, 4, 5, 6, 8, 10, 20$ that they think, when the sample mean weight is taken, will result in their best estimate of μ for that value of n. There is considerable variation in the weights of the agates and the human eye has difficulty judging weight or volume, so this task is more difficult than one might suspect. During one of the more recent lab sessions, with results comparable to those on many previous occasions, $M = 22$ students recorded the total weight of agates for each sample that was selected and, as a subsequent lab exercise, students calculated the corresponding mean weight ($\hat{\mu}$) for each sample selected. The students were then informed of the population mean weight ($\mu = 14.843$ g), and for each n were asked to calculate the following performance measures for their collective judgment sampling effort. Letting $\hat{\mu}_j$ denote the sample mean weight obtained by student judge j, $j = 1, 2, \ldots, M$: (1) average value $= \sum_{j=1}^{M} \hat{\mu}_j / M$ (analogous to expected value), (2) bias $=$ average value $- \mu$, (3) proportional bias $=$ bias$/\mu$, (4) variance $= \sum_{j=1}^{M} (\hat{\mu}_j - $ average value$)^2 / M$ (analogous to sampling variance), and (5) variance $+$ bias^2 (analogous to mean square error). The analogies are a bit imperfect, of course, because the lab participants are only a small number of individuals out of billions of potential judges and it is risky to generalize from the performance of so few students.

As we will learn in Chapter 3, if SRS is used to select samples of size n from a population of size N, then the sample mean is an unbiased estimator of μ, with sampling variance equal to $\left(\frac{N-n}{N}\right) \frac{\sigma^2}{n}$, where σ^2 is the finite population variance (a measure of the variation in agate weights, defined in Chapter 3), here equal to 307.383. With SRS, the bias and proportional bias of the sample mean is zero, and therefore its mean square error is the same as its sampling variance. But how might SRS perform compared to judgment sampling assuming the sample mean is used to estimate μ for each method of selection?

Figure 2.6 summarizes the collective performance of judgment sampling for the 22 students compared to that for SRS and its dependence on n. Panel (a) displays the sample mean weights obtained by the individual students at each value of n, along with a horizontal dashed line referencing the population mean value of $\mu = 14.843$ g. Note that the bias for judgment sampling [panel (b)] is at first very high (proportional bias is more than 90% for $n = 1$) and gradually declines with increasing n, but is negligible only for $n = 20$. In contrast, with SRS the sample mean is unbiased for all values of n. Although for small samples sizes ($n \leq 2$) the sampling variance for judgment sampling [panel (c)] is less than the sampling variance for SRS, for larger sample sizes the sampling variance for SRS is always less than that for judgment sampling. Finally, the mean square error for SRS is less than that for judgment sampling for all values of n [panel (d)]. Thus, the accuracy with SRS is greater than with judgment sampling when using the sample mean as an estimator of μ, at least for the values of n evaluated.

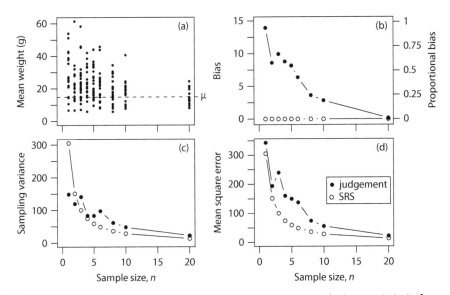

Fig. 2.6 Performance of 22 student judges in estimating the mean weight ($\mu = 14.843$ g) of agates in a small population ($N = 125$) with considerable variation in agate weights ($\sigma^2 = 307.383$), compared to performance for SRS. Both approaches used the sample mean as the estimator of μ. See text for definition of performance metrics.

Overall it appears that the only potential advantage of judgment sampling over SRS might be for very small sample sizes, when the variance among judges is less than the sampling variance for SRS. If judges could improve their skills in selection of agates, so as to dramatically reduce their bias, then judgment sampling might be recommended for the very smallest sample sizes. Otherwise, the performance of judgment sampling seems clearly inferior to SRS. It is also important to note that all units of the population were on full display in this sampling exercise. This is the most favorable kind of setting that one might possibly imagine for application of judgment sampling. When a field ecologist makes a judgment selection of a "representative" patch of vegetation or section of stream, etc., the choice is made based on his/her past experience which generally does not include a simultaneous view of the physical attributes of the entire population or even a view of these same physical attributes over the course of a professional career. Against this backdrop, it is all the more remarkable that the performance for SRS—in which units are selected purely at random, without regard to the y values—compares so favorably to that for judgment sampling. Wouldn't you think that an average student judge selecting a single agate could do a much better job purposely selecting a *single representative agate* than a randomized selection procedure that is just as likely to select the heaviest agate or the lightest agate as to select an "average" weight agate? Imagine how much more difficult it might be to purposely select a representative patch of a vegetation type or a representative section of a stream.

CHAPTER 3

Equal probability sampling

Fig. 3.1 Western lily, *Lilium occidentale*, an endangered species with distribution limited to a few isolated locations in northern California and southern Oregon. Abundance of this species might be estimated using equal probability strategies described in this chapter, but adaptive sampling (Chapter 11) might be a better strategy (Tout 2009). Photo credit: Dave Imper.

Equal probability samples may be selected **without replacement** or **with replacement**. We begin our formal treatment of random selection methods with a consideration of equal probability without replacement sampling because we will use this method of sampling as a "baseline" from which to evaluate the performance of alternative selection methods.

3.1 Without replacement sampling

When a person or paper states that "sample units were selected at random", they usually, but not always, mean that they have used **simple random sampling (SRS)** for selection of population units from a simple frame. With SRS we select n units from the N units

Sampling Theory: For the Ecological and Natural Resource Sciences. David G. Hankin, Michael S. Mohr, and Ken B. Newman, Oxford University Press (2019). © David G. Hankin, Michael S. Mohr, and Ken B. Newman. DOI: 10.1093/oso/9780198815792.001.0001

in the population with *equal probability without replacement*. To select a simple random sample, we select a first unit "at random" (i.e., with equal probability) from the N units in the population. Because SRS is a without replacement selection method, we retain this first selected unit and then select our second unit with equal probability from the $N-1$ remaining units. We retain this second selected unit and continue this process until we have selected a total of n units. (Section 2.6 illustrates how an SRS may be selected using R code.) The set of units that appears in an SRS is thus a set-valued **random variable** (Section A.6.1), which we label as an uppercase italic S, and it will have a *variable* outcome over repeated uses of the SRS unit selection procedure. We use lowercase italic s to identify and distinguish a *particular, realized set of sample units* that is selected by SRS (or some other random selection procedure) from the unspecified random outcome, S.

When sample units are selected by SRS, several important properties of this selection scheme emerge.

Property 1. The total number of possible distinct samples (without regard to order of selection, Section A.1.3), T, can be calculated as

$$T = \binom{N}{n} = \frac{N!}{n!(N-n)!} \tag{3.1}$$

Property 2. Each of the T possible distinct samples is equally likely to be selected and the probability of selecting any particular sample $S = s$ ($s = 1, 2, \ldots, T$) is

$$\Pr\{S = s\} = \frac{1}{T} = \frac{1}{\binom{N}{n}} \tag{3.2}$$

Property 3. The probability that any particular unit i ($i = 1, 2, \ldots, N$) is included in a simple random sample of size n selected from N is denoted by π_i and termed the **first order inclusion probability**. In SRS, all π_i are equal and

$$\pi_i = \frac{n}{N} \tag{3.3}$$

The first two properties may be motivated from the following argument (Cochran 1977). Consider a particular sample of size n, containing n distinct units, that has been selected from N by SRS. When units are selected with equal probability without replacement, the probability that any one of these sample units was selected on the first draw is n/N ($= \pi_i$). Given that one of the n sample units was selected on the first draw, the probability that any one of the remaining units in the sample was selected on the second draw would be $(n-1)/(N-1)$. Given that two of the n sample units have been selected, the probability that any one of the remaining sample units is selected on the third draw is $(n-2)/(N-2)$. This logic continues until the very last (n^{th}) unit in the sample is selected on the nth draw with probability $[n-(n-1)]/[N-(n-1)] = 1/(N-n+1)$. The probability that our SRS sample contains a specific set of n distinct units selected from N is thus equal to the product of: (a) the probability of selecting any one of the sample units on the first draw, (b) the probability of selecting a second sample unit on the second draw given that one of the sample units has been selected on the first draw, (c) the probability of selecting a third sample unit on the third draw given that two sample units have been selected on the previous two draws and so on, so that

Pr{any distinct SRS sample of size n drawn from N}

$$= \Pr\{S = s\} = \frac{n}{N} \cdot \frac{(n-1)}{(N-1)} \cdot \frac{(n-2)}{(N-2)} \cdot \frac{(n-3)}{(N-3)} \cdots \frac{1}{(N-n+1)} = \frac{n!}{\frac{N!}{(N-n)!}} = \frac{1}{\binom{N}{n}}$$

This logic does not specify the labels (unit indexes) for any of the units that appear in a particular sample of size n selected from N by SRS, so $\Pr\{S = s\}$ must be the same for all possible samples $T = \binom{N}{n}$ that can be selected. Note that because each possible SRS has the same probability, $1/T = 1/\binom{N}{n}$, and the total number of possible samples is $T = \binom{N}{n}$, then the probabilities of all the distinct SRS samples must sum to exactly one.

The total number of distinct simple random samples increases at a remarkable rate as n and N increase. For example, for $n = 2$ and $N = 4$, $\binom{N}{n} = 6$; for $n = 4$ and $N = 20$, $\binom{N}{n} = 4{,}845$; and for $n = 10$ and $N = 40$, $\binom{N}{n} = 847{,}660{,}528$, nearly one billion distinct samples.

3.1.1 Estimation of the population mean, proportion, and total

As noted in Chapter 2, the objective of sampling is usually a simple one, namely estimation of a **population parameter**: a mean, a total, or a proportion. In a finite population of size N, we label the fixed target variable values attached to particular units by y_i $(i = 1, 2, \ldots, N)$ and we define the fixed targets of estimation as

population mean

$$\mu = \frac{\sum\limits_{i=1}^{N} y_i}{N} \tag{3.4}$$

population proportion $(y = 0 \text{ or } 1)$

$$\pi = \frac{\sum\limits_{i=1}^{N} y_i}{N} \tag{3.5}$$

population total

$$\mathcal{T} = \sum\limits_{i=1}^{N} y_i = N\mu \tag{3.6}$$

Note that the expression for the population proportion, π, is identical to that for the population mean, μ. Thus, a proportion can be viewed as a special case of a population mean, where y is a binary (0, 1) variable. We use π to denote the population proportion to be consistent with our use of Greek letters to denote all other population parameters (e.g., the mean, μ, and the total, $\mathcal{T}$).[1] When samples are selected by SRS, we often estimate the values of the population mean, proportion, and total using the following **mean-per-unit estimators**

$$\hat{\mu}_{\text{mpu}} = \frac{\sum\limits_{i \in S} y_i}{n} \tag{3.7}$$

$$\hat{\pi}_{\text{mpu}} = \frac{\sum\limits_{i \in S} y_i}{n} \tag{3.8}$$

$$\hat{\mathcal{T}}_{\text{mpu}} = N \sum\limits_{i \in S} y_i / n = N\hat{\mu}_{\text{mpu}} \tag{3.9}$$

---

[1] In the sampling theory literature, the letter p is often used to denote a proportion. Our usage of π also avoids potential notational confusion between p, if it were used to denote a proportion, and $p(s)$, a shorthand which we use to denote the probability of selecting a particular sample s: $p(s) = \Pr\{S = s\}$. However, our usage of π does overlap with our inclusion probability notation, π_i.

Note that these estimators are all defined over the population units that appear in the random sample S ($i \in S$) and represent general calculation schemes or formulas. As noted in Chapter 2, for a particular realized sample, s, an estimator will take on a specific realized numeric value which we call an *estimate*, and is defined over the units that appear in the realized sample, s ($i \in s$). The notation that we use in Equations (3.10)–(3.12) serves to clearly distinguish these realized sample estimates from the estimators with which they are associated.

$$\hat{\mu}_{\text{mpu}}(s) = \frac{\sum\limits_{i \in s} y_i}{n} \tag{3.10}$$

$$\hat{\pi}_{\text{mpu}}(s) = \frac{\sum\limits_{i \in s} y_i}{n} \tag{3.11}$$

$$\hat{T}_{\text{mpu}}(s) = N \sum_{i \in s} y_i / n = N\hat{\mu}_{\text{mpu}}(s) \tag{3.12}$$

In the sampling theory literature, the estimator of the mean is often denoted by $\bar{y}$. In this text we instead use $\hat{\mu}$ as an estimator of μ, consistent with the generalized use of $\hat{\theta}$ as an estimator of θ, where θ is some arbitrary target of estimation, and we use the notation $\hat{\theta}(s)$ to indicate the value of an estimator for a particular realized sample, s. In some cases, we will use subscripts below estimators to indicate the particular sampling design that leads to the particular form of the estimator as in, e.g., $\hat{\mu}_{\text{mpu}}$ as an estimator of the mean under SRS with a simple sampling frame (where mpu is shorthand for *mean-per-unit*). We may omit the subscript on an estimator when context makes the subscript implicitly obvious.

One of the many reasons that SRS and mean-per-unit estimation is used as a sampling strategy is that it is guaranteed to produce *unbiased* estimation of the population mean, proportion, and total. That is, $E(\hat{\mu}_{\text{mpu}}) = \mu$, $E(\hat{\pi}_{\text{mpu}}) = \pi$, and $E(\hat{T}_{\text{mpu}}) = T$, where $E(\hat{\theta})$ denotes expected value of an arbitrary estimator $\hat{\theta}$. It takes a bit of work to formally prove unbiasedness of the mpu estimator under SRS, but it is easy to generate simple numerical illustrations of this fact for small populations and small sample sizes. Indeed, throughout this text we will use small population and small sample size settings to illustrate the fundamental properties of estimators. (Note, however, that we will prove the unbiasedness of SRS/mean-per-unit estimation in Chapter 8 as a special case of Horvitz–Thompson (Horvitz and Thompson 1952) estimation.)

To numerically illustrate the unbiasedness of mpu estimators of the mean and total, we construct a small population consisting of $N = 4$ stones. The target variable of interest is the weight of a stone

Population unit (i):	1	2	3	4
Stone weight (y_i):	2	5	7	14

For this population of stone weights, $\mu = 7$ and $T = 28$. Table 3.1 provides a listing of each of the $T = \binom{4}{2} = 6$ possible simple random samples of $n = 2$ stones that can be selected from the $N = 4$ stones in the population. For each such sample s, the mpu estimates of the population mean, $\hat{\mu}(s)$, and total, $\hat{T}(s)$, are also listed.

Throughout this text we use the notation $\sum_{s \in S}$ to denote summation over the **sample space**, S, the set of all possible distinct samples ($s \in S$) of size n that can be selected from N according to some sampling design (sampling frame and selection method). The set of values (sample-specific estimates) that an estimator $\hat{\theta}$ takes on over all possible samples,

Table 3.1 Sample space illustration of the behavior of $\hat{\mu}_{mpu}$ and $\hat{T}_{mpu}$ for SRS samples of size $n = 2$ selected from the $N = 4$ population of stone weights.

ID	Units (s)	Values ($y_i, i \in s$)	$\hat{\mu}_{mpu}(s)$	$\hat{T}_{mpu}(s)$
1	1, 2	2, 5	3.5	14
2	1, 3	2, 7	4.5	18
3	1, 4	2, 14	8.0	32
4	2, 3	5, 7	6.0	24
5	2, 4	5, 14	9.5	38
6	3, 4	7, 14	10.5	42
Sum:			42.0	168

along with the probabilities of obtaining the individual samples, $\Pr\{S = s\}$, can be used to specify the *sampling distribution* of an estimator $\hat{\theta}$. As noted in Chapter 2, the sampling distribution can be characterized by its mean and its variance. The expected value, $E(\hat{\theta})$, of some estimator $\hat{\theta}$ is the mean of the estimator's sampling distribution and is equal to the following sum over the sample space

$$E(\hat{\theta}) = \sum_{s \in \mathcal{S}} \hat{\theta}(s) \Pr\{S = s\} = \sum_{s \in \mathcal{S}} \hat{\theta}(s) p(s)$$

where $p(s)$ will be used, throughout the remainder of this text, as a convenient shorthand for $\Pr\{S = s\}$. For $\hat{\mu}_{mpu}$

$$E(\hat{\mu}_{mpu}) = \sum_{s \in \mathcal{S}} \hat{\mu}_{mpu}(s) p(s) = \sum_{s \in \mathcal{S}} \hat{\mu}_{mpu}(s) \frac{1}{T} = \frac{1}{T} \sum_{s \in \mathcal{S}} \hat{\mu}_{mpu}(s)$$

In the Table 3.1 sample space example, $T = 6$, so $p(s) = 1/6$. Thus, we may numerically verify the unbiasedness of $\hat{\mu}_{mpu}$

$$E(\hat{\mu}_{mpu}) = \frac{1}{6} \sum_{s \in \mathcal{S}} \hat{\mu}_{mpu}(s) = \frac{1}{6} \cdot 42 = 7 = \mu$$

Similarly, we may numerically verify unbiasedness of $\hat{T}_{mpu}$

$$E(\hat{T}_{mpu}) = \sum_{s \in \mathcal{S}} \hat{T}_{mpu}(s) p(s) = \frac{1}{T} \sum_{s \in \mathcal{S}} \hat{T}_{mpu}(s) = \frac{1}{6} \cdot 168 = 28 = T$$

3.1.2 *Sampling variance*

In Chapter 2, we noted that sampling variance is a measure of the reliability or precision of an estimator. If an estimator is very precise (very reliable), then it has low sampling variance, whereas an estimator that is very imprecise (very unreliable) has high sampling variance. Sampling variance measures the variance of the sampling distribution of an estimator. Formally, we define sampling variance, $V(\hat{\theta})$, of some estimator $\hat{\theta}$ as

$$V(\hat{\theta}) = E[\hat{\theta} - E(\hat{\theta})]^2 = \sum_{s \in \mathcal{S}} [\hat{\theta}(s) - E(\hat{\theta})]^2 p(s) \tag{3.13}$$

For an unbiased estimator, $E(\hat{\theta}) = \theta$ and so Equation (3.13) would be equivalent to

$$V(\hat{\theta}) = \sum_{s \in \mathcal{S}} (\hat{\theta}(s) - \theta)^2 p(s) \tag{3.14}$$

Table 3.2 tabulates sample-specific values of $(\hat{\theta}(s) - \theta)^2$ and allows us to numerically calculate the sampling variance for the unbiased estimator $\hat{\mu}_{\text{mpu}}$ for our sample space example. We calculate sampling variance as

$$V(\hat{\mu}_{\text{mpu}}) = \frac{1}{6} \sum_{s \in \mathcal{S}} (\hat{\mu}_{\text{mpu}}(s) - \mu)^2 = \frac{1}{6} \cdot 39 = 6.5$$

Similarly, Table 3.3 allows us to numerically calculate sampling variance for the unbiased estimator $\hat{T}_{\text{mpu}}$

$$V(\hat{T}_{\text{mpu}}) = \frac{1}{6} \sum_{s \in \mathcal{S}} (\hat{T}_{\text{mpu}}(s) - T)^2 = \frac{1}{6} \cdot 624 = 104$$

A tabulation of all sample estimates $[\hat{\theta}(s)]$, all squared differences between sample estimates and the estimator's expected value $([\hat{\theta}(s) - E(\hat{\theta})]^2)$, and all corresponding sample probabilities $[p(s)]$ over the sample space can, in principle, be used to calculate $E(\hat{\theta})$ and $V(\hat{\theta})$ for almost any estimator $\hat{\theta}$. We use this device (presentation of an example sample

Table 3.2 Sample space tabulation for calculating the sampling variance of $\hat{\mu}_{\text{mpu}}$ for SRS samples of size $n = 2$ selected from the $N = 4$ population of stone weights (as listed in Table 3.1).

Sample ID	$\hat{\mu}_{\text{mpu}}(s)$	$\hat{\mu}_{\text{mpu}}(s) - 7$	$(\hat{\mu}_{\text{mpu}}(s) - 7)^2$
1	3.5	−3.5	12.25
2	4.5	−2.5	6.25
3	8.0	+1.0	1.00
4	6.0	−1.0	1.00
5	9.5	+2.5	6.25
6	10.5	+3.5	12.25
Sum:		0.0	39.00

Table 3.3 Sample space tabulation for calculating the sampling variance of $\hat{T}_{\text{mpu}}$ for SRS samples of size $n = 2$ from the $N = 4$ population of stone weights (as listed in Table 3.1).

Sample ID	$\hat{T}_{\text{mpu}}(s)$	$\hat{T}_{\text{mpu}}(s) - 28$	$(\hat{T}_{\text{mpu}}(s) - 28)^2$
1	14	−14	196
2	18	−10	100
3	32	+4	16
4	24	−4	16
5	38	+10	100
6	42	+14	196
Sum:		0	624

space) throughout this text for four purposes: (1) to provide *worked examples* of sample-specific calculations for many estimators, (2) to give simple tangible concrete meaning to the concepts of sample space and sampling variance for alternative sampling strategies, (3) to allow numerical verification of important estimator properties (e.g., unbiasedness), and (4) to allow numerical verification that the results of sample space calculations are equivalent to those that can (in most settings) instead be generated using efficient analytic expressions.

There are, fortunately, simple explicit analytic expressions for the sampling variances of $\hat{\mu}_{\text{mpu}}$, $\hat{\pi}_{\text{mpu}}$, and $\hat{T}_{\text{mpu}}$,

$$V(\hat{\mu}_{\text{mpu}}) = V(\hat{\pi}_{\text{mpu}}) = \left(\frac{N-n}{N}\right)\frac{\sigma^2}{n} = (1-f)\frac{\sigma^2}{n} \tag{3.15}$$

and

$$V(\hat{T}_{\text{mpu}}) = N^2\left(\frac{N-n}{N}\right)\frac{\sigma^2}{n} = N^2 V(\hat{\mu}_{\text{mpu}}) \tag{3.16}$$

where

$$\sigma^2 = \frac{\sum\limits_{i=1}^{N}(y_i - \mu)^2}{(N-1)} \tag{3.17}$$

and $f = n/N$. We refer to σ^2 as the **finite population variance**. It can be interpreted, for large N, as the average mean-corrected variation in y among the population units. The divisor of $N-1$ rather than N ensures that many important results in sampling theory have an "attractive" form, and is a convention that has been adopted by most sampling theorists.[2] The term $(1-f)$ is the complement of f which is known as the **sampling fraction**, the fraction of the N population units that appear in the sample of size n. When $y_i = 0, 1$, only, it can be easily shown that $\sigma^2 = \frac{N}{N-1}\pi(1-\pi)$. Therefore, $V(\hat{\pi}_{\text{mpu}})$ can alternatively be written as

$$V(\hat{\pi}_{\text{mpu}}) = \left(\frac{N-n}{N-1}\right)\frac{\pi(1-\pi)}{n} \tag{3.18}$$

Inspection of Equation (3.15) and Equation (3.16) reveals that $V(\hat{\mu}_{\text{mpu}})$, $V(\hat{\pi}_{\text{mpu}})$, and $V(\hat{T}_{\text{mpu}})$ all depend directly on the underlying variation in y, namely on σ^2. Also, sampling variances of $\hat{\mu}_{\text{mpu}}$, $\hat{\pi}_{\text{mpu}}$, and $\hat{T}_{\text{mpu}}$ all decline with increasing sample size, n, in two respects: (a) by the factor $1/n$ in the σ^2/n term, and (b) by the factor $(N-n)/N = 1-f$. The factor $(1-f)$ is termed the **finite population correction** (fpc) factor and causes sampling variance (when sampling from a finite population) to decrease more than would be expected as a function of sample size alone. In fact, if $n = N$, then $f = 1$, and the sampling variance reduces to 0, regardless of the value of σ^2.

The formulas for sampling variance of $V(\hat{\mu}_{\text{mpu}})$, $V(\hat{\pi}_{\text{mpu}})$, and $V(\hat{T}_{\text{mpu}})$ allow us to calculate sampling variance without having to construct the equivalent of Table 3.2 and Table 3.3. If σ^2 is known, we can instead calculate $V(\hat{\mu}_{\text{mpu}})$, $V(\hat{\pi}_{\text{mpu}})$, and $V(\hat{T}_{\text{mpu}})$ directly from σ^2, n, and N. For example, for our population of stone weights, $\sigma^2 = 26$. If $n = 2$ were selected from $N = 4$, then $f = 2/4 = 0.5$ and $V(\hat{\mu}_{\text{mpu}}) = (1-f)\sigma^2/n = (1-0.5) \cdot 26/2 = 6.5$, and $V(\hat{T}_{\text{mpu}}) = N^2 V(\hat{\mu}_{\text{mpu}}) = 4^2 \cdot 6.5 = 104$, as calculated previously. If the population y

---

[2] In the sampling theory literature, the symbol S^2 is often used to denote the finite population variance, again with a divisor of $N-1$.

values were instead binary (0, 1), then similar calculations could be made for $V(\hat{\pi}_{\text{mpu}})$ using either Equation (3.15) or Equation (3.18) and would give identical results.

3.1.3 Estimation of sampling variance

In practice, the sampling variance of an estimator is rarely known exactly, as in the preceding sample space examples, because in ordinary circumstances (a) σ^2 is unknown and (b) we typically select just one (SRS) sample, s, from the sample space, $\mathcal{S}$. From just a single sample, however, we can use an unbiased estimator for σ^2, namely $\hat{\sigma}^2_{\text{mpu}}$, to unbiasedly estimate sampling variance for the mpu estimators of the population mean, proportion, or total

$$\hat{V}(\hat{\mu}_{\text{mpu}}) = \hat{V}(\hat{\pi}_{\text{mpu}}) = \left(\frac{N-n}{N}\right)\frac{\hat{\sigma}^2_{\text{mpu}}}{n} = (1-f)\frac{\hat{\sigma}^2_{\text{mpu}}}{n} \tag{3.19}$$

and

$$\hat{V}(\hat{T}_{\text{mpu}}) = N^2\left(\frac{N-n}{N}\right)\frac{\hat{\sigma}^2_{\text{mpu}}}{n} = N^2\hat{V}(\hat{\mu}_{\text{mpu}}) \tag{3.20}$$

where

$$\hat{\sigma}^2_{\text{mpu}} = \frac{\sum\limits_{i\in S}(y_i - \hat{\mu}_{\text{mpu}})^2}{n-1} \tag{3.21}$$

is the **sample variance**, the (approximate) average mean-corrected variation among the population y values that appear in the sample.[3] When y is a binary (0, 1) variable, then $\hat{\sigma}^2_{\text{mpu}} = \frac{n}{n-1}\hat{\pi}_{\text{mpu}}(1 - \hat{\pi}_{\text{mpu}})$ and $\hat{V}(\hat{\pi}_{\text{mpu}})$ can alternatively be expressed as

$$\hat{V}(\hat{\pi}_{\text{mpu}}) = \left(\frac{N-n}{N}\right)\frac{\hat{\pi}_{\text{mpu}}(1 - \hat{\pi}_{\text{mpu}})}{n-1} \tag{3.22}$$

Continuing with our stone population example, the Table 3.4 sample space tabulation allows us to numerically illustrate that, for the mean-per-unit estimator, $E(\hat{\sigma}^2_{\text{mpu}}) = \sigma^2$, $E[\hat{V}(\hat{\mu}_{\text{mpu}})] = V(\hat{\mu}_{\text{mpu}})$, and $E[\hat{V}(\hat{T}_{\text{mpu}})] = V(\hat{T}_{\text{mpu}})$.

$$E(\hat{\sigma}^2_{\text{mpu}}) = \sum_{s\in\mathcal{S}}\hat{\sigma}^2_{\text{mpu}}(s)p(s) = \frac{1}{6}\sum_{s=1}^{6}\hat{\sigma}^2_{\text{mpu}}(s) = \frac{1}{6}\cdot 156 = 26 = \sigma^2$$

$$E[\hat{V}(\hat{\mu}_{\text{mpu}})] = \sum_{s\in\mathcal{S}}\hat{V}(\hat{\mu}_{\text{mpu}})(s)p(s) = \frac{1}{6}\sum_{s=1}^{6}\hat{V}(\hat{\mu}_{\text{mpu}})(s) = \frac{1}{6}\cdot 39 = 6.5 = V(\hat{\mu}_{\text{mpu}})$$

$$E[\hat{V}(\hat{T}_{\text{mpu}})] = \sum_{s\in\mathcal{S}}\hat{V}(\hat{T}_{\text{mpu}})(s)p(s) = \frac{1}{6}\sum_{s=1}^{6}\hat{V}(\hat{T}_{\text{mpu}})(s) = \frac{1}{6}\cdot 624 = 104 = V(\hat{T}_{\text{mpu}})$$

For the previous numerical verifications of unbiased mpu estimation under SRS, it is worth noting that the estimators $\hat{V}(\hat{\mu}_{\text{mpu}})$ and $\hat{V}(\hat{T}_{\text{mpu}})$ are, for known n and N, of the form of a constant known factor multiplied by $\hat{\sigma}^2_{\text{mpu}}$, the sample variance. Therefore, rather

---

[3] Throughout this text we denote an estimator of σ^2 as $\hat{\sigma}^2$, consistent with the general convention that an estimator of a parameter θ is denoted by $\hat{\theta}$. In the sampling theory literature, however, the finite population variance is often denoted as S^2, as noted previously, and its estimator (the sample variance) denoted by s^2. In yet another twist, Thompson (2012) uses σ^2 to denote the finite population variance, but uses s^2 to denote its estimator.

Table 3.4 Sample space tabulation for calculating the expected values of $\hat{\sigma}^2_{mpu}$, $\hat{V}(\hat{\mu}_{mpu})$, and $\hat{V}(\hat{T}_{mpu})$ for SRS samples of size $n = 2$ selected from the $N = 4$ population of stone weights (as listed in Table 3.1).

Sample ID	$\hat{\sigma}^2_{mpu}(s)$	$\hat{V}(\hat{\mu}_{mpu})(s)$	$\hat{V}(\hat{T}_{mpu})(s)$
1	4.5	1.125	18
2	12.5	3.125	50
3	72.0	18.000	288
4	2.0	0.500	8
5	40.5	10.125	162
6	24.5	6.125	98
Sum:	156.0	39.000	624

than having numerically shown that $E[\hat{V}(\hat{\mu}_{mpu})] = V(\hat{\mu}_{mpu})$ and $E[\hat{V}(\hat{T}_{mpu})] = V(\hat{T}_{mpu})$, we could instead have relied on a very simple property of expected values of estimators (Section A.3.4). For any constant a

$$E(a\hat{\theta}) = \sum_{s \in \mathcal{S}} a\hat{\theta}(s)p(s) = a \sum_{s \in \mathcal{S}} \hat{\theta}(s)p(s) = aE(\hat{\theta}) \tag{3.23}$$

That is, the expected value of the product of a constant and an estimator is equal to the constant multiplied by the expected value of the estimator. Therefore, the key calculation result is the verification that $E(\hat{\sigma}^2_{mpu}) = \sigma^2$.[4]

Given this result, it should be clear that $\hat{V}(\hat{\mu}_{mpu})$ and $\hat{V}(\hat{T}_{mpu})$ must also be unbiased. That is, for example,

$$E[\hat{V}(\hat{\mu}_{mpu})] = E\left[\left(\frac{N-n}{N}\right)\frac{\hat{\sigma}^2_{mpu}}{n}\right] = \left(\frac{N-n}{Nn}\right)E(\hat{\sigma}^2_{mpu}) = \left(\frac{N-n}{Nn}\right)\sigma^2 = V(\hat{\mu}_{mpu})$$

3.1.4 *Bernoulli sampling*

In some field application settings, when there is no pre-existing map or listing of population units, it may make sense to select units "on the fly", with equal probability. For example, suppose that one wishes to estimate the total area of pools in a reach of a small stream, but there is no pre-existing map of habitat units and N is unknown. **Bernoulli sampling** proceeds as follows. Choose a desired sampling fraction, f, and then, working upstream, select individual pool units independently with *probability f*, say $f = 0.2$. This can be accomplished very easily in the field with a pocket calculator— just select a continuous uniform(0,1) random variate, $U = u$, and select a unit for sampling whenever $u \leq f$, say $u \leq 0.2$. Alternatively, in advance of fieldwork, generate a listing of a large number (well in excess of the unknown N) of independent selections of U, with outcomes 0 or 1 corresponding to whether the values of $u > f$ or $u \leq f$, respectively. Field personnel could then just "check off" the list as they encountered pool units in

---

[4] The unbiasedness of $\hat{\sigma}^2$ for σ^2 is in part a consequence of the sampling theory convention of defining the finite population variance, σ^2, with a divisor of $N-1$, rather than a divisor of N (Section A.5). If σ^2 was defined with a divisor of N, then $\hat{\sigma}^2$ would not be an unbiased estimator of σ^2.

their upstream field work. Whenever the current outcome is "1", then the area of a unit is measured. When the survey has been completed, N will be known as will the realized number of measured pool units, n. Bernoulli sampling, named after the Bernoulli probability distribution (Section A.4.2), is an equal probability selection method.

In Bernoulli sampling, the sample size n is not fixed; its a random variable. For the binomial distribution (which describes a series of N independent Bernoulli trials with probability of success f, Section A.4.4), the expected sample size is $E(n) = Nf$, and the variance in sample size is $V(n) = Nf(1-f)$. When using Bernoulli sampling, we recommend use of an estimator analogous to the mean-per-unit estimator of the total in SRS, $\hat{T}_{\text{mpu}}$, namely

$$\hat{T}_B = N \sum_{i \in S} y_i / n_s \tag{3.24}$$

where n_s is the *realized* sample size rather than a fixed sample size n (as in SRS).

The notion of (unconditional) sampling variance for this estimator, with random sample size, can be understood by imagining repetition of a Bernoulli sampling experiment a very large number of times (e.g., 10^6 times). For each independent experiment, given the realized but variable sample size, estimate the population total using Equation (3.24). Then, numerically calculate the sampling variance over this large number of experiments with variable sample size. Särndal et al. (1992 p. 260) provide an approximate expression for this unconditional[5] sampling variance

$$V(\hat{T}_B) \approx N^2 \left(\frac{1-f}{Nf} \right) \left(1 + \frac{1}{Nf} \right) \sigma^2 = V(\hat{T}_{\text{mpu}}) \left[1 + \frac{1}{Nf} \right] \tag{3.25}$$

because $f = n/N$ for an SRS fixed n design. The factor $[1 + 1/(Nf)]$ is generally not much greater than 1. Thus, there is usually a relatively small price to pay for variable sample size when using Bernoulli sampling with $\hat{T}_B$ compared to SRS and $\hat{T}_{\text{mpu}}$. When estimating sampling variance from a single sample, Särndal et al. [1992 Equation (7.10.6)] recommend use of the *conditionally unbiased*[6] variance estimator

$$\hat{V}(\hat{T}_B) = N^2 \left(\frac{N - n_s}{N} \right) \frac{\hat{\sigma}_B^2}{n_s} \tag{3.26}$$

where $\hat{\sigma}_B^2 = \sum_{i \in S} [(y_i - \hat{T}_B/N)^2]/(n_s - 1)$. By chance, the realized sample size, n_s, will be smaller or larger than $E(n)$, and the conditional variance estimator will directly account for that fact.

In Chapter 4, we provide an alternative equal probability selection method, systematic sampling, which could have been used in this same field context where N is not known in advance. This method has the desirable property that sample size is approximately fixed (fixed exactly when Nf is an integer), but suffers from other issues that are considered in Chapter 4.

3.2 With replacement sampling

Rather than using equal probability without replacement (SRS) selection of units, we could instead use equal probability **sampling with replacement (SWR)**. In SWR, or

---

[5] The unconditional sampling variance is the variation in $\hat{T}_B$ among all possible realized samples, regardless of the value of n.

[6] The conditional sampling variance is the variation in $\hat{T}_B$ among all possible realized samples of size $n = n_s$.

binomial sampling, successive sample units are selected with equal probability $(1/N)$ on an independent draw-by-draw basis. Using the marbles in a jar analogy, if a particular marble is selected on a particular draw, then it is returned to the jar and may be selected on any one or more of the remaining draws. Thus, the same population unit (assuming a simple frame) may appear more than once in a SWR sample. The total number of distinct SWR samples that can be selected, without regard to order of selection, is $\binom{n+N-1}{n}$ (Section A.1.3).

For example, suppose we were to select $n = 2$ sample units from a population of size $N = 4$ with population units identified by the capital letters A, B, C, and D. Without regard to the order in which units were selected, there are ten possible SWR samples that could be selected.

Samples with distinct units: AB, AC, AD, BC, BD, CD

Samples with repeated units: AA, BB, CC, DD

In contrast to SRS, however, these 10 distinct SWR samples are *not* all equally likely. A sample containing the same unit twice, say AA, can only result from selection of that unit on draw 1 *and* draw 2, whereas a sample containing two distinct units, say A and B, can be selected in two different ways: (a) by selecting A on draw 1 and then B on draw 2, *or* (b) by selecting B on draw 1 and then A on draw 2. If each unit is equally likely to be selected on a given draw, then the probability of selecting any particular unit on any given draw is always $1/N$ or, for $N = 4$, 1/4.

Thus, under SWR, individual sample probabilities $[p(s)]$ are not equal. Instead, for $n = 2$ from $N = 4$,

for repeat units $\{i, i\}$:

$$p(s) = \text{Pr}\{\text{select unit } i \text{ on first draw, and unit } i \text{ on second draw}\}$$
$$= \frac{1}{N} \cdot \frac{1}{N} = \frac{1}{4} \cdot \frac{1}{4} = \frac{1}{16}$$

for distinct units $\{i, j\}$:

$$p(s) = \text{Pr}\{\text{select unit } i \text{ on first draw, and unit } j \text{ on second draw}\} +$$
$$\text{Pr}\{\text{select unit } j \text{ on first draw, and unit } i \text{ on second draw}\}$$
$$= \frac{1}{4} \cdot \frac{1}{4} + \frac{1}{4} \cdot \frac{1}{4} = \frac{1}{8}$$

Also, in an SWR sample of size n selected from N, the first order inclusion probability for unit i $(i = 1, 2, \ldots, N)$ is not equal to n/N, as in SRS. The probability that unit i appears x times in a SWR sample of size n selected from N can be calculated from the binomial probability distribution (Section A.4.4) as

$$p(x) = \binom{n}{x} q^x (1-q)^{n-x}, \quad x = 0, 1, 2, \ldots, n$$

where $q = 1/N$ is the draw-by-draw selection probability. The probability that unit i appears in a SWR sample is equal to the sum of the probabilities of all of the sample outcomes that would lead to inclusion of unit i in the sample, $\pi_i = \sum_{x=1}^{n} p(x)$. For example, if $n=4$, and $N = 8$, then π_i would be the sum of the binomial probabilities that unit i appeared one, two, three, or four times in a sample of size 4. As the only other possible outcome is that unit i does not appear in the sample, $p(0)$, it is more efficient to calculate the first order SWR inclusion probabilities as

$$\pi_i = 1 - p(0) = 1 - \binom{n}{0} \left(\frac{1}{N}\right)^0 \left(1 - \frac{1}{N}\right)^n = 1 - \left(1 - \frac{1}{N}\right)^n, \quad i = 1, 2, \ldots, N$$

For $n = 4$ selected from $N = 10$ by SWR, $\pi_i = 1 - 0.9^4 = 1 - 0.6561 = 0.3439$, considerably less than $\pi_i = 4/10$ for an SRS of size 4 selected from $N = 10$ without replacement. Only when n is small compared to N will the inclusion probabilities for SWR and SRS be approximately the same. For example, if $n = 10$ is selected from $N = 1{,}000$ by SRS, then $\pi_i = 10/1000 = 0.01$, but if $n = 10$ were selected from $N = 1{,}000$ by SWR, then $\pi_i = 0.00995512$.

3.2.1 Estimation of the population mean, proportion, and total

Estimators for the population mean, proportion ($y = 0$ or 1), total, and the finite population variance under SWR are identical in form to those for SRS selection and mpu estimation

$$\hat{\mu}_{swr} = \hat{\pi}_{swr} = \sum_{i \in S} y_i/n \tag{3.27}$$

$$\hat{T}_{swr} = N \sum_{i \in S} y_i/n = N\hat{\mu}_{swr} \tag{3.28}$$

$$\hat{\sigma}^2_{swr} = \frac{\sum_{i \in S}(y_i - \hat{\mu}_{swr})^2}{n - 1} \tag{3.29}$$

but, unlike SRS, the random sample S may include repetitions of individual units (Section A.6.1).

Using the same small ($N = 4$) population of stone weights ($\mu = 7$, $T = 28$, $\sigma^2 = 26$) as for the previous SRS/mpu examples, sample space entries in Table 3.5 can be used to verify that the SWR estimators of the population mean and population total are unbiased, but the SWR estimator of finite population variance is negatively biased. Relative magnitude of this bias $(B(\hat{\sigma}^2_{swr})/\sigma^2)$ will be small, however, when n is a small fraction of N. From the fundamental definition of the expected value of an estimator, $E(\hat{\theta}) = \sum_{s \in S} \hat{\theta}(s)p(s)$, we find

Table 3.5 Listing of all possible distinct SWR samples of size $n = 2$ selected from the stone population of size $N = 4$ and associated sample probabilities [$p(s)$], with stone weight population parameter values $\mu = 7$, $T = 28$, and $\sigma^2 = 26$. Associated $\hat{\mu}_{swr}$, $\hat{T}_{swr}$, and $\hat{\sigma}^2_{swr}$ are tabulated, along with $(\hat{\mu}_{swr} - \mu)^2$ for calculation of sampling variance.

	Sample						
ID	Units (s)	Values ($y_i, i \in s$)	$p(s)$	$\hat{\mu}_{swr}(s)$	$\hat{T}_{swr}(s)$	$\hat{\sigma}^2_{swr}(s)$	$(\hat{\mu}_{swr}(s) - \mu)^2$
1	1, 2	2, 5	1/ 8	3.5	14	4.5	12.25
2	1, 3	2, 7	1/ 8	4.5	18	12.5	6.25
3	1, 4	2, 14	1/ 8	8.0	32	72.0	1.00
4	2, 3	5, 7	1/ 8	6.0	24	2.0	1.00
5	2, 4	5, 14	1/ 8	9.5	38	40.5	6.25
6	3, 4	7, 14	1/ 8	10.5	42	24.5	12.25
7	1, 1	2, 2	1/16	2.0	8	0.0	25.00
8	2, 2	5, 5	1/16	5.0	20	0.0	4.00
9	3, 3	7, 7	1/16	7.0	28	0.0	0.00
10	4, 4	14, 14	1/16	14.0	56	0.0	49.00

$$E(\hat{\mu}_{swr}) = \frac{1}{8}[3.5 + 4.5 + 8.0 + 6.0 + 9.5 + 10.5] + \frac{1}{16}[2.0 + 5.0 + 7.0 + 14.0] = 7.0 = \mu$$

$$E(\hat{T}_{swr}) = \frac{1}{8}[14 + 18 + 32 + 24 + 38 + 42] + \frac{1}{16}[8 + 20 + 28 + 56] = 28 = T$$

$$E(\hat{\sigma}^2_{swr}) = \frac{1}{8}[4.5 + 12.5 + 72.0 + 2.0 + 40.5 + 24.5] + \frac{1}{16}[0 + 0 + 0 + 0] = 19.5 \neq \sigma^2 = 26$$

In general under SWR, $E(\hat{\sigma}^2_{swr}) = \sigma^2 \left(\frac{N-1}{N}\right)$. For the population of stone weights, $E(\hat{\sigma}^2_{swr}) = 26 \cdot \frac{3}{4} = 19.5$, as obtained from the sample space listings in Table 3.5.

3.2.2 Sampling variance and variance estimation

Sampling variance for $\hat{\mu}_{swr}$ can be calculated from the sample space entries for $p(s)$ and $(\hat{\mu}_{swr}(s) - \mu)^2$ listed in Table 3.5. Because $E(\hat{\mu}_{swr}) = \mu$ under SWR,

$$V(\hat{\mu}_{swr}) = \sum_{s \in \mathcal{S}} (\hat{\mu}_{swr}(s) - \mu)^2 p(s)$$

$$= \frac{1}{8}(12.25 + 6.25 + \cdots + 6.25 + 12.25) + \frac{1}{16}(25.0 + 4.0 + 0.0 + 49.0) = 9.75$$

In general, under SWR

$$V(\hat{\mu}_{swr}) = \left(\frac{N-1}{N}\right)\frac{\sigma^2}{n} \tag{3.30}$$

and

$$V(\hat{T}_{swr}) = N^2 V(\hat{\mu}_{swr}) = N^2 \left(\frac{N-1}{N}\right)\frac{\sigma^2}{n} \tag{3.31}$$

For our numerical example: $V(\hat{\mu}_{swr}) = \frac{3}{4} \cdot \frac{26}{2} = 9.75$, as calculated from the sample space. Convince yourself that, under SWR

$$\hat{V}(\hat{\mu}_{swr}) = \hat{\sigma}^2_{swr}/n \tag{3.32}$$

must be an unbiased estimator of $V(\hat{\mu}_{swr})$.

3.2.3 Rao–Blackwell theorem

The Rao–Blackwell theorem (Blackwell 1947, Rao 1945) can be used to create a more efficient estimator when sampling is with replacement (Basu 1958, Cassel et al. 1977 Section 2.3).

Theorem 3.1. *If an existing estimator is conditioned on a sufficient statistic, then taking its expected value provides an alternative estimator with a mean square error that is no more than, and perhaps substantially less than, that for the original estimator. Moreover, the expected value of the two estimators is equivalent.*

A sufficient statistic is "sufficient" in the sense that it contains all of the information in the sample data that is relevant to estimating the parameter of interest. In the context of SWR, $y_i, i \in S_d$ is a sufficient statistic for estimating μ and T (Cassel et al. 1977), where S_d is the set of *distinct* units in the random sample S. That is, repeat measurements on the same sample unit provide no additional information relevant to the estimation of μ or T.

Table 3.6 Sample space for SWR samples of size $n = 3$ selected from the stone population of size $N = 4$ with $\mathcal{T} = 28$ and associated sample probabilities $[p(s)]$. Sample-specific $\hat{\mathcal{T}}_{\text{swr}}$ and $\hat{\mathcal{T}}_{\text{RB}}$ estimates of the population total are indicated by $\hat{\mathcal{T}}_{\text{swr}}(s)$ and $\hat{\mathcal{T}}_{\text{RB}}(s)$, respectively.

	Sample				
ID	Units (s)	Values $(y_i, i \in s)$	$p(s)$	$\hat{\mathcal{T}}_{\text{swr}}(s)$	$\hat{\mathcal{T}}_{\text{RB}}(s)$
1	1, 1, 1	2, 2, 2	0.015625	8.000	8.000
2	2, 2, 2	5, 5, 5	0.015625	20.000	20.000
3	3, 3, 3	7, 7, 7	0.015625	28.000	28.000
4	4, 4, 4	14, 14, 14	0.015625	56.000	56.000
5	1, 1, 2	2, 2, 5	0.046875	12.000	14.000
6	1, 1, 3	2, 2, 7	0.046875	14.666	18.000
7	1, 1, 4	2, 2, 14	0.046875	24.000	32.000
8	2, 2, 1	5, 5, 2	0.046875	16.000	14.000
9	2, 2, 3	5, 5, 7	0.046875	22.666	24.000
10	2, 2, 4	5, 5, 14	0.046875	32.000	38.000
11	3, 3, 1	7, 7, 2	0.046875	21.333	18.000
12	3, 3, 2	7, 7, 5	0.046875	25.333	24.000
13	3, 3, 4	7, 7, 14	0.046875	37.333	42.000
14	4, 4, 1	14, 14, 2	0.046875	40.000	32.000
15	4, 4, 2	14, 14, 5	0.046875	44.000	38.000
16	4, 4, 3	14, 14, 7	0.046875	46.666	42.000
17	1, 2, 3	2, 5, 7	0.093750	18.666	18.660
18	1, 2, 4	2, 5, 14	0.093750	28.000	28.000
19	1, 3, 4	2, 7, 14	0.093750	30.666	30.666
20	2, 3, 4	5, 7, 14	0.093750	34.666	34.666

Applying the theorem to $\hat{\mathcal{T}}_{\text{swr}}$ and this sufficient statistic, Cassel et al. (1977) derive an alternative estimator, $\hat{\mathcal{T}}_{\text{RB}} = N \sum_{i \in S_d} y_i / n_d$, where n_d is the number of units in S_d, which is also unbiased with $V(\hat{\mathcal{T}}_{\text{RB}}) < V(\hat{\mathcal{T}}_{\text{swr}})$.

Table 3.6 presents the sample space for an SWR sample size of $n = 3$ selected from the $N = 4$ population of stone weights, and the associated $\hat{\mathcal{T}}_{\text{swr}}$ and $\hat{\mathcal{T}}_{\text{RB}}$ estimates. Convince yourself that both $\hat{\mathcal{T}}_{\text{swr}}$ and $\hat{\mathcal{T}}_{\text{RB}}$ are unbiased for $\mathcal{T} = 28$, and that the sampling variance for $\hat{\mathcal{T}}_{\text{RB}}$ is 91, less than $V(\hat{\mathcal{T}}_{\text{swr}}) = 104$. The reduction in sampling variance from use of the improved estimator $\hat{\mathcal{T}}_{\text{RB}}$, is, however, small compared to the reduction that would be achieved if a fixed sample of size $n = 3$ had instead been selected by SRS. For $n = 3$ selected by SRS, $V(\hat{\mathcal{T}}_{\text{mpu}}) = 34.667$. Nevertheless, this is an important theorem to be aware of and it sees useful application in the theory of adaptive sampling (Thompson and Seber, 1996).

3.3 Relative performance of alternative sampling strategies

3.3.1 *Measures of relative performance*

Throughout this text, we will be interested in assessing the relative performance of one sampling strategy compared to another. Borrowing loosely from Jessen (1978), we define three different measures of relative performance: relative precision, relative efficiency, and net relative efficiency. First, define the **precision** (P) of strategies A and B for estimating a parameter θ as

$$P(\hat{\theta}_A) = 1/V(\hat{\theta}_A)$$

$$P(\hat{\theta}_B) = 1/V(\hat{\theta}_B)$$

where $V(\hat{\theta}_A)$ and $V(\hat{\theta}_B)$ denote sampling variances under strategies A and B, respectively. Then, the **relative precision** (RP) of strategy A compared to strategy B is

$$RP(\hat{\theta}_A, \hat{\theta}_B) = \frac{P(\hat{\theta}_A)}{P(\hat{\theta}_B)} = \frac{1/V(\hat{\theta}_A)}{1/V(\hat{\theta}_B)} = \frac{V(\hat{\theta}_B)}{V(\hat{\theta}_A)} \tag{3.33}$$

Two sampling strategies would be equally precise if the relative precision had value 1.0.

   Because relative precision does not require that sample sizes or costs of sampling are the same for the two sampling strategies that are compared, it is often more useful to fix sample sizes for the two schemes and then compare the two sampling variances, i.e., set $n_A = n_B$. This kind of comparison is termed the **relative efficiency** (RE) of strategy A compared to strategy B

$$RE(\hat{\theta}_A, \hat{\theta}_B) = RP(\hat{\theta}_A, \hat{\theta}_B | n_A = n_B) \tag{3.34}$$

If the sampling variance of strategy B were to exceed the sampling variance of strategy A for equal sample size n, then the relative efficiency of strategy A compared to strategy B would exceed 1. Even when sample sizes are equal, however, the costs of implementing strategy A may be considerably less than or more than the costs of implementing strategy B.[7] Therefore, it is often most instructive to consider both the costs and sampling variances of two alternative strategies. One reasonable way to do this is to fix the total cost for two competing sampling strategies and to compare sampling variances of the two strategies given equal cost. In general, sample sizes for alternative strategies may be different for this comparison. We refer to such a comparison as **net relative efficiency** (NRE)

$$NRE(\hat{\theta}_A, \hat{\theta}_B) = RP(\hat{\theta}_A, \hat{\theta}_B | C_A = C_B) \tag{3.35}$$

We present calculated values and analytic expressions for relative efficiency and net relative efficiency throughout this text. They are key to evaluation of competing sampling strategies.

3.3.2 An example: SRS/mean-per-unit estimation versus SWR

Suppose that we fixed sample size when SRS and SWR selection are used to select units from a simple sampling frame and estimate the population mean. From Equations (3.15) and (3.30), the relative efficiency of the SRS mean-per-unit strategy compared to the SWR strategy for estimating μ would be

$$RE(\hat{\mu}_{mpu}, \hat{\mu}_{swr}) = \frac{V(\hat{\mu}_{swr})}{V(\hat{\mu}_{mpu})} = \frac{[(N-1)/N](\sigma^2/n)}{[(N-n)/N](\sigma^2/n)} = \frac{N-1}{N-n} \tag{3.36}$$

Note that the relative efficiency of the SRS mean-per-unit strategy compared to SWR would equal 1 only for $n = 1$. For larger sample sizes, $V(\hat{\mu}_{mpu})$ is always less than $V(\hat{\mu}_{swr})$ and the relative efficiency of the SRS mean-per-unit strategy compared to SWR will always

---

[7] For example, the expected cost of implementing an unequal probability without replacement sampling design will typically exceed the expected cost of implementing an equal probability without replacement sampling design with identical sample size. Unequal probability sampling designs are considered in Chapter 8.

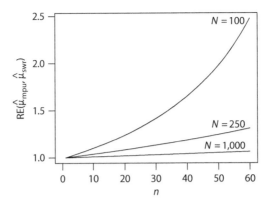

Fig. 3.2 Relative efficiency of SRS mean-per-unit strategy compared to SWR for estimating μ for three different population sizes and sample sizes ranging from 1 through 60.

exceed 1. For this reason, SRS mean-per-unit estimation is almost always used in preference to SWR—it is more efficient for $n > 1$. If N is large, however, then the relative efficiency of the SRS mean-per-unit strategy compared to SWR will be close to 1 if sample sizes are small compared to N (e.g., see the $N = 1{,}000$ curve in Figure 3.2).

3.4 Sample size to achieve desired level of precision

One of the most frequent questions that young graduate students ask a statistician is: "What should my sample size be?" As we will see next, this kind of question is generally unanswerable without some critical information concerning (a) desired precision of estimation, and (b) some preliminary data that can shed light on the likely level of variation in the target variable for the population from which the sample will be selected.

The desired precision of estimation is most frequently specified by a desired bounds on the error of estimation, which typically requires that one invoke some notion of confidence interval construction around the estimated value. Thus, the first topic that must be considered is construction of confidence intervals in the context of finite population sampling. (A more formal treatment of this topic is given in Section A.7.4.)

3.4.1 *Approximate normality of sampling distributions*

There are two key theorems for infinite populations that provide the essential logic behind construction of the symmetric confidence intervals that are routinely constructed in sampling theory. These theorems assume that sample units have been drawn independently and may be loosely phrased as follows:

Theorem 3.2. *If Y is a normally distributed random variable with mean μ and standard deviation σ, then, for any sample size n, the sampling distribution of the sample mean, $\hat{\mu} = \sum_i y_i / n$, will be normal with mean μ and standard deviation $\sigma / \sqrt{n}$.*

Theorem 3.3. *If Y possesses a distribution with mean μ and standard deviation σ, then, for large n, the sampling distribution of the sample mean, $\hat{\mu}$, will be approximately normal with mean μ and standard deviation $\sigma / \sqrt{n}$. The approximation to normality will improve as n gets larger.*

The first theorem states that if the target variable for the population from which the sample is drawn is normally distributed, then the sampling distribution of sample mean will also be normally distributed, for any sample size n. The second theorem, known as the **central limit theorem (CLT)**, states that even if the target variable is not normally distributed, the limiting shape of the sampling distribution of the sample mean will still be approximately normal. The approximation to normality will become better and better for large sample sizes (typically, $n > 30$ is "large" enough for the normal approximation to be a good one). Although these theorems strictly apply only to an infinite population and independent sampling, they apply reasonably well for finite populations and other probability-based selection methods, where the square root of the sampling variance replaces $\sigma/\sqrt{n}$. Most important, the approximate normality of the sampling distribution of the sample mean (and of other mean-based estimators used in sampling theory, e.g., $\hat{\pi}$ and $\hat{T}$) is key to the motivation and logic for construction of symmetric confidence intervals for many of the estimators presented in this text.

For large N and n, the number of possible samples that can be selected quickly becomes so large that it is generally not feasible to directly determine the sampling distribution of an estimator by constructing all possible samples, calculating the associated sample-specific estimates, and displaying the results as an appropriately scaled histogram. For example, for $N = 1,000$ and $n = 5$, $\binom{N}{n} = 8.250291 \times 10^{12}$, so there are more than 8 trillion possible samples! Instead, we most often *approximate* the sampling distribution of estimators via **Monte Carlo simulation** of the selection of a large number, Q, of independent samples, selected according to the sampling design, calculating sample-specific estimates for each of these samples and then displaying these results as a histogram (Section A.7.2). If the histogram bin frequencies are divided by Q, this would give the approximate probability that the estimator would produce an estimate within each of these bin intervals, and thus provides an approximation of the estimator sampling distribution. If Q is sufficiently large (say, $Q \geq 10^4$), then the shape of the simulated sampling distribution and its expected value and variance, will closely match those of the actual sampling distribution, and better matches can be achieved with larger values of Q, restricted only by the computational capabilities of a computer.

Monte Carlo simulations of sampling distributions can be used to assess the degree to which Theorems 3.2 and 3.3 apply to a small ($N = 200$) finite population and a variety of distributions for the target variable. Figure 3.3 (top row) displays five different distributions of the target variable that were obtained by generating 200 independent random numbers from each of the following distributions: normal, uniform, exponential, binormal, and binomial. The remaining rows in the figure show the scaled histograms (approximate sampling distributions) of the sample-specific values $\hat{\mu}_{\mathrm{mpu}}(s)$ for sample sizes of 2, 5, 15, and 30, respectively, based on Monte Carlo simulations of $Q = 10^4$ independently selected SRS samples for each sample size and variable distribution. From inspection of these sampling distributions, we draw the following *very* crude "rule of thumb" conclusions.

1. If the variable has an approximately normal frequency distribution, then the sampling distribution of $\hat{\mu}_{\mathrm{mpu}}$ will be approximately normal in shape for all sample sizes, consistent with Theorem 3.2.
2. If the variable has an approximately uniform frequency distribution, then the sampling distribution of $\hat{\mu}_{\mathrm{mpu}}$ will be approximately symmetric and "triangular" in shape for very small sample sizes ($n = 2$) and will become approximately normal at quite modest sample sizes (say, $n \geq 10$).

3. If the variable has an approximately exponential frequency distribution, then the shape of the sampling distribution of $\hat{\mu}_{mpu}$ will be similar to the exponential distribution at $n = 2$, will have substantial skewness consistent with the exponential distribution at $n = 5$, but becomes approximately normal in shape somewhere between $n = 15$ and $n = 30$.

4. If the variable has an approximately binormal frequency distribution, then the shape of the sampling distribution of $\hat{\mu}_{mpu}$ will strongly reflect the distribution of population values at $n = 2$, but becomes approximately normal in shape somewhere between $n = 15$ and $n = 30$.

5. If the population y values are binary (0, 1), and $0.05 < \mu < 0.95$, then the sampling distribution of $\hat{\mu}_{mpu}$ does not approach normality until approximately $n = 30$.

There are, of course, many alternative variable distributions, and a more refined analysis would examine the shape of the sampling distribution at a far larger number of sample sizes, but the Monte Carlo simulations of sampling distributions displayed in Figure 3.3 do provide some direct reassurance that the central limit theorem does generally apply when SRS mean-per-unit estimation is used in a small finite population. The shape of the sampling distribution of $\hat{\mu}_{mpu}$ seems nearly normal so long as $n \geq 15$, with the exception of the case when population y values are binary (0, 1). For binary population values, where $\hat{\mu}_{mpu}$ is equivalent to an estimator of a proportion, $\hat{\pi}_{mpu}$, it seems clear that n must be considerably larger (at least 30). If the proportion is close to 0 or to 1, then it can take a sample much larger than $n = 30$ for the normal approximation to be considered a good one.

3.4.2 *Confidence interval construction*

If the shapes of the sampling distributions of $\hat{\mu}$, $\hat{T}$, or $\hat{\pi}$ are approximately normal, then we can form symmetric confidence intervals about our sample estimates using

$$\hat{\theta} \pm t_{(1-\alpha/2),(n-1)} \sqrt{\hat{V}(\hat{\theta})} \tag{3.37}$$

where $\hat{\theta} = \hat{\mu}$, $\hat{T}$, or $\hat{\pi}$ and are assumed unbiased. The term $t_{(1-\alpha/2),(n-1)}$ is the $(1-\alpha/2)$ quantile of a t distribution with $(n-1)$ degrees of freedom. When $1 - \alpha = 0.95$ and n is "large", then $t_{(1-\alpha/2),(n-1)} \approx 2$, and we therefore often form an (approximate) **95% confidence interval** as $\hat{\theta} \pm 2\sqrt{\hat{V}(\hat{\theta})}$. We refer to $\sqrt{\hat{V}(\hat{\theta})} = \widehat{SE}(\hat{\theta})$ as the *estimated standard error* of $\hat{\theta}$, where the standard error of $\hat{\theta}$ is $SE(\hat{\theta}) = \sqrt{V(\hat{\theta})}$, and we refer to $2\sqrt{\hat{V}(\hat{\theta})}$ as the large sample **95% bounds on the error of estimation**, B.[8] If the normality and large sample assumptions are met, then in *repeated sampling experiments* we would expect that (approximately) 95% of the confidence intervals constructed using $\hat{\theta} \pm 2\sqrt{\hat{V}(\hat{\theta})}$ would contain the target value θ.

3.4.3 *Sample size determination*

The precision of $\hat{\mu}_{mpu}$ $[1/V(\hat{\mu}_{mpu})]$ or $\hat{\mu}_{swr}$ $[1/V(\hat{\mu}_{swr})]$ depends on sample size, in addition to N and σ^2. In a survey planning context, the reverse is true: required sample

---

[8] The factor of 2 is close to the limiting value of t, 1.96, as the degrees of freedom ($n-1$ in mpu estimation) becomes large. Even for relatively small degrees of freedom, the approximate value of 2 is often not much less than the actual value of t. For example, for $n = 11$ (10 degrees of freedom in mpu estimation), $t = 2.2$.

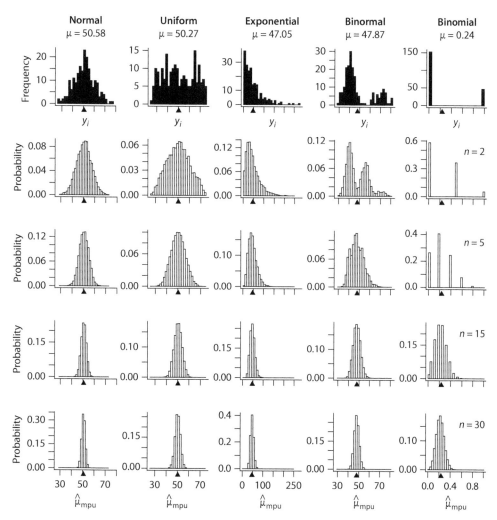

Fig. 3.3 Monte Carlo approximations of sampling distributions of $\hat{\mu}_{mpu}$ for sample sizes of $n = 2, 5$, 15, and 30 selected from populations of size $N = 200$ with target variable values randomly generated from normal, uniform, exponential, binormal, and binomial ($y = 0$ or 1) distributions. Variable value distributions (solid black), with population means, are displayed in the top row; simulated sampling distributions (white bars) are displayed in the remaining rows for each value of n, respectively. Approximate sampling distributions are based on 10^4 independent SRS samples selected from each population for each sample size.

size depends on the desired (prespecified) level of precision. Given a desired sampling variance, $V(\hat{\mu}_{mpu})$, and known, estimated or assumed values for N and σ^2, one may solve Equation (3.15) for n to obtain

$$n = \frac{\sigma^2}{V(\hat{\mu}_{mpu}) + \sigma^2/N} \tag{3.38}$$

Because Equations (3.15) and (3.16) have the same general form, the solution for n may be generalized for estimating $\theta = \mu, \pi,$ or $\mathcal{T}$ as

$$n = \frac{\sigma^2}{\mathcal{K}(\hat{\theta}_{\mathrm{mpu}}) + \sigma^2/N}, \qquad \mathcal{K}(\hat{\theta}) = \begin{cases} V(\hat{\theta}), & \text{for } \theta = \mu, \pi \\ V(\hat{\theta})/N^2, & \text{for } \theta = \mathcal{T} \end{cases} \tag{3.39}$$

Similarly, for $\hat{\mu}_{\mathrm{swr}}$, Equations (3.30) and (3.31) lead to the following general solution for sample size for equal probability with replacement sampling

$$n = \left(\frac{N-1}{N}\right)\frac{\sigma^2}{\mathcal{K}(\hat{\theta}_{\mathrm{swr}})} \tag{3.40}$$

with $\mathcal{K}(\hat{\theta})$ defined as in Equation (3.39).

Desired estimator uncertainty is rarely specified directly in terms of sampling variance, however. Instead, it is usually expressed as a desired standard error, confidence interval half-width (bounds on error of estimation), or **coefficient of variation** [CV$(\hat{\theta}) =$ SE$(\hat{\theta})/E(\hat{\theta})$]. (For an unbiased estimator $\hat{\theta}$, CV$(\hat{\theta}) = $ SE$(\hat{\theta})/\theta$, and an estimator for the coefficient of variation is $\widehat{\mathrm{CV}}(\hat{\theta}) = \widehat{\mathrm{SE}}(\hat{\theta})/\hat{\theta}$.) The coefficient of variation compares the standard error to the target of estimation, is often expressed as a percentage (instead of a proportion), and is a very useful measure of *relative uncertainty* that is often adopted by resource managers. In contrast, sampling variance or a desired standard error, by themselves, provide no notion of relative uncertainty without knowledge of the magnitude of the target of estimation or the units of measurement. For example, a sampling variance of 4,000,000 (SE = 2,000) when compared to a target θ value of 10,000 (CV = 20%) has quite a different meaning than when compared to a target θ value of 1,000 (CV = 200%).

For a large sample 95% bounds on the error of estimation, re-express $V(\hat{\theta})$ in terms of $B = 2\mathrm{SE}(\hat{\theta})$ so that $V(\hat{\theta}) = B^2/4$. For a specified coefficient of variation, assuming $\hat{\theta}$ is unbiased (and positive valued), re-express $V(\hat{\theta})$ as $\mathrm{CV}^2(\hat{\theta})\theta^2$. Either of these re-expressions for $V(\hat{\theta})$ can then be substituted in Equations (3.39) or (3.40) as appropriate.

In practice, σ^2 is not a known value, so use of these solutions for n requires some preliminary survey data from which to generate an estimate for σ^2. In the case of a proportion, an a priori notion of, or likely range for, the value of π is all that is required because for a binary (0, 1) variable $\sigma^2 = N\pi(1-\pi)/(N-1)$. Thus, the necessary n for a specified $V(\hat{\pi})$, B, or CV$(\hat{\pi})$ depends only on N and π.

3.5 Nonresponse and oversampling

In practice, it is not uncommon that observation or measurement of y cannot be made for one or more of the selected units in an SRS or SWR sample of size n. The consequences of such events, which are generically referred to as **nonresponse** events and nonresponse units, depends on whether the reason for the nonresponse is independent of the unit's y value. For example, suppose that the objective of a sample survey is to allow estimation of the number of an endangered amphibian species in a watershed of small lakes and that the population units consist of these small lakes. Within each selected lake, a nighttime vocalization survey will be used to enumerate the number of individuals present. However, permission for property access must be granted prior to surveying the selected lakes. Imagine two very different scenarios for nonresponse. First, a landowner is notified but does not respond to the access request. Second, a landowner denies permission knowing that the endangered species is present on their property and is concerned that the survey data might lead to further restrictions on land use. In the first case, it may be reasonable to suppose that nonresponse resulted from *pure chance*, i.e., that it is unrelated to the value of y for the affected units. That is, the landowner may have not responded to the notification due to some sort of clerical/mail handling error, or is currently unreachable, etc. (Other examples of *pure chance* nonresponse include such things as "field crew inadvertently skipped a sample location in the survey", "the variable measurement for a unit was compromised in some way, or incorrectly recorded", etc.) For the second case however, there is good reason to believe that the nonresponse is related to the value of y. If a landowner knows that a survey of the variable of interest on their land is likely to lead to greater restrictions on their property use, they may be less likely to grant access.

For the *pure chance* case of nonresponse, let S_r denote the random sample of *response units* for which y can be measured or observed, and let n_r be the associated sample size, with $n_r \leq n$. Then, for both SRS and SWR selection of n population units, the usual mpu and swr estimators, calculated using the *random sample of n_r response units, S_r*, remain unbiased estimators of the population parameters T, μ, and π

$$\hat{\mu} = \hat{\pi} = \sum_{i \in S_r} \frac{y_i}{n_r}, \qquad \hat{T} = N\hat{\mu}$$

Corresponding estimators of sampling variance, based on the S_r sample of size n_r, are unbiased estimators of *conditional* sampling variance of the analogous target population parameters *given the reduced sample size n_r*. For example, $\hat{V}(\hat{T}_{mpu}) = N^2 [(N - n_r)/N] \hat{\sigma}^2_{mpu}/n_r$ is an unbiased estimator of $V(\hat{T}_{mpu}) = N^2 [(N - n_r)/N] \sigma^2/n_r$, where $\hat{\sigma}^2_{mpu} = \sum_{i \in S_r} (y_i - \hat{\mu}_{mpu})^2/(n_r - 1)$. Thus, if nonresponse is the result of *pure chance*, then the usual equal probability estimators calculated over the reduced sample S_r remain unbiased.

If instead the nonresponse depends on y, then the estimators calculated over the reduced sample S_r will be biased. The degree of bias will depend on the relative reduction in sample size and the degree of association between the y values and nonresponse. For example, if only one nonresponse unit is encountered in a sample of size $n = 100$, then the degree of bias is likely to be relatively small, even when that nonresponse is associated with the value of y. But if one nonresponse unit is encountered in a sample of size 10 and nonresponse is strongly associated with the value of y, then the bias may be substantial. In some field survey contexts, nonresponse may be encountered at relatively high frequency (e.g., 20% of selected units may be nonresponse units) and in such cases the potential for serious estimation bias should be a cause for concern.

If nonresponse is anticipated to be a major problem or it is critical to achieve a realized sample of fixed size n and some nonresponse is anticipated, then one may instead select an initial SRS **oversample** of size $n' = \gamma n$, where γ is sufficiently large so as to nearly guarantee that $n' - n_{nr} \geq n$, where n_{nr} is the number of nonresponse units. For example, if the nonresponse rate is anticipated to be about 20%, the desired sample size is 10, and $\gamma = 2$, then the probability that the number of response units is at least 10 is, in R, `sum(dbinom(x=10:20, size=20, prob=0.8))`, which gives the result 0.9994, close to certainty. Select the first n response units in the oversample to achieve the desired sample of size n. Problem 3.10 asks you to show, using R, that the usual mean-per-unit estimators will remain unbiased for this oversampling scheme (assuming that all nonresponse is *pure chance*) and that the conditional sampling variance given n will be the same as for a sample of fixed size n for which there is no nonresponse. The number of potential sample units which will have to be *evaluated* for nonresponse, however, is a random variable and there may be an additional cost for this evaluation as compared to selection of a sample of fixed size n with no nonresponse. The unbiasedness of this oversampling approach is a consequence of the fact that *a simple random sample of a simple random sample is also a simple random sample!* This oversampling device is used in GRTS, a spatially representative selection method considered in Chapter 12.

3.6 Sampling in R

3.6.1 *SRS and SWR*

Unit labels for individual SRS or SWR samples of size n selected from N can be generated using the following R code

```
s <- sample(1:N, size=n, replace=FALSE)   # SRS sample
s <- sample(1:N, size=n, replace=TRUE)    # SWR sample
```

Example 3.1. *Select SRS and obtain associated y values.*

```
N <- 10; n <- 3
y <- seq(from=10, to=100, by=10)
s <- sample(1:N, size=n, replace=FALSE)

y[s]
[1] 60 90 10
```

3.6.2 *Sample spaces*

R package iterpc (Lai 2018) can be used to obtain a listing of all possible samples (sample space) for with and without replacement sampling. The listing is in the form of a matrix, with rows containing the unit labels of each sample. The package must be downloaded and installed on your computer the first time it is used. To do this type

```
install.packages("iterpc")
```

and select a nearby server from the presented list. Once the package is installed, you must load the package in an R session prior to using it. To do this type

```
library("iterpc")
```

Once the iterpc package has been loaded, the following R code can be used to generate a $\binom{N}{n} \times n$ (SRS) or $\binom{n+N-1}{n} \times n$ (SWR) matrix containing all possible sample unit selections (distinct samples of size n selected from N without regard to order of selection)

```
SS <- getall(iterpc(N, n, replace=FALSE))   # SRS sample space
SS <- getall(iterpc(N, n, replace=TRUE))    # SWR sample space
```

Example 3.2. *SWR sample space and associated y values.*

```
N  <- 3; n <- 2
y  <- seq(from=10, to=30, by=10)
SS <- getall(iterpc(N, n, replace=TRUE))

SS
      [,1] [,2]
[1,]    1    1
[2,]    1    2
[3,]    1    3
[4,]    2    2
[5,]    2    3
```

```
[6,]    3    3

matrix(y[SS], nrow=nrow(SS))
        [,1] [,2]
[1,]    10   10
[2,]    10   20
[3,]    10   30
[4,]    20   20
[5,]    20   30
[6,]    30   30
```

3.7 Chapter comments

Although it might be argued that with replacement selection of n units makes no sense because $V(\hat{\mu}_{swr}) > V(\hat{\mu}_{mpu})$ unless $n = 1$, there are at least two reasons to consider with replacement selection of units. First, there is an unequal probability extension of with replacement selection (Section 8.2) which has several positive features. Second, and perhaps more important, in some settings it is much more appropriate to view the sampling process as a with replacement process. For example, if hook-and-line catch-and-release fishing is used to collect individual fish to determine the average age in a population, then this is clearly a with replacement process; the same fish might be captured more than once.

In Appendix A, we provide a more general and more formal treatment of many of the fundamental principles that have been introduced in this chapter for the special case of equal probability selection of units from a finite population and estimation using $\hat{\theta}_{mpu}$ or $\hat{\theta}_{swr}$. Readers with more advanced mathematics and statistics backgrounds may wish to carefully study all material in Appendix A before proceeding to the following chapters, but other readers may wish instead to use Appendix A as a reference source and review the appendix material only as it is cross-referenced elsewhere in the text. All readers should be aware, however, that Appendix A provides the mathematical foundation of finite population sampling theory.

We briefly discuss the issue of nonresponse also in Chapter 4 and also in Chapter 8 which provides a review of the more general setting where sample units are selected with unequal probability. The topics of nonresponse and what to do about it are, generally, beyond the level of this introductory text. An excellent though advanced introduction to this subject area is Särndal et al. (1992 Chapter 15); Särndal and Lundström (2005) is an entire text devoted to this topic.

Problems

Problem 3.1. The coefficient of variation of an estimator is defined as $CV(\hat{\theta}) = SE(\hat{\theta})/E(\hat{\theta})$. (a) For sample sizes $n = 10$ and $n = 20$, calculate the coefficients of variation of $\hat{\mu}_{mpu}$, $\hat{\mu}_{swr}$, and $\hat{\tau}_{mpu}$ for each of the sets of $N = 100$ variable values listed in the following table. The R function seq() can be used to quickly generate the population values. Enter your calculated CVs in a table like the one below. (b) Briefly discuss the relevance of your results.

Estimator ($\hat{\theta}$)	Variable values ($y_i, i = 1, 2, \ldots, N$)	CV($\hat{\theta}$)	
		$n = 10$	$n = 20$
$\hat{\mu}_{\text{mpu}}$	$1, 2, \ldots, 100$	_____	_____
$\hat{\mu}_{\text{mpu}}$	$0.01, 0.02, \ldots, 1.00$	_____	_____
$\hat{\mu}_{\text{swr}}$	$1, 2, \ldots, 100$	_____	_____
$\hat{\mu}_{\text{swr}}$	$0.01, 0.02, \ldots, 1.00$	_____	_____
$\hat{T}_{\text{mpu}}$	$1, 2, \ldots, 100$	_____	_____
$\hat{T}_{\text{mpu}}$	$0.01, 0.02, \ldots, 1.00$	_____	_____

Problem 3.2. For a linear combination of independent estimates $\hat{\theta}_i$, $\hat{V}[\sum(a_i\hat{\theta}_i)] = \sum a_i^2 \hat{V}(\hat{\theta}_i)$. Let $\hat{\theta}_1$ and $\hat{\theta}_2$ be independent unbiased estimates of the same target value θ. Then, any weighted average of these estimates will also be unbiased. That is, if $\hat{\theta}^* = \alpha\hat{\theta}_1 + (1-\alpha)\hat{\theta}_2$, for $0 \leq \alpha \leq 1$, then $E(\hat{\theta}^*) = \theta$. Suppose that you were given the following two independent and unbiased estimates of the total number of marbled murrelets, *Brachyramphus marmoratus*, a small (threatened status) seabird which nests in coastal redwood forests, in Humboldt County, California, along with their estimated sampling variances: $\hat{\theta}_1 = 732$, $\hat{V}(\hat{\theta}_1) = 49{,}672$, and $\hat{\theta}_2 = 1{,}332$, $\hat{V}(\hat{\theta}_2) = 78{,}443$. Find (numerically or using calculus) the value of α that will minimize the estimated sampling variance of the total number of murrelets and calculate the associated *best estimate* of the total number of murrelets.

Problem 3.3. Suppose that we were to select samples by SRS from a population of size $N = 5$, with $y_1 = 3$, $y_2 = 9$, $y_3 = 14$, $y_4 = 6$, and $y_5 = 13$. (a) Construct all possible distinct SRS samples of size $n = 3$. (b) For each sample, calculate $\hat{T}_{\text{mpu}}$, $\hat{\sigma}^2_{\text{mpu}}$, $\hat{\sigma}_{\text{mpu}} = \sqrt{\hat{\sigma}^2_{\text{mpu}}}$, and $\hat{V}(\hat{T}_{\text{mpu}})$. (c) Show numerically that $V(\hat{T}_{\text{mpu}})$ calculated over the sample space is equal to $N^2((N-n)/N)\sigma^2/n$, that $\hat{T}_{\text{mpu}}$, $\hat{\sigma}^2_{\text{mpu}}$, and $\hat{V}(\hat{T}_{\text{mpu}})$ are unbiased estimators of T_y, σ_y^2, and $V(\hat{T}_{\text{mpu}})$, respectively, but that $\hat{\sigma}_{\text{mpu}}$ is not an unbiased estimator of $\sigma = \sqrt{\sigma^2}$.

Problem 3.4. Suppose that we were to select samples by SWR from a population of size $N = 4$, with $y_1 = 3$, $y_2 = 9$, $y_3 = 14$ and $y_4 = 6$. (a) Construct all possible distinct SWR samples of size $n = 3$ (i.e., without regard to order of selection of units) and calculate the corresponding sample probabilities, $p(s)$. (b) For each sample s, calculate $\hat{\mu}_{\text{swr}}(s)$ and $\hat{\sigma}^2_{\text{swr}}(s)$. (c) Using the list of all possible samples, corresponding sample estimates, and sample probabilities, find $E(\hat{\mu}_{\text{swr}})$, $E(\hat{\sigma}^2_{\text{swr}})$, and $V(\hat{\mu}_{\text{swr}})$. Is $\hat{\sigma}^2_{\text{swr}}$ an unbiased estimator of σ^2? (d) For each sample, estimate the sampling variance using $\hat{V}(\hat{\mu}_{\text{swr}}) = \hat{\sigma}^2/n$. Is $\hat{V}(\hat{\mu}_{\text{swr}})$ an unbiased estimator of $V(\hat{\mu}_{\text{swr}})$? (e) Suppose that SRS had instead been used to select samples of size $n = 3$. How would $V(\hat{\mu}_{\text{mpu}})$ compare with $V(\hat{\mu}_{\text{swr}})$?

Problem 3.5. Suppose that $N = 4$ and that the target variable values are $y = \{0, 1, 1, 0\}$. (a) Construct the SRS sample space for $n = 2$. (b) Numerically verify that $E(\hat{\sigma}^2) = \sigma^2$, $V(\hat{\pi}) = \left(\frac{N-n}{N-1}\right)\frac{\pi(1-\pi)}{n}$, and $E[\hat{V}(\hat{\pi})] = V(\hat{\pi})$, where $\hat{V}(\hat{\pi}) = \left(\frac{N-n}{N}\right)\hat{\pi}(1-\hat{\pi})/(n-1)$. (c) Show algebraically that $\sigma_y^2 = N\pi(1-\pi)/(N-1)$ when y is a binary (0, 1) variable.

Problem 3.6. It may often be of interest to use SRS to estimate a mean or total over some "domain of interest" which is a subset of the population. For example, in an SRS of n 100 m^2 forest plots selected from N, one might be interested in estimating the total biomass of trees with diameter at breast height exceeding 0.5 m that are on plots with northern exposure. That is, we desire an estimate of T_d, where d designates the domain of interest, and N_d is the size of that domain. Consider the following two alternative unbiased estimators

$$\hat{T}_{d,1} = N_d \sum_{\substack{i \in S \\ i \in d}} y_i / n_d, \text{ for } N_d \text{ known;} \qquad \hat{T}_{d,2} = \frac{N}{n} \sum_{\substack{i \in S \\ i \in d}} y_i, \text{ for } N_d \text{ unknown}$$

Note that $\hat{T}_{d,1}$ is equivalent to $N_d \hat{\mu}_d$ and that $\hat{\mu}_d$ is equal to the mean of the sample y that fall in the *random* sample size n_d (sample units that fall in domain d). (a) Write a simulation program in R that evaluates the performance of these two alternative estimators for $n = (10, 20, 30, 40, 50, 100)$. Assume that $N = 200$, $N_d = 100$, and base your simulations on the following code.

```
set.seed(1000)                                  # ensures fixed y
y.d    <- round(rnorm(100, mean= 180, sd= 10))  # in domain
y.not <- round(rnorm(100, mean=1000, sd=100))   # not in domain
y      <- c(y.d, y.not)                          # all y values
```

(b) Numerically confirm that $E(\hat{\mu}_d) = \mu_d$. (c) Calculate and contrast/discuss the simulated sampling variances for $\hat{T}_{d,1}$ and $\hat{T}_{d,2}$ based on 10^6 independent SRS sample selections.

Problem 3.7. Given that $E(\hat{\sigma}^2_{swr}) = \sigma^2 (N-1)/N$, show algebraically that $\hat{\sigma}^2_{swr}/n$ must be an unbiased estimator of $V(\hat{\mu}_{swr}) = [(N-1)/N]\sigma^2/n$.

Problem 3.8. Suppose that finite population variance was defined as $\sigma^{*2} = \sum_{i=1}^{N}(y_i - \mu)^2/N$ rather than $\sigma^2 = \sum_{i=1}^{N}(y_i - \mu)^2/(N-1)$. (a) Find algebraic expressions for $V(\hat{\mu}_{mpu})$ and $V(\hat{\mu}_{swr})$ in terms of σ^{*2} rather than σ^2. (b) In SRS and SWR, would $\hat{\sigma}^2 = \sum_{i \in S}(y_i - \hat{\mu})^2/(n-1)$ be unbiased estimators of σ^{*2}?

Problem 3.9. (a) Using a without replacement sample space of $n = 3$ and $N = 4$, show algebraically that $\sum_{s \in S} \sum_{i \in s} y_i = \binom{3}{2} \sum_{i=1}^{N} y_i$. (b) More generally, try to prove that, when samples are selected without replacement, the number of samples that contain unit i is equal to $\binom{N-1}{n-1}$. (Note: you might find it useful to refer to Section A.1.3.)

Problem 3.10. Suppose that the probability of *pure chance* nonresponse is 0.75 and that it is desired to achieve an SRS sample size of $n = 10$ when sampling from a population of size $N = 100$. Write a simulation program in R that implements the SRS oversampling procedure described in Section 3.5. Choose y so that $\Pr\{n_r \geq 10\} > 0.999$, where n_r is the number of response units in the oversample of size $n' = \gamma n$. Verify that this oversampling procedure leads to unbiased estimation of T with sampling variance equal to $V(\hat{T}_{mpu})$ for the fixed sample size of $n = 10$, assuming that all nonresponse is due to pure chance.

Problem 3.11. In SWR, the first order inclusion probability is $\pi_i = 1 - [1 - (1/N)]^n$ for all i, and the second order inclusion probability (the probability that units i and j appear together in a sample of size n from N) is $\pi_{ij} = 1 - 2[1 - (1/N)]^n + [1 - (2/N)]^n$ for all $i \neq j$ [Särndal et al. 1992 Equation (2.9.4)]. (a) Numerically verify that these expressions give the correct π_i and π_{ij} for the sample space summarized in Table 3.5. (b) Justify these expressions algebraically. (You should find it helpful to review Equation (A.14).)

CHAPTER 4

Systematic sampling

Fig. 4.1 Dungeness crabs, *Cancer magister*, stacked in boxes on the stern of a small commercial vessel in northern California. Selection of an SRS of crabs from these boxes would be a practical impossibility, whereas systematic sampling would be entirely feasible. Photo credit: D. Hankin.

In many instances it may be impractical or impossible to select a sample by SRS. For example, consider the practical sampling problem of estimating the average carapace width of Dungeness crabs that are landed by a particular vessel on a particular day (Figure 4.1). Selecting an SRS sample of these crabs would require, for example, that one first line up all N crabs, and then select n positions at random with equal probability without replacement from the $1, 2, \ldots, N$ positions in the line. What fisher would allow use of such a sampling method on board his or her vessel? Imagine instead that one wished to estimate the average carapace width of crabs at a processing plant where many vessel owners sold their catch. Would the processing plant allow many thousands of these crabs to be "lined up" so that a valid SRS sample could be selected? In both of these examples, the practical difficulty of selecting an SRS sample would likely rule out its use. (The issue that live crabs move would also have to be overcome!) Clearly, a more practical sampling method would be valuable, especially one that does not require advance knowledge of N.

Sampling Theory: For the Ecological and Natural Resource Sciences. David G. Hankin, Michael S. Mohr, and Ken B. Newman, Oxford University Press (2019). © David G. Hankin, Michael S. Mohr, and Ken B. Newman. DOI: 10.1093/oso/9780198815792.001.0001

In the above examples, selection by SRS would not be "impossible," but it would be "infeasible". In other settings, selection by SRS would simply be impossible. Consider the following two examples. First, imagine that you wished to estimate the total kill of birds on a game refuge on a particular date during which a limited kill was allowed. Hunters enter and leave the area all day, at various times, but there are only four access points through which they may enter/exit the refuge. How could you make a list (in advance) of all those hunters who will hunt on that day and then draw an SRS sample of those hunters? Such a list would not be available until all hunting had ceased for that day and, by that time, it would no longer be possible to select an SRS sample of hunters onsite as they would have already left the refuge with their birds. Second, imagine that you have been asked to conduct an oral interview SRS sample of 10% of all individuals who were observing waterfowl at a different refuge on a no-hunting day. Again, because the number of individuals (N) participating in the activity of interest would not be known in advance, it would be impossible to prespecify the sample size ($n = 0.1 \times N$).

In cases like these, where it is clearly impractical or simply impossible to select an SRS, it may be feasible to instead select an equal probability *systematic* sample. In **linear systematic sampling**, one first chooses a random start, r, on the integers 1 through k. One then selects unit r and every k^{th} unit thereafter. Even if N is not known in advance, the resulting sample size will be $n \approx N/k$ and the sampling fraction will be $f \approx 1/k$. Systematic sampling has many obvious advantages. First, it clearly provides a practical alternative to SRS selection for each of the settings described above. One could easily select a "1 in k" sample of crabs from a box, or of hunters as they left each of the four access points. Second, selecting a sample systematically through time or space should guarantee a temporal or spatial "representativeness"[1] that cannot be guaranteed by SRS selection (see, e.g., Figure 4.2). For example, suppose that hunter success varies according to time of day. The systematic sampling procedure would guarantee that hunters would be selected "evenly" at each access site across the full spectrum of actual hunting times. An SRS selection procedure, if it were feasible, might by chance result in hunters primarily being

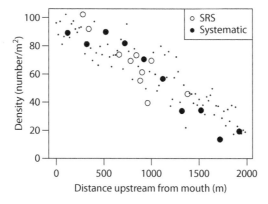

Fig. 4.2 Density of juvenile steelhead (anadromous rainbow trout, *Oncorhynchus mykiss*) within 20 m sections of the lower 2,000 m of a stream ($N = 100$ units). Location of sample units in a particular systematic sample (solid circles) compared to a particular simple random sample (open circles) of the same size ($n = 10$). Note the difference in degree of spatial regularity for the systematic as compared to the SRS unit selections.

---

[1] We will use the terms representativeness, regularity, evenness, and balance interchangeably when referring to this property of systematic samples in time or space.

sampled in the early part of the day. With SRS selection, especially for small sample sizes, there is no assurance that a sample will be "well-balanced" with respect to time, space or any other auxiliary variable that may influence the target variable values.

There are, of course, penalties that are associated with these obvious desirable characteristics of systematic sampling. First, sample size is often not fixed (pre-specified) for this selection method, although sample size is approximately "conditionally fixed" given the true N (sometimes unknown before the sample has been selected). Indeed, the sample size that results from a 1 in k sample may prove too large or too costly when N is not known in advance. Second, and often of much greater concern, we shall see that it is nearly impossible to obtain an unbiased estimate of sampling variance from a single systematic sample of size n. Having made an initial random start on the integers 1 through k, the remaining sample units are selected at a fixed interval (every k^{th} unit) thereafter, whereas with SRS the selection of *every* sample unit is subject to chance. Finally, there are certain circumstances (regular periodic fluctuations in unit variable values through time or space) in which systematic selection of samples can lead to seriously erroneous estimates. Such cases are pathological rather than ordinary, however, and can typically be anticipated and prevented. On balance, the practical advantages of systematic sampling (simplicity, feasibility, and desirable spatial or temporal "balance" when unit labels have an important natural order) and its typically solid performance characteristics (negligible bias and modest sampling variance) often outweigh the concern regarding variance estimation.

4.1 Linear systematic sampling

4.1.1 *N/k is integer-valued*

As noted in the previous section, in **linear systematic sampling**, one first selects an initial random start, r, on the integers 1 through k. One then selects unit r and every k^{th} unit thereafter. If the total number of units in the population is N, and N/k is integer-valued, then sample size is fixed at $n = N/k$. Table 4.1 lists all possible samples for this type of design. The population mean is estimated using a mean-per-unit estimator, $\hat{\mu}_{\text{lsys}} = \sum_{i \in s} y_i / n$, and the population total is estimated as $\hat{T}_{\text{lsys}} = N \hat{\mu}_{\text{lsys}}$.

Each of the k possible linear systematic samples that can be selected is equally likely because the initial random start is drawn with equal probability on the integers 1 through k. Therefore, the probability of selecting any particular sample, $p(s)$, is the same for all samples and equal to $1/k$. For a simple numerical example, let's continue with the same $N = 4$ population of stones, for which $\mu = 7$ and $\sigma^2 = 26$

Table 4.1 Generic listing of all possible linear systematic samples for a 1 in k design where N/k is integer-valued. There are k possible samples, each of size $n = N/k$.

Sample ID	Random start (r)	Sample unit values ($y_i, i \in s$)				
1	1	y_1	y_{k+1}	y_{2k+1}	$\cdots$	$y_{(n-1)k+1}$
2	2	y_2	y_{k+2}	y_{2k+2}	$\cdots$	$y_{(n-1)k+2}$
$\vdots$	$\vdots$	$\vdots$	$\vdots$	$\vdots$	$\vdots$	
k	k	y_k	y_{2k}	y_{3k}	$\cdots$	$y_{nk} (= y_N)$

Population unit (i):	1	2	3	4
Variable value (y_i):	2	5	7	14

Suppose that we want to select linear systematic samples of size $n = 2$, so we need to set $k = N/n = 2$. The two possible random starts, associated sample units and sample means are

ID	Units (s)	Values ($y_i, i \in s$)	$p(s)$	$\hat{\mu}_{\text{lsys}}(s)$
		Sample		
1	1, 3	2, 7	1/2	4.5
2	2, 4	5, 14	1/2	9.5

The expected value of $\hat{\mu}_{\text{lsys}}$ would be

$$E(\hat{\mu}_{\text{lsys}}) = \sum_{s \in \mathcal{S}} \hat{\mu}_{\text{lsys}}(s)p(s) = \frac{1}{2}(4.5 + 9.5) = 7 = \mu$$

and the associated sampling variance of $\hat{\mu}_{\text{lsys}}$ would be

$$V(\hat{\mu}_{\text{lsys}}) = \sum_{s \in \mathcal{S}} [\hat{\mu}_{\text{lsys}}(s) - E(\hat{\mu}_{\text{lsys}})]^2 p(s) = \frac{1}{2}\left[(4.5 - 7)^2 + (9.5 - 7)^2\right] = 6.25$$

Thus, for N/k giving an integer result, $\hat{\mu}_{\text{lsys}}$ is unbiased and its sampling variance (here equal to 6.25) may compare favorably to sampling variance of $\hat{\mu}_{\text{mpu}}$ under SRS (equal to 6.5 for this same $N = 4$ population of stones).

Define the average variation within all possible linear systematic samples

$$\sigma^2_{\text{lsys}} = \frac{\sum\limits_{s \in \mathcal{S}} \sigma^2_{\text{lsys}}(s)}{k}, \quad \text{where } \sigma^2_{\text{lsys}}(s) = \frac{\sum\limits_{i \in s}\left[y_i(s) - \hat{\mu}_{\text{lsys}}(s)\right]^2}{n - 1}$$

and $y_i(s)$ denotes the target variable value of unit i that is included in a particular realized sample, $S = s, i \in s$. For this numerical example, the within sample variances $[\sigma^2_{\text{lsys}}(s)]$ are

Sample 1: $\sigma^2_{\text{lsys}}(s) = \left[(2 - 4.5)^2 + (7 - 4.5)^2\right]/(2 - 1) = 12.5$

Sample 2: $\sigma^2_{\text{lsys}}(s) = \left[(5 - 9.5)^2 + (14 - 9.5)^2\right]/(2 - 1) = 40.5$

Therefore, the average variation within systematic samples is $\sigma^2_{\text{lsys}} = (12.5 + 40.5)/2 = 26.5$.

Relative efficiency

The sampling variance of $\hat{\mu}_{\text{lsys}}$ can be expressed as (Cochran 1977 Section 8.3)

$$V(\hat{\mu}_{\text{lsys}}) = \frac{(N - 1)}{N}\sigma^2 - \frac{k(n - 1)}{N}\sigma^2_{\text{lsys}} \tag{4.1}$$

For the $N = 4$ population of stones and $n = 2$, Equation (4.1) gives

$$V(\hat{\mu}_{\text{lsys}}) = \frac{(4 - 1)}{4}26 - \frac{2(2 - 1)}{4}26.5 = 19.5 - 13.25 = 6.25$$

as directly calculated over the sample space. The relative efficiency of $\hat{\mu}_{\text{lsys}}$ compared to $\hat{\mu}_{\text{mpu}}$ is

$$RE(\hat{\mu}_{lsys}, \hat{\mu}_{mpu}) = \frac{V(\hat{\mu}_{mpu})}{V(\hat{\mu}_{lsys})} = \frac{\left(\frac{N-n}{N}\right)\frac{\sigma^2}{n}}{\frac{(N-1)}{N}\sigma^2 - \frac{k(n-1)}{N}\sigma^2_{lsys}}$$

Substituting N/n for k (assuming N/n is integer-valued), simplifying and rearranging gives

$$RE(\hat{\mu}_{lsys}, \hat{\mu}_{mpu}) = \frac{\frac{\sigma^2}{n} - \frac{\sigma^2}{N}}{\left(\frac{\sigma^2_{lsys}}{n} - \frac{\sigma^2}{N}\right) + \left(\sigma^2 - \sigma^2_{lsys}\right)} \tag{4.2}$$

Convince yourself that Equation (4.2) leads to a conclusion that $RE(\hat{\mu}_{lsys}, \hat{\mu}_{mpu}) > 1$ whenever $\sigma^2_{lsys} > \sigma^2$. That is, $\hat{\mu}_{lsys}$ will outperform $\hat{\mu}_{mpu}$ *whenever the average variation within the systematic samples exceeds the finite population variance.* In simple random sampling, of course, the average variation within samples is exactly equal to the finite population variance, σ^2, because $E(\hat{\sigma}^2) = \sigma^2$.

4.1.2 N/k is not integer-valued

If N/k does not generate an integer result, then not all linear systematic samples will have the same size and $\hat{\mu}_{lsys}$ is not unbiased. There are two quite different ways that one may respond to this problem if one is extremely concerned about estimator bias. First, one may construct alternative estimators for linear systematic sampling that are always unbiased, regardless of the values of N and k. Second, one may achieve unbiasedness by varying the systematic sample selection method itself so as to ensure that all systematic samples are of equal size.

Unbiased estimation

Unbiased estimators of the population mean and total in linear systematic sampling when N/k is non-integer are (Murthy 1967)

$$\hat{\mu}_{lsys,u} = \frac{\sum_{i \in S} y_i}{N/k} \tag{4.3}$$

$$\hat{T}_{lsys,u} = k \sum_{i \in S} y_i = N\hat{\mu}_{lsys,u} \tag{4.4}$$

Unbiasedness can be established as

$$E(\hat{\mu}_{lsys,u}) = \sum_{s \in \mathcal{S}} \hat{\mu}_{lsys}(s)p(s) = \frac{1}{k}\sum_{s \in \mathcal{S}}\frac{\sum_{i \in s} y_i}{N/k} = \frac{1}{N}\sum_{s \in \mathcal{S}}\sum_{i \in s} y_i = \frac{1}{N}\sum_{i=1}^{N} y_i = \mu$$

(Note that $\sum_{s \in \mathcal{S}} \sum_{i \in s} y_i = \sum_{i=1}^{N} y_i$ because each population unit appears in only one of the possible systematic samples.)

As the following numerical example illustrates, however, there may often be a price paid to achieve unbiasedness: the sampling variances of these unbiased estimators will generally be greater than the sampling variances of the usual (biased) estimators. Consider a population of $N = 7$ units having the following target variable values.

Population unit (i):	1	2	3	4	5	6	7
Variable value (y_i):	1	3	4	6	11	4	6

For this set of variable values, $\mu = 5$. For $k = 2$, there will be two linear systematic samples, one of size $n = 4$ and one of size $n = 3$.

ID	Sample Units (s)	Values $(y_i, i \in s)$	$p(s)$	$\hat{\mu}_{lsys}(s)$	$\hat{\mu}_{lsys,u}(s)$
1	1, 3, 5, 7	1, 4, 11, 6	1/2	22/4	22/3.5
2	2, 4, 6	3, 6, 4	1/2	13/3	13/3.5

For this example, the expected value of the usual linear systematic sampling estimator of the mean would be $E(\hat{\mu}_{lsys}) = \sum_{s \in S} \hat{\mu}_{lsys}(s) p(s) = \frac{1}{2}\left(\frac{22}{4} + \frac{13}{3}\right) = 4.917$ so that its bias would be $B(\hat{\mu}_{lsys}) = E(\hat{\mu}_{lsys}) - \mu = 4.917 - 5 = -0.083$. We also note that $V(\hat{\mu}_{lsys}) = 0.340$ (calculations not shown). For the unbiased estimator, $E(\hat{\mu}_{lsys,u}) = \sum_{s \in S} \hat{\mu}_{lsys,u}(s) p(s) = \frac{1}{2}\left(\frac{22}{3.5} + \frac{13}{3.5}\right) = 5 = \mu$ so that $B(\hat{\mu}_{lsys,u}) = 0$. Note, however, that the *range* of sample estimates increased from $(22/4 - 13/3) = 1.167$ for $\hat{\mu}_{lsys}$ to $(22/3.5 - 13/3.5) = 2.571$ for $\hat{\mu}_{lsys,u}$. As a consequence, $V(\hat{\mu}_{lsys,u}) = 1.653$ (calculations not shown) is approximately $4.9 \times V(\hat{\mu}_{lsys})$.

Bias of $\hat{\mu}_{lsys}$

The algebraic argument below leads to an expression for the bias of $\hat{\mu}_{lsys}$ when N/k does not generate an integer result.

$$B(\hat{\mu}_{lsys}) = E(\hat{\mu}_{lsys}) - \mu = E(\hat{\mu}_{lsys}) - E(\hat{\mu}_{lsys,u}) = E(\hat{\mu}_{lsys} - \hat{\mu}_{lsys,u})$$

Substituting the expressions for $\hat{\mu}_{lsys}$ and $\hat{\mu}_{lsys,u}$ in the above expectation gives

$$B(\hat{\mu}_{lsys}) = E\left[\frac{\sum_{i \in S} y_i}{n} - \frac{\sum_{i \in S} y_i}{N/k}\right] = E\left[\frac{\sum_{i \in S} y_i}{n}\left(1 - \frac{nk}{N}\right)\right]$$

The expectation clearly has value zero whenever $N/k = n$ (integer result). For large n and N the expectation will approach zero even when N/k is non integer. For example, suppose that $N = 204$ and $k = 10$. In that case, n may take on the values 20 or 21 (only). For $n = 20$, the factor $(1 - \frac{nk}{N}) = +0.0196$, whereas for $n = 21$, the factor $(1 - \frac{nk}{N}) = -0.0294$. Because for fixed n the estimator $\sum_{i \in S} y_i/n$ is unbiased for μ, this means that the *conditional* bias of the linear systematic estimator would be about $+.02\mu$ for $n = 20$, about $-.03\mu$ for $n = 21$, and the *unconditional* bias would be a weighted average of these two conditional biases. In contrast, for $N = 5$ and $k = 2$, conditional bias would be much larger: $+.2\mu$ for $n = 2$ and -0.2μ for $n = 3$.

4.2 Selection methods that guarantee fixed n

In this section we describe two alternative systematic selection methods that ensure a fixed sample size and allow for unbiased estimation of the population mean and total.

4.2.1 Circular systematic sampling

In **circular systematic sampling**, given a specified integer value of k, total sample size is fixed and equal to the nearest integer exceeding N/k. For example, if $N = 7$ and $k = 2$, then $n = 4$. The initial random start is selected with equal probability on the integers

1 through N rather than 1 through k. Thus, there will be a total of N possible circular systematic samples and each sample will have probability $1/N$. The method is termed *circular* because it is most easily understood by taking a line segment with the N unit numbers regularly spaced along its length, and connecting it end-to-end so as to produce a circular arrangement of the unit numbers. Thus, unit $N+i$ on the circle is the same unit as unit i. From the random start, one proceeds around the circle selecting every k^{th} unit until a total of n units have been selected. Figure 4.3 provides an illustration of this construction, and displays all possible circular systematic samples for the example of $N = 7$ and $k = 2$.

As for linear systematic sampling, the population mean is estimated using a mean-per-unit estimator, $\hat{\mu}_{csys} = \sum_{i \in s} y_i / n$, the population total is estimated as $\hat{T}_{csys} = N\hat{\mu}_{csys}$, and both estimators are unbiased. Let us take up again the same set of $N = 7$ population units and associated variable values previously considered. Table 4.2 provides a listing of all possible circular systematic samples of size $n = 4$ from this $N = 7$ population corresponding to the samples graphically displayed in Figure 4.3.

For this example, the expected value of $\hat{\mu}_{csys}$ would be

$$E(\hat{\mu}_{csys}) = \frac{1}{7}\left[\frac{22 + 14 + 24 + 15 + 26 + 20 + 19}{4}\right] = \frac{140}{28} = 5 = \mu$$

and thus $\hat{\mu}_{csys}$ would be unbiased. We also note that $V(\hat{\mu}_{csys}) = 1.054$ (calculations not shown).

4.2.2 *Fractional interval random start*

Another systematic sampling approach is to select a real-valued random start with uniform probability on the continuous interval 0 to $k = N/n$, where k is typically a real-valued number. This has been called the **fractional interval method** (Särndal et al. 1992). The random start, r, is a randomly selected value from the continuous uniform$(0,k)$ distribution, and the usual systematic sequence is then formed as $r, r+k, r+2k, \ldots, r+$

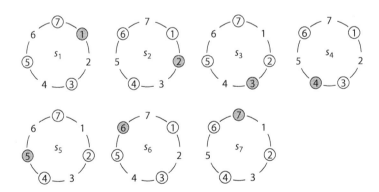

Fig. 4.3 Circular systematic sample construction for $N = 7$ population units and a sampling interval of $k = 2$, which results in a sample size of $n = 4$. There are $N = 7$ possible random starting units which lead to the seven possible samples $(s_1, s_2, \ldots, s_7)$ displayed. For each possible sample, the starting unit number is ringed and shaded, and every other unit after that moving clockwise around the circle is selected until a total of n units have been selected. The ringed numbers thus identify the units to be sampled.

$(n-1)k$. The nearest integers *above* these values are the unit numbers to be sampled.[2] For example, continuing with our same previous example population of $N = 7$ units, to fix the sample size at $n = 4$ we set $k = N/n = 7/4 = 1.75$. Suppose that the starting value selected at random from 0 to 1.75 is $r = 1.4765$. The systematic sequence of values would

Table 4.2 Sample space for circular systematic samples of size $n = 4$ from the $N = 7$ example population. Random starts are made with equal probability among the integers 1 through N. Sample units and associated values $y_i \in s$ are listed in order of selection following the initial random start. See Figure 4.3 for a graphical depiction of the sample construction process.

		Sample		
ID	Units (s)	Values ($y_i, i \in s$)	$p(s)$	$\hat{\mu}_{csys}(s)$
1	1, 3, 5, 7	1, 4, 11, 6	1/7	22/4
2	2, 4, 6, 1	3, 6, 4, 1	1/7	14/4
3	3, 5, 7, 2	4, 11, 6, 3	1/7	24/4
4	4, 6, 1, 3	6, 4, 1, 4	1/7	15/4
5	5, 7, 2, 4	11, 6, 3, 6	1/7	26/4
6	6, 1, 3, 5	4, 1, 4, 11	1/7	20/4
7	7, 2, 4, 6	6, 3, 6, 4	1/7	19/4

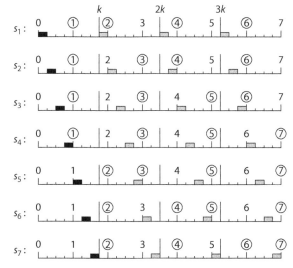

Fig. 4.4 Fractional interval systematic sample construction for $N = 7$, $n = 4$, and $k = N/n = 1.75$. Horizontal lines depict the real number line. Labels 1, 2, ..., 7 denote both real line units and population units. Suppose a start value of $0 < r \leq 0.25$, within the range spanned by the s_1 black rectangle. Successive values in the systematic sequence would be $r + k$, $r + 2k$, and $r + 3k$ and would fall within the s_1 grey rectangle ranges. Rounding any values within these rectangle ranges up to the nearest integer would result in the s_1 sample of circled unit numbers (units 1, 2, 4, 6). Other ranges possible for r lead to the seven samples ($s_1, s_2, \ldots, s_7$) indicated. The range of r values leading to each s is equal width (0.25), and thus the samples are equally likely: $p(s) = 0.25/1.75 = 1/7$.

[2] The fractional interval selection method described here can be shown to be a special case of a more general selection method that is often used in unequal probability systematic sampling (Section 8.5).

Table 4.3 Sample space for fractional interval systematic samples of size $n = 4$ from the $N = 7$ population, where $k = N/n = 1.75$. A random start, r, selected with uniform probability on the continuous interval $(0, k)$ is first generated, and then the sequence of values $r, r + k, r + 2k, r + 3k$ is used to select sample units. See Figure 4.4 for a graphical depiction of the sample construction process.

ID	Sample Units (s)	Values ($y_i, i \in s$)	$p(s)$	$\hat{\mu}_{fsys}(s)$
1	1, 2, 4, 6	1, 3, 6, 4	1/7	14/4
2	1, 3, 4, 6	1, 4, 6, 4	1/7	15/4
3	1, 3, 5, 6	1, 4, 11, 4	1/7	20/4
4	1, 3, 5, 7	1, 4, 11, 6	1/7	22/4
5	2, 3, 5, 7	3, 4, 11, 6	1/7	24/4
6	2, 4, 5, 7	3, 6, 11, 6	1/7	26/4
7	2, 4, 6, 7	3, 6, 4, 6	1/7	19/4

then be 1.4765, 3.2265, 4.9765, 6.7265, resulting in the sample unit numbers 2, 4, 5, 7. Figure 4.4 displays all possible fractional interval samples for $N = 7$ and $k = 1.75$, and how they arise depending on the value of the random start, r. As for linear and circular systematic sampling, the population mean is estimated using a mean-per-unit estimator, $\hat{\mu}_{fsys} = \sum_{i \in S} y_i / n$, the population total is estimated as $\hat{T}_{fsys} = N\hat{\mu}_{fsys}$, and both estimators are unbiased.

For our example ($N = 7, k = 4$), the sample space consists of seven distinct samples, each of equal probability ($p(s) = 1/7$), as listed in Table 4.3. Note that, for this example, the sample space for the circular and fractional interval selection methods is identical (unit ordering varies within the listed samples but that is inconsequential). Therefore, both the bias and the sampling variance of $\hat{\mu}_{fsys}$ are identical to that of $\hat{\mu}_{csys}$ for this example: $B(\hat{\mu}_{fsys}) = 0$ and $V(\hat{\mu}_{fsys}) = 1.054$.

In general, however, the sample space for the fractional interval selection method is not the same at that for circular systematic sampling. In circular systematic sampling there are always N possible samples (N possible integer random starts on the integers 1 through N), but for the fractional interval method, the number of possible samples is not always equal to N. For example, for $N = 20$ and $n = 6$ selected by the fractional interval scheme, there are only 10 distinct samples.

Is there a strong reason to use one of the unbiased estimators $\hat{\mu}_{lsys,u}$, $\hat{\mu}_{csys}$, or $\hat{\mu}_{fsys}$ and their associated sampling designs instead of the biased estimator $\hat{\mu}_{lsys}$ (when N/k is non-integer) and linear systematic selection? The answer to this question depends, in part, on how concerned one is about the bias of an estimator compared to the overall accuracy of an estimator. Recall that in Chapter 1 we introduced the concept of mean square error (MSE) as the most appropriate measure of the accuracy of an estimator, and that $\text{MSE}(\hat{\theta}) = B(\hat{\theta})^2 + V(\hat{\theta})$. For an unbiased estimator, the mean square error is thus equivalent to the sampling variance. For our numerical example, $\text{MSE}(\hat{\mu}_{lsys}) = (-0.083)^2 + 0.340 = 0.347$, which is substantially less than $V(\hat{\mu}_{csys}) = V(\hat{\mu}_{fsys}) = 1.054$, which are both less than $V(\hat{\mu}_{lsys,u}) = 1.653$. Thus, for this example, the usual linear systematic sampling estimator ($\hat{\mu}_{lsys}$) would be the estimator of choice, even though it has small negative bias.

4.3 Estimation of sampling variance

The main difficulty associated with use of systematic sampling is not that the linear systematic estimator may have modest bias when N/k is non-integer. Instead, the primary difficulty is that there is no unbiased estimator of sampling variance for any of the systematic selection schemes. There are a number of different intuitive explanations for this fact. The easiest to understand is based on a realization that only the very first unit in a systematic sample is subject to chance (via the random start r selected on the integers 1 through k); all other sample units are predetermined given that initial random start. Thus, a single systematic sample is in some respects like selection of a random sample of size one from which no estimate of variance can be made. In Chapter 6 we consider **cluster sampling**, where the basic *sampling unit* consists of a group or cluster of units. If we select just one of these clusters, then unbiased variance estimation is not possible in that context either. Thus, the sample space of systematic samples can be considered analogous to a set of k cluster samples. In Chapter 8 we will present an alternative and more formal perspective, based on the general theory of probability sampling, on why unbiased variance estimation is impossible for a single systematic sample.

4.3.1 *Biased estimation*

SRS proxy

There are many different approaches for estimation of sampling variance in systematic sampling (Wolter 2007 Chapter 8). The most common approach is to use $\hat{V}(\hat{\mu}_{mpu})$ for variance estimation. In this subsection and the next, we use $\hat{\mu}_{sys}$ to refer to any one of the three systematic selection method estimators ($\hat{\mu}_{lsys}$, $\hat{\mu}_{csys}$, or $\hat{\mu}_{fsys}$). Under situations where $\hat{\mu}_{sys}$ should outperform $\hat{\mu}_{mpu}$, use of $\hat{V}(\hat{\mu}_{mpu})$ as an estimator of $V(\hat{\mu}_{sys})$ will typically result in substantial positive bias and is therefore considered a *conservative* variance estimator [i.e., on average estimates will be larger than the true, but unknown, $V(\hat{\mu}_{sys})$]. The logic behind the use of $\hat{V}(\hat{\mu}_{mpu})$ for variance estimation in systematic sampling is based on the following arguments or contexts.

1. If the target variable y is *randomly* ordered with respect to the population unit label $i, i = 1, 2, \ldots, N$, then the *expected* sampling variance of $\hat{\mu}_{sys}$ should be equivalent to that of $\hat{\mu}_{mpu}$. That is, in this setting $E(\sigma_{sys}^2) = \sigma^2$, and $\hat{V}(\hat{\mu}_{mpu})$ should therefore provide approximately unbiased estimation of sampling variance under systematic sampling.

2. If there is a linear trend (increasing or descreasing) in the target variable y with respect to the population unit label, $i, i = 1, 2, \ldots, N$, then this will generate greatly reduced sampling variance for $\hat{\mu}_{sys}$ as compared to $\hat{\mu}_{mpu}$ so that $\sigma_{sys}^2 > \sigma^2$. Therefore, use of $\hat{V}(\hat{\mu}_{mpu})$ would tend to result in positively biased estimates of sampling variance. Although biased estimation of sampling variance is certainly undesirable, there is less risk in concluding that an estimator has lower accuracy than it actually has, than in concluding that an estimator has greater accuracy than it actually has.

Estimation in presence of linear trend

Another approach for estimation of $V(\hat{\mu}_{lsys})$ should be considered *only* when one has detected a strong linear trend in the target variable with respect to the population unit label as revealed, for example, by plotting the sample y_i against the unit label, $i, i \in S$. In such cases, $V(\hat{\mu}_{lsys})$ should be much less than $V(\hat{\mu}_{mpu})$ and it would be worth using an estimator of $V(\hat{\mu}_{lsys})$ with an expected value closer to this sampling variance than would be achieved using $\hat{V}(\hat{\mu}_{mpu})$. In this case, we suggest the following "mean square successive differences" estimator be used. To appy this estimator, sort the sample unit labels $i \in S$ into ascending order and denote these sorted label values as $i_1, i_2, \ldots, i_n$. $V(\hat{\mu}_{lsys})$ can then be estimated based on the squared differences between the neighboring y values defined by this sort order

$$\hat{V}_{lt}(\hat{\mu}_{lsys}) = \frac{(N-n)}{Nn} \sum_{j=1}^{n-1} \frac{(y_{i_j} - y_{i_{j+1}})^2}{2(n-1)} \tag{4.5}$$

where the V subscript "lt" denotes linear trend.

Consider the following set of $N = 8$ population units which exhibit a perfect linear trend between y_i and the unit label, i: $y_i = 2i, i = 1, 2, \ldots, N$.

Population unit (i):	1	2	3	4	5	6	7	8
Variable value (y_i):	2	4	6	8	10	12	14	16

For this population $\mu = 9$. Suppose that we were to set $k = 2$ so that there would be two possible linear systematic samples of size $n = 4$. The results for this sample space are listed in the table below.

ID	Sample Units (s)	Values ($y_i, i \in s$)	$p(s)$	$\hat{\mu}_{lsys}(s)$	$\hat{V}(\hat{\mu}_{mpu})(s)$	$\hat{V}_{lt}(\hat{\mu}_{lsys})(s)$
1	1, 3, 5, 7	2, 6, 10, 14	1/2	8	10/3	1
2	2, 4, 6, 8	4, 8, 12, 16	1/2	10	10/3	1

For this example,

$$E(\hat{\mu}_{lsys}) = \sum_{s \in S} \hat{\mu}_{lsys}(s)p(s) = \frac{1}{2}(8 + 10) = 9 = \mu$$

$$V(\hat{\mu}_{lsys}) = \sum_{s \in S} [\hat{\mu}_{lsys}(s) - E(\hat{\mu}_{lsys})]^2 p(s) = \frac{1}{2}\left[(8-9)^2 + (10-9)^2\right] = 1$$

Note that the linear trend estimator of sampling variance is unbiased in this *perfect* setting $[E[\hat{V}_{lt}(\hat{\mu}_{lsys})] = 1 = V(\hat{\mu}_{lsys})]$ whereas reliance on $\hat{V}(\hat{\mu}_{mpu})$ would lead to substantial overestimation of sampling variance $(E[\hat{V}(\hat{\mu}_{mpu})] = 3.333 > V(\hat{\mu}_{lsys}) = 1)$. If a linear trend were present but imperfect, then the linear trend estimator of sampling variance would no longer be unbiased, but it might still be a considerable improvement over using $\hat{V}(\hat{\mu}_{mpu})$. Perhaps even more important than such variance estimation issues is the fact that systematic sampling over a linear trend should greatly outperform $\hat{\mu}_{mpu}$. For a fixed sample size of $n = 4$ units selected by SRS from the same population of y values, $V(\hat{\mu}_{mpu}) = 3$ (three times the sampling variance for $\hat{\mu}_{lsys}$).

4.3.2 *Unbiased estimation*

If unbiased estimation of sampling variance is essential, then one may draw two or more systematic samples, either *independently* (i.e., multiple random starts *with replacement*) or *without replacement*, and the variation among samples estimates can provide an unbiased estimate of sampling variance for $\hat{\mu}_{sys}$ (or for any other method of sample selection and estimation). Wolter (2007 Chapter 2) refers to this kind of approach generally as the *method of random groups* and notes that the idea was first introduced by Mahalanobis (1940), and often referred to as **interpenetrating subsamples**. Murthy (1967 Section 5.8d) applied the interpenetrating subsamples approach to variance estimation in systematic sampling. At first glance, this approach may seem like a good resolution to the problem of generating an unbiased estimate of sampling variance in systematic sampling. However, it has two shortcomings. First, in order to draw two or more systematic samples, the sample size for each of these samples must usually be reduced compared to a single systematic sample because total survey costs are typically limited. Thus, one might have the choice of drawing a single large systematic sample or two smaller systematic samples that are each only half as large. Second, although the resulting variance estimators are unbiased, the sampling variance of these variance estimators may be large. Therefore, they may generate unbiased, but unreliable, estimates of sampling variance.

In the numerical examples that follow, we assume that N/k is integer-valued so that the resulting sample size, $n = N/k$, for a single linear systematic sample would be *fixed*. For each of the alternatives considered next, we will be selecting $m = 2$ systematic samples of size $n^* = n/2$ so that the total sample size remains the same when we compare the approaches. For these examples we apply linear systematic selection to the same population used for the previous illustration of sampling along a perfect linear trend, with $N = 8$ and $\mu = 9$.

m samples selected independently

For this approach, we draw m *independent* systematic samples of size $n^* = n/m$ by taking m independent random starts on the integers 1 through k^* where $k^* = mk$. Thus, for $m = 2$ and $k = 2$, $k^* = mk = 2 \cdot 2 = 4$. That is, we would select $m = 2$ independent systematic samples by taking two independent (with replacement) random starts on the integers 1 through 4. For $N = 8$, our resulting sample sizes would be $n^* = N/k^* = 2$ as compared to the original sample size $n = N/k = 8/2 = 4$. The total combined size of the two independent samples would equal that of the original sample. The $k^* = 4$ possible systematic samples of size two that could be selected for any one of the m systematic samples would be

	Sample			
ID	Units (s)	Values ($y_i, i \in s$)	$p(s)$	$\hat{\mu}_{lsys}(s)$
1	1, 5	2, 10	1/4	6
2	2, 6	4, 12	1/4	8
3	3, 7	6, 14	1/4	10
4	4, 8	8, 16	1/4	12

The following two theorems are broadly applicable to estimation of a population parameter based on a combination of independent, individually-unbiased estimators of that same parameter, and apply here in the context of estimating a population mean based on the sample means resulting from a set of independent systematic samples.

Theorem 4.1. *Let $\hat{\theta}_j$ denote an unbiased estimator of θ, and let $\bar{\hat{\theta}} = \sum_{j=1}^{m} \hat{\theta}_j/m$, the mean of m ($j = 1, 2, \ldots, m$) independent unbiased estimators. Then, $E(\bar{\hat{\theta}}) = \theta$; that is, the mean of m unbiased estimators is also an unbiased estimator.*

Theorem 4.2. *Given m independent unbiased estimators $\hat{\theta}_j$, an unbiased estimator of sampling variance for the mean of these estimators, $\bar{\hat{\theta}}$, is*

$$\hat{V}(\bar{\hat{\theta}}) = \frac{\sum_{j=1}^{m}(\hat{\theta}_j - \bar{\hat{\theta}})^2}{m(m-1)} \tag{4.6}$$

Therefore, for estimates $\hat{\mu}_{sys}$ originating from m independent systematic samples, let $\bar{\hat{\mu}}_{sys} = \sum_{j=1}^{m}\hat{\mu}_{sys,j}/m$ be the arithmetic mean of m unbiased estimates of the mean, $\hat{\mu}_{sys,j}$, $j = 1, 2, \ldots, m$, generated from m independent systematic samples. By Theorem 4.1, $\bar{\hat{\mu}}_{sys}$ will also be an unbiased estimator of the mean, and by Theorem 4.2 the sampling variance of $\bar{\hat{\mu}}_{sys}$ can be unbiasedly estimated as

$$\hat{V}(\bar{\hat{\mu}}_{sys}) = \frac{\sum_{j=1}^{m}(\hat{\mu}_{sys,j} - \bar{\hat{\mu}}_{sys})^2}{m(m-1)} \tag{4.7}$$

Returning to our example, for the $k^{\star} = 4$ possible systematic samples that can be selected using the four possible random starts on the integers 1 through $k^{\star}$, Table 4.4 lists the 10 possible pairs of distinct samples that could be selected, along with the associated $\bar{\hat{\mu}}_{sys}$ and $\hat{V}(\bar{\hat{\mu}}_{sys})$. We thus find

$$E(\bar{\hat{\mu}}_{lsys}) = \frac{1}{16}[6+8+10+12] + \frac{1}{8}[7+8+9+9+10+11] = 9 = \mu$$

$$V(\bar{\hat{\mu}}_{lsys}) = \frac{1}{16}\left[(6-9)^2 + \ldots + (12-9)^2\right] + \frac{1}{8}\left[(7-9)^2 + \ldots + (11-9)^2\right] = 20/8$$

$$E[\hat{V}(\bar{\hat{\mu}}_{lsys})] = \frac{1}{16}[0+0+0+0] + \frac{1}{8}[1+4+9+1+4+1] = 20/8 = V(\bar{\hat{\mu}}_{lsys})$$

Therefore, both $\bar{\hat{\mu}}_{lsys}$ and $\hat{V}(\bar{\hat{\mu}}_{lsys})$ are unbiased estimators. Note, however, that four of the systematic sample pairs, with a total probability of 1/4, would generate $\hat{V}(\bar{\hat{\mu}}_{lsys})(s) = 0$.

Table 4.4 Sample space for two independent linear systematic samples of size $n = 2$ selected *with replacement* from a population of size $N = 8$.

Samples						
ID$_1$	ID$_2$	$p(s)$	$\hat{\mu}_{lsys,1}$	$\hat{\mu}_{lsys,2}$	$\bar{\hat{\mu}}_{lsys}$	$\hat{V}(\bar{\hat{\mu}}_{lsys})$
1	1	1/16	6	6	6	0
2	2	1/16	8	8	8	0
3	3	1/16	10	10	10	0
4	4	1/16	12	12	12	0
1	2	1/ 8	6	8	7	1
1	3	1/ 8	6	10	8	4
1	4	1/ 8	6	12	9	9
2	3	1/ 8	8	10	9	1
2	4	1/ 8	8	12	10	4
3	4	1/ 8	10	12	11	1

m samples selected without replacement

Suppose that we instead select m systematic samples of size $n^\star = n/m$ *without replacement* by taking m without replacement random starts on the integers 1 through $k^\star$ where $k^\star = mk$, and we use $\bar{\hat{\mu}}_{\text{sys}}$ to estimate μ. In contrast to independent (with replacement) selection of the systematic samples, selection without replacement rules out selecting the same systematic sample more than once. In this case, $\bar{\hat{\mu}}_{\text{sys}}$ is again an unbiased estimator of μ, and an unbiased estimator of $V(\bar{\hat{\mu}}_{\text{sys}})$ is[3] [Wolter 2007 Equation (8.2.17)]

$$\hat{V}(\bar{\hat{\mu}}_{\text{sys}}) = \left(\frac{k^\star - m}{k^\star}\right)\frac{\sum_{j=1}^{m}(\hat{\mu}_{\text{sys},j} - \bar{\hat{\mu}}_{\text{sys}})^2}{m(m-1)} \tag{4.8}$$

Returning to our example, for the $k^\star = 4$ possible systematic samples that can be selected using the four possible random starts on the integers 1 through $k^\star$, there are now only six possible pairs of two distinct systematic samples that could be selected (Table 4.5). The with replacement sample pairs j,j are not possible when selection is without replacement. We thus find

$$E(\bar{\hat{\mu}}_{\text{lsys}}) = \frac{1}{6}[7+8+9+9+10+11] = 9 = \mu$$

$$V(\bar{\hat{\mu}}_{\text{lsys}}) = \frac{1}{6}\left[(7-9)^2 + (8-9)^2 + (9-9)^2 + (9-9)^2 + (10-9)^2 + (11-9)^2\right] = 10/6$$

$$E[\hat{V}(\bar{\hat{\mu}}_{\text{lsys}})] = \frac{1}{6}[4/8 + 16/8 + 36/8 + 4/8 + 16/8 + 4/8] = 10/6 = V(\bar{\hat{\mu}}_{\text{lsys}})$$

Thus, both $\bar{\hat{\mu}}_{\text{lsys}}$ and $\hat{V}(\bar{\hat{\mu}}_{\text{lsys}})$ are again unbiased estimators, but there are no $\hat{V}(\bar{\hat{\mu}}_{\text{lsys}}) = 0$ estimates because samples are selected without replacement. Note also that the sampling variance of the without replacement method of selecting the multiple systematic samples (10/6) is considerably less than the sampling variance for the with replacement selection method (20/8) and compares favorably with the sampling variance (1) of the original single systematic sample of size $n = 4$. (See also Problem 4.5.)

Table 4.5 Illustrative sample space for unbiased estimation of $\hat{\mu}_{\text{lsys}}$ when two linear systematic samples of size $n = 2$ are selected *without replacement* from a population of size $N = 8$.

Samples						
ID$_1$	ID$_2$	$p(s)$	$\hat{\mu}_{\text{lsys},1}$	$\hat{\mu}_{\text{lsys},2}$	$\bar{\hat{\mu}}_{\text{lsys}}$	$\hat{V}(\bar{\hat{\mu}}_{\text{lsys}})$
1	2	1/6	6	8	7	4/8
1	3	1/6	6	10	8	16/8
1	4	1/6	6	12	9	36/8
2	3	1/6	8	10	9	4/8
2	4	1/6	8	12	10	16/8
3	4	1/6	10	12	11	4/8

---

[3] Selection of $m \geq 2$ systematic samples with or without replacement can be viewed as drawing an SWR or SRS sample of $m \geq 2$ units from a population consisting of target variable values that correspond to the means of the k possible systematic samples. Thus, the expressions for unbiased estimation of sampling variance are directly analogous to the corresponding formulas for $\hat{V}(\hat{\mu}_{\text{swr}})$ and $\hat{V}(\hat{\mu}_{\text{srs}})$, respectively.

4.4 Unpredictable trend in sampling variance with *n*

Cochran (1946 Table 1) long ago provided numerical illustration of the fact that the pattern of change in sampling variance with increasing sample size for $\hat{\mu}_{lsys}$ may be highly erratic, in contrast to the steady decline in sampling variance with sample size achieved with $\hat{\mu}_{mpu}$ or $\hat{\mu}_{st}$ (Chapter 5). The pattern of decline in $V(\hat{\mu}_{lsys})$ with increasing sample size depends on the particular arrangement of units in the population. Figure 4.5 displays sampling variances of $\hat{\mu}_{lsys}$ for two random arrangements of $N = 240$ units (for which N/k is integer for many choices of k) and associated unit weights taken from the agate population used in the judgment sampling class experiment (Chapter 2). In Figure 4.5(a), there are four instances where an increase in sample size resulted in an increase rather than a decrease in sampling variance, but in this case the changes are fairly moderate and the overall decline in sampling variance with sample size is fairly smooth. Figure 4.5(b) depicts a rather extreme example, in which there are seven instances of an increase in sampling variance following an increase in sample size, and some of these increases are substantial. Other random arrangements of the $N = 240$ units resulted in variance patterns that were intermediate between these two cases. As noted by Murthy (1967 section 5.13), the unpredictable pattern of $V(\hat{\mu}_{lsys})$ as a function of sample size means that determination of sample size to achieve an expected precision objective can only be done using simulations assuming that some auxiliary variable, available for all units, is closely proportional to the values of the target variable.

4.5 Warning: Pathological settings

We would be remiss if we failed to note certain "pathological" settings where systematic sampling can only be safely applied using common sense to avoid certain unacceptable outcomes. The classic setting for pathological behavior involves target variable values which vary in a regular periodic pattern with respect to some temporal or spatial variable which might be used in specifying the order of the population unit labels. For example, suppose that the variable of interest is dissolved oxygen (DO) concentration and it is desired to obtain an estimate of the mean DO concentration at a depth of six inches below the surface of a small freshwater pond that is covered with filamentous algae. The DO concentration is to be measured at a particular location of the pond over the course of a week. Assume that it is not possible to continuously monitor DO concentration.

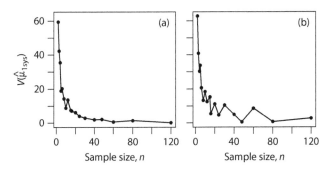

Fig. 4.5 Sampling variance of $\hat{\mu}_{lsys}$ as a function of sample size for two random rearrangements of $N = 240$ units selected from the population of agates considered in Chapter 2. Plotted points are those for which N/k gives an integer-valued n so that sample size is fixed and $\hat{\mu}_{lsys}$ is unbiased.

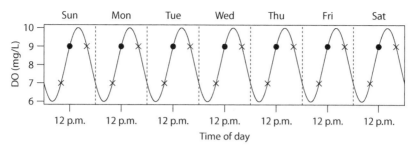

Fig. 4.6 Dissolved oxygen (DO) concentration six inches below the surface of a small freshwater pond over the course of one week. DO concentration is sinusoidal with a period of 24 hours, and at its minimum at 4 a.m. and maximum at 4 p.m. Mean DO concentration is $\mu = 8$. Black dots indicate observations for a systematic sample having a 24 hour sampling interval and a random start of Sunday 12 p.m.; X's indicate observations for a systematic sample having a 12 hour sampling interval and a random start of Sunday 8 a.m.

Due to photosynthetic activity during the day and the lack thereof at night, DO concentration in the pond will exhibit a diurnal pattern, generally reaching a maximum a few hours before sunset, and a minimum a few hours before sunrise. For simplicity, imagine that the variation in DO concentration is perfectly periodic, having a sine wave form with a period of 24 hours and a mean value of $\mu = 8$ mg/L (Figure 4.6). Thus, if the sampling interval is set equal to 24 hours, $\hat{\mu}_{\text{sys}}$ will be based on seven identical observed values, and though the estimator would be unbiased over the possible random start times, it would have high sampling variance even though each of the possible samples would yield $\hat{\sigma}^2 = 0$. For example, a Sunday 12 p.m. random start with a 24 hour interval would give the sample depicted by the dots in Figure 4.6, resulting in seven DO measurements equal to 9 and $\hat{\mu}_{\text{sys}} = 9$. For other random starts with a 24 hour sampling interval, $\hat{\mu}_{\text{sys}}$ ranges from 6 to 10 but in all cases $\hat{\sigma}^2 = 0$. At the other extreme, if the sampling interval is set equal to half the period length, 12 hours, all possible random starts will lead to samples yielding $\hat{\mu}_{\text{sys}} = \mu$ so that $V(\hat{\mu}_{\text{sys}}) = 0$. For example, a Sunday 8 a.m. random start with a 12 hour interval would give the sample depicted by the X's in Figure 4.6, resulting in seven DO measurements equal to 7 and seven DO measurements equal to 9 for a mean of $\hat{\mu}_{\text{sys}} = 8 = \mu$. In practice, this sort of pathological situation is unusual. If present, and one is aware of this a priori, then the sampling interval can be set to take advantage of the situation. If one is not aware of this a priori, and in the process of taking a systematic sample one is obtaining nearly equal observed values, some additional units should be selected off-interval to evaluate for the possibility of a periodic process.

4.6 Nonresponse and oversampling

As for SRS and SWR (Section 3.5), if one or more units in a linear systematic sample of size n are *pure chance* nonresponse units, then the usual linear systematic sampling estimators (assuming N/k is integer-valued) of the mean and total will remain unbiased when calculated over the sample of response units of size n_r, with $n_r \leq n$.

If it is critical to ensure that exactly n response units are selected, then oversampling may be used to guarantee that the desired sample size is realized. Let $n' = \gamma n$ denote the size of the oversample that should ensure with near certainty that n_r, the number of response units in n', is at least as large as the desired n, and let $k' = k/\gamma$ denote the associated sampling interval. Unbiased estimation based on the desired n should be maintained assuming that (a) nonresponse is due to pure chance, and (b) N/k and N/k' both give integer results. The steps in the oversample procedure are as follows. First, select a linear systematic sample of n' units such that $\Pr\{n_r \geq n\} > 0.999$ given an assumed response rate. Second, randomize the order of the selected units. Third, select the first n response units in this randomized list for the final sample. Calculate the usual linear systematic estimates over the n response units. The second step (randomization of unit order in the oversample) is intended to preserve much of the desirable spatial properties of systematic samples when taken over a list of units that are ordered with respect to location or some other auxiliary variable. Without such randomization, the n sample units would all be from the first $(n + n_{nr})/n'$ portion of the unit order $i = 1, 2, \ldots, N$, where n_{nr} is the number of nonresponse units encountered prior to achieving the desired n. Note that while this randomization step attempts to maintain the spatial "spread" of a systematic sample, it also eliminates the fixed interval sampling period normally associated with a systematic sampling design.

4.7 Chapter comments

As noted in Section 1.5, the benefits of systematic sampling have been recognized at least since Bowley (1926). If individual plants are sampled along a soil moisture gradient, pools are selected from reach of stream, or a temporal process is sampled at equal intervals of time (assuming intervals are short compared to any periodic behavior), then systematic selection of units will deliver excellent spatial or temporal coverage and will, in this respect, be superior to SRS selection of units. Whenever there is an increasing or decreasing trend in y_i with the unit label, $i, i = 1, 2, \ldots, N$, the sampling variance of estimators relying on systematic selection of units should be less than for those relying on SRS selection. The important logistical advantages of systematic sampling also cannot be overly stressed. First, no pre-existing map of unit locations or of the total number of population units is required. Instead, one may simply take a random start on the integers 1 through k, and then sample every k^{th} unit thereafter until the Nth unit is encountered. Second, systematic sampling can feasibly be implemented in many settings where selection of an SRS may be out of the question (e.g., selecting fish for measurement on a working fishing vessel). Third, the simplicity of the selection approach (just one random unit selected, the remaining at fixed intervals) means that field execution is greatly simplified and less prone to flaws in selection of units *intended* for inclusion in the sample.

One objection to linear systematic sampling, that unbiased estimation is only possible when N/k is integer-valued, can be easily overcome via circular systematic sampling and other devices, but this objection is a minor one that does not fundamentally undermine the positive aspects of systematic sampling. Unfortunately, systematic sampling relies on the most restricted randomization that might be imagined. Only selection of the first unit (r, the initial random start) is subject to chance, whereas remaining units are selected with certainty given r. As a consequence of this extremely restricted randomization, second order inclusion probabilities (π_{ij}) are equal to zero for all i, j pairs that fall in different linear systematic samples. Linear systematic sampling is therefore an example of a **non-measureable** design[4] and unbiased estimation of sampling variance is impossible. The

---

[4] Särndal et al. (1992 Section 14.3) refer to any sampling design for which one or more $\pi_{ij} = 0$ as a non-measureable design. Unbiased variance estimation is not possible for non-measureable designs, whereas for a **measureable** design, for which all $\pi_{ij} > 0$, unbiased or nearly unbiased variance estimation is possible. See also Chapter 8.

typical device of using $\hat{V}(\hat{\mu}_{mpu})$ for estimation of $V(\hat{\mu}_{lsys})$, approximately valid when population unit labels reflect a randomized order with respect to the value of y, has the ironic effect of resulting in substantial overestimation of sampling variance when systematic sampling is most effective because in such cases the average variation within systematic samples is far greater than the finite population variance. To a large extent, the dilemma remains as expressed by Madow and Madow (1944): " ... the statistician ... has the alternative of recommending a systematic sampling procedure for which no theory exists, or a random sampling procedure that may yield worse results than the systematic procedure."

Because systematic sampling has such obvious practical application advantages and because it poses exceptional difficulties for variance estimation, statisticians have devoted considerable effort to the exploration of how the idea of systematic sampling might be improved upon. Iachan (1982) provides a critical review of research on systematic sampling through 1981, and Mostafa and Ahmad (2018) provide a review of more recent developments. Some recent attention has been given to creation of *quasi-systematic* sampling methods that have fixed sample size, good spatial properties, and are also measureable designs, thus allowing unbiased variance estimation using Equation (8.10). One such method is the *partially systematic* circular sampling procedure developed by Leu and Tsui (1996). For this selection method, a random start, r, is taken on the integers 1 through N, and then a small SRS sample of size a (e.g., $a = 2$) is taken over the relatively short range of units, $(r, r + u - 1)$, where $u = N - (n - a)k$, and k is the closest integer to $N/(n - 1)$ (or $k = N/n$ if integer). Thereafter, a systematic 1 in k sample of size $n - a$ is taken over the remaining units through $N + r + u - 1$ (i.e., continuing "around the circle"), where unit $N + i$ is defined to be the same unit as unit i. The full partially systematic sample consists of the union of the initial SRS sample of size a and the following systematic sample of size $n - a$. First order inclusion probabilities are all n/N for this procedure and the sample mean is an unbiased estimator of μ. Explicit formulas are available for all second order inclusion probabilities and, though rather complex, can certainly be implemented in R or a similar programming environment. The disadvantage of such partially systematic procedures, however, is loss of the practical field advantages resulting from the simplicity with which conventional linear or circular systematic samples can be selected. Nevertheless, we believe that such partially-systematic selection schemes have substantial promise.

Finally, we note that Amode Sen once *winked* at two of the authors of this text when discussing the difficulties of variance estimation for systematic sampling. In effect, he posed the same question implied by Madow and Madow (1944): Which would you rather have: a good (accurate) estimate with a poor (and probably falsely pessimistic) notion of its accuracy, or a poor (sometimes much less accurate) estimate, but with a good notion of its accuracy? We will wink at you when we have an authoritative answer to this question!

Problems

Problem 4.1. Using the R code displayed immediately below: (A) Generate the following sets of (1) $N = 72$ approximately uniformly distributed auxiliary variable values, and (2) associated target variable values that are linearly related to the auxiliary variable values. The first three values of x and y should be (112, 102, 160) and (549, 535, 755), respectively, and Cor(x, y) should be 0.8646.

```
set.seed(100); N <- 72
x <- round(runif(N, min=50, max=250))
e <- rnorm(N, mean=0, sd=sqrt(2*var(x)))
y <- round(200 + (3*x) + e)
```

(B) Construct all possible linear systematic samples for $n = 9$. For the target variable y calculate the associated $\hat{T}_{lsys}$ for each sample, and calculate the sampling variance $V(\hat{T}_{lsys})$. (C) Using the analytic expression for $V(\hat{\mu}_{lsys})$ [Equation (4.1)], verify that $N^2 V(\hat{\mu}_{lsys})$ is identical to the $V(\hat{T}_{lsys})$ value calculated directly over the sample space. Note: in R, given k and a particular random start $r, r = 1, 2, \ldots, k$, the associated sample units s and target variable values y can be obtained using the following syntax:

```
s <- seq(from=r, to=N, by=k);   y.s <- y[s]
```

Problem 4.2. Using the same simulated values for the auxiliary and target variables as for Problem 4.1: (A) For sample sizes $n = (2, 3, 4, 6, 9, 12, 18, 24, 36)$, *assuming that units are labeled according to their natural (simulated) order*, (1) construct all possible linear systematic samples for each sample size, (2) calculate the corresponding expected values and sampling variances of $\hat{\mu}_{lsys}$ for the target variable y, and (3) calculate the expected values of $\hat{V}(\hat{\mu}_{mpu})$ as an estimator of $V(\hat{\mu}_{lsys})$. (B) Suppose instead that the population units are re-labeled (sorted) according to the value of x (low to high). Given this new ordering of population units, repeat the calculations described in steps (1) and (2), but use both $\hat{V}(\hat{\mu}_{mpu})$ and $\hat{V}_{lt}(\hat{\mu}_{lsys})$ [Equation (4.5)] as alternative estimators for $V(\hat{\mu}_{lsys})$. Note: R code helpful for determining the sort order of x and the correspondingly ordered values of x and y is

```
j <- order(x)   # indicies of sorted x
x.sort <- x[j]; y.sort <- y[j]
```

(C) Plot the sampling variance of $\hat{\mu}_{lsys}$ against sample size for the unordered and ordered settings. (D) For each of the sample sizes, calculate (1) relative efficiency of $\hat{\mu}_{lsys}$ compared to $\hat{\mu}_{mpu}$ (original natural order and sorted on x), and (2) evaluate the performance of the estimators of sampling variance for the unordered and ordered settings. Discuss your findings.

Problem 4.3. Suppose that $N = 10$ and $n = 4$. (A) Construct all possible equally likely systematic samples that can be selected using the fractional interval method (Section 4.2.2) and determine the first order sample inclusion probability, π_i, for each population unit. (B) Is the sample space equivalent to that of circular systematic sampling for $n = 4$?

Problem 4.4. Show algebraically that Equation (4.2) leads to a conclusion that linear systematic sampling (with $\hat{\mu}_{lsys}$) outperforms SRS (with $\hat{\mu}_{mpu}$) when $\sigma^2_{sys} > \sigma^2$.

Problem 4.5. In R: (A) Develop a program that will (1) calculate the sampling variance of a linear systematic sample ($k=5$, integer-valued $n = N/k$) selected from a population of size $N = 150$, (2) calculate the sampling variance for the mean of $m = 3$ SRS linear systematic samples (selected without replacement) of size $n^* = n/3$ ($k^* = mk = 15$), and (3) verify that unbiased estimation of sampling variance is achieved using Equation (4.8) when multiple systematic samples are selected without replacement. Note: Use the following R code to generate values of the target variable

```
y <- rnorm(N, mean=150, sd=50)
```

(B) Run the program repeatedly (say, 10 times, with new y values generated for each run), and record key sampling variance results for each repeated simulation. (C) Discuss the apparent significance of your simulation results.

Problem 4.6. Write an R program that allows calculation of the sampling variance of the circular systematic estimator of the population mean. Use it to calculate the sampling variance for a setting where y_i is linearly related to unit label (position), i, using the following syntax:

```
set.seed(100); N <- 80
i <- seq(from=10, to=89)
e <- rnorm(N, mean=0, sd=25)
y <- 10 + (2*i) + e
```

What is the relatively efficiency of the circular systematic sampling estimator of the mean compared to the SRS mean-per-unit estimator for $n = 12$?

Problem 4.7. The chapter comments section describes a *partially systematic* sampling approach that achieves $\pi_i = n/N$ for all i, but also guarantees that all second order inclusion probabilities exceed zero. (A) Write an R program to implement this sample selection approach for $a = 2$. (B) Simulate selection of a large number (say, 10^5) of independent samples of the Problem 4.6 population y values. (C) For $n = 12$, numerically verify that the partially systematic estimator of the population mean, $\hat{\mu}_{ps} = \sum_{i \in S} y_i/n$, is, within the limits of simulation accuracy, (1) unbiased, that (2) all $\pi_i = 12/80 = 0.15$, and that (3) all $\pi_{ij} > 0$.

Problem 4.8. Write a program in R that numerically verifies that the oversampling procedure described in Section 4.6 delivers unbiased estimation given an assumption that all nonresponse is pure chance.

Problem 4.9. There is an alternative exact expression for the sampling variance of the linear systematic estimator of the mean, $\hat{\mu}_{lsys}$, assuming integer-valued N/k (Cochran 1977 Theorem 8.2):

$$V(\hat{\mu}_{lsys}) = \frac{N-1}{N}\frac{\sigma^2}{n}(1 + (n-1)\text{ICCor}), \quad \text{where ICCor} = \frac{2\sum_{i=1}^{k}\sum_{j>j'}(y_{ij} - \mu)(y_{ij'} - \mu)}{(n-1)(N-1)\sigma^2}$$

(4.9)

is the *intraclass correlation*. The summation is over the $i = 1, 2, \ldots, k$ possible systematic samples, and j and j' refer to distinct unit positions in a particular systematic sample. (ICCor is closely related to the intracluster correlation, ICor, considered in Section 6.2.2.)

Suppose that $k = 2$ and that the y values for two populations are defined in R as

```
y1 <- c(2,  7, 9, 16, 19, 23, 28, 40)
y2 <- c(9, 28, 7, 40,  2, 23, 16, 19)
```

(A) Construct the sample spaces and determine $V(\hat{\mu}_{lsys})$ for both populations. (B) Calculate ICCor for both sample spaces and numerically verify that Equation (4.9) gives the exact sampling variance for the two sample spaces. (C) Wolter (2007) states that if $\hat{V}(\hat{\mu}_{mpu})$ is used as an estimator of $V(\hat{\mu}_{lsys})$, then it will be biased upward or downward as ICCor is less than or greater than $-1/(N-1)$. Do your results support Wolter's statement?

CHAPTER 5

Stratified sampling

Fig. 5.1 Classic pool (flat water)/riffle (turbulent surface) sequence in the Van Duzen River, northern California. Stratification by habitat type may lead to better understanding of fish habitat preferences and improved estimation of abundance. Photo credit: Mark Allen.

In **stratified sampling**, the N units in the sampling population are partitioned into L strata of size $N_h, h = 1, 2, \ldots, L$, such that $\sum_{h=1}^{L} N_h = N$. Strata may be "natural" [e.g., male and female strata created from a mixed population of male and female animals; pool, riffle and run habitat type strata in a stream (Figure 5.1)] or strata may be constructed based on auxiliary information (e.g., age, elevation, aspect, vegetation type). Samples are then selected *independently* from each of the L strata. Within any particular stratum, the target of estimation is typically the stratum mean, proportion, or total; any probability sampling strategy may be used to obtain an estimate of this stratum-specific population parameter. If the within-strata estimators are all unbiased, then properly weighted stratified estimators of the corresponding overall (across all strata) population parameters will also be unbiased.

Given a favorable setting and intelligent construction of strata, the sampling variance of a stratified estimator may be considerably less than the sampling variance of an unstratified estimator. Consider an unstratified population with variable values restricted to the integers 10, 20, 30, and 40 and imagine that each distinct value is associated with an auxiliary color variable (say, red, blue, green, yellow, respectively). If we were to stratify the population units into four strata according to their colors, then the values would all be 10 within the "red" stratum, 20 within the "blue" stratum, and so on. In this case, because there is no variation in values within these constructed strata, sampling within each stratum would lead to error-free estimates of the stratum mean, proportion, or total, and then the stratified estimate of the corresponding population parameter would also be error-free! Thus, stratification partitions the finite population variance, σ^2, into variation *within strata* and variation *between strata*. If there is no variation in variable values within

Sampling Theory: For the Ecological and Natural Resource Sciences. David G. Hankin, Michael S. Mohr, and Ken B. Newman, Oxford University Press (2019). © David G. Hankin, Michael S. Mohr, and Ken B. Newman. DOI: 10.1093/oso/9780198815792.001.0001

strata, then there will be no errors of estimation in stratified sampling. Of course, this is an extreme illustration of the most favorable setting for stratified sampling; in practice one constructs the strata so that the variable values within strata are as homogeneous as possible.

With stratified sampling, independent samples of size n_h are selected from strata of size N_h, and the total sample size is $n = \sum_{h=1}^{L} n_h$. We label the variable values of the units within strata as y_{hj}, where h denotes stratum and $j = 1, 2, \ldots, N_h$ denotes a particular unit within stratum h. In many applications SRS is used within strata, a sampling design we label as **stratified SRS**. Many of the fundamental results in stratified sampling that we present in this chapter (e.g., general formulas for sampling variance and estimation of sampling variance) allow for alternative methods of selecting samples within strata, but some of the results that we present are specific to stratified SRS (e.g., optimal allocation results). We therefore attempt to derive and motivate results in a manner that makes it clear how similar results could be generated for other sampling designs that might instead be used within strata.

5.1 Estimation of the population mean

In a stratified population, we may express the overall population mean as

$$\mu = \frac{\sum_{i=1}^{N} y_i}{N} = \frac{\sum_{h=1}^{L} \sum_{j=1}^{N_h} y_{hj}}{N} = \frac{\sum_{h=1}^{L} T_h}{N} = \frac{\sum_{h=1}^{L} N_h \mu_h}{N} = \sum_{h=1}^{L} W_h \mu_h$$

where T_h is a stratum total, and μ_h is a stratum mean. Thus, μ can be formulated as a weighted average of the stratum means where the **stratum weights**, $W_h = N_h/N$, are the fractions of the total number of population units contained in stratum h. In general, the stratum means can be estimated using any sampling strategy(ies) that produces unbiased estimation of μ_h (e.g., SRS with $\hat{\mu}_{mpu} = \hat{\mu}_h$ within strata) and the stratified estimator of the population mean, μ, is

$$\hat{\mu}_{st} = \sum_{h=1}^{L} W_h \hat{\mu}_h \tag{5.1}$$

With stratified SRS, the individual stratum means are estimated using the mpu estimator, $\hat{\mu}_h = \sum_{j \in S_h} y_{hj}/n_h$, where S_h is the random set of sample units selected from stratum h.

5.1.1 Expected value

The expected value of $\hat{\mu}_{st}$ can be found using the fact that the expectation of a linear combination of random variables (here the $\hat{\mu}_h$ values) is equal to the linear combination of the random variable expected values [Equation (A.40)].

$$E(\hat{\mu}_{st}) = E\left(\sum_{h=1}^{L} W_h \hat{\mu}_h \right) = \sum_{h=1}^{L} W_h E(\hat{\mu}_h)$$

For SRS and mpu estimation used within strata, and for many other sampling strategies, $E(\hat{\mu}_h) = \mu_h$, in which case $\hat{\mu}_{st}$ is also unbiased

$$E(\hat{\mu}_{st}) = \sum_{h=1}^{L} W_h \mu_h = \mu \tag{5.2}$$

5.1.2 *Sampling variance*

The variance of a linear combination of *independent* random variables is equal to the linear combination of the random variable variances, but with the scalar coefficients squared [Equation (A.56)]. Thus, because samples are selected independently from strata in stratified sampling

$$V(\hat{\mu}_{st}) = V\left(\sum_{h=1}^{L} W_h \hat{\mu}_h\right) = \sum_{h=1}^{L} W_h^2 V(\hat{\mu}_h) \tag{5.3}$$

Equation (5.3) is the general expression for sampling variance of a stratified estimator of the population mean. The within stratum sampling strategies used to estimate the stratum means determine the form of the $V(\hat{\mu}_h)$. For example, sampling within strata could be SWR, SRS, systematic, unequal probability with or without replacement (see Chapter 8), or some other method. Each of these alternative selection methods and associated estimators of the population mean would have distinct $V(\hat{\mu}_h)$ which could be substituted in Equation (5.3) to give the sampling variance when that sampling strategy is used within strata. When SRS and mpu estimation are used within strata, then

$$V(\hat{\mu}_h) = \left(\frac{N_h - n_h}{N_h}\right) \frac{\sigma_h^2}{n_h}$$

where $\sigma_h^2 = \sum_{j=1}^{N_h} (y_{hj} - \mu_h)^2/(N_h - 1)$ is the **stratum variance**, the variance of the y_{hj} within stratum h, so that

$$V(\hat{\mu}_{st}) = \sum_{h=1}^{L} W_h^2 \left(\frac{N_h - n_h}{N_h}\right) \frac{\sigma_h^2}{n_h} \tag{5.4}$$

5.1.3 *Numerical examples*

In this section we present four small population examples illustrating the dependence of $V(\hat{\mu}_{st})$ on strata membership and sample size allocation, and examine the relative efficiency of stratified SRS ($\hat{\mu}_{st}$) in these settings compared to SRS ($\hat{\mu}_{mpu}$). We also compare the results of applying Equation (5.4) with calculation of $V(\hat{\mu}_{st})$ based directly on the sampling distribution of $\hat{\mu}_{st}$.

The examples are based on a $N = 6$ unit population and a sample size of $n = 4$. The population y values are 10, 9, 12, 6, 25, 16, for which $\mu = 13$, and $\sigma^2 = 45.6$. For SRS selection with mpu estimation of the mean, $V(\hat{\mu}_{mpu}) = [(N - n)/N]\sigma^2/n = [(6 - 4)/6] \cdot 45.6/4 = 3.8$. The examples consider two alternative stratifications of the units into two equal size strata so that $N_1 = N_2 = 3$ and $W_1 = W_2 = 1/2$ (Table 5.1). Stratification 1 is a haphazard grouping of the units, whereas Stratification 2 is a sorted grouping of the units with the three smallest y values in Stratum 1 and the three largest y values in Stratum 2. Notice the differences in μ_h and σ_h^2 between strata and across the stratifications. Of course, these are just two of the $\binom{6}{3} = 20$ possible $N_1 = N_2 = 3$ stratifications of these units.

Example 5.1. For Stratification 1 (haphazard) with $n_1 = n_2 = 2$, Table 5.2 lists all (nine) possible stratified SRS samples that could be selected under this arrangement. Because these samples are all equally likely, one can directly calculate the expected value and sampling variance of $\hat{\mu}_{st}$ as

$$E(\hat{\mu}_{st}) = \sum_{s \in \mathcal{S}} \hat{\mu}_{st}(s)p(s) = \frac{1}{9} \sum_{s \in \mathcal{S}} \hat{\mu}_{st}(s) = \frac{1}{9} \cdot 117 = 13 = \mu$$

$$V(\hat{\mu}_{st}) = \sum_{s \in \mathcal{S}} \left(\hat{\mu}_{st}(s) - \mu\right)^2 p(s) = \frac{1}{9} \sum_{s \in \mathcal{S}} \left(\hat{\mu}_{st}(s) - 13\right)^2 = \frac{1}{9} \cdot 34.75 = 3.861$$

Table 5.1 Example stratifications of a $N = 6$ unit population with population y values 10, 9, 12, 6, 25, 16 into two equal size strata ($h = 1, 2$). Stratification 1 is *haphazard* whereas Stratification 2 is based on the *order* of y values.

	Stratification 1		Stratification 2	
Quantity	Stratum 1	Stratum 2	Stratum 1	Stratum 2
y_{h1}	10	6	6	12
y_{h2}	9	25	9	16
y_{h3}	12	16	10	25
μ_h	10.3333	15.6667	8.3333	17.6667
σ_h^2	2.3333	90.3333	4.3333	44.3333

Table 5.2 Listing of all possible stratified samples and corresponding sample estimates for Example 5.1. Sample identifiers and sample probabilities are denoted by s and $p(s)$, respectively. Summed values can be used to directly calculate $E(\hat{\mu}_{st})$ and $V(\hat{\mu}_{st})$.

	Stratum 1	Stratum 2					
Sample ID	$y_{1j}, j \in s_1$	$y_{2j}, j \in s_2$	$p(s)$	$\hat{\mu}_1(s)$	$\hat{\mu}_2(s)$	$\hat{\mu}_{st}(s)$	$(\hat{\mu}_{st}(s) - \mu)^2$
1	10, 9	6, 25	1/9	9.5	15.5	12.50	0.2500
2	10, 9	6, 16	1/9	9.5	11.0	10.25	7.5625
3	10, 9	25, 16	1/9	9.5	20.5	15.00	4.0000
4	10, 12	6, 25	1/9	11.0	15.5	13.25	0.0625
5	10, 12	6, 16	1/9	11.0	11.0	11.00	4.0000
6	10, 12	25, 16	1/9	11.0	20.5	15.75	7.5625
7	9, 12	6, 25	1/9	10.5	15.5	13.00	0.0000
8	9, 12	6, 16	1/9	10.5	11.0	10.75	5.0625
9	9, 12	25, 16	1/9	10.5	20.5	15.50	6.2500
Sum:			1			117.00	34.7500

Equation (5.4) gives the identical result for the sampling variance

$$V(\hat{\mu}_{st}) = \sum_{h=1}^{L} W_h^2 \left(\frac{N_h - n_h}{N_h} \right) \frac{\sigma_h^2}{n_h} = \left[\frac{1}{4} \cdot \frac{1}{3} \cdot \frac{2.3333}{2} \right] + \left[\frac{1}{4} \cdot \frac{1}{3} \cdot \frac{90.3333}{2} \right] = 3.861$$

For this example, the relative efficiency of $\hat{\mu}_{st}$ compared to $\hat{\mu}_{mpu}$ is

$$RE(\hat{\mu}_{st}, \hat{\mu}_{mpu}) = V(\hat{\mu}_{mpu})/V(\hat{\mu}_{st}) = 3.8/3.861 \approx 1.0$$

and thus the stratified SRS with $\hat{\mu}_{st}$ strategy is no more precise than the SRS with $\hat{\mu}_{mpu}$ strategy. As the following two examples illustrate, however, other stratifications and/or alternative allocations of n across strata can create settings for which $V(\hat{\mu}_{st})$ is substantially less than $V(\hat{\mu}_{mpu})$.

Example 5.2. *Continuation.* Now consider Stratification 2 (ordered), again with $n_1 = n_2 = 2$. In this case

$$V(\hat{\mu}_{st}) = \left[\frac{1}{4} \cdot \frac{1}{3} \cdot \frac{4.3333}{2} \right] + \left[\frac{1}{4} \cdot \frac{1}{3} \cdot \frac{44.3333}{2} \right] = 2.028$$

resulting in a relative efficiency of $3.8/2.028 \approx 1.9$.

Example 5.3. *Continuation.* Now assume Stratification 1 (haphazard), but with unequal stratum sample sizes of $n_1 = 1$ and $n_2 = 3$. In this case the stratum-specific fpc values are $(N_1 - n_1)/N_1 = 2/3$ and $(N_2 - n_2)/N_2 = 0/3$ so that

$$V(\hat{\mu}_{st}) = \left[\frac{1}{4} \cdot \frac{2}{3} \cdot \frac{2.3333}{1}\right] + \left[\frac{1}{4} \cdot \frac{0}{3} \cdot \frac{90.3333}{3}\right] = 0.389$$

resulting in a relative efficiency of $3.8/0.389 \approx 9.8$. The contribution to the sampling variance from Stratum 2, the more variable stratum, is 0 because of the complete census taken in Stratum 2 (i.e., $n_2 = N_2$).

Example 5.4. *Continuation.* Assume again Stratification 1 (haphazard), but switch the stratum sample sizes of the previous example so that $n_1 = 3$ and $n_2 = 1$. In this case

$$V(\hat{\mu}_{st}) = \left[\frac{1}{4} \cdot \frac{0}{3} \cdot \frac{2.3333}{3}\right] + \left[\frac{1}{4} \cdot \frac{2}{3} \cdot \frac{90.3333}{1}\right] = 15.056$$

resulting in a relative efficiency of $3.8/15.056 \approx 0.3$. This example shows that, if ill-conceived, it is possible for stratified SRS to perform poorly compared to SRS!

5.2 Estimation of the population proportion

As noted previously, a proportion can be viewed as a special case of a population mean, where y is a binary (0, 1) variable. Thus, the results for estimating a population mean using a stratified sample transfer directly over to estimating a population proportion

$$\hat{\pi}_{st} = \sum_{h=1}^{L} W_h \hat{\pi}_h \tag{5.5}$$

where $\hat{\pi}_h$ is an estimator of the proportion for stratum h, with the following general expression for sampling variance

$$V(\hat{\pi}_{st}) = \sum_{h=1}^{L} W_h^2 V(\hat{\pi}_h) \tag{5.6}$$

When SRS is used within strata, $\hat{\pi}_h = \sum_{j \in S_h} y_{hj}/n_h$ and

$$V(\hat{\pi}_{st}) = \sum_{h=1}^{L} W_h^2 \left(\frac{N_h - n_h}{N_h}\right) \frac{\sigma_h^2}{n_h} \tag{5.7}$$

where

$$\sigma_h^2 = \left(\frac{N_h}{N_h - 1}\right) \pi_h(1 - \pi_h) \tag{5.8}$$

As noted previously, many other methods of sampling could be used within strata and would generate different, possibly stratum-specific, expressions for $V(\hat{\pi}_h)$, but Equation (5.6) would continue to apply.

5.3 Estimation of the population total

The population total, $\mathcal{T}$, can be estimated in stratified sampling by just "scaling up" a stratified estimator of the population mean, or by summing up estimated stratum totals

$$\hat{T}_{st} = N\hat{\mu}_{st} = \sum_{h=1}^{L} \hat{T}_h \qquad (5.9)$$

These alternative forms of the estimator of the total give identical results since

$$\hat{T}_{st} = N\hat{\mu}_{st} = N\sum_{h=1}^{L} W_h\hat{\mu}_h = \sum_{h=1}^{L} N_h\hat{\mu}_h = \sum_{h=1}^{L} \hat{T}_h$$

If estimates of the stratum means (or totals) are unbiased, as would be true for SRS selection within strata, then $\hat{T}_{st}$ will also be unbiased

$$E(\hat{T}_{st}) = E(N\hat{\mu}_{st}) = NE(\hat{\mu}_{st}) = N\mu = T$$

Because the samples from within strata are selected independently of one another, the $\hat{T}_h$ are also independent across strata and the general expression for sampling variance for a stratified estimator of the total is

$$V(\hat{T}_{st}) = V\left(\sum_{h=1}^{L} \hat{T}_h\right) = \sum_{h=1}^{L} V(\hat{T}_h) \qquad (5.10)$$

If SRS and mpu estimation are used to estimate stratum totals, then

$$V(\hat{T}_{st}) = \sum_{h=1}^{L} N_h^2 \left(\frac{N_h - n_h}{N_h}\right) \frac{\sigma_h^2}{n_h} \qquad (5.11)$$

Convince yourself that the sampling variance for the stratified SRS estimator of the total can also be expressed as $V(\hat{T}_{st}) = N^2 V(\hat{\mu}_{st})$. If some other sampling strategy were used within strata, then the corresponding expression for $V(\hat{T}_h)$ would be substituted into Equation (5.10).

5.4 Estimation of sampling variance

Estimation of $V(\hat{\mu}_h)$ is straightforward when the stratum samples are selected independently of one another. In this case, the sampling variance can be estimated as a weighted sum of the estimated stratum-specific sampling variances. The general expression for estimation of sampling variance for a stratified estimator, with unbiased estimation within strata, is:

$$\hat{V}(\hat{\mu}_{st}) = \sum_{h=1}^{L} W_h^2 \hat{V}(\hat{\mu}_h) \qquad (5.12)$$

When SRS and mpu estimation are used within strata, an unbiased estimator for $V(\hat{\mu}_h)$ is

$$\hat{V}(\hat{\mu}_h) = \left(\frac{N_h - n_h}{N_h}\right) \frac{\hat{\sigma}_h^2}{n_h}$$

where $\hat{\sigma}_h^2 = \sum_{j \in S_h} (y_{hj} - \hat{\mu}_h)^2/(n_h - 1)$. Thus, $V(\hat{\mu}_{st})$ can be estimated as

$$\hat{V}(\hat{\mu}_{st}) = \sum_{h=1}^{L} W_h^2 \left(\frac{N_h - n_h}{N_h}\right) \frac{\hat{\sigma}_h^2}{n_h} \qquad (5.13)$$

It is not difficult to verify that this estimator is also unbiased

$$E\left[\hat{V}(\hat{\mu}_{st})\right] = E\left[\sum_{h=1}^{L} W_h^2 \hat{V}(\hat{\mu}_h)\right] = \sum_{h=1}^{L} W_h^2 E[\hat{V}(\hat{\mu}_h)] = \sum_{h=1}^{L} W_h^2 V(\hat{\mu}_h) = V(\hat{\mu}_{st})$$

Similarly, general expressions for estimation of sampling variance of an estimator of a population proportion, π, or total, $\mathcal{T}$, would be

$$\hat{V}(\hat{\pi}_{st}) = \sum_{h=1}^{L} W_h^2 \hat{V}(\hat{\pi}_h) \tag{5.14}$$

$$\hat{V}(\hat{\mathcal{T}}_{st}) = \sum_{h=1}^{L} \hat{V}(\hat{\mathcal{T}}_h) \tag{5.15}$$

With SRS and mpu estimation used within strata

$$\hat{V}(\hat{\pi}_{st}) = \sum_{h=1}^{L} W_h^2 \left(\frac{N_h - n_h}{N_h}\right) \frac{\hat{\sigma}_h^2}{n_h} \tag{5.16}$$

$$\hat{V}(\hat{\mathcal{T}}_{st}) = \sum_{h=1}^{L} N_h^2 \left(\frac{N_h - n_h}{N_h}\right) \frac{\hat{\sigma}_h^2}{n_h} \tag{5.17}$$

where for a binary (0, 1) variable

$$\hat{\sigma}_h^2 = \left(\frac{n_h}{n_h - 1}\right) \hat{\pi}_h (1 - \hat{\pi}_h) \tag{5.18}$$

If some alternative sampling strategy(ies) were used within strata, then the corresponding strategy-specific expressions for $\hat{V}(\hat{\mu}_h)$, $\hat{V}(\hat{\pi}_h)$, or $\hat{V}(\hat{\mathcal{T}}_h)$ would be substituted in Equations (5.12), (5.14), or (5.15), respectively. The general expressions for estimation of sampling variance will be unbiased whenever the $\hat{V}(\hat{\mu}_h)$, $\hat{V}(\hat{\pi}_h)$, or $\hat{V}(\hat{\mathcal{T}}_h)$, respectively, are unbiased. Therefore, selection of single linear systematic samples within strata would not allow unbiased estimation of sampling variance because within stratum sampling variance could not be unbiasedly estimated.

5.5 Allocation of the sample across strata

As Examples 5.2 and 5.3 illustrate, stratified SRS can achieve unbiased estimation with greatly reduced sampling variance when compared to SRS with mpu estimation. The sampling variance of $\hat{\mu}_{st}$ depends primarily upon (a) the allocation of n across strata via the sample **allocation weights**, $w_h = n_h/n$, and (b) the degree of homogeneity among unit y values within strata. Note that the allocation weights, w_h, must all be positive (i.e., every stratum must be sampled) and that they must sum to one, $\sum_{h=1}^{L} w_h = 1$. Throughout this section we assume that SRS and mpu estimation are used within strata, but the general concepts of optimal allocation that are motivated in this section apply regardless of the sampling strategy(ies) used within strata.

Optimal allocation results are typically based on either (a) minimization of the sampling variance of $\hat{\mu}_{st}$ given a specified budget available for the survey, or (b) minimization of total survey cost given a target precision level (target sampling variance) for $\hat{\mu}_{st}$. For both cases, a *cost function* is required that specifies total survey cost as a function of the stratum-specific sample sizes and stratum-specific costs per unit of measuring variables in stratum samples. We assume the following simple linear cost function (others are certainly possible)

$$C = c_0 + \sum_{h=1}^{L} c_h n_h \qquad (5.19)$$

where c_h is the cost per unit in stratum h, c_0 is the survey "overhead" cost (e.g., deployment and transportation costs for survey personnel that are not directly related to costs of taking sample observations), and C is the total survey cost.

5.5.1 *Optimal allocation: Graphical analysis*

Before deriving the optimal allocation analytical results, it is helpful to develop a graphical perspective of the objective and its solution. To simplify matters, we'll assume here that $c_h = c$ is a constant for all strata, and that the overall budget, C, has been specified for the survey. Thus, with the linear cost function, the total sample size available for the survey is $n = (C - c_0)/c$, and our objective is to allocate the available n across strata so as to minimize the sampling variance $V(\hat{\mu}_{st})$. Consider first a simple case of two strata of size $N_1 = N_2 = 200$, with stratum variances of $\sigma_1^2 = 200$ and $\sigma_2^2 = 40$, and a total available sample size of $n = n_1 + n_2 = 60$. All strata must be sampled at some level, so there are $n - 1 = 59$ possible allocations of the total sample size across the two strata: $(n_1, n_2) = (1, 59), (2, 58), ..., (59, 1)$. For each such pair (n_1, n_2) we can calculate $V(\hat{\mu}_{st})$ using Equation (5.4). Figure 5.2(a) illustrates how $V(\hat{\mu}_{st})$ varies according to the (n_1, n_2) allocation of the total sample size $n = 60$. (Note that for the two strata case, given n, $V(\hat{\mu}_{st})$ is actually a function of a single variable, n_1, because $n_2 = n - n_1$.) If only a very small sample size is selected from Stratum 1 (e.g., $n_1 = 5$), then $V(\hat{\mu}_{st})$ is about 10 and much larger than if n_1 were relatively large. The *optimal allocation* of n appears to be $n_1 \approx 40$ and $n_2 \approx 60 - 40 = 20$. Note that the graph of $V(\hat{\mu}_{st})$ against n_1 is relatively flat over the range $30 \leq n_1 \leq 50$.

This same concept of optimal allocation applies in higher dimensional ($L > 2$) settings. For example, if there are three strata, given the available n, $V(\hat{\mu}_{st})$ is a function of two variables (n_1 and n_2, with $n_3 = n - n_1 - n_2$). Suppose that $N_1 = N_2 = N_3 = 200$, $\sigma_1^2 = \sigma_2^2 = \sigma_3^2 = 100$, and $n = 60$. Then $V(\hat{\mu}_{st})$ when plotted against (n_1, n_2) appears as a purse-like surface (Figure 5.2b). The optimal allocation of n would be the (n_1, n_2) combination corresponding to the low point of the purse. A $V(\hat{\mu}_{st})$ contour plot on the (n_1, n_2) plane could also be used to identify the optimal allocation. Figure 5.2(c) indicates that the optimal allocation for our "purse" example (where all N_h and all σ_h^2 are equal with $n = 60$) is $n_{1,opt} = n_{2,opt} = n_{3,opt} = 20$, not a surprise! But if instead, for example, $\sigma_1^2 = 240$, $\sigma_2^2 = 80$, $\sigma_3^2 = 20$ (Figure 5.2d), the optimal allocation shifts toward the more variable strata: $n_{1,opt} = 32$, $n_{2,opt} = 19$, $n_{3,opt} = 9$.

5.5.2 *Optimal allocation: Analytical analysis*

In this section we formally derive optimal allocation results for an arbitrary number of strata, L, each of which must be sampled at some level, $n_h = w_h n$, with $w_h > 0$. Before we can identify the "best possible" choices for the n_h, we must impose an additional side condition (*constraint*) that specifies either the overall cost of the survey or the target precision level. If we simply wanted the most precise estimate possible, that would be accomplished by conducting a complete census, $n_h = N_h, h = 1, 2, ..., L$ which, presumably, would not be a realistic option. As mentioned previously, either one of the two following constraints can be used to determine the optimal allocation results for stratified SRS.

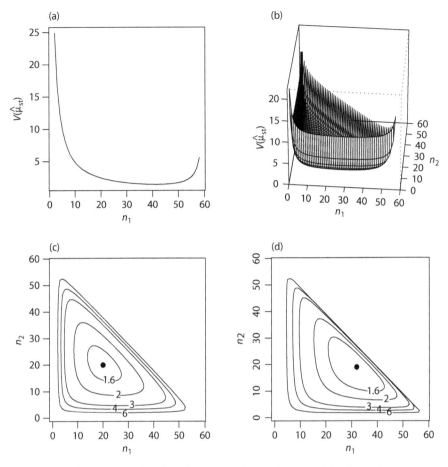

Fig. 5.2 Optimal allocation graphical analysis. Sampling variance and its dependence on stratum sample sizes. (a) Two stratum case with $N_1 = N_2 = 200$, $\sigma_1^2 = 200$, and $\sigma_2^2 = 40$. Total sample size is fixed at $n = 60$, with $n_2 = n - n_1$; (b) perspective plot, three stratum case with $N_1 = N_2 = N_3 = 200$ and $\sigma_1^2 = \sigma_2^2 = \sigma_3^2 = 100$. Total sample size is fixed at $n = 60$, with $n_3 = n - n_1 - n_2$; (c) contour plot, three stratum case, same parameters as for panel (b), resulting in $n_{1,opt} = n_{2,opt} = n_{3,opt} = 20$ (solid dot); (d) contour plot, three stratum case, same N_h parameters as for panel (c), but with $\sigma_1^2 = 240$, $\sigma_2^2 = 80$, $\sigma_3^2 = 20$, resulting in $n_{1,opt} = 32$, $n_{2,opt} = 19$, $n_{3,opt} = 9$ (solid dot).

(a) Minimize $V(\hat{\mu}_{st})$ subject to a fixed total survey cost, C: C is the constraint.
(b) Minimize total survey cost, C, subject to achieving a specified $V(\hat{\mu}_{st})$: $V(\hat{\mu}_{st})$ is the constraint.

For the linear cost function [Equation (5.19)], the solutions that one obtains for the $w_h = n_h/n$ are identical under these two alternative constraints (Cochran 1977). The solutions for total sample size, n_{opt}, and sample sizes within strata, $n_{h,opt} = w_{h,opt}n_{opt}$, however, do depend on which constraint is imposed. We use condition (a) to derive the optimal allocation results.

Use of Lagrange multipliers

In stratified SRS, $V(\hat{\mu}_{st}) = \sum_h W_h^2[(N_h - n_h)/N_h]\sigma_h^2/n_h$ is a function of the *variables* n_h and the *constants* N_h, N, and σ_h^2. $V(\hat{\mu}_{st})$ will thus vary according to our choices of n_h. We want to minimize $V(\hat{\mu}_{st})$ through our choices of n_h subject to the constraint that the total cost of the survey is equal to a specified value C. That is, we want to produce the most accurate estimator that we can given the amount of money that we have available to carry out the survey.

Our problem is thus a minimization (calculus) problem, but subject to a constraint. This kind of problem can be solved using the method of *Lagrange multipliers*. Section A.9 provides an illustrative application of this method to minimization of the simple function $f(x_1, x_2) = 2x_1^2 + x_2^2$, subject to the constraint $g(x_1, x_2) = x_1 + x_2 - 10 = 0$. We now apply this method to find the optimal allocation of a stratified SRS sample.

We seek to minimize $V(\hat{\mu}_{st}) = \sum_h W_h^2[(N_h - n_h)/N_h]\sigma_h^2/n_h = f(\mathbf{n})$, where $\mathbf{n} = (n_1, n_2, \ldots, n_L)$, subject to the constraint $g(\mathbf{n}) = c_0 + \sum_h c_h n_h - C = 0$ for a specified total survey cost, C. Following the steps outlined in Section A.9, we first form the Lagrange function

$$\mathcal{L}(\mathbf{n}, \lambda) = f(\mathbf{n}) + \lambda g(\mathbf{n}) = \sum_{h=1}^{L} W_h^2 \left(\frac{N_h - n_h}{N_h}\right) \frac{\sigma_h^2}{n_h} + \lambda \left(c_0 + \sum_{h=1}^{L} c_h n_h - C\right)$$

$$= \sum_{h=1}^{L} \frac{W_h^2 \sigma_h^2}{n_h} - \sum_{h=1}^{L} \frac{W_h^2 \sigma_h^2}{N_h} + \lambda \left(c_0 + \sum_{h=1}^{L} c_h n_h - C\right)$$

Now find the partial derivatives of $\mathcal{L}(\mathbf{n}, \lambda)$ with respect to the $n_h, h = 1, 2, \ldots, L$, and λ, set them equal to zero, and solve for n_h in terms of λ

$$\frac{\partial \mathcal{L}}{\partial n_h} = -\frac{W_h^2 \sigma_h^2}{n_h^2} + \lambda c_h = 0, \text{ so that } n_h = \frac{1}{\sqrt{\lambda}} \frac{W_h \sigma_h}{\sqrt{c_h}}, h = 1, 2, \ldots, L$$

$$\frac{\partial \mathcal{L}}{\partial \lambda} = c_0 + \sum_{h=1}^{L} c_h n_h - C = 0 \quad \text{(the constraint, } g(\mathbf{n}) = 0)$$

Substituting the solution for n_h into the final (constraint) equation and solving for $1/\sqrt{\lambda}$ gives

$$\frac{1}{\sqrt{\lambda}} = \frac{C - c_0}{\sum_{h=1}^{L} W_h \sigma_h \sqrt{c_h}}$$

and substituting this equation for $1/\sqrt{\lambda}$ back into the n_h equation gives

$$n_{h,\text{opt}}|C = \frac{(C - c_0) N_h \sigma_h / \sqrt{c_h}}{\sum_{h=1}^{L} N_h \sigma_h \sqrt{c_h}} \tag{5.20}$$

with

$$n_{\text{opt}}|C = \sum_{h=1}^{L} n_{h,\text{opt}}|C = \frac{(C - c_0) \sum_{h=1}^{L} N_h \sigma_h / \sqrt{c_h}}{\sum_{h=1}^{L} N_h \sigma_h \sqrt{c_h}} \tag{5.21}$$

The **optimal allocation weights** are therefore

$$w_{h,\text{opt}} = \frac{n_{h,\text{opt}}|C}{n_{\text{opt}}|C} = \frac{N_h \sigma_h / \sqrt{c_h}}{\sum_{h=1}^{L} N_h \sigma_h / \sqrt{c_h}} \tag{5.22}$$

5.5.3 *Comments on optimal allocation*

Note that although $w_{h,\text{opt}}$ was derived to minimize $V(\hat{\mu}_{\text{st}})$ subject to an overall cost constraint, $w_{h,\text{opt}}$ also minimizes $V(\hat{T}_{\text{st}})$ subject to that same cost constraint because $V(\hat{T}_{\text{st}})$ is a constant multiple of $V(\hat{\mu}_{\text{st}})$, $V(\hat{T}_{\text{st}}) = N^2 V(\hat{\mu}_{\text{st}})$. Note also that $w_{h,\text{opt}}$ is independent of the total survey cost, C, and that the optimal allocation weights can be determined using *relative costs* per unit rather than actual costs. Specifically, if c_h is proportional to some other measure c'_h so that $c_h = \alpha c'_h$ for all h with α being a proportionality constant, then if $\alpha c'_h$ is substituted for c_h in Equation (5.22), α factors out of the equation leaving c'_h in place of c_h. For example, if $\{c_h\} = \{100, 200, 500\}$, the same $w_{h,\text{opt}}$ will result from using $\{c'_h\} = \{1, 2, 5\}$, with $c_h = 100c'_h$. This can be important for survey planning, when the actual costs per unit in strata are not accurately known but the relative costs seem fairly well defined.

Although $w_{h,\text{opt}}$ was derived to minimize $V(\hat{\mu}_{\text{st}})$ subject to a specified total survey cost, C, minimization of total survey cost subject to a specified sampling variance $V(\hat{\mu}_{\text{st}}) = V$ leads to the identical solution for $w_{h,\text{opt}}$ (see Problem 5.2). However, the optimal total sample size differs

$$n_{\text{opt}}|V = \frac{\left(\sum\limits_{h=1}^{L} N_h \sigma_h \sqrt{c_h}\right)\left(\sum\limits_{h=1}^{L} N_h \sigma_h / \sqrt{c_h}\right)}{N^2 V(\hat{\mu}_{\text{st}}) + \sum\limits_{h=1}^{L} N_h \sigma_h^2} \tag{5.23}$$

and the corresponding $n_{h,\text{opt}}|V = w_{h,\text{opt}} \cdot n_{\text{opt}}|V$.

The solution for $w_{h,\text{opt}}$ [Equation (5.22)] has important implications for stratified SRS in general: greater sampling effort should be given to those strata that

(a) are larger (larger N_h)
(b) have greater within stratum variation (larger σ_h^2)
(c) have lower cost per sampled unit (smaller c_h)

which suggests the following three "special cases" of the optimal allocation solution in terms of the $w_{h,\text{opt}}$.

(1) All c_h are equal so that $w_{h,\text{opt}} = (N_h \sigma_h)/(\sum_h N_h \sigma_h)$, termed **Neyman allocation** after Jerzy Neyman for his seminal paper (Neyman 1934) on stratified sampling and sample allocation.
(2) All σ_h^2 are equal so that $w_{h,\text{opt}} = (N_h/\sqrt{c_h})/(\sum_h N_h/\sqrt{c_h})$, which we term **constant variance allocation**.
(3) All c_h are equal and all σ_h^2 are equal so that $w_{h,\text{opt}} = N_h/N = W_h$, termed **proportional allocation**.

If estimates of σ_h^2 and c_h are not available during the planning stages of a stratified survey, then proportional allocation is a good choice for allocation of the total sample. In practical applications, the w_h are calculated using $\sqrt{\hat{\sigma}_h^2}$ in place of σ_h.

The calculated overall sample size, n_{opt}, is real-valued and should be rounded up (to the nearest integer) in the case of $n_{\text{opt}}|V$, or rounded down in the case of $n_{\text{opt}}|C$, to satisfy the V or C constraint, respectively. Secondly, given the integer-valued n_{opt}, the calculated sample sizes within strata, $n_{h,\text{opt}} = w_{h,\text{opt}} n_{\text{opt}}$, are also real-valued, and a method known as *controlled rounding* is often used to round these real values to a set of integers that sum to n_{opt}.

With controlled rounding, for any set of w_h, how many of the n_h are to be rounded up is given by the sum of the n_h fractional parts. Those n_h with the largest fractional parts are then rounded up and the rest are rounded down. For example, if $n = 20$ and the initially calculated $\{n_h\} = \{w_h n\} = \{4.8, 8.3, 6.9\}$, two (= 0.8 + 0.3 + 0.9) of the n_h are to be rounded up and the other down, so that $\{n_h\} = \{5, 8, 7\}$. Controlled rounding often results in the true, integer-valued optimal allocation of n_{opt}, but sometimes the resulting allocation may only be nearly optimal (Wright 2012). If this is an important consideration, Wright's (2012, 2014, 2016) algorithms may be used to determine the exact, integer-valued optimal allocation.

Optimal allocation formulas may also yield one or more $n_{h,opt} > N_h$, which is impossible when sampling without replacement. Whenever this occurs, $n_{h,opt}$ should be set equal to N_h for these strata and the formulas for $w_{h,opt}$, n_{opt}, and $n_{h,opt}$ should be applied to the remaining strata. Recall that the contribution to the sampling variance is zero for those strata in which $n_{h,opt} = N_h$.

5.6 Sample size determination

Section 5.5 addressed the distribution of sample size across strata via the optimal allocation weights, $w_{h,opt}$. It also provided the overall sample size necessary, n_{opt}, given either a survey cost constraint C, or a sampling variance target of $V(\hat{\mu}_{st})$. In this section, the $n|C$ and $n|V$ formulas, Equations (5.21) and (5.23), are generalized to: (a) permit an arbitrary set of specified w_h values, (b) accommodate any one of the three estimators ($\hat{\mu}$, $\hat{\pi}$, $\hat{T}$), and (c) allow for specified measures of target uncertainty other than sampling variance (e.g., standard error, coefficient of variation).

The key to developing general formulas for $n|C$ and $n|V$ is to first re-express C and V in terms of n rather than n_h. Writing our linear cost function, Equation (5.19), in terms of n rather than n_h gives $C = c_0 + \sum_h n_h c_h = c_0 + n \sum_h w_h c_h$, because $n_h = w_h n$. Solving this expression for n gives the solution

$$n|C = \frac{C - c_0}{\sum_{h=1}^{L} w_h c_h} \tag{5.24}$$

Thus, given a specified C and values for c_0, c_h, and w_h, the corresponding n can be calculated and rounded down to the nearest integer (to satisfy the cost constraint), and sample sizes within strata can then be determined from $n_h = w_h n$ using controlled rounding as described in Section 5.5.3. Note that $n|C$ is independent of the population parameter (μ, π, T) being estimated. Note also that if the optimal allocation weights $w_{h,opt}$ [Equation (5.22)] are substituted for w_h in Equation (5.24), this yields our previous result for $n_{opt}|C$ [Equation (5.21)].

The case of V is a bit more involved. Assume first that we are interested in estimating μ and want to know the n necessary to achieve a specified value of $V(\hat{\mu}_{st})$. We re-express $V(\hat{\mu}_{st})$ in terms of n rather than n_h, recognizing that $n_h = w_h n$

$$V(\hat{\mu}_{st}) = \sum_{h=1}^{L} W_h^2 \left(\frac{N_h - n_h}{N_h} \right) \frac{\sigma_h^2}{n_h} = \frac{1}{N^2} \left[\frac{1}{n} \sum_{h=1}^{L} N_h^2 \frac{\sigma_h^2}{w_h} - \sum_{h=1}^{L} N_h \sigma_h^2 \right]$$

Solving this expression for n gives

$$n|V = \frac{\sum_{h=1}^{L} N_h^2 \sigma_h^2 / w_h}{N^2 V(\hat{\mu}_{st}) + \sum_{h=1}^{L} N_h \sigma_h^2} \tag{5.25}$$

Thus, given a specified $V(\hat{\mu}_{\mathrm{st}})$ and values for N_h, σ_h^2, and w_h, the corresponding n can be calculated using Equation (5.25) and rounding up to the nearest integer (to satisfy the variance constraint). Sample sizes within strata are determined from $n_h = w_h n$ using controlled rounding as described in the previous section. Again, note that if the optimal allocation weights $w_{h,\mathrm{opt}}$ [Equation (5.22)] are substituted for w_h in Equation (5.25), this yields our previous result for $n_{\mathrm{opt}}|V$ [Equation (5.23)].

Because the sampling variance equations for $\hat{\mu}$, $\hat{\pi}$, and $\hat{T}$ (Equations (5.4), (5.7), and (5.11), respectively) have the same general form, Equation (5.25) for $n|V$ can be generalized for estimating $\theta = \mu$, π, or T as

$$n|V = \frac{\sum\limits_{h=1}^{L} N_h^2 \sigma_h^2 / w_h}{\mathcal{K}(\hat{\theta}_{\mathrm{st}}) + \sum\limits_{h=1}^{L} N_h \sigma_h^2}, \qquad \mathcal{K}(\hat{\theta}) = \begin{cases} N^2 V(\hat{\theta}), & \text{for } \theta = \mu, \pi \\ V(\hat{\theta}), & \text{for } \theta = T \end{cases} \qquad (5.26)$$

As noted in Section 3.4.3, desired estimator uncertainty is rarely specified directly in terms of sampling variance. Instead, it is usually specified as a desired standard error, $1 - \alpha$ level confidence interval half-width (bounds on error of estimation), or coefficient of variation. For a specified (large sample) 95% bounds on error of estimation, $B = 2\mathrm{SE}(\hat{\theta})$, re-express $V(\hat{\theta})$ in terms of B giving $V(\hat{\theta}) = B^2/4$. For a specified coefficient of variation, $\mathrm{CV}(\hat{\theta}) = \mathrm{SE}(\hat{\theta})/\theta$, assuming $\hat{\theta}$ is unbiased (and positive valued), re-express $V(\hat{\theta})$ as $\mathrm{CV}^2(\hat{\theta})\theta^2$. Either of these re-expressions for $V(\hat{\theta})$ can then be substituted in Equation (5.26).

5.7 Relative efficiency

The general expression for the Relative Efficiency (RE) of stratified SRS ($\hat{\mu}_{\mathrm{st}}$) as compared to SRS with mpu estimation ($\hat{\mu}_{\mathrm{mpu}}$) is

$$\mathrm{RE}(\hat{\mu}_{\mathrm{st}}, \hat{\mu}_{\mathrm{mpu}}) = \frac{V(\hat{\mu}_{\mathrm{mpu}})}{V(\hat{\mu}_{\mathrm{st}})}$$

There is not, however, any simple general result that describes the relative efficiency of $\hat{\mu}_{\mathrm{st}}$ compared to $\hat{\mu}_{\mathrm{mpu}}$, whereas there is a simple general expression for, say, the relative efficiency of $\hat{\mu}_{\mathrm{mpu}}$ compared to $\hat{\mu}_{\mathrm{swr}}$ in an unstratified setting. As demonstrated previously in small population examples, however, for a fixed total sample size, n, two factors have primary influence over the relative size of $V(\hat{\mu}_{\mathrm{st}})$ [Equation (5.4)] compared to $V(\hat{\mu}_{\mathrm{mpu}})$: (a) allocation of n across strata (the w_h), and (b) assignment of units to strata because the unit y values within strata determine the σ_h^2. We can, however, gain considerable general insight into the relative efficiency of stratified SRS ($\hat{\mu}_{\mathrm{st}}$) as compared to SRS ($\hat{\mu}_{\mathrm{mpu}}$) by adopting an analysis of variance (ANOVA) perspective and making an assumption of proportional allocation.

5.7.1 Proportional allocation

With proportional allocation

$$w_h = W_h = \frac{N_h}{N}, \text{ so that } n_h = \frac{N_h}{N} n$$

In this case, the finite population correction factor for stratum h is

$$\left(\frac{N_h - n_h}{N_h}\right) = \left(\frac{N_h - (N_h/N)n}{N_h}\right) = \left(\frac{N - n}{N}\right)$$

Thus, under proportional allocation with $n_h = W_h n$, and $N_h = W_h N$

$$V(\hat{\mu}_{st,prop}) = \sum_{h=1}^{L} W_h^2 \left(\frac{N_h - n_h}{N_h}\right) \frac{\sigma_h^2}{n_h} = \left(\frac{N-n}{N}\right) \sum_{h=1}^{L} W_h \frac{\sigma_h^2}{n} \tag{5.27}$$

and the relative efficiency of $\hat{\mu}_{st,prop}$ compared to $\hat{\mu}_{mpu}$ is

$$RE(\hat{\mu}_{st,prop}, \hat{\mu}_{mpu}) = \frac{V(\hat{\mu}_{mpu})}{V(\hat{\mu}_{st,prop})} = \frac{\sigma^2}{\sum_{h=1}^{L} W_h \sigma_h^2}$$

The interpretation of this result becomes more readily apparent if it is recast in terms of mean squares from an ANOVA. We begin by noting that the sums of squares identity for a stratified population is

$$\sum_{h=1}^{L} \sum_{j=1}^{N_h} (y_{hj} - \mu)^2 = \underbrace{\sum_{h=1}^{L} \sum_{j=1}^{N_h} (y_{hj} - \mu_h)^2}_{} + \underbrace{\sum_{h=1}^{L} \sum_{j=1}^{N_h} (\mu_h - \mu)^2}_{} \tag{5.28}$$

$$\underbrace{\phantom{\sum_{h=1}^{L} \sum_{j=1}^{N_h} (y_{hj} - \mu)^2}}_{SS(T)} \quad \underbrace{\phantom{\sum_{h=1}^{L} \sum_{j=1}^{N_h} (y_{hj} - \mu_h)^2}}_{SS(W)} \quad \underbrace{\phantom{\sum_{h=1}^{L} \sum_{j=1}^{N_h} (\mu_h - \mu)^2}}_{SS(B)}$$

where SS(T), SS(W), and SS(B) are the total sum of squares and the sum of squares within and between strata, respectively. By definition, $MS(T) = \sum_{h=1}^{L} \sum_{j=1}^{N_h} (y_{hj} - \mu)^2 / (N-1) = \sigma^2$, the mean square total variation. In a stratified population, $\sum_{h=1}^{L} W_h \sigma_h^2$ is approximately equal to the mean square within strata variation

$$MS(W) = \frac{\sum_{h=1}^{L} \sum_{j=1}^{N_h} (y_{hj} - \mu_h)^2}{\sum_{h=1}^{L} (N_h - 1)} = \frac{\sum_{h=1}^{L} (N_h - 1)\sigma_h^2}{\sum_{h=1}^{L} (N_h - 1)} \approx \sum_{h=1}^{L} W_h \sigma_h^2$$

assuming that the N_h are all "large" so that $(N_h - 1)/\sum_{h=1}^{L}(N_h - 1) \approx W_h$. Therefore

$$RE(\hat{\mu}_{st,prop}, \hat{\mu}_{mpu}) \approx \frac{\sigma^2}{MS(W)} = \frac{MS(T)}{MS(W)}$$

Thus, with proportional allocation, if the average variation within strata, MS(W), is less than the average variation in the overall population, $MS(T) = \sigma^2$, then stratified SRS will outperform SRS. MS(W) can be reduced to well below σ^2 by constructing strata so that as much of the total variation as possible is attributable to between stratum variation rather than within stratum variation. Thus, it is desirable for the y values within strata to be as homogeneous as possible and for stratum means to be substantially different from one another.

5.7.2 *Estimation of finite population variance*

Having selected a sample using stratified SRS, whether selected using proportional allocation, some form of optimal allocation, or any other sample allocation, it is often of interest to estimate the finite population variance, σ^2, in the unstratified population. From such an estimate one could calculate the sampling variance that would have been expected if instead SRS or SWR (rather than stratified SRS) had been used to select units. An unbiased estimator for σ^2 (Wakimoto 1970), based on data collected from a stratified SRS, is derived next, following Cochran (1977 Theorem 5A.1).

We begin by expressing the population total sum of squares as

$$(N-1)\sigma^2 = \sum_{i=1}^{N}(y_i - \mu)^2 = \sum_{i=1}^{N} y_i^2 - N\mu^2 \tag{5.29}$$

(convince yourself of the last equality). If we could find stratified SRS sample-based quantities whose expectations were equal to $\sum_{i=1}^{N} y_i^2$ and μ^2, respectively, then we could substitute these quantities into Equation (5.29) and rearrange for an unbiased estimator of σ^2.

Consider first $\sum_{i=1}^{N} y_i^2 = \sum_{i=1}^{N} z_i = T_z$, where $z_i = y_i^2$. We know that for an arbitrary population variable Z, $\hat{T}_{z,st} = \sum_{h=1}^{L} N_h \hat{\mu}_{z,h}$ is an unbiased estimator for T_z so that

$$E\left(\sum_{h=1}^{L} N_h \frac{\sum_{j \in S_h} y_{hj}^2}{n_h} \right) = \sum_{i=1}^{N} y_i^2$$

Now consider μ^2. Recall that $V(\hat{\theta}) = E(\hat{\theta}^2) - [E(\hat{\theta})]^2$, for any estimator $\hat{\theta}$. Because $\hat{\mu}_{st}$ is unbiased for μ,

$$V(\hat{\mu}_{st}) = E(\hat{\mu}_{st}^2) - \left[E(\hat{\mu}_{st})\right]^2 = E(\hat{\mu}_{st}^2) - \mu^2$$

and since $\hat{V}(\hat{\mu}_{st})$ is unbiased for $V(\hat{\mu}_{st})$, this implies that

$$E\left(\hat{\mu}_{st}^2 - \hat{V}(\hat{\mu}_{st}) \right) = \mu^2$$

Substituting these unbiased estimators for $\sum_{i=1}^{N} y_i^2$ and μ^2 into Equation (5.29), and dividing both sides by $N-1$ gives the following unbiased estimator for σ^2

$$\hat{\sigma}^2 = \frac{N}{N-1} \left[\left(\sum_{h=1}^{L} W_h \frac{\sum_{j \in S_h} y_{hj}^2}{n_h} \right) - \hat{\mu}_{st}^2 + \hat{V}(\hat{\mu}_{st}) \right] \tag{5.30}$$

Example 5.5. We demonstrate the unbiased property of $\hat{\sigma}^2$ by returning to the Example 5.1 setting with $\sigma^2 = 45.6$, and extending Table 5.2 to include, for each of the possible samples, the various quantities involved in Equation (5.30) including $\hat{\sigma}^2$ itself (Table 5.3).

Because each of the samples are equally likely, the expected value of $\hat{\sigma}^2$ can be calculated by simply averaging the sample-specific $\hat{\sigma}^2(s)$ values

$$E(\hat{\sigma}^2) = \sum_{s \in S} \hat{\sigma}^2(s) p(s) = \frac{1}{9} \sum_{s \in S} \hat{\sigma}^2(s) = \frac{1}{9} \cdot 410.40 = 45.6 = \sigma^2$$

Using an estimate of σ^2 from a stratified SRS sample, an estimate of the relative efficiency of $\hat{\mu}_{st}$ compared to $\hat{\mu}_{mpu}$ can then be calculated as

$$\widehat{RE}(\hat{\mu}_{st}, \hat{\mu}_{mpu}) = \frac{\hat{V}(\hat{\mu}_{mpu})}{\hat{V}(\hat{\mu}_{st})} = \frac{\left(\frac{N-n}{N}\right)\frac{\hat{\sigma}^2}{n}}{\sum_{h=1}^{L} W_h^2 \hat{V}(\hat{\mu}_h)}$$

where $\hat{V}(\hat{\mu}_h) = [(N_h - n_h)/N_h]\hat{\sigma}_h^2/n_h$. This estimator $\widehat{RE}(\hat{\mu}_{st}, \hat{\mu}_{mpu})$ is not, however, unbiased, even though both numerator and denominator are unbiased estimates of the respective

Table 5.3 Listing of all possible stratified samples and corresponding sample estimates for Example 5.5. Sample identifier and probability is denoted by s and $p(s)$, respectively.

Sample ID	$p(s)$	$\hat{\mu}_{st}(s)$	$\hat{\mu}_{st}^2(s)$	$\hat{V}(\hat{\mu}_{st})(s)$	$\sum_h W_h \frac{\sum_{j \in s_h} y_{hj}^2}{n_h}(s)$	$\hat{\sigma}^2(s)$
1	1/9	12.50	156.2500	7.541667	210.50	74.15
2	1/9	10.25	105.0625	2.104167	118.25	18.35
3	1/9	15.00	225.0000	1.708333	265.50	50.65
4	1/9	13.25	175.5625	7.604167	226.25	69.95
5	1/9	11.00	121.0000	2.166667	134.00	18.20
6	1/9	15.75	248.0625	1.770833	281.25	41.95
7	1/9	13.00	169.0000	7.708333	221.50	72.25
8	1/9	10.75	115.5625	2.270833	129.25	19.15
9	1/9	15.50	240.2500	1.875000	276.50	45.75
Sum:	1					410.40

sampling variances. The ratio of two unbiased estimators does not, in general, yield an unbiased estimator. This phenomenon will feature prominently in Chapter 7 which considers *ratio* estimation.

5.8 Effective degrees of freedom

With SRS, the appropriate degrees of freedom for construction of a Student's t-based confidence interval for μ is $n - 1$, the degrees of freedom associated with the estimated sampling variance, $\hat{V}(\hat{\mu})$, and estimated finite population variance, $\hat{\sigma}^2 = \sum_{i \in S}(y_i - \hat{\mu})^2/(n-1)$. For stratified SRS, however, $\hat{V}(\hat{\mu}_{st})$ is a linear combination of the independent, stratum-specific, estimated variances

$$\hat{V}(\hat{\mu}_{st}) = \sum_{h=1}^{L} a_h \hat{\sigma}_h^2, \quad a_h = W_h^2 \left(\frac{N_h - n_h}{N_h}\right) \frac{1}{n_h}$$

and its sampling distribution is not so easily characterized. For normally distributed data, however, Satterthwaite (1946) found that the (scaled) distribution of $\hat{V}(\hat{\mu}_{st})$ is well approximated by a chi-square distribution, χ_v^2, with "effective degrees of freedom", v, specified so that the expectation and variance of the approximating distribution is equal to that of the actual distribution. That is

$$\frac{v \cdot \hat{V}(\hat{\mu}_{st})}{E[\hat{V}(\hat{\mu}_{st})]} = \frac{v \cdot \sum_{h=1}^{L} a_h \hat{\sigma}_h^2}{\sum_{h=1}^{L} a_h \sigma_h^2} \dot{\sim} \chi_v^2$$

with v given by

$$v = \frac{\left(\sum_{h=1}^{L} a_h \sigma_h^2\right)^2}{\sum_{h=1}^{L} \frac{(a_h \sigma_h^2)^2}{v_h}}, \quad v_h = n_h - 1$$

The value of v will be greater than the smallest v_h, but no more than the sum of the v_h

$$\min\{v_1, v_2, \ldots, v_L\} < v \leq v_1 + v_2 + \ldots + v_L$$

depending on the values of the N_h, σ_h^2, and n_h.

From sample data, v can be estimated by substituting $\hat{\sigma}_h^2$ for σ_h^2 in the previous equation for v

$$\hat{v} = \frac{\left(\sum\limits_{h=1}^{L} a_h \hat{\sigma}_h^2\right)^2}{\sum\limits_{h=1}^{L} \dfrac{\left(a_h \hat{\sigma}_h^2\right)^2}{v_h}}, \quad v_h = n_h - 1 \tag{5.31}$$

and truncating down to the next smallest integer. You might recognize the similarity of this equation for the degrees of freedom to that used for the degrees of freedom associated with a two-sample t-test or confidence interval when the two underlying σ_h^2 are believed not to be equal. Note that the formulas for v and $\hat{v}$ are also applicable to $\hat{V}(\hat{T}_{st})$ since the N term contained in a_h ($W_h = N_h/N$) is a constant that factors out of the numerator and denominator of both of these quantities.

Example 5.6. Suppose that there are three equally sized and equally variable strata with $N_1 = N_2 = N_3 = 100$, and $\sigma_1^2 = \sigma_2^2 = \sigma_3^2 = 200$. If $n_h = 10$ units are selected from each stratum, then $v = 27 = \sum_{h=1}^{L} v_h$. This holds true more generally: if stratum sizes, variances, and sample sizes are all constant, then v will equal its maximum possible value.

Example 5.7. *Continuation.* Suppose now that there are three strata with $N_1 = 100$, $N_2 = 200$, $N_3 = 50$, and $\sigma_1^2 = 200$, $\sigma_2^2 = 500$, $\sigma_3^2 = 4000$. Assuming an overall sample size of $n = 40$, the value of v (truncated) is provided in the following table for six particular allocations of n (of the $\binom{n-1}{2} = 741$ possible), numerically illustrating that the value of v

Allocation	n_1	n_2	n_3	v
1	20	18	2	1
2	25	10	5	11
3	20	10	10	16
4	10	10	20	13
5	5	5	30	5
6	11	23	6	12

does indeed depend on the allocation of n. Allocation 6 reflects proportional allocation. This would be the optimal allocation if stratum variances and stratum costs per unit (c_h) were all equal. In this case, however, stratum variances are highly unequal. If $n = 40$ were instead optimally allocated (assuming equal c_h: Neyman allocation), then the corresponding sample sizes would be $n_1 = 6$, $n_2 = 20$, and $n_3 = 14$ and the effective degrees of freedom would be $v = 36$. This is very close to the maximum possible of $\sum_{h=1}^{L} v_h = 37$ and substantially more than any of the allocations in the preceding table.

5.9 Post-stratification

In post-stratification, an initial SRS of size n is selected from N with equal probability without replacement. *After* this SRS sample is collected, it is then *post-stratified* into strata of the analyst's choosing, thereby allowing estimation of the mean, proportion, or total in the various strata and, ultimately, application of a stratified estimator for the overall population mean, proportion, or total. This procedure differs from stratified SRS because the sample is taken as a single SRS from all population units (not as a set of independently drawn, within-stratum SRS samples), and as a result the within-strata sample sizes are now *random variables* rather than fixed (pre-specified) quantities. Note also that this procedure does not necessarily guarantee that units will have been selected from all strata, depending on how they are defined.

The **post-stratified estimator** of the mean has the same form as the stratified SRS estimator

$$\hat{\mu}_{post} = \sum_{h=1}^{L} W_h \hat{\mu}_h \tag{5.32}$$

5.9.1 Unconditional sampling variance

Denote the strata sample size random vector as $\mathbf{n} = (n_1, n_2, \ldots, n_L)$, with fixed total sample size $n = \sum_{h=1}^{L} n_h$. Given n, the strata sample size vector $\mathbf{n}$ is distributed as a multivariate hypergeometric random variable (see Section A.4.7)

$$p(\mathbf{n}) = \frac{\binom{N_1}{n_1}\binom{N_2}{n_2}\cdots\binom{N_L}{n_L}}{\binom{N}{n}}$$

Because $\mathbf{n}$ is a random variable with post-stratification, this introduces a very large number of additional possible samples in the sample space over those which would be possible under stratified SRS with fixed $\mathbf{n}$.

For sample size planning purposes, and to better understand the costs and benefits of post-stratification, it is the *unconditional* properties (given n only) of the estimator that are relevant.[1] The unconditional expectation and variance of $\hat{\mu}_{post}$ are most easily derived using the *laws of total expectation and variance* (Section A.3.7),

$$E(\hat{\mu}_{post}) = E\left[E(\hat{\mu}_{post} \mid \mathbf{n})\right]$$
$$V(\hat{\mu}_{post}) = E\left[V(\hat{\mu}_{post} \mid \mathbf{n})\right] + V\left[E(\hat{\mu}_{post} \mid \mathbf{n})\right]$$

where the outside $E[\cdot]$ and $V[\cdot]$ operators are with respect to all possible random configurations of the n-sample across strata (i.e., variation in $\mathbf{n}$), and the inside $E(\cdot)$ and $V(\cdot)$ operators are with respect to a particular sample size vector $\mathbf{n}$. Because $\hat{\mu}_{post}$ is conditionally unbiased, $E(\hat{\mu}_{post} \mid \mathbf{n}) = \mu$, and does not vary with $\mathbf{n}$. Thus, $E(\hat{\mu}_{post}) = E[\mu] = \mu$, and the second term in the variance equation is $V[\mu] = 0$. The first term of the variance equation is the expected value of $V(\hat{\mu}_{post} \mid \mathbf{n})$ over all possible $\mathbf{n}$ configurations. Therefore

$$E(\hat{\mu}_{post}) = \mu \tag{5.33}$$

$$V(\hat{\mu}_{post}) = \sum_{\mathbf{n}} p(\mathbf{n}) V\left(\hat{\mu}_{post} \mid \mathbf{n}\right) = \sum_{\mathbf{n}} \left[\frac{\binom{N_1}{n_1}\binom{N_2}{n_2}\cdots\binom{N_L}{n_L}}{\binom{N}{n}} \sum_{h=1}^{L} W_h^2 \left(\frac{N_h - n_h}{N_h}\right) \frac{\sigma_h^2}{n_h}\right] \tag{5.34}$$

---

[1] Note that similar issues are raised in the context of Bernoulli sampling, Chapter 3. In Bernoulli sampling, the unconditional sampling variance [Equation (3.25)] would be relevant for survey planning purposes.

where the summation is over all possible configurations of **n**. (This summation is analogous to a summation over a sample space, and the multivariate hypergeometric probability of any particular **n** configuration, $p(\mathbf{n})$, is analogous to the probability of a particular sample.)

Equation (5.34) for the unconditional sampling variance of $\hat{\mu}_{post}$, while exact, is not especially convenient or illuminating. Various simplifying approximations to this formula have been developed, and we present three

$$
V(\hat{\mu}_{post}) \approx \begin{cases} \dfrac{N-n}{N} \sum\limits_{h=1}^{L} W_h \dfrac{\sigma_h^2}{n} + \dfrac{1}{n} \sum\limits_{h=1}^{L} (1-W_h) \dfrac{\sigma_h^2}{n} & \text{(Cochran 1977)} \\[3ex] \dfrac{N-n}{N} \sum\limits_{h=1}^{L} W_h \dfrac{\sigma_h^2}{n} + \dfrac{1}{n} \dfrac{N-n}{N} \sum\limits_{h=1}^{L} (1-W_h) \dfrac{\sigma_h^2}{n} & \text{(Särndal et al. 1992)} \\[3ex] \dfrac{N-n}{N} \sum\limits_{h=1}^{L} W_h \dfrac{\sigma_h^2}{n} + \dfrac{1}{n} \dfrac{N-n}{N} \dfrac{N}{N-1} \sum\limits_{h=1}^{L} (1-W_h) \dfrac{\sigma_h^2}{n} & \text{(Thompson 2012)} \end{cases}
$$

In these expressions, the first terms are all identical and are equal to the sampling variance that one would obtain under stratified SRS with proportional allocation [Equation (5.27)]. Note that the *expected* **n** configuration with post-stratification is equal to proportional allocation: $E[n_h] = n(N_h/N), h = 1,2,\dots,L$. This follows directly from the expectation formula for a multivariate hypergeometric random variable, but it also follows from the fact that with an overall SRS, the first order inclusion probabilities for all population units are equal to n/N, so that one expects, on average, to select $(n/N)N_h$ units from stratum h. The second term in these approximations is sometimes referred to as a "penalty" (relative to proportional allocation) for having stratified an SRS sample *after* it was collected rather than using stratified SRS with pre-specified (and proportionally allocated) stratum sample sizes.

The magnitude of the penalty term, relative to the sampling variance for stratified SRS with proportional allocation, is typically quite small. This, combined with the observation that the sampling variance for stratified SRS with proportional allocation is often not much greater than with optimal allocation, argues that post-stratification may often be a very worthwhile strategy. In addition, if the goal of a sample survey includes the characterization of several different target variables for which the optimal stratification of units may differ significantly, an overall SRS may be used followed by separate post-stratifications for each target variable.

5.9.2 Conditional sampling variance

When estimating sampling variance from a post-stratified SRS sample, we support the position that variance estimation should be *conditioned* on the realized **n**, and that the usual (pre-specified) fixed sample size formulas should be used to estimate sampling variance [e.g., Equation (5.12)]. That is, given the realized vector of post-stratified stratum-specific sample sizes, $\mathbf{n}^*$, we treat $\mathbf{n}^*$ *as if* it had been pre-specified, and the inferred sampling variance is over all possible samples given the realized $\mathbf{n}^*$ rather than over all possible **n**. Moreover, given $\mathbf{n}^*$, and assuming $n_h \geq 1, h = 1,2,\dots,L$, we can also treat the within-stratum samples as independently drawn SRS selections in the sense that, given $\mathbf{n}^*$, all possible samples within strata were both possible and equally likely. Therefore, the $E(\hat{\mu}_{st})$, $V(\hat{\mu}_{st})$, and $\hat{V}(\hat{\mu}_{st})$ results and formulas [Equations (5.2), (5.3), (5.12)] apply equally to $\hat{\mu}_{post}$ given $\mathbf{n} = \mathbf{n}^*$.

Holt and Smith (1979) and Smith (1991) provide compelling justification for such use of conditioning for estimation of post-stratified sampling variance. They note that the magnitude of the unconditional sampling variance can be useful for pre-survey specification of total sample size (when post-stratification is a strategy planned in advance), but that the conditional sampling variance of the post-stratified estimator should be used for variance estimation and construction of confidence intervals about a post-stratified estimate.

5.10 Chapter comments

Stratification can be a very successful strategy for reducing sampling variance of an estimator of the overall population mean, proportion or total, while at the same time providing stratum-specific estimates which may be of substantial interest in themselves. Optimal allocation theory allows calculation of stratum-specific sample sizes (and sampling fractions) that should achieve an objective of minimizing sampling variance subject to a fixed total survey cost constraint or minimizing total survey cost subject to a sampling variance constraint. Proportional allocation often results in a sampling variance only slightly greater than optimal allocation and is a good place to start if there is no pre-existing information concerning stratum-specific variances.

Some surveys may have multiple targets of estimation. A stratification of units that is good for one target variable may not be good for another target variable, and optimal allocations for one variable will not be optimal for a different variable given the same stratification of units. When optimal allocation is adopted, it will lead to variation in stratum-specific sampling fractions which in turn implies that first order inclusion probabilities (over the entire population) will be highly variable across strata. This variation in first order inclusion probabilities can lead to substantial complications if one of the survey objectives involves post-survey analyses of many measured survey variables (e.g., relationships among response variables in the context of a survey questionnaire administered across strata). Appropriate analyses of relations among these variables should account for such variability in inclusion probabilities (Korn and Graubard 1999).

In expectation, post-stratification of an unstratified SRS sample should generate the equivalent of simple proportional allocation, often a "near optimal" allocation strategy, and with only a very modest sampling variance penalty relative to pre-specified proportional allocation. Post-stratification can also be carried out with different stratifications of units for different estimation targets if more than one target variable is measured for each sampled unit. Overton and Stehman (1996) praise the use of equal probability selection with post-stratification for long-term monitoring, in part because it allows flexibility in identification of target variables which may change through time as a function of resource status or changes in resource management policies.

On balance, we recommend consideration of stratified sampling in contexts where there is a single target variable of interest and when there is also interest in generating stratum-specific estimates of means, proportions, or totals for this single target variable. When there are multiple target variables of interest and/or when there is interest in post-survey analysis of relations among survey variables, we hesitate to strongly recommend use of stratified sampling. If stratified sampling is used in such contexts, we encourage adoption of identical sampling fractions for all strata (i.e., proportional allocation) so that first order inclusion probabilities are equal across all strata and typical analyses of post-survey relations among variables are not complicated by unequal inclusion probabilities (Overton and Stehman (1996)). The alternative of post-stratification of an initial SRS, without initial stratification, appears to have substantial merit whenever collection of sample survey data has primarily analytic as opposed to primarily descriptive objectives.

Problems

Problem 5.1. Use the optimal allocation formulas to confirm the solutions to the optimal allocation graphical analyses presented in Section 5.5.1: (a) two strata example, (b) three strata equal σ_h^2 example, and (c) three strata unequal σ_h^2 example.

Problem 5.2. Using the method of Lagrange multipliers, verify that Equation (5.23) gives the solution for sample size to achieve a specified variance, V, at minimum cost $C = c_0 + \sum_h c_h n_h$, and that the corresponding $w_{h,\mathrm{opt}}$ is given by Equation (5.22).

Problem 5.3. With proportional allocation ($w_h = W_h$), stratified SRS can be described as *self-weighting*, in the sense that each population unit has the same probability of being selected in the overall sample, regardless of its stratum membership. In this case: (a) Show algebraically that $\hat{\mu}_{\mathrm{st}} = \sum_{h=1}^{L} \sum_{j \in S_h} y_{hj}/n$. (b) What form does $V(\hat{\mu}_{\mathrm{st}})$ take, and how can it be interpreted?

Problem 5.4. Show algebraically that the total sum of squares [SS(T)] can be neatly partitioned into the sums of squares within [SS(W)] and between [SS(B)] strata [see Equation (5.28)].

Problem 5.5. For sample size determination with proportional allocation, show that if N is large Equation (5.25) reduces to $n|V \approx (\sum_h W_h \sigma_h^2)/V(\hat{\mu}_{\mathrm{st}})$.

Problem 5.6. Consider a small population with two strata of size $N_h = 3$ with y_{hj} taking on binary values $(0, 1)$, with Stratum 1 values $\{y_{1j}\} = \{0, 1, 1\}$ and Stratum 2 values $\{y_{2j}\} = \{0, 0, 1\}$. Suppose that stratified SRS is used with $n_1 = n_2 = 2$. (a) Construct the sample space (as in Table 5.2) and numerically show that $\hat{\sigma}^2 = [N/(N-1)][\hat{\pi}_{\mathrm{st}}(1 - \hat{\pi}_{\mathrm{st}}) + \hat{V}(\hat{\pi}_{\mathrm{st}})]$ is an unbiased estimator of σ^2 for binary $(0, 1)$ values. (b) Derive this estimator for σ^2 in the case of binary $(0, 1)$ values from Equation (5.30).

Problem 5.7. Consider a small population consisting of two strata with $\{y_{1j}\} = \{3, 4, 5\}$ and $\{y_{2j}\} = \{7, 10, 12, 14\}$. (a) Construct all possible stratified samples of size $n = 4$ that can be selected when $n_1 = n_2 = 2$ and sampling within strata is equal probability with replacement (i.e., SWR within strata). (b) Calculate the corresponding sample probabilities $[p(s)]$ and sample estimates $\hat{\mu}_{\mathrm{st,swr}} = \sum_h W_h \hat{\mu}_{h,\mathrm{swr}}$. (c) Calculate the sampling variance of $\hat{\mu}_{\mathrm{st,swr}}$ over the sample space using the general expression $V(\hat{\theta}) = \sum_s [\hat{\theta}(s) - E(\hat{\theta})]^2 p(s)$. (d) Derive an algebraic expression for the sampling variance of $\hat{\mu}_{\mathrm{st,swr}}$ using Equation (5.3). (e) Use this algebraic expression to calculate $V(\hat{\mu}_{\mathrm{st,swr}})$, and check that it agrees with the value of $V(\hat{\mu}_{\mathrm{st,swr}})$ that was calculated directly over the sample space.

Problem 5.8. With stratified sampling, estimated CVs of the estimated stratum totals $(\widehat{SE}(\hat{T}_h)/\hat{T}_h)$ may exceed the estimated CV of the estimated overall total $(\widehat{SE}(\hat{T}_{\mathrm{st}})/\hat{T}_{\mathrm{st}})$. For each of the three following example sets of stratified population values, can you find any stratified samples (possibly none or at least one) that could generate this surprising result when $n_h = 2$ for each stratum?

Stratification	Population values	
	Stratum 1	Stratum 2
1	5, 15, 8	12, 20, 40
2	6, 7, 8	9, 10, 11
3	5, 20, 80	8, 40, 70

Problem 5.9. At a large Pacific salmon hatchery in northern California, a systematic sampling procedure was used to estimate the total number of rearing juvenile Chinook salmon (*Oncorhynchus tshawytscha*) present in a raceway (shallow cement channel) by transferring the fish to an adjacent raceway using buckets, each bucket containing approximately 10 pounds of fish. A 1 in k systematic sample of these buckets was taken and the number of fish in each selected bucket was counted. The number of fish in the last (partial) bucket used to transfer the fish was also counted.

In the following table are some counts of fish that were generated from a 1 in $k = 15$ systematic sample, with a random start of 10, taken over a seven day period. The count for the last, partial bucket (bucket number 246) was $c = 849$ fish.

April 1		April 6		April 7	
Bucket	Count	Bucket	Count	Bucket	Count
10	1895	100	1715	175	1658
25	1845	115	1636	190	1615
40	1803	130	1680	205	1618
55	1943	145	1665	220	1625
70	1870	160	1603	235	1631
85	1887				

The original plan was to estimate the number of fish present in the raceway as $\hat{T}_{sys} = N\hat{\mu}_{sys} + c$, where N = total number of 10-pound buckets; $\hat{\mu}_{sys} = \sum_{i \in s} y_i / n$, where y_i = count for bucket i, and n = number of sampled buckets. Assuming no trend in number or size of fish with the order of bucket selection, $\hat{V}(\hat{\mu}_{mpu})$ can be used to estimate the error of estimation of $\hat{T}_{sys}$.

Note, however, that the average count of fish per bucket was considerably greater on April 1 than on April 6–7, implying that fish had grown rapidly over an (unplanned) five day gap prior to the resumption of sampling. Thus, an estimator post-stratified by sample date might provide an improved estimate of the total number of fish present with a lower sampling variance. For the post-stratified approach, assume that n_1 of N_1 buckets were selected on April 1, and that n_2 buckets were selected from $N_2 = N - N_1$ buckets on April 6–7. (a) Estimate the total number of fish present in the raceway and estimate its standard error and coefficient of variation using (1) the original single systematic sample approach, and (2) the proposed post-stratified approach. State any assumptions that you had to make in doing so. (b) Discuss your numerical results. Did post-stratification significantly reduce the error of estimation in this case?

Problem 5.10. A Master's student sought the advice of a distinguished statistician concerning allocation of sampling effort in a stratified random sampling design applied to vegetation analysis. The student, without a background in sampling theory, asked: "What should my sample size be?" The statistician, in a crabby mood, barked: "The bigger it is, the more sampling effort you should devote." Three years later the student presented his completed MS thesis to his graduate committee. Committee members reluctantly approved the thesis research, but all noted that they had never seen such an ineffective application of stratified sampling. The following table presents a summary of data presented in the student's thesis.

Stratum (h)	N_h	n_h	$\hat{\sigma}_h^2$	$\hat{\mu}_h$	$\hat{T}_h$
1	2,000	420	100	17.0	34,000
2	20,400	35	600	1.4	28,560
3	16,000	50	1,000	2.0	32,000
4	1,600	495	244	20.0	32,000

(a) How had the student interpreted the statistician's advice in allocating his sampling effort? (b) Estimate T and $\text{CV}(\hat{T})$ based on the sample results. (c) If the student had instead correctly interpreted the statistician's advice, how might the estimates of T and $\text{CV}(\hat{T})$ been affected? (Assume $\hat{\sigma}_h^2$ and $\hat{\mu}_h$ would remain unchanged.)

Problem 5.11. A Pacific Northwest state has 700 licensed Christmas tree growers with farms ranging in size from 7 to 7,000 acres. For sample planning purposes, a representative for the Christmas tree growers selected a preliminary stratified random sample (six farms selected by SRS from each of four farm size strata) and contacted growers for documented evidence of the number of trees they had sold in the most recent season. The resulting data are summarized in the following table.

Acreage class (h)	N_h	Number of trees sold ($y_{hj}, j \in s_h$)					
0–50	513	1,435	2,542	3,904	4,988	4,320	5,214
51–200	168	14,378	19,661	16,276	5,903	7,792	9,806
201–500	11	28,737	33,732	32,149	30,803	49,601	22,640
>500	8	417,442	281,101	574,854	319,735	524,042	560,325

(a) Based on these preliminary sample data, estimate the total number of trees sold and an associated bounds on the error of estimation of $2 \cdot \widehat{\text{SE}}(\hat{T}_{st})$. (b) For the subsequent survey, the desired bounds on the error of estimation is $2 \cdot \text{SE}(\hat{T}_{st}) = 400,000$ trees. Assume that the cost per farm of collecting and reviewing the documentation needed to substantiate the number of trees sold by a farm is proportional to the cube root of the stratum-specific mean number of trees sold. Estimate the associated optimal allocation weights, optimal overall sample size, and optimal stratum-specific sample sizes required for the follow-up survey. (Note that the optimal allocation formulas presented in this chapter can yield values of $n_{h,opt} > N_h$.)

Problem 5.12. A small population consisting of six individuals, three males (M) and three females (F), has the following variable values for age (yr), body weight (lb), and systolic blood pressure (mmHg).

Individual	Sex	Age	Body weight	Blood pressure
1	M	80	180	105
2	M	20	200	140
3	M	20	160	155
4	F	25	140	145
5	F	55	100	125
6	F	75	120	110

Suppose that a sample of $n = 4$ individuals were selected from this population by (1) SRS with mpu estimation, or (2) stratified SRS with $n_h = 2$ and strata constructed based on sex. (a) Calculate the sampling variances of estimators of mean body weight and mean blood pressure for these two alternative sampling strategies. (b) Calculate the (unconditional) sampling variances of SRS post-stratified estimators of mean body weight and mean blood pressure if the post-stratification were made on the basis of (1) sex, or (2) age (≤ 25, > 25). (c) Discuss the implications of this problem for stratified SRS as compared to SRS post-stratification when there are multiple targets of estimation.

Problem 5.13. Commercial fishermen often sort sablefish (*Anoplopoma fimbria*) by size prior to delivering to fish processors because the price per pound typically increases with increasing fish size. Suppose that you were given the following data summary based on lengths (y_{hj}, in cm) of sablefish randomly selected from each of four market size categories from a commercial fishing vessel.

Size class (h)	N_h	n_h	$\sum_{j\in s_h} y_{hj}$	$\sum_{j\in s_h} y_{hj}^2$
Small	2,432	152	5,284	185,532
Small-medium	1,656	92	3,817	158,953
Medium	2,268	63	3,033	146,357
Large	665	35	2,027	118,169

From this data summary: (a) Construct a 95% confidence interval for the mean length of sablefish on board the vessel. (b) Use Equation (5.30) to estimate $V(\hat{\mu}_{mpu})$ with $n = \sum n_h$. (c) Estimate the relative efficiency of stratified SRS as compared to SRS.

CHAPTER 6

Single-stage cluster sampling: Clusters of equal size

Fig. 6.1 Purse seine commercial fishing for chum salmon, *Oncorhynchus keta*, in Lynn Canal near Juneau, Alaska. Field collection of fish or birds is often based on use of gear (nets, traps) that collect individuals as groups or "clusters". Photo credit: John Hyde.

For the SRS, SWR, and systematic sample selection methods that we have thus far examined, we have selected n units from the N units in a *simple* sampling frame, or, in the case of stratified sampling, we have selected L independent samples of n_h units from the N_h units within each stratum, $h = 1, 2, \ldots, L$, from a *complex* stratified sampling frame for which $N = \sum_{h=1}^{L} N_h$. With cluster sampling, the population units are again defined in a *complex* sampling frame consisting of aggregated groupings or *clusters* of population units (similar to strata). An individual cluster i consists of M_i population units, and N now denotes the number of clusters, so that the total number of population units is $\sum_{i=1}^{N} M_i$. The estimation targets are still defined at the population unit level (e.g., the mean or total of the target variable over all population units), but it may not be feasible or cost-

Sampling Theory: For the Ecological and Natural Resource Sciences. David G. Hankin, Michael S. Mohr, and Ken B. Newman, Oxford University Press (2019). © David G. Hankin, Michael S. Mohr, and Ken B. Newman. DOI: 10.1093/oso/9780198815792.001.0001

effective to select a simple random sample of the population units. Instead, n clusters are selected from the N clusters in the frame according to some specified method of selection. In **single-stage cluster sampling**, y_{ij}, the target variable for the jth unit within cluster i, is then measured for *all* of the units in the selected clusters.

Clusters often arise naturally—a carton of eggs, a family or household, a fishing or hunting party consisting of several individuals, a brood of goslings produced by a pair of Canada geese—but clusters may also be deliberately constructed to generate sampling frame units that make execution of a survey more efficient—all households on a city block, all pools in a short reach of stream. Often the y values *within* natural clusters are relatively similar to one another. For example, members of a family with a single set of parents may have similar heights due to genetic relatedness and trait inheritance. The y values within constructed clusters also are often similar, as for example with household incomes on a city block.

We begin our consideration of cluster sampling by investigating a special case of single-stage cluster sampling—**equal size cluster sampling**—in which all clusters have the same number of units ($M_i = M, i = 1, 2, \ldots, N$).[1] In this case, the total number of population units is NM. The sum of the y values within cluster i, the *cluster total*, is given by $T_i = \sum_{j=1}^{M} y_{ij}$, and the mean value per unit in cluster i, the *cluster mean*, by $\mu_i = T_i/M$.

6.1 Estimation of the population mean

In equal size cluster sampling, the overall mean per population unit can be expressed as

$$\mu = \frac{\sum_{i=1}^{N}\sum_{j=1}^{M} y_{ij}}{NM} = \frac{\sum_{i=1}^{N} T_i}{NM} = \frac{\sum_{i=1}^{N} M\mu_i}{NM} = \frac{\sum_{i=1}^{N} \mu_i}{N}$$

Thus, when clusters are of equal size, the overall mean per unit is equivalent to the simple average of the cluster means. If n clusters are selected from N clusters by SRS and y is measured or recorded for every unit in each of the n clusters that are sampled, the cluster sampling estimator of the mean per population unit is

$$\hat{\mu}_c = \frac{\sum_{i \in S} \mu_i}{n} \tag{6.1}$$

the simple average of cluster means in the sample.

Consider the following small population of $N = 4$ clusters of $M = 3$ units each.

Cluster (i)	Variable values $(y_{ij}, j = 1, 2, 3)$	T_i	μ_i
1	1, 4, 7	12	4
2	3, 10, 5	18	6
3	2, 8, 2	12	4
4	4, 9, 2	15	5
Sum:		57	

---

[1] The general case, where the numbers of units within clusters may vary, is considered in Chapter 9.

The population mean and finite population variance are

$$\mu = \frac{\sum\limits_{i=1}^{N}\sum\limits_{j=1}^{M} y_{ij}}{NM} = \frac{\sum\limits_{i=1}^{N} T_i}{NM} = \frac{57}{4\cdot3} = 4.75$$

$$\sigma^2 = \frac{\sum\limits_{i=1}^{N}\sum\limits_{j=1}^{M} (y_{ij} - \mu)^2}{NM-1} = \frac{102.25}{(4\cdot3)-1} = 9.2955$$

Suppose that $n = 2$ clusters are selected by SRS from the $N = 4$ clusters. The sample space and associated estimates are as follows

ID	Sample Clusters (s)	$p(s)$	$\hat{\mu}_c(s)$	$(\hat{\mu}_c(s) - \mu)^2$
1	1, 2	1/6	5.0	0.0625
2	1, 3	1/6	4.0	0.5625
3	1, 4	1/6	4.5	0.0625
4	2, 3	1/6	5.0	0.0625
5	2, 4	1/6	5.5	0.5625
6	3, 4	1/6	4.5	0.0625
Sum:		1	28.5	1.3750

Expected value and sampling variance for $\hat{\mu}_c$ in this case would be

$$E(\hat{\mu}_c) = \sum_{s\in\mathcal{S}} \hat{\mu}_c(s)p(s) = (1/6)\sum_{s\in\mathcal{S}} \hat{\mu}_c(s) = (1/6)\cdot 28.5 = 4.75 = \mu$$

$$V(\hat{\mu}_c) = \sum_{s\in\mathcal{S}} [\hat{\mu}_c(s) - E(\hat{\mu}_c)]^2 p(s) = (1/6)\sum_{s\in\mathcal{S}} [\hat{\mu}_c(s) - \mu]^2 = (1/6)\cdot 1.375 = 0.2292$$

Note that if instead $n = 6$ population units were selected from the $NM = 12$ units by SRS, then sampling variance for $\hat{\mu}_{mpu}$ would have been

$$V(\hat{\mu}_{mpu}) = \left(\frac{NM-n}{NM}\right)\frac{\sigma^2}{n} = \left(\frac{12-6}{12}\right)\frac{9.2955}{6} = 0.7746$$

Thus, for this numerical example, $\hat{\mu}_c$ had lower sampling variance than $\hat{\mu}_{mpu}$ when the same number of population units were included in the SRS and cluster samples.

6.2 Sampling variance

There are two general approaches taken to characterize the sampling variance of a single-stage equal size cluster sample. The more easily understood approach relies on an ANOVA sum of squares analogy, similar to the approach taken for stratified sampling. The less intuitive approach relies on definition of an unusual *correlation*. Both approaches generate important insights concerning when this simplest kind of cluster sampling may be more efficient than SRS.

Table 6.1 ANOVA mean squares for a population grouped into N clusters of M units each.

Source of variation	Sum of squares	df	Mean square
Total	SS(T)	$NM - 1$	$MS(T) = \sigma^2$
Within clusters	SS(W)	$N(M - 1)$	MS(W)
Between clusters	SS(B)	$N - 1$	MS(B)

6.2.1 *ANOVA/mean squares approach*

As for stratified sampling, we can partition the total squared variation in the population into two components—variation *within* clusters and variation *between* clusters (Problem 6.1).

$$\underbrace{\sum_{i=1}^{N}\sum_{j=1}^{M}(y_{ij} - \mu)^2}_{SS(T)} = \underbrace{\sum_{i=1}^{N}\sum_{j=1}^{M}(y_{ij} - \mu_i)^2}_{SS(W)} + \underbrace{\sum_{i=1}^{N}\sum_{j=1}^{M}(\mu_i - \mu)^2}_{SS(B)} \tag{6.2}$$

Note that SS(B) can be rewritten as $M\sum_{i=1}^{N}(\mu_i - \mu)^2$. Mean squares are defined in Table 6.1. For our Section 6.1 example population, the ANOVA mean squares are

Source of variation	Sum of squares	df	Mean square
Total	102.25	11	9.2955
Within clusters	94.00	8	11.7500
Between clusters	8.25	3	2.7500

In terms of these ANOVA quantities, it can be shown that

$$V(\hat{\mu}_c) = \left(\frac{N - n}{N}\right)\frac{MS(B)}{nM} \tag{6.3}$$

which, for this example, gives

$$V(\hat{\mu}_c) = \left(\frac{4 - 2}{4}\right)\frac{2.75}{2 \cdot 3} = 0.2292$$

as previously calculated. Note that Equation (6.3) implies that sampling variance in single-stage equal size cluster sampling will be minimized when MS(B) is as small as possible, i.e., when the N cluster means are each as close to μ as possible.

Substituting MS(B) = SS(B)/$(N - 1)$ into Equation (6.3) and simplifying gives

$$V(\hat{\mu}_c) = \left(\frac{N - n}{N}\right)\frac{\sum_{i=1}^{N}(\mu_i - \mu)^2}{n(N - 1)} \tag{6.4}$$

Equation (6.4) is directly analogous to the expression for the sampling variance of $\hat{\mu}_{mpu}$. In single-stage equal size cluster sampling using SRS, one can view the design as drawing an SRS sample of n cluster means from a population consisting of N cluster means. The finite population correction should thus be $(N - n)/N$, as for $\hat{\mu}_{mpu}$. If a population consisted of N cluster means, then the finite population variance would be $\sigma_c^2 = \sum_{i=1}^{N}(\mu_i - \mu)^2/(N - 1)$, so that the sampling variance for an mpu estimator based on the n sample means would be $V(\hat{\mu}_c) = [(N - n)/N]\sigma_c^2/n$, equivalent to Equation (6.4). Convince yourself that $\sigma_c^2 = MS(B)/M$ (Problem 6.5).

6.2.2 Intracluster correlation approach

The **intracluster correlation** measures the average degree of homogeneity among the units present within clusters and provides an alternative perspective on sampling variance in single-stage equal size cluster sampling. Cochran (1977) defines the intracluster correlation, ICor, as

$$ICor = \frac{E[(y_{ij} - \mu)(y_{ik} - \mu)]}{E[(y_{ij} - \mu)^2]}$$

where $j \neq k$ are distinct units within cluster i. For samples of n clusters selected from N by SRS, the numerator of the expression for ICor is the average crossproduct of differences between the y_{ij} and the overall population mean per unit, μ, for distinct pairs of units ($j \neq k$) within selected clusters. The denominator is the average squared difference between y_{ij} in selected clusters and the overall population mean per unit, μ. Cochran (1977) shows that[2]

$$ICor = \frac{\sum\limits_{i=1}^{N} \sum\limits_{j \neq k}^{M} (y_{ij} - \mu)(y_{ik} - \mu)}{(M-1)(NM-1)\sigma^2} \tag{6.5}$$

As will be shown in Equation (6.6), the sign of ICor affects the sampling variance of $\hat{\mu}_c$, and having a general understanding of the conditions where ICor is positive, negative, or approximately zero can help to explain the relative efficiency of $\hat{\mu}_c$ compared to other estimators, e.g., $\hat{\mu}_{mpu}$.

ICor > 0. If the y values within clusters are quite similar to one another (e.g., most are greater than the overall mean or most are less than the overall mean), then the signs of the differences between the various pairs ($y_{ij} - \mu$) and ($y_{ik} - \mu$) within the same cluster will tend to be the same. Thus, the products will tend to be positive, thereby leading to a positive intracluster correlation.

ICor ≈ 0. If the y values within clusters have a degree of variation similar to σ^2 and cluster means are similar to one another, then it is just as likely for the signs of the various pairs ($y_{ij} - \mu$) and ($y_{ik} - \mu$) within the same cluster to be positive or negative, and for the magnitude of the products to be relatively similar across the pairings. In that case, the intracluster correlation will tend to be close to zero.

ICor < 0. If the average heterogeneity within clusters exceeds σ^2 and the cluster means are similar to μ, then signs of the products of pairs within clusters will more often be negative and the magnitude of negative products larger than positive products. As a result, the intracluster correlation will be negative.

Using the intracluster correlation approach, the sampling variance of $\hat{\mu}_c$ under equal size cluster sampling with SRS can be shown to be (Cochran 1977)

$$V(\hat{\mu}_c) = \left(\frac{N-n}{N}\right)\left[\frac{NM-1}{M(N-1)}\right]\frac{\sigma^2}{nM}[1 + (M-1) \cdot ICor] \tag{6.6}$$

For large N and M, an approximate expression for the sampling variance is

$$V(\hat{\mu}_c) \approx \left(\frac{N-n}{N}\right)\frac{\sigma^2}{nM}[1 + (M-1) \cdot ICor] \tag{6.7}$$

---

[2] Note that $\sum_{j \neq k}(y_{ij} - \mu)(y_{ik} - \mu) = 2\sum_{j<k}(y_{ij} - \mu)(y_{ik} - \mu)$. For example, if $M = 3$, then the sum over $j \neq k$ would give the pairs [(1,2),(1,3),(2,1),(2,3),(3,1),(3,2)], whereas the sum over $j < k$ would give the pairs [(1,2),(1,3),(2,3)]. Each pair product appears twice in the sum over $j \neq k$ but only once in the sum over $i < j$. We will revisit this type of notation in Chapter 8.

In contrast, if a SRS sample of size nM were selected from a population of size NM, the sampling variance would be

$$V(\hat{\mu}_{\mathrm{mpu}}) = \left(\frac{NM - nM}{NM}\right)\frac{\sigma^2}{nM} = \left(\frac{N - n}{N}\right)\frac{\sigma^2}{nM} \tag{6.8}$$

Thus, for large N and an equal number of sampled population units (nM), the intracluster correlation formulation suggests that $V(\hat{\mu}_c) < V(\hat{\mu}_{\mathrm{mpu}})$ whenever $\mathrm{ICor} < 0$.

For our Section 6.1 example population, we find

Cluster (i)	Variable values $(y_{ij}, j = 1, 2, 3)$	$\sum_{j \neq k}^{3}(y_{ij} - \mu)(y_{ik} - \mu)$	$\sum_{j=1}^{3}(y_{ij} - \mu)^2$
1	1, 4, 7	−14.625	19.6875
2	3, 10, 5	−16.625	30.6875
3	2, 8, 2	−20.625	25.6875
4	4, 9, 2	−25.625	26.1875
Sum:		−77.500	102.2500

From the preceding table $\sum_{i=1}^{N}\sum_{j \neq k}^{M}(y_{ij} - \mu)(y_{ik} - \mu) = -77.5$, and because $N = 4$, $M = 3$, and $\sigma^2 = 9.2955$

$$\mathrm{ICor} = \frac{\sum_{i=1}^{N}\sum_{j \neq k}^{M}(y_{ij} - \mu)(y_{ik} - \mu)}{(M - 1)(NM - 1)\sigma^2} = \frac{-77.5}{2 \cdot 11 \cdot 9.2955} = -0.3790$$

and,

$$V(\hat{\mu}_c) = \left(\frac{N - n}{N}\right)\left[\frac{NM - 1}{M(N - 1)}\right]\frac{\sigma^2}{nM}[1 + (M - 1) \cdot \mathrm{ICor}]$$

$$= \left(\frac{4 - 2}{4}\right)\left[\frac{12 - 1}{3(4 - 1)}\right]\frac{9.2955}{2 \cdot 3}[1 + (3 - 1)(-0.379)] = 0.2292$$

as calculated previously over the sample space.

Finally, it is also worth noting a connection between the mean squares approach and the intracluster correlation approach

$$\mathrm{MS(W)} = \left(\frac{NM - 1}{NM}\right)\sigma^2(1 - \mathrm{ICor})$$

For our numerical example, $\mathrm{MS(W)} = (11/12) \cdot 9.2955 \cdot [1 - (-0.379)] = 11.75$. For large N and M, $\mathrm{MS(W)} \approx \sigma^2(1 - \mathrm{ICor})$, which indicates that the average variation within clusters will exceed the average variation in the population (σ^2) whenever $\mathrm{ICor} < 0$. Thus, when $\mathrm{MS(W)}$ is large, $\mathrm{MS(B)}$ will be small and $\hat{\mu}_c$ will outperform $\hat{\mu}_{\mathrm{mpu}}$.

6.3 Estimation of the population total and proportion

Because $\hat{\mu}_c$ is an unbiased estimator of μ, we can simply "scale up" the estimator of the mean, as for $\hat{\mu}_{\mathrm{mpu}}$, to obtain an unbiased estimator of the total

$$\hat{T}_c = NM\hat{\mu}_c \tag{6.9}$$

Because the scaling factor NM is known with certainty, the sampling variance of $\hat{T}_c$ is

$$V(\hat{T}_c) = V(NM\hat{\mu}_c) = N^2M^2V(\hat{\mu}_c) = N^2M^2\left(\frac{N - n}{N}\right)\frac{\mathrm{MS(B)}}{nM} \tag{6.10}$$

or

$$V(\hat{T}_c) = N^2 M^2 \left(\frac{N-n}{N}\right) \frac{\sum_{i=1}^{N}(\mu_i - \mu)^2}{n(N-1)} \tag{6.11}$$

Estimation of the proportion is directly analogous to estimation of the mean. In this case, y is a binary (0, 1) variable, and the cluster i total, $T_i = \sum_{j=1}^{M} y_{ij}$, is the total number of units in cluster i that possess a particular characteristic or attribute. Thus, the proportion of population units for which $y = 1$ is

$$\pi = \frac{\sum_{i=1}^{N}\sum_{j=1}^{M} y_{ij}}{NM} = \frac{\sum_{i=1}^{N} T_i}{NM} = \frac{\sum_{i=1}^{N} \pi_i}{N}$$

and an unbiased estimator of π is

$$\hat{\pi}_c = \frac{\sum_{i\in S} T_i}{nM} = \frac{\sum_{i\in S} \pi_i}{n} \tag{6.12}$$

with sampling variance

$$V(\hat{\pi}_c) = \left(\frac{N-n}{N}\right) \frac{\sum_{i=1}^{N}(\pi_i - \pi)^2}{n(N-1)} \tag{6.13}$$

6.4 Estimation of sampling variance

Mean squares can be calculated for individual samples and are useful for estimation of sampling variance and for other purposes. Sample mean squares are defined as follows where we use SS(t), SS(w), and SS(b) to denote the sample analogues of the population quantities SS(T), SS(W), and SS(B), respectively, and likewise for the MS terms.

Source of variation	Sum of squares	df	Mean square
Total	$SS(t) = \sum_{i\in S}\sum_{j=1}^{M}(y_{ij} - \hat{\mu}_c)^2$	$nM - 1$	MS(t)
Within clusters	$SS(w) = \sum_{i\in S}\sum_{j=1}^{M}(y_{ij} - \mu_i)^2$	$n(M-1)$	MS(w)
Between clusters	$SS(b) = \sum_{i\in S}\sum_{j=1}^{M}(\mu_i - \hat{\mu}_c)^2$	$n - 1$	MS(b)

For our Section 6.1 example population, we provide a numerical illustration (Table 6.2) of calculated sums of squares and their associated mean squares over the sample space. Because each of the six possible cluster samples are equally likely and have probability 1/6, we can calculate the expected values of the sample mean squares by simply multiplying the appropriate column sum from Table 6.2 by 1/6.

$$E[MS(b)] = \frac{1}{6} \cdot 16.5 = 2.75 = MS(B)$$

$$E[MS(w)] = \frac{1}{6} \cdot 70.5 = 11.75 = MS(W)$$

$$E[MS(t)] = \frac{1}{6} \cdot 65.85 = 10.975 \neq \sigma^2 = 9.2955$$

Table 6.2 Sample space sums of squares and mean squares for the Section 6.1 example population.

Sample			Sum of squares			Mean square		
ID	Clusters (s)	$p(s)$	SS(t)	SS(w)	SS(b)	MS(t)	MS(w)	MS(b)
1	1, 2	1/6	59.00	44	6.0	11.80	11.0	6.0
2	1, 3	1/6	45.75	42	0.0	9.15	10.5	0.0
3	1, 4	1/6	47.00	44	1.5	9.40	11.0	1.5
4	2, 3	1/6	62.75	50	6.0	12.55	12.5	6.0
5	2, 4	1/6	64.00	52	1.5	12.80	13.0	1.5
6	3, 4	1/6	50.75	50	1.5	10.15	12.5	1.5
Sum:						65.85	70.5	16.5

Note that although MS(w) and MS(b) are unbiased estimators of MS(W) and MS(B), respectively, MS(t) is *not* an unbiased estimator of σ^2 in single-stage equal size cluster sampling.

Because MS(b) is an unbiased estimator of MS(B), an unbiased estimator of $V(\hat{\mu}_c)$ is

$$\hat{V}(\hat{\mu}_c) = \left(\frac{N-n}{N}\right)\frac{\text{MS(b)}}{nM} = \left(\frac{N-n}{N}\right)\frac{M\sum_{i \in S}(\mu_i - \hat{\mu}_c)^2/(n-1)}{nM}$$

or

$$\hat{V}(\hat{\mu}_c) = \left(\frac{N-n}{N}\right)\frac{\sum_{i \in S}(\mu_i - \hat{\mu}_c)^2}{n(n-1)} \tag{6.14}$$

Unbiased estimators for $V(\hat{T}_c)$ and $V(\hat{\pi}_c)$ are

$$\hat{V}(\hat{T}_c) = N^2M^2\left(\frac{N-n}{N}\right)\frac{\sum_{i \in S}(\mu_i - \hat{\mu}_c)^2}{n(n-1)} \tag{6.15}$$

and

$$\hat{V}(\hat{\pi}_c) = \left(\frac{N-n}{N}\right)\frac{\sum_{i \in S}(\pi_i - \hat{\pi}_c)^2}{n(n-1)} \tag{6.16}$$

6.5 Estimation of finite population variance

After a sample has been collected using an equal size cluster sampling design, it may be of interest to know what the sampling variance would have been if instead a simple random sample of size nM population units had been selected from a population of NM units (i.e., with no cluster structure). An estimator of σ^2 can be derived as follows.

First, in a clustered population, we may express σ^2 in terms of the sums of squares between and within clusters

$$\sigma^2 = \text{SS(T)}/(NM - 1) = [\text{SS(B)} + \text{SS(W)}]/(NM - 1)$$

Next, substitute for SS(B) and SS(W) in terms of MS(B) and MS(W) (see Table 6.1), giving

$$\sigma^2 = \frac{(N-1)\text{MS(B)} + N(M-1)\text{MS(W)}}{NM - 1} \tag{6.17}$$

Because $E[\text{MS(b)}] = \text{MS(B)}$ and $E[\text{MS(w)}] = \text{MS(W)}$, an unbiased estimator of σ^2 is

$$\hat{\sigma}_c^2 = \frac{(N-1)\text{MS(b)} + N(M-1)\text{MS(w)}}{NM-1} \tag{6.18}$$

6.6 Sample size determination

Write the sampling variance of $\hat{\mu}_c$ as

$$V(\hat{\mu}_c) = \left(\frac{N-n}{N}\right)\frac{\text{MS(B)}}{nM} = \left(\frac{1}{n} - \frac{1}{N}\right)\frac{\text{MS(B)}}{M}$$

Given a desired sampling variance $V(\hat{\mu}_c)$, this expression can be solved for n to obtain

$$n = \frac{\text{MS(B)}/M}{V(\hat{\mu}_c) + \text{MS(B)}/(NM)} \tag{6.19}$$

Because $V(\hat{\mu}_c)$ and $V(\hat{T}_c)$ have the same general form [Equation (6.10)], Equation (6.19) for n can be generalized for estimating $\theta = \mu, \pi$, or T as

$$n = \frac{\text{MS(B)}/M}{\mathcal{K}(\hat{\theta}_c) + \text{MS(B)}/(NM)}, \qquad \mathcal{K}(\hat{\theta}) = \begin{cases} V(\hat{\theta}), & \text{for } \theta = \mu, \pi \\ V(\hat{\theta})/(NM)^2, & \text{for } \theta = T \end{cases} \tag{6.20}$$

If a 95% bound on the error of estimation, B, is specified, replace $V(\hat{\theta})$ with $B^2/4$. If a CV constraint is specified, replace $V(\hat{\theta})$ with $\text{CV}(\hat{\theta})^2\theta^2$. In practice, a preliminary estimate of MS(B) and possibly also of θ (for CV specification) would be required for calculation of sample size using Equation (6.20).

6.7 Relative efficiency

For an SRS of size nM selected from a population of size NM, the sampling variance of $V(\hat{\mu}_{\text{mpu}})$ is given by Equation (6.8). Thus, from Equation (6.3) the relative efficiency of $\hat{\mu}_c$ compared to $\hat{\mu}_{\text{mpu}}$ for an equal number of sample units is

$$\text{RE}(\hat{\mu}_c, \hat{\mu}_{\text{mpu}}) = \frac{V(\hat{\mu}_{\text{mpu}})}{V(\hat{\mu}_c)} = \frac{\left(\dfrac{N-n}{N}\right)\dfrac{\sigma^2}{nM}}{\left(\dfrac{N-n}{N}\right)\dfrac{\text{MS(B)}}{nM}} = \frac{\sigma^2}{\text{MS(B)}}$$

The relative efficency will therefore exceed 1 $[V(\hat{\mu}_c) < V(\hat{\mu}_{\text{mpu}})]$ whenever $\sigma^2 > \text{MS(B)}$. Equivalently, referring to the approximate sampling variance of $\hat{\mu}_c$ expressed in terms of ICor [Equation (6.7)], $\text{RE}(\hat{\mu}_c, \hat{\mu}_{\text{mpu}}) > 1$ whenever ICor < 0. These conditions will be realized when (a) y values within clusters are very heterogeneous, thus making MS(W) large, and when (b) cluster means are similar to one another (so that MS(B) is small). Ideally, each cluster should capture the full variation present in the population and clusters should be as similar as possible to one another. Note that these characteristics are essentially the opposite of those characteristics that are most desirable for stratified SRS ($\hat{\mu}_{\text{st}}$). In stratified SRS, one attempts to minimize within strata variation and maximize between strata variation; in equal size cluster sampling, we want to maximize within cluster variation and minimize between cluster variation.

Unfortunately, it seems rare for clusters to have these ideal properties. Indeed, as previously noted, natural and constructed clusters often tend to contain similar y values and between cluster variation may often be large. For example, suppose that a social science survey concerns attitudes towards police. If the family or household is the basic cluster unit, attitudes within a given household (as for individual heights) will tend to be relatively homogeneous and variation in attitudes between households will tend to be relatively large.

6.8 Chapter comments

This chapter has briefly introduced the simplest setting for application of cluster sampling, namely when (a) clusters are all of equal size, and (b) all subunits within clusters are selected for inclusion in the sample. We will visit the more general setting for application of **unequal size cluster sampling** in Chapter 9 which is devoted to multi-stage sampling. In multi-stage sampling, the **primary units** may consist of clusters of subunits, not necessarily with the same number of subunits for each cluster, and sampling takes place in two or more stages. For example, in two-stage sampling, the first stage of sampling consists of selection of n primary units from N, and the second stage of sampling consists of selecting m_i subunits from the M_i subunits within the ith cluster selected at the first stage. It is impossible to understand or fully appreciate the wide variety of sampling strategies that can be used in this general multi-stage setting without knowledge of (a) ratio and regression estimation (covered in Chapter 7) and (b) unequal probability selection and associated estimators (covered in Chapter 8). For that reason, we have only introduced single-stage equal size cluster sampling in this chapter.

It is important to recognize that many field collections of data in fisheries and wildlife settings often are best represented by analogy to cluster sampling, though clusters are typically of unequal sizes. For example, nets and traps are routinely used to capture fish in fisheries field research, resulting in collection of groups or clusters of individuals (e.g., Figure 6.1). Similarly, in wildlife research, mist nets may be used to collect birds which are typically also captured in groups or clusters, or a collection of live traps may be set overnight in specific (often randomly selected) sites resulting in collections of groups or clusters of animals at each of these specific sites. Clustering is also an important notion to be acutely aware of in genetic sampling. For example, in small streams it has been shown that genetic relatedness among juvenile fish collected from a single pool is usually much greater than for fish collected from a large number of pools. Therefore, it is unsound to estimate population-level allele frequencies, say, from cluster type data collected at just one or two locations which may instead reflect genetic variation primarily within families rather than within the entire population (Hansen et al. 1997). In each of these settings, there are good biological reasons to suspect that the individuals within sample clusters may share similar attributes, an undesirable though often typical feature of *natural* clusters. Many other examples of ecological field research that results in collection of clusters of individuals could be identified.

It is worth emphasizing that cluster samples are often taken for reasons of convenience rather than for greater statistical efficiency. Indeed, in many cases the relative efficiency of $\hat{\mu}_c$ as compared to $\hat{\mu}_{\text{mpu}}$ will be < 1, for reasons given earlier, but selecting a SRS would be a considerably more difficult task. However, the cost of obtaining measurements in the nM units that appear in an equal size cluster sample may be less than for an SRS of size nM. In the two-stage cluster sampling context (Chapter 9), the total cost of a survey is assumed to reflect two different kinds of costs: (a) travel costs to clusters, and (b) costs of measurements of (sub)units within clusters. For an easily understood setting within which one can imagine travel costs as distinguished from measurement costs, imagine the (travel) cost savings if social scientists interviewed M households in a city block as compared to M independently selected households separated by possibly large travel distances. Whenever it seems obvious that data or individuals have been collected as clusters, then statistical analysis of data should account for this underlying cluster structure (Nelson 2014). Material presented in this chapter and also in Chapter 9 provides explicit guidance for use of cluster sampling data to estimate simple population parameters.

Problems

Problem 6.1. For equal size cluster sampling, SS(T) can be expanded as

$$\sum_{i=1}^{N}\sum_{j=1}^{M}(y_{ij}-\mu)^2 = \sum_{i=1}^{N}\sum_{j=1}^{M}(y_{ij}-\mu_i+\mu_i-\mu)^2$$

$$= \sum_{i=1}^{N}\sum_{j=1}^{M}(y_{ij}-\mu_i)^2 + 2\sum_{i=1}^{N}\sum_{j=1}^{M}(y_{ij}-\mu_i)(\mu_i-\mu) + \sum_{i=1}^{N}\sum_{j=1}^{M}(\mu_i-\mu)^2$$

(A) Show numerically that $2\sum_{i=1}^{N}\sum_{j=1}^{M}(y_{ij}-\mu_i)(\mu_i-\mu)=0$ for the equal size cluster example population in Section 6.1. (B) Show algebraically that $2\sum_{i=1}^{N}\sum_{j=1}^{M}(y_{ij}-\mu_i)(\mu_i-\mu)=0$, which thus leads to Equation (6.2).

Problem 6.2. Using this chapter's small clustered population and the sample mean squares tabulated in Table 6.4, verify numerically that $\hat{\sigma}_c^2 = [(N-1)\text{MS(b)} + N(M-1)\text{MS(w)}]/(NM-1)$ is an unbiased estimator of the finite population variance, σ^2.

Problem 6.3. A large supermarket gets a delivery of 1,200 cartons of eggs (12 eggs per carton) every week. The store manager has been concerned that the incidence of cracked eggs may be higher than acceptable, so he instructed staff to select a "random" sample of 25 cartons from a recent delivery. Staff found that 21 cartons had no broken eggs, 2 cartons had one broken egg, 1 carton had 2 broken eggs, and 1 carton had 10 broken eggs. (A) Calculate the cluster sampling estimate of the proportion of broken eggs ($\hat{\mu}_c = \hat{\pi}_c$) and the corresponding estimate of standard error. (B) Calculate the expected number of cartons that would need to be examined in order to achieve a 95% bounds on error of estimation of 0.02. (C) Assuming that eggs could instead be sampled individually by SRS, what SRS sample size should achieve the same bounds on error of estimation assuming use of $\hat{\mu}_{\text{mpu}}$? (D) Contrast/explain the difference in *total numbers of eggs examined* for your answers to parts (B) and (C). Is the comparison of cluster sampling to SRS selection of practical relevance? Would a more relevant target of estimation be the proportion of cartons that had at least one broken egg?

Problem 6.4. The relative efficiency of $\hat{\mu}_c$ compared to $\hat{\mu}_{\text{st,prop}}$ is $V(\hat{\mu}_{\text{st,prop}})/V(\hat{\mu}_c)$. Show algebraically that $\text{RE}(\hat{\mu}_c, \hat{\mu}_{\text{st,prop}}) > 1$ when $\text{MS(B)} < \sum_{h=1}^{L} W_h^2 \sigma_h^2$. Is this a particularly useful result? Discuss.

Problem 6.5. Define $\sigma_c^2 = \sum_{i=1}^{N}(\mu_i-\mu)^2/(N-1)$. Show algebraically that $\sigma_c^2/n = \text{MS(B)}/(nM)$.

Problem 6.6. Show algebraically that $\sum_{i=1}^{N}(y_i - \mu)^2 = \sum_{i=1}^{N} y_i^2 - N\mu^2$. (Note: This result is relied upon in many different contexts in this text and, prior to computer-based vector calculations, was historically an important result for numerical calculations, including ANOVA.)

Problem 6.7. Suppose that $N = 200$, $M = 20$, and that a preliminary sample of $n = 10$ clusters generated the following sample cluster totals: $\{T_i(s)\} = \{102, 314, 186, 90, 205, 178, 163, 86, 144, 189\}$. Estimate the number of clusters that would need to be sampled to achieve $CV(\hat{\mu}_c) = 0.05$.

CHAPTER 7

Ratio and regression estimation

Fig. 7.1 Close up photo of juvenile coho salmon, *Oncorhynchus kisutch*, from a small northern California stream. Ratio estimation may be used to link visual counts of fish made in all (or a large fraction of) population units with more expensive "exhaustive removals" made in an SRS of population units (Problem 10.1). Photo credit: Darren Ward.

For the sampling strategies that have been considered thus far, values of only a single target variable, y, have been recorded or measured for sample units selected from a finite population. Many other variables may be associated with the units in a population, of course. Values of some of these variables may be much cheaper to record or much more readily available than values of the target variable. When such **auxiliary** variables are highly correlated with the target variable, they can be explicitly incorporated in estimators that have dramatically improved performance (reduced sampling variance) compared to estimators based only on the sample y values. (In Chapter 8, we will learn that these auxiliary variables can also be used to develop unequal selection probabilities which can

Sampling Theory: For the Ecological and Natural Resource Sciences. David G. Hankin, Michael S. Mohr, and Ken B. Newman, Oxford University Press (2019). © David G. Hankin, Michael S. Mohr, and Ken B. Newman. DOI: 10.1093/oso/9780198815792.001.0001

also greatly reduce sampling variance when the auxiliary and target variables are highly correlated.)

In this chapter we introduce **ratio estimation**—the simplest example of using an auxiliary variable to aid in estimation of a target variable population parameter—and also **regression estimation**, a related approach that is very similar to the linear regression prediction technique that is covered in introductory statistics courses and texts.

We also briefly consider the use of *models* in sampling theory. As noted in Chapter 1, some sampling theorists (e.g., Brewer and Gregoire 2009) consider design-based sampling theory to be a branch of *non-parametric* statistics because concepts like design-unbiasedness of an estimator do not invoke any assumed model relationships among population variables or any parametric form for the distribution of y values in a finite population. Statistical models of conjectured relationships within and among variables form the foundation of parametric statistics, however, and models of relationships may help us to better understand the conditions under which certain sampling strategies may outperform others. This kind of use of statistical models—for generating insight into the relative performance of sampling strategies—has been termed **model-assisted sampling** (Särndal et al. 1992). Alternatively, models may be used directly to develop a **prediction-based** approach to sampling theory which is rooted in parametric statistics. We introduce the notions of model-assisted and prediction-based estimation in this chapter, following our presentation of the design-based theory of ratio and regression estimation.

7.1 Estimation of the mean and total

At a conceptual level, ratio and regression estimation are closely related to one another. Superior performance of each of these approaches requires identification of an auxiliary variable that is highly correlated with the target variable.

In ratio estimation, we typically assume (for estimation of the mean or total) that values of an auxiliary variable, x_i, are available for all of the population units, $i = 1, 2, \ldots, N$, or that at least the mean or total of the auxiliary variable is known.[1] Suppose that we select a sample of size n from N by SRS. Sample means are then calculated or recorded for the target variable, $\hat{\mu}_y$, and for the auxiliary variable, $\hat{\mu}_x$. The sample ratio, $\hat{R} = \hat{\mu}_y/\hat{\mu}_x$, provides an estimate of the population ratio $R = T_y/T_x = \mu_y/\mu_x$, and can be used to generate estimators of the mean or total of the target variable, y.

$$\hat{R} = \hat{\mu}_y/\hat{\mu}_x \tag{7.1}$$

$$\hat{\mu}_{y,\mathrm{rat}} = \mu_x \hat{R} = \frac{\mu_x}{\hat{\mu}_x}\hat{\mu}_y \tag{7.2}$$

$$\hat{T}_{y,\mathrm{rat}} = T_x \hat{R} = N\hat{\mu}_{y,\mathrm{rat}} \tag{7.3}$$

In regression estimation, we select a sample of size n from N by SRS and calculate sample means for the auxiliary and target variables, as for ratio estimation, but base estimation on

---

[1] Consider, for example, a population consisting of N ripe ovaries of domesticated female Russian sturgeon, *Acipenser gueldenstaedtii*, from which fine caviar can be made. Suppose that the auxiliary variable is the weight of an ovary, and the target variable is the total number of eggs (fecundity) in an ovary. Fecundity ranges from about 85,000–835,000 eggs in this species. The total for the auxiliary variable could be obtained by taking a single weight measurement of all ovaries and the mean for the auxiliary variable would simply be that total divided by N. Thus, in this case it would not be necessary to assume that the x values were known for all of the population units in order to satisfy an assumption that the mean or total of the auxiliary variable were known.

$$\hat{\beta} = \frac{\sum\limits_{i \in S}(x_i - \hat{\mu}_x)(y_i - \hat{\mu}_y)}{\sum\limits_{i \in S}(x_i - \hat{\mu}_x)^2} \tag{7.4}$$

rather than $\hat{R}$, and estimate the mean and total as

$$\hat{\mu}_{y,\text{reg}} = \hat{\mu}_y + \hat{\beta}(\mu_x - \hat{\mu}_x) \tag{7.5}$$

$$\hat{T}_{y,\text{reg}} = N\hat{\mu}_{y,\text{reg}} \tag{7.6}$$

Your may recognize $\hat{\beta}$ as identical in form to the ordinary least-squares estimator of the slope of a straight line (with intercept) with constant variation around the regression line.[2] [$\hat{\beta}$ is also derived later in this chapter, just prior to Equation (7.44).]

7.1.1 Graphical representation

For both ratio and regression estimation, the sample mean (mean-per-unit estimate) of the target variable, $\hat{\mu}_y$, is *adjusted* to a higher or lower value for the estimate based on a comparison of the auxiliary variable population and sample means. It is clear from the forms of Equations (7.2) and (7.5) that if $\mu_x > \hat{\mu}_x$ then $\hat{\mu}_y$ is adjusted upwards, and if $\mu_x < \hat{\mu}_x$ then $\hat{\mu}_y$ is adjusted downwards. As Figure 7.2 illustrates, these adjustments make good sense when the target variable is linearly related to the auxiliary variable.

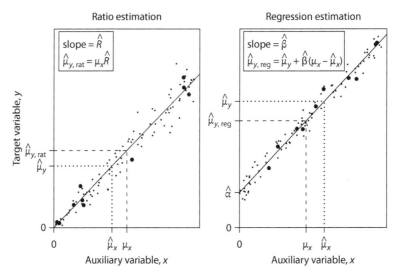

Fig. 7.2 Graphical illustration of ratio and regression estimation. In both panels, the dots depict a population of 100 units from which an SRS of 10 units has been selected (larger dots), $(\hat{\mu}_x, \hat{\mu}_y)$ are the sample means (mean-per-unit estimates), and (μ_x, μ_y) are the population means of the auxiliary and target variables, respectively. Ratio estimation: the sample data are used to calculate $\hat{R} = \hat{\mu}_y/\hat{\mu}_x$, which is the slope of the solid line passing through $(0,0)$ and $(\hat{\mu}_x, \hat{\mu}_y)$. The ratio estimate, $\hat{\mu}_{y,\text{rat}}$, is the y value on this line at $x = \mu_x$. It is an increase over $\hat{\mu}_y$ because $\mu_x > \hat{\mu}_x$ (it would result in a decrease if $\mu_x < \hat{\mu}_x$). Regression estimation: the sample data are used to calculate $\hat{\beta}$, which is the slope of the solid line passing through $(0,\hat{\alpha})$ and $(\hat{\mu}_x, \hat{\mu}_y)$, where $\hat{\alpha} = \hat{\mu}_y - \hat{\beta}\hat{\mu}_x$. The regression estimate, $\hat{\mu}_{y,\text{reg}}$, is the y value on this line at $x = \mu_x$. It is a decrease over $\hat{\mu}_y$ because $\mu_x < \hat{\mu}_x$ (it would result in an increase if $\mu_x > \hat{\mu}_x$).

---

[2] The usual slope-intercept representation for a linear regression can be seen by defining $\hat{\alpha} = \hat{\mu}_y - \hat{\beta}\hat{\mu}_x$. Then $\hat{\mu}_{y,\text{reg}} = \hat{\alpha} + \hat{\beta}\mu_x$.

7.1.2 Sample space illustration

Consider a population of size $N = 5$ with the following auxiliary and target variable values

Population unit (i):	1	2	3	4	5
Auxiliary variable (x_i):	10	20	30	40	50
Target variable (y_i):	2	5	7	14	20

The corresponding population parameters are

μ_x	T_x	μ_y	T_y	R	σ_x^2	σ_y^2
30	150	9.6	48	0.32	250	53.3

Suppose that an SRS of size $n = 3$ is selected from this population, and that ratio and regression estimators are both used to estimate μ_y. Table 7.1 lists the resulting sample space and corresponding estimates.

Because each of the 10 SRS samples is equally likely, expected values can be calculated by multiplying the column totals by $p(s) = 0.10$. Thus

$$E(\hat{R}) = 0.31457; \quad E(\hat{\mu}_{y,\text{rat}}) = 9.43697; \quad E(\hat{\mu}_{y,\text{reg}}) = 9.10293$$

Note that these estimators are all biased

$$B(\hat{R}) = 0.31457 - 0.32 = -0.00543$$
$$B(\hat{\mu}_{y,\text{rat}}) = 9.43697 - 9.6 = -0.16303$$
$$B(\hat{\mu}_{y,\text{reg}}) = 9.10293 - 9.6 = -0.49707$$

Note also that the estimator $\hat{R}$ is biased even though both $\hat{\mu}_x$ and $\hat{\mu}_y$ are unbiased estimators: $E(\hat{\mu}_x) = 30 = \mu_x$, and $E(\hat{\mu}_y) = 9.6 = \mu_y$. That is

$$E(\hat{R}) = E\left(\frac{\hat{\mu}_y}{\hat{\mu}_x}\right) \neq \frac{E(\hat{\mu}_y)}{E(\hat{\mu}_x)}$$

Table 7.1 Sample space for SRS samples of size $n = 3$ from the $N = 5$ example population, and corresponding estimates. Population unit-specific values $\{(i, x_i, y_i),$ $i = 1, 2, 3, 4, 5\}$ of auxiliary and target variables are: $\{(1, 10, 2), (2, 20, 5), (3, 30, 7), (4, 40, 14), (5, 50, 20)\}$.

ID	Sample Units (s)	$p(s)$	$\hat{\mu}_x(s)$	$\hat{\mu}_y(s)$	$\hat{R}(s)$	$\hat{B}(s)$	$\hat{\mu}_{y,\text{rat}}(s)$	$\hat{\mu}_{y,\text{reg}}(s)$
1	1, 2, 3	0.1	20.00	4.66	0.2333	0.2500	7.0000	7.1666
2	1, 2, 4	0.1	23.33	7.00	0.3000	0.4071	9.0000	9.7143
3	1, 2, 5	0.1	26.66	9.00	0.3375	0.4615	10.1250	10.5385
4	1, 3, 4	0.1	26.66	7.66	0.2875	0.3786	8.6250	8.9286
5	1, 3, 5	0.1	30.00	9.66	0.3222	0.4500	9.6666	9.6666
6	1, 4, 5	0.1	33.33	12.00	0.3600	0.4385	10.8000	10.5385
7	2, 3, 4	0.1	30.00	8.66	0.2888	0.4500	8.6666	8.6666
8	2, 3, 5	0.1	33.33	10.66	0.3200	0.5214	9.6000	8.9286
9	2, 4, 5	0.1	36.66	13.00	0.3545	0.4929	10.6363	9.7143
10	3, 4, 5	0.1	40.00	13.66	0.3417	0.6500	10.2500	7.1666
Sum:		1.0	300.00	96.00	3.1457	4.5000	94.3697	91.0293

Indeed, it is generally true that the expectation of a ratio of random variables is not equal to the ratio of the expectations of the random variables (Springer 1979).

Given that the ratio and regression estimators of the mean are biased, the corresponding estimators of the total will be biased too. The expected value of the ratio estimator of the total is

$$E(\hat{T}_{y,\text{rat}}) = E(N\hat{\mu}_{y,\text{rat}}) = N \cdot E(\hat{\mu}_{y,\text{rat}}) = 5(9.43697) = 47.18485$$

so that $B(\hat{T}_{y,\text{rat}}) = 47.18485 - 48 = -0.81515$, and the expected value of the regression estimator of the total is

$$E(\hat{T}_{y,\text{reg}}) = E(N\hat{\mu}_{y,\text{reg}}) = N \cdot E(\hat{\mu}_{y,\text{rat}}) = 5(9.10293) = 45.51465$$

so that $B(\hat{T}_{y,\text{reg}}) = 45.51465 - 48 = -2.48535$. The bias of the T_y estimator in each case is equal to the bias of the μ_y estimator scaled by N.

$$B(\hat{T}_{y,\text{rat}}) = E(\hat{T}_{y,\text{rat}}) - T_y = N \cdot B(\hat{\mu}_{y,\text{rat}}) = 5(-0.16303) = -0.81515$$
$$B(\hat{T}_{y,\text{reg}}) = E(\hat{T}_{y,\text{reg}}) - T_y = N \cdot B(\hat{\mu}_{y,\text{reg}}) = 5(-0.49707) = -2.48535$$

For the sampling variances of $\hat{R}$, $\hat{T}_{y,\text{rat}}$, and $\hat{T}_{y,\text{reg}}$, we need to calculate the squared difference $[\hat{\theta}(s) - E(\hat{\theta})]^2$ for each $s \in \mathcal{S}$, as tabulated in Table 7.2. Multiplying the column totals for $[\hat{\theta}(s) - E(\hat{\theta})]^2$ (for $\hat{R}$, $\hat{\mu}_{y,\text{rat}}$, and $\hat{\mu}_{y,\text{reg}}$) by $p(s) = 0.10$ gives

$$V(\hat{R}) = 0.001321396, \quad V(\hat{\mu}_{y,\text{rat}}) = 1.189256635, \text{ and } \quad V(\hat{\mu}_{y,\text{reg}}) = 1.293617115$$

$V(\hat{\mu}_{y,\text{rat}})$ could alternatively be obtained as $V(\mu_x\hat{R}) = \mu_x^2 V(\hat{R}) = 30^2 \cdot 0.001321396 = 1.189256 \approx 1.189256635$ (agreement would be exact if additional decimal places were reported in Table 7.2). Also, sampling variances for estimators of the total can be easily obtained from sampling variances for estimators of the mean as

$$V(\hat{T}_{y,\text{rat}}) = V(N\hat{\mu}_{y,\text{rat}}) = N^2 V(\hat{\mu}_{y,\text{rat}}) = 25(1.189256) = 29.7314$$
$$V(\hat{T}_{y,\text{reg}}) = V(N\hat{\mu}_{y,\text{reg}}) = N^2 V(\hat{\mu}_{y,\text{reg}}) = 25(1.293617) = 32.34043$$

Table 7.2 Tabulation of sample-specific squared differences $[\hat{\theta}(s) - E(\hat{\theta})]^2$ for ratio estimators $\hat{R}$, $\hat{\mu}_{y,\text{rat}}$ and regression estimator $\hat{\mu}_{y,\text{reg}}$ based on the sample estimates shown in Table 7.1.

Sample ID	$p(s)$	$\hat{\theta} = \hat{R}$	$\hat{\theta} = \hat{\mu}_{y,\text{rat}}$	$\hat{\theta} = \hat{\mu}_{y,\text{reg}}$
			$[\hat{\theta}(s) - E(\hat{\theta})]^2$	
1	0.1	0.00659869	5.93882130	3.74911726
2	0.1	0.00021216	0.19094252	0.37375532
3	0.1	0.00052598	0.47338570	2.06074964
4	0.1	0.00073255	0.65929479	0.03040105
5	0.1	0.00005862	0.05276070	0.31779858
6	0.1	0.00206428	1.85785161	2.06074964
7	0.1	0.00065930	0.59336676	0.19032605
8	0.1	0.00002953	0.02657888	0.03040105
9	0.1	0.00159838	1.43854582	0.37375532
10	0.1	0.00073447	0.66101827	3.74911726
Sum:	1.0	0.01321396	11.89256635	12.93617115

Because ratio and regression estimators are not unbiased, the appropriate measure of estimator accuracy is mean square error ($\text{MSE}(\hat{\theta}) = V(\hat{\theta}) + [B(\hat{\theta})]^2$) rather than sampling variance. For this small sample space, mean square errors of the ratio and regression estimators are

$$\text{MSE}(\hat{R}) = V(\hat{R}) + [B(\hat{R})]^2 = 0.001321396 + (-0.00543)^2 = 0.001350881$$

$$\text{MSE}(\hat{\mu}_{y,\text{rat}}) = V(\hat{\mu}_{y,\text{rat}}) + [B(\hat{\mu}_{y,\text{rat}})]^2 = 1.189256635 + (-0.16303)^2 = 1.215835$$

$$\text{MSE}(\hat{\mu}_{y,\text{reg}}) = V(\hat{\mu}_{y,\text{reg}}) + [B(\hat{\mu}_{y,\text{reg}})]^2 = 1.293617115 + (-0.49707)^2 = 1.540696$$

Thus, although ratio and regression estimators are modestly biased, the contribution of the squared bias to the overall MSE was either minor (2.2% for $\hat{R}$ and $\hat{\mu}_{y,\text{rat}}$) or modest (16.0% for $\hat{\mu}_{y,\text{reg}}$).

Finally, if an SRS of size $n = 3$ had been selected from this population and only the y values had been used for estimation, then $V(\hat{\mu}_{y,\text{mpu}}) = [(N-n)/N]\sigma_y^2/n = [(5-3)/5] \cdot 53.3/3 = 7.11$. Thus, the MSE of both $\hat{\mu}_{y,\text{rat}}$ and $\hat{\mu}_{y,\text{reg}}$ (approximately 1.22 and 1.54, respectively) were substantially less than the sampling variance of $\hat{\mu}_{y,\text{mpu}}$. Incorporation of an auxiliary variable in estimators can, under favorable circumstances, dramatically reduce sampling variance and overall mean square error.

7.1.3 Bias

For the ratio estimator $\hat{R}$, Hartley and Ross (1954) provides an exact expression for bias

$$B(\hat{R}) = \frac{-\text{Cov}(\hat{R}, \hat{\mu}_x)}{\mu_x} \tag{7.7}$$

where $\text{Cov}(\hat{R}, \hat{\mu}_x) = E[(\hat{R} - E(\hat{R}))(\hat{\mu}_x - E(\hat{\mu}_x))]$ and measures the *covariation* of the estimators $\hat{R}$ and $\hat{\mu}_x$ over the sample space. For the sample space example in Table 7.1

$$\begin{aligned}
\text{Cov}(\hat{R}, \hat{\mu}_x) &= E\left[[\hat{R} - E(\hat{R})][\hat{\mu}_x - E(\hat{\mu}_x)]\right] \\
&= \sum_{s \in \mathcal{S}}\left[[\hat{R}(s) - E(\hat{R})][\hat{\mu}_x(s) - E(\hat{\mu}_x)]\right] \cdot p(s) \\
&= (1/10)\sum_{s \in \mathcal{S}}\left[(\hat{R}(s) - 0.31457)(\hat{\mu}_x(s) - 30)\right] = (1/10) \cdot 1.6303 = 0.16303
\end{aligned}$$

Thus, $B(\hat{R}) = -0.16303/\mu_x = -0.16303/30 = -0.00543$, as previously calculated over the sample space. Note that this exact expression for $B(\hat{R})$ also implies that $B(\hat{\mu}_{y,\text{rat}}) = -\text{Cov}(\hat{R}, \hat{\mu}_x) = -0.16303$, because $B(\hat{\mu}_{y,\text{rat}}) = B(\mu_x\hat{R}) = \mu_x B(\hat{R}) = -\text{Cov}(\hat{R}, \hat{\mu}_x)$.

Lohr (2010 p. 139) provides an exact expression for the bias of the regression estimator

$$B(\hat{\mu}_{y,\text{reg}}) = -\text{Cov}(\hat{\beta}, \hat{\mu}_x) \tag{7.8}$$

where $\text{Cov}(\hat{\beta}, \hat{\mu}_x) = E\left[[\hat{\beta} - E(\hat{\beta})][\hat{\mu}_x - E(\hat{\mu}_x)]\right]$ measures the covariation of the estimators $\hat{\beta}$ and $\hat{\mu}_x$ over the sample space. For the sample space example in Table 7.1

$$\begin{aligned}
\text{Cov}(\hat{\beta}, \hat{\mu}_x) &= E\left[(\hat{\beta} - E(\hat{\beta}))(\hat{\mu}_x - E(\hat{\mu}_x))\right] \\
&= \sum_{s \in \mathcal{S}}\left[(\hat{\beta}(s) - E(\hat{\beta}))(\hat{\mu}_x(s) - E(\hat{\mu}_x))\right] \cdot p(s) \\
&= (1/10)\sum_{s \in \mathcal{S}}\left[(\hat{\beta}(s) - 0.45)(\hat{\mu}_x(s) - 30)\right] = (1/10) \cdot 4.970696 = 0.4970696
\end{aligned}$$

Therefore, $B(\hat{\mu}_{y,\text{reg}}) = -0.4970696$ which is equal (except for minor rounding differences) to the value $B(\hat{\mu}_{y,\text{reg}}) = -0.49707$ previously calculated over the sample space.

7.1.4 Sampling variance

There are no exact expressions for sampling variance of the ratio and regressions estimators, but there are excellent approximations to sampling variance. These approximations are based on application of the delta method (Section A.8, Seber 1982 Section 1.3.3, Wolter 2007 Chapter 6). Suppose that $\hat{\theta}$ is a function of two estimators $\hat{\theta}_1$ and $\hat{\theta}_2$, i.e., $\hat{\theta} = f(\hat{\theta}_1, \hat{\theta}_2)$. According to the delta method, an approximate expression for the sampling variance of $\hat{\theta}$ [denoted by $V^*(\hat{\theta})$] is [Equation (A.150)]

$$V(\hat{\theta}) \approx V^*(\hat{\theta}) = \left(\frac{\partial\hat{\theta}}{\partial\hat{\theta}_1}\right)^2 V(\hat{\theta}_1) + \left(\frac{\partial\hat{\theta}}{\partial\hat{\theta}_2}\right)^2 V(\hat{\theta}_2) + 2\left(\frac{\partial\hat{\theta}}{\partial\hat{\theta}_1}\right)\left(\frac{\partial\hat{\theta}}{\partial\hat{\theta}_2}\right)\text{Cov}(\hat{\theta}_1, \hat{\theta}_2) \qquad (7.9)$$

where $\partial\hat{\theta}/\partial\hat{\theta}_i$ is the partial derivative of $\hat{\theta}$ with respect to $\hat{\theta}_i$ $(i = 1, 2)$ and these partial derivatives are evaluated at the expected values of $\hat{\theta}_1$ and $\hat{\theta}_2$, and $\text{Cov}(\hat{\theta}_1, \hat{\theta}_2)$ is the *covariance* (Section A.3.6) of $\hat{\theta}_1$ and $\hat{\theta}_2$.

Ratio estimator

Application of Equation (7.9) to the estimator $\hat{R} = \hat{\mu}_y/\hat{\mu}_x$ gives the specific result

$$V^*(\hat{R}) = \left(\frac{\partial\hat{R}}{\partial\hat{\mu}_x}\right)^2 V(\hat{\mu}_x) + \left(\frac{\partial\hat{R}}{\partial\hat{\mu}_y}\right)^2 V(\hat{\mu}_y) + 2\left(\frac{\partial\hat{R}}{\partial\hat{\mu}_x}\right)\left(\frac{\partial\hat{R}}{\partial\hat{\mu}_y}\right)\text{Cov}(\hat{\mu}_x, \hat{\mu}_y) \qquad (7.10)$$

The partial derivatives of $\hat{R}$ with respect to $\hat{\mu}_x$ and $\hat{\mu}_y$ are $\partial\hat{R}/\partial\hat{\mu}_x = -\hat{\mu}_y/\hat{\mu}_x^2$ and $\partial\hat{R}/\partial\hat{\mu}_y = 1/\hat{\mu}_x$. If units are selected by SRS, then $V(\hat{\mu}_x) = \left(\frac{N-n}{N}\right)\frac{\sigma_x^2}{n}$ and $V(\hat{\mu}_y) = \left(\frac{N-n}{N}\right)\frac{\sigma_y^2}{n}$. Similarly, for SRS, $\text{Cov}(\hat{\mu}_x, \hat{\mu}_y)$ depends on the finite population covariance of x and y, $\sigma_{x,y}$,

$$\text{Cov}(\hat{\mu}_x, \hat{\mu}_y) = \left(\frac{N-n}{N}\right)\frac{\sigma_{x,y}}{n}, \quad \text{where } \sigma_{x,y} = \sum_{i=1}^{N}(x_i - \mu_x)(y_i - \mu_y)/(N-1) \qquad (7.11)$$

Substitution for $V(\hat{\mu}_x)$, $V(\hat{\mu}_y)$, $\text{Cov}(\hat{\mu}_x, \hat{\mu}_y)$ and the partial derivatives of $\hat{R}$ with respect to $\hat{\mu}_x$ and $\hat{\mu}_y$ evaluated at $E(\hat{\mu}_x) = \mu_x$ and $E(\hat{\mu}_y) = \mu_y$, respectively, into Equation (7.10) gives

$$V^*(\hat{R}) = \left(\frac{-\mu_y}{\mu_x^2}\right)^2 \left(\frac{N-n}{N}\right)\frac{\sigma_x^2}{n} + \left(\frac{1}{\mu_x}\right)^2 \left(\frac{N-n}{N}\right)\frac{\sigma_y^2}{n} + 2\left(\frac{-\mu_y}{\mu_x^2}\right)\left(\frac{1}{\mu_x}\right)\left(\frac{N-n}{N}\right)\frac{\sigma_{x,y}}{n}$$

$$= \frac{1}{\mu_x^2}\left(\frac{N-n}{N}\right)\frac{R^2\sigma_x^2 + \sigma_y^2 - 2R\sigma_{x,y}}{n}$$

Alternatively, $\sigma_{x,y}$ in the formula immediately above can be re-expressed in terms of the *correlation* (Section A.3.6) between x and y, $\text{Cor}(x, y) = \sigma_{x,y}/(\sigma_x\sigma_y)$ [Equation (A.119)], which may provide additional insight into the conditions under which the ratio estimator can be expected to perform well.

$$V^*(\hat{R}) = \frac{1}{\mu_x^2}\left(\frac{N-n}{N}\right)\frac{R^2\sigma_x^2 + \sigma_y^2 - 2R\text{Cor}(x, y)\sigma_x\sigma_y}{n} \qquad (7.12)$$

Correlation ranges from a minumum value of -1 (perfect negative correlation between x and y) to a maximum value of +1 (perfect positive correlation between x and y) and depends directly on the covariance between x and y. If small and large x values tend to be associated with small and large y values, respectively, then there is evidence of *positive covariance* between x and y and $\text{Cor}(x,y)$ will be positive. If instead small and large x values tend to be associated with large and small y values, respectively, then there is evidence of *negative covariance* between x and y and $\text{Cor}(x,y)$ will be negative. If the values of x and y are independent of one another, then $\text{Cov}(x,y) = 0$ and $\text{Cor}(x,y) = 0$.

Similar approximate expressions for the sampling variances of the ratio estimators $\hat{\mu}_{y,\text{rat}} = \mu_x \hat{R}$ and $\hat{T}_{y,\text{rat}} = T_x \hat{R} = N\hat{\mu}_{y,\text{rat}}$ follow from $V^*(\hat{R})$ given that $V(c\hat{\theta}) = c^2 V(\hat{\theta})$ for any constant c [Equation (A.52)]. Therefore

$$V^*(\hat{\mu}_{y,\text{rat}}) = \mu_x^2 V^*(\hat{R}) = \left(\frac{N-n}{N}\right) \frac{R^2 \sigma_x^2 + \sigma_y^2 - 2R\text{Cor}(x,y)\sigma_x\sigma_y}{n} \tag{7.13}$$

and

$$V^*(\hat{T}_{y,\text{rat}}) = N^2 V^*(\hat{\mu}_{y,\text{rat}}) = N^2 \left(\frac{N-n}{N}\right) \frac{R^2 \sigma_x^2 + \sigma_y^2 - 2R\text{Cor}(x,y)\sigma_x\sigma_y}{n} \tag{7.14}$$

Finally, an alternative and simpler expression for the sampling variance of $\hat{\mu}_{y,\text{rat}}$, but identical in value to Equation (7.13), is

$$V^*(\hat{\mu}_{y,\text{rat}}) = \left(\frac{N-n}{N}\right) \frac{\sum_{i=1}^{N}(y_i - Rx_i)^2/(N-1)}{n} \tag{7.15}$$

Thus, $V^*(\hat{\mu}_{y,\text{rat}})$ will become smaller as (a) n increases, (b) the sampling fraction n/N increases, and (c) when x and y are positively correlated *and* y is (approximately) proportional to x. If satisfied, factor (c) would make the sum of the squared differences, $\sum_{i=1}^{N}(y_i - Rx_i)^2$, small.

The bias of $V^*(\hat{\mu}_{y,\text{rat}})$ for $V(\hat{\mu}_{y,\text{rat}})$ is of order $1/n$ (Cochran 1977), so it becomes an increasingly more accurate approximation as sample size increases. If Equation (7.15) is applied to the set of population values used for illustration in Table 7.1, then $V^*(\hat{\mu}_{y,\text{rat}}) = 0.92$, which is a bit less than $V(\hat{\mu}_{y,\text{rat}}) = 1.189$, but still a useful approximation for many applications even at this very small sample size ($n = 3$). $V^*(\hat{\mu}_{y,\text{rat}})/V(\hat{\mu}_{y,\text{rat}})$ will be much closer to 1 for larger n and N. (The performance of $V^*(\hat{\mu}_{y,\text{rat}})$ as an approximation for $V(\hat{\mu}_{y,\text{rat}})$ is explored for some simulated populations in Section 7.5.)

Regression estimator

As for ratio estimation, there is no exact expression for sampling variance of $\hat{\mu}_{y,\text{reg}}$. Application of the delta method to the estimator $\hat{\mu}_{y,\text{reg}}$ provides a large sample approximation to sampling variance equivalent to the formula derived by Lohr (2010 p. 139)

$$V(\hat{\mu}_{y,\text{reg}}) \approx V^*(\hat{\mu}_{y,\text{reg}}) = \left(\frac{N-n}{N}\right) \frac{\sigma_y^2[1 - \text{Cor}(x,y)^2]}{n} \tag{7.16}$$

As for the ratio estimator of μ_y, $V^*(\hat{\mu}_{y,\text{reg}})$ will become smaller as (a) n increases, (b) the sampling fraction n/N increases, and (c) $|\text{Cor}(x,y)|$ approaches 1. Indeed, if $|\text{Cor}(x,y)| = 1$, then $V^*(\hat{\mu}_{y,\text{reg}}) = 0$.

The bias of $V^*(\hat{\mu}_{y,\text{reg}})$ for $V(\hat{\mu}_{y,\text{reg}})$ is also of order $1/n$ (Cochran 1977). If Equation (7.16) is applied to the $N = 5$ example population values, then $V^*(\hat{\mu}_{y,\text{reg}}) = 0.3567$, considerably less than $V(\hat{\mu}_{y,\text{reg}}) = 1.294$ calculated directly over the sample space. Indeed, $V^*(\hat{\mu}_{y,\text{reg}})$ is generally *not to be trusted* for small n, but performance improves greatly as n exceeds 10 or so. (The performance of $V^*(\hat{\mu}_{y,\text{reg}})$ as an approximation for $V(\hat{\mu}_{y,\text{reg}})$ is explored for some simulated populations in Section 7.5.)

7.1.5 *Estimation of sampling variance*

Estimators for sampling variance of the ratio and regression estimators are typically based on substitution of sample estimates of the quantities that appear in the approximate expressions $V^*(\hat{\theta}_y)$ for sampling variance. These estimators are not unbiased, but for large n and N they typically have expected values that are very close to $V^*(\hat{\theta}_y)$ and are reasonably close (say, $\pm 10\%$) to actual sampling variances.

For the ratio estimator of μ_y, variance estimators are of two forms. Analogous to Equation (7.15) is the computationally simple estimator

$$\hat{V}(\hat{\mu}_{y,\text{rat}}) = \left(\frac{N-n}{N}\right)\frac{\sum\limits_{i\in S}(y_i - \hat{R}x_i)^2/(n-1)}{n} \tag{7.17}$$

where $\hat{R} = \hat{\mu}_y/\hat{\mu}_x$. An alternative form of this estimator, analogous to Equation (7.13), is

$$\hat{V}(\hat{\mu}_{y,\text{rat}}) = \left(\frac{N-n}{N}\right)\frac{\hat{R}^2\hat{\sigma}_x^2 + \hat{\sigma}_y^2 - 2\hat{R}\widehat{\text{Cor}}(x,y)\hat{\sigma}_x\hat{\sigma}_y}{n} \tag{7.18}$$

where

$$\widehat{\text{Cor}}(x,y) = \frac{\sum\limits_{i\in S}(x_i - \hat{\mu}_x)(y_i - \hat{\mu}_y)/(n-1)}{\hat{\sigma}_x\hat{\sigma}_y} = \frac{\hat{\sigma}_{x,y}}{\hat{\sigma}_x\hat{\sigma}_y}$$

and $\hat{\sigma}_x = \sqrt{\hat{\sigma}_x^2}$, $\hat{\sigma}_y = \sqrt{\hat{\sigma}_y^2}$. For a constant a, $\hat{V}(a\hat{\theta}) = a^2\hat{V}(\hat{\theta})$, so that estimators for the sampling variance of $\hat{R}$ and $\hat{T}_{y,\text{rat}}$ are

$$\hat{V}(\hat{R}) = \frac{1}{\hat{\mu}_x^2}\hat{V}(\hat{\mu}_{y,\text{rat}}) \tag{7.19}$$

$$\hat{V}(\hat{T}_{y,\text{rat}}) = N^2\hat{V}(\hat{\mu}_{y,\text{rat}}) \tag{7.20}$$

For the regression estimator, an estimator of sampling variance analogous to Equation (7.17) is

$$\hat{V}(\hat{\mu}_{y,\text{reg}}) = \left(\frac{N-n}{N}\right)\frac{\sum_{i\in S}(y_i - \hat{y}_i)^2/(n-2)}{n} \tag{7.21}$$

where $\hat{y}_i = \hat{\mu}_y + \hat{\beta}(x_i - \hat{\mu}_x)$. The $\{\hat{y}_i\}$ can be viewed as *predicted* values under a linear relation between y and x. Note that the Equation (7.21) divisor of the sum of the squared *residuals*, $\{y_i - \hat{y}_i\}$, is $n-2$ rather than the divisor of $n-1$ for Equation (7.17) owing to two parameters having been estimated for the regression estimator (the slope $\hat{\beta}$ and the intercept $\hat{\alpha} = \hat{\mu}_y - \hat{\beta}\hat{\mu}_x$) versus the one parameter having been estimated for the zero-intercept ratio estimator ($\hat{R}$). The corresponding estimator for $V(\hat{T}_{y,\text{reg}})$ is

$$\hat{V}(\hat{T}_{y,\text{reg}}) = N^2\hat{V}(\hat{\mu}_{y,\text{reg}}) \tag{7.22}$$

7.1.6 *Sample size determination*

Ratio estimation

For the ratio estimator of μ_y

$$V^*(\hat{\mu}_{y,\text{rat}}) = \left(\frac{N-n}{N}\right)\frac{R^2\sigma_x^2 + \sigma_y^2 - 2R\text{Cor}(x,y)\sigma_x\sigma_y}{n} = \left(\frac{1}{n} - \frac{1}{N}\right)Z$$

where $Z = \left(R^2 \sigma_x^2 + \sigma_y^2 - 2R\mathrm{Cor}(x,y)\sigma_x\sigma_y\right)$. Given a desired sampling variance, $V^*(\hat{\mu}_{y,\mathrm{rat}})$, this epression can be solved for n to obtain

$$n = \frac{Z}{V^*(\hat{\mu}_{y,\mathrm{rat}}) + Z/N} \tag{7.23}$$

Because $V^*(\hat{T}_{y,\mathrm{rat}}) = N^2 V^*(\hat{\mu}_{y,\mathrm{rat}})$, the expression for n can be generalized for estimating $\theta = \mu, \pi,$ or T as

$$n = \frac{Z}{\mathcal{K}(\hat{\theta}_{y,\mathrm{rat}}) + Z/N}, \qquad \mathcal{K}(\hat{\theta}) = \begin{cases} V^*(\hat{\theta}), & \text{for } \theta = \mu, \pi \\ V^*(\hat{\theta})/N^2, & \text{for } \theta = T \end{cases} \tag{7.24}$$

If a 95% bound on the error of estimation, B, is specified, replace $V^*(\hat{\theta})$ with $B^2/4$. If a CV constraint is specified, replace $V^*(\hat{\theta})$ with $\mathrm{CV}(\hat{\theta})^2\theta^2$ (assuming that n is large enough so that $B(\hat{\theta})$ is negligible).

In practice, use of Equation (7.24) would require preliminary sample estimates of R, σ_y^2, σ_x^2, $\mathrm{Cor}(x,y)$, and possibly also $\hat{\mu}_y$ (for CV specification). Note that the preliminary sample estimate of σ_x^2 should be used for calculation of Z rather than the known σ_x^2 (assuming all x values are known) (Problem 7.6).

Regression estimation

For the regression estimator of μ_y

$$V^*(\hat{\mu}_{y,\mathrm{reg}}) = \left(\frac{N-n}{N}\right)\sigma_y^2(1 - \mathrm{Cor}(x,y)^2)/n = \left(\frac{1}{n} - \frac{1}{N}\right)Z$$

where $Z = \sigma_y^2(1 - \mathrm{Cor}(x,y)^2)$. This has the same general form as for the ratio estimator, and thus n can be solved for in the same manner to give

$$n = \frac{Z}{\mathcal{K}(\hat{\theta}_{y,\mathrm{reg}}) + Z/N}, \qquad \mathcal{K}(\hat{\theta}) = \begin{cases} V^*(\hat{\theta}), & \text{for } \theta = \mu, \pi \\ V^*(\hat{\theta})/N^2, & \text{for } \theta = T \end{cases} \tag{7.25}$$

If a 95% bound on the error of estimation, B, is specified, replace $V^*(\hat{\theta})$ with $B^2/4$. If a CV constraint is specified, replace $V^*(\hat{\theta})$ with $\mathrm{CV}(\hat{\theta})^2\theta^2$. Note that solutions for sample size for the regression estimator are not to be trusted if the calculated n is small because $V^*(\hat{\mu}_{y,\mathrm{reg}})$ may be substantially less than $V(\hat{\mu}_{y,\mathrm{reg}})$ at small n (Section 7.5). In practice, preliminary estimates of $\mathrm{Cor}(x,y)$ and σ_y^2 would be substituted into Equation (7.25) to get an approximate solution for the required sample size.

7.1.7 Relative efficiency

Ratio estimator

Assuming that sample size is large enough so that $V^*(\hat{\mu}_{y,\mathrm{rat}})$ provides an accurate approximation to $V(\hat{\mu}_{y,\mathrm{rat}})$, we can use this expression to determine the conditions under which $V^*(\hat{\mu}_{y,\mathrm{rat}}) < V(\hat{\mu}_{y,\mathrm{mpu}})$ (i.e., to determine when $\mathrm{RE}(\hat{\mu}_{\mathrm{rat}}, \hat{\mu}_{\mathrm{mpu}}) > 1$). First, from the structure of Equation (7.15), it is immediately apparent that $V^*(\hat{\mu}_{y,\mathrm{rat}}) < V(\hat{\mu}_{y,\mathrm{mpu}})$ whenever

$$\frac{\sum\limits_{i=1}^{N}(y_i - Rx_i)^2}{N-1} < \frac{\sum\limits_{i=1}^{N}(y_i - \mu_y)^2}{N-1} = \sigma_y^2$$

If, on average, the values Rx_i are closer to y_i than the fixed value μ_y, then the sampling variance of the ratio estimator will be less than that of $\hat{\mu}_{y,\text{mpu}}$. This would happen if y were approximately proportional to x, by the factor R, and $\text{Cor}(x,y)$ were *large* and positive. In that case, one would expect that $\sum_{i=1}^{N}(y_i - Rx_i)^2$ would be small compared to $\sum_{i=1}^{N}(y_i - \mu_y)^2$.

Using Equation (7.13), we could alternatively calculate the relative efficiency of the two sampling strategies as

$$\text{RE}(\hat{\mu}_{y,\text{rat}}, \hat{\mu}_{y,\text{mpu}}) \approx \frac{V(\hat{\mu}_{y,\text{mpu}})}{V^*(\hat{\mu}_{y,\text{rat}})} = \frac{\left(\frac{N-n}{N}\right)\sigma_y^2/n}{\left(\frac{N-n}{N}\right)\left[\sigma_y^2 + R^2\sigma_x^2 - 2R\text{Cor}(x,y)\sigma_x\sigma_y\right]/n}$$

$$= \frac{\sigma_y^2}{\sigma_y^2 + R^2\sigma_x^2 - 2R\text{Cor}(x,y)\sigma_x\sigma_y}$$

Thus, whenever $\left[\sigma_y^2 + R^2\sigma_x^2 - 2R\text{Cor}(x,y)\sigma_x\sigma_y\right] < \sigma_y^2$, ratio estimation will outperform SRS with mean-per-unit estimation (RE > 1). That is, whenever $\left[R^2\sigma_x^2 - 2R\text{Cor}(x,y)\sigma_x\sigma_y\right] < 0$ or $R\sigma_x/\sigma_y < 2\text{Cor}(x,y)$. With $R = \mu_y/\mu_x$ and $\sigma_x/\sigma_y = \sqrt{V(x)}/\sqrt{V(y)}$, where $V(y) = \sigma_y^2$ $[(N-1)/N]$ is the population variance of y [Equations (A.113), (A.114)], $R\sigma_x/\sigma_y = [\sqrt{V(x)}/\mu_x]/[\sqrt{V(y)}/\mu_y] = \text{CV}(x)/\text{CV}(y)$ [Equation (A.115)], thus giving the following condition for $V(\hat{\mu}_{y,\text{rat}}) < V(\hat{\mu}_{y,\text{mpu}})$

$$\left(\frac{1}{2}\right)\frac{\text{CV}(x)}{\text{CV}(y)} < \text{Cor}(x,y)$$

This result leads to the conclusion that, when $\text{CV}(x) = \text{CV}(y)$, then $V^*(\hat{\mu}_{y,\text{rat}}) < V(\hat{\mu}_{y,\text{mpu}})$ whenever $\text{Cor}(x,y) > 0.5$. Thus, if one is willing to assume that the coefficients of variation of the x and y variables have similar values, then ratio estimation should outperform SRS with mean-per-unit estimation whenever $\text{Cor}(x,y) > 0.5$.

Regression estimator

The relative efficiency of regression estimation as compared to SRS with mean-per-unit estimation is

$$\text{RE}(\hat{\mu}_{y,\text{reg}}, \hat{\mu}_{y,\text{mpu}}) \approx \frac{V(\hat{\mu}_{y,\text{mpu}})}{V^*(\hat{\mu}_{y,\text{reg}})} = \frac{\left(\frac{N-n}{N}\right)\sigma_y^2/n}{\left(\frac{N-n}{N}\right)\sigma_y^2\left[(1 - \text{Cor}(x,y)^2)\right]/n} = \frac{1}{1 - \text{Cor}(x,y)^2}$$

Thus, so long as $|\text{Cor}(x,y)| > 0$, $\hat{\mu}_{y,\text{reg}}$ should theoretically outperform $\hat{\mu}_{y,\text{mpu}}$. It is important to note that the theoretical superiority of regression estimation ($|\text{Cor}(x,y)| > 0$ generates RE > 1) as compared to ratio estimation ($\text{Cor}(x,y) > 0.50$ generates RE > 1) may not apply when n is small (Section 7.5) because $V^*(\hat{\mu}_{y,\text{reg}})$ may be considerably less than $V(\hat{\mu}_{y,\text{reg}})$ for small n.

Finally, it is very important to recognize that the expressions concerning relative efficiency of ratio and regression estimation implicitly assume that the x values are available prior to executing a survey and tacitly assume that the costs of obtaining the x values are trivial compared to the costs associated with obtaining the y values.

Whenever there are substantial costs associated with obtaining the x values, then the presented formulas for relative efficiency will be misleading. In such cases a more appropriate measure of relative performance would be the net relative efficiency [NRE, Equation (3.35)], which is the relative precision of the two strategies given an equivalent overall budget.

7.2 Ratio estimation of a proportion

Suppose that x and y are both binary (0, 1) variables. Let $x = 1$ when the auxiliary variable indicates (correctly or not) that $y = 1$, and let $x = 0$ otherwise. Let $y = 1$ if a unit is a member of the class of interest, and let $y = 0$ otherwise. Then, the ratio estimator for μ_y, Equation (7.2), can be applied directly to estimation of the population proportion for the target variable, $\pi_y = \sum_{i=1}^{N} y_i/N = \mu_y$, and

$$\hat{\pi}_{y,\text{rat}} = \frac{\sum_{i \in S} y_i}{\sum_{i \in S} x_i} \pi_x \tag{7.26}$$

where $\pi_x = \sum_{i=1}^{N} x_i/N = \mu_x$. Equations (7.15) and (7.17) could be used as expressions for approximate sampling variance and for estimation of sampling variance, respectively, for $\hat{\pi}_{y,\text{rat}}$. In the following paragraphs, however, we develop an alternative approach to expression of sampling variance for the ratio estimator of a proportion. This alternative approach can generate greater insight into the behavior of this estimator.

Intuition suggests that $\hat{\pi}_{y,\text{rat}}$ should have small sampling variance when there is a high proportion of $y = 1$ units for which $x = 1$, and when there is a high proportion of $y = 0$ units for which $x = 0$. These two notions are captured by the following quantities

$$Sensitivity = P_{\text{sens}} = \mathcal{P}(x = 1|y = 1) = 1 - \mathcal{P}(\text{false negative}) \tag{7.27}$$
$$Specificity = P_{\text{spec}} = \mathcal{P}(x = 0|y = 0) = 1 - \mathcal{P}(\text{false positive}) \tag{7.28}$$

where $\mathcal{P}(\cdot)$ denotes the proportion of units meeting the parenthetical criteria stated. Thus, sensitivity is the proportion of units for which the auxiliary variable equals 1 given that the target variable equals 1, whereas specificity is the proportion of units for which the auxiliary variable equals 0 given that the target variable equals 0. Sensitivity and specificity have long been of interest in medical contexts where the auxiliary variable might be the outcome (yes or no) of an imperfect but rapid preliminary diagnosis of disease or psychosis generated from an inexpensive diagnostic test, whereas direct and essentially error-free measurements of the target variable (definitive disease or psychosis presence diagnosis) might require a battery of very expensive tests. For sensitivity and specificity to be high, the proportions of false negatives, $\mathcal{P}(x = 0|y = 1)$, and false positives, $\mathcal{P}(x = 1|y = 0)$, respectively, must be low. Estimators for sensitivity and specificity are

$$\hat{P}_{\text{sens}} = \frac{\sum_{i \in S} x_i y_i}{\sum_{i \in S} y_i}, \quad \text{and } \hat{P}_{\text{spec}} = \frac{\sum_{i \in S}(1 - x_i)(1 - y_i)}{\sum_{i \in S}(1 - y_i)}$$

For ratio estimation of a proportion, Equation (7.15) can be re-expressed in terms of π_y, π_x, P_{sens}, and P_{spec} (Section 10.2 and Hankin et al. 2009) giving

$$V(\hat{\pi}_{y,\text{rat}}) = \left(\frac{N-n}{N-1}\right)\left(\frac{\pi_y}{\pi_x}\right)\frac{1 - \pi_y P_{\text{sens}} - (1 - \pi_y)P_{\text{spec}}}{n} \tag{7.29}$$

where

$$\pi_x = \pi_y P_{\text{sens}} + (1 - \pi_y)(1 - P_{\text{spec}})$$

That is, π_x, the proportion of $x = 1$ units, can be expressed as the proportion of units that either are (a) a member of the class of interest and correctly classified as such by the auxiliary variable, or (b) a not a member of the class of interest but incorrectly classified as one by the auxiliary variable.

From Equation (3.18), the sampling variance for $\hat{\pi}_{y,\text{mpu}}$ is $V(\hat{\pi}_{y,\text{mpu}}) = \left(\frac{N-n}{N-1}\right)\frac{\pi_y(1-\pi_y)}{n}$. Therefore, the relative efficiency of SRS with ratio estimation as compared to mean-per-unit estimation of a proportion is

$$\text{RE}(\hat{\pi}_{y,\text{rat}}, \hat{\pi}_{y,\text{mpu}}) \approx \frac{\left(\frac{N-n}{N-1}\right)\frac{\pi_y(1-\pi_y)}{n}}{\left(\frac{N-n}{N-1}\right)\left(\frac{\pi_y}{\pi_x}\right)\frac{1-\pi_y P_{\text{sens}}-(1-\pi_y)P_{\text{spec}}}{n}} = \frac{(1-\pi_y)\pi_x}{1-\pi_y P_{\text{sens}}-(1-\pi_y)P_{\text{spec}}} \quad (7.30)$$

Equation (7.30) can be used to determine those circumstances under which the ratio estimator of a proportion would outperform the mean-per-unit estimator of a proportion with simple random sampling. Figure 7.3 displays $\text{RE}(\hat{\pi}_{y,\text{rat}}, \hat{\pi}_{y,\text{mpu}})$ for a series of π_y values ranging from 0.05–0.95, plotted against P_{sens} for a range of P_{spec} values. Overall, the relative efficiency increases as the sensitivity and specificity are increased. Relative efficiencies greater than 1 require high levels of P_{sens} (> 0.7) for all values of π_y, and the minimum P_{sens} required to achieve RE > 1 increases with the value of π_y. The most favorable settings for ratio estimation are for "intermediate" π_y values (e.g., $0.3 < \pi_y < 0.6$). In this favorable setting, $\text{RE}(\hat{\pi}_{y,\text{rat}}, \hat{\pi}_{y,\text{mpu}})$ can often exceed 2. Application of ratio estimation for extremely small or extremely large proportions seems generally problematic, however. For example, for $\pi_y = 0.05$ [panel (a)], RE > 1.5 cannot be achieved unless $P_{\text{spec}} > 0.9$, regardless of the value of P_{sens}. For $\pi_y = 0.95$ [panel (f)], RE > 1.5 cannot be achieved unless P_{sens} exceeds 0.95, regardless of the value of P_{spec}. These insights into the relative efficiency of $\hat{\pi}_{y,\text{rat}}$ as an

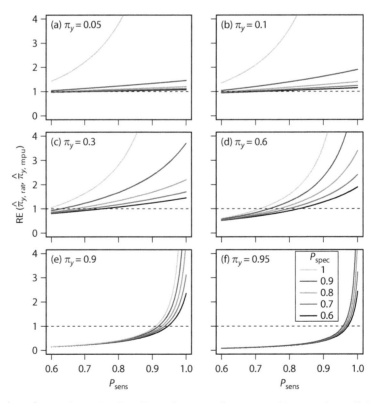

Fig. 7.3 $\text{RE}(\hat{\pi}_{y,\text{rat}}, \hat{\pi}_{y,\text{mpu}})$ [Equation (7.30)] as a function of sensitivity (P_{sens}) and specificity (P_{spec}) for $\pi_y = 0.05, 0.1, 0.3, 0.6, 0.9, 0.95$ (panels (a)–(f), respectively).

estimator of π_y would not have been possible if Equation (7.15) had not been re-expressed in terms of π_y, π_x, P_{spec} and P_{sens} [i.e., re-expressed as Equation (7.29)].

7.3 Ratio estimation with stratified sampling

Ratio estimation can be used in stratified sampling in two different ways. First, stratified estimates of T_y and T_x can be used to generate a single combined stratified estimate of $R = T_y/T_x$ which can in turn be used to estimate the population total, $T_y = RT_x$. Second, separate independent ratio estimates of stratum totals can be made in all strata and then these separate ratio estimates can be summed across strata to get an estimate of the population total, $T_y = \sum_{h=1}^{L} T_{yh}$.

7.3.1 Combined estimator

Label the strata from $h = 1$ through L, select independent SRS samples of size n_h from N_h within all strata, and define $T_x = \sum_{h=1}^{L} T_{xh}$ (the sum of the stratum-specific totals for x). Estimate the totals T_y and T_x using the ordinary stratified sampling procedures: $\hat{T}_{y,st} = \sum_{h=1}^{L} \hat{T}_{yh}$, and $\hat{T}_{x,st} = \sum_{h=1}^{L} \hat{T}_{xh}$, where $\hat{T}_{yh} = N_h \sum_{j\in S_h} y_{hj}/n_h$ and $\hat{T}_{xh} = N_h \sum_{j\in S_h} x_{hj}/n_h$.

The **combined stratified ratio estimator** of T_y is

$$\hat{T}_{y,comb.rat} = \hat{R}_{st} T_x = \frac{\hat{T}_{y,st}}{\hat{T}_{x,st}} T_x \tag{7.31}$$

where $\hat{R}_{st} = \hat{T}_{y,st}/\hat{T}_{x,st}$. Cochran (1977 Section 6.11) motivates the following approximation for sampling variance of $\hat{T}_{y,comb.rat}$, valid for large $n = \sum_{h=1}^{L} n_h$

$$V^*(\hat{T}_{y,comb.rat}) = \sum_{h=1}^{L} N_h^2 \left(\frac{N_h - n_h}{N_h}\right) \frac{R^2 \sigma_{xh}^2 + \sigma_{yh}^2 - 2RCor(x_h, y_h)\sigma_{yh}\sigma_{xh}}{n_h} \tag{7.32}$$

where σ_{yh}^2, σ_{xh}^2 and $Cor(x_h, y_h)$ are stratum-specific finite population variances and correlations. An estimate of $V^*(\hat{T}_{y,comb.rat})$ can be calculated by substituting $\hat{R}_{st}$ for R, and stratum-specific estimates of σ_{yh}^2, σ_{xh}^2, σ_{yh}, σ_{xh}, and $Cor(x_h, y_h)$ in Equation (7.32).

7.3.2 Separate estimator

For this approach, separate ratio estimates of T_{yh} are made in each of the L strata using

$$\hat{T}_{yh,rat} = \hat{R}_h T_{xh} = \frac{\sum\limits_{j\in S_h} y_{hj}}{\sum\limits_{j\in S_h} x_{hj}} T_{xh}$$

where $\hat{R}_h = \hat{T}_{yh}/\hat{T}_{xh}$ is an estimator of $R_h = T_{yh}/T_{xh}$. The **separate stratified ratio estimator** of T_y is then obtained by summing these stratum-specific estimates

$$\hat{T}_{y,sep.rat} = \sum_{h=1}^{L} \hat{T}_{yh,rat} \tag{7.33}$$

Because strata are sampled independently, we make the obvious extension of Equation (7.14) to express the approximate sampling variance within each strata and then sum across strata for the sampling variance of $\hat{T}_{y,sep.rat}$.

$$V^*(\hat{T}_{y,\text{sep.rat}}) = \sum_{h=1}^{L} V^*(\hat{T}_{yh,\text{rat}}) = \sum_{h=1}^{L} N_h^2 \left(\frac{N_h - n_h}{N_h} \right) \frac{R_h^2 \sigma_{xh}^2 + \sigma_{yh}^2 - 2R_h \text{Cor}(x_h, y_h)\sigma_{yh}\sigma_{xh}}{n_h}$$

(7.34)

An estimate of $V^*(\hat{T}_{y,\text{sep.rat}})$ can be calculated by substituting stratum-specific estimates of R_h, σ_{yh}^2, σ_{xh}^2, σ_{yh}, σ_{xh}, and $\text{Cor}(x_h, y_h)$ in Equation (7.34).

Note that Equations (7.32) and (7.34) differ only by the use of $R = \mu_y/\mu_x$ for the combined stratified ratio estimator as compared to the use of $R_h = \mu_{yh}/\mu_{xh}$ for the separate stratified ratio estimator.

7.3.3 *Choosing between combined and separate estimators*

If there is reason to believe that the $R_h = T_{yh}/T_{xh}$ are all very similar to $R = T_y/T_x$, then there is a clear preference for use of the combined stratified estimator because overall estimator bias should be small, as for an unstratified ratio estimator. If, however, there is reason to suspect that the R_h vary substantially across strata, then sampling variance of $\hat{T}_{y,\text{comb.st}}$ may be large. In this setting, the separate stratified ratio estimator would have smaller sampling variance. However, because the separate ratio estimator sums independent ratio estimates of T_{yh} across L strata, estimation bias, $B(\hat{T}_{yh,\text{rat}})$, also sums across strata. Thus, overall estimator bias, $B(\hat{T}_{y,\text{sep.rat}}) = \sum_{h=1}^{L} B(\hat{T}_{yh,\text{rat}})$, may make a substantial contribution to mean square error for the separate stratified ratio estimator unless n_h are large for all strata.

7.4 A model-based perspective

In this and the following section we introduce a model-based perspective on estimation in finite populations. Our objectives are threefold: (1) to introduce the model-based perspective, including derivation of best linear unbiased (BLU) estimators of linear model parameters, (2) to show how a prediction approach, relying on BLU estimators of linear model parameters, can be used to estimate finite population parameters, and (3) to provide insights into those situations where we might expect design-based ratio and regression estimators to perform well. Before working through this section, we recommend that readers review Sections A.3 and A.9.

From the model-based perspective, the sample units s and auxiliary variable values x are regarded as fixed, but the target variable values (lower case) y are regarded as realized outcomes of random variables (upper case) Y generated from some presumed stochastic process. That is, the $Y = y$ values observed for a set of units on a given occasion are measureable, but are just one realization of a random process that may have resulted in different Y values for those units. Population parameters, such as the population total, $T_y = \sum_{i=1}^{N} Y_i$, are therefore also regarded as random variables.

In contrast, in design-based sampling theory the variables x and y are both regarded as fixed (non-random) values associated with specific population units. The population parameters $T_y = \sum_{i=1}^{N} y_i$, $\mu_y = \pi_y = T_y/N$, $R = T_y/T_x$ are thus also regarded as fixed (and very *real*) and are the explicit targets of estimation. From the design-based perspective, the randomness that arises is due to the sample selection process that yields a random sample S. A design-based estimator, $\hat{\theta}$, is thus a random variable over the sample space induced by the probability sampling design that is imposed on the sampling frame.

Thus, the key distinction between the model-based and design-based perspectives is where the randomness originates from and how it propogates through to estimation uncertainty. In the model-based case, the target variable, Y_i, associated with population unit i, is a random variable. Given a particular *realization*, the target variable values for the s units are known (measureable), but to estimate the population mean or total the non-sample unit values for that realization must be *predicted* based on the presumptive model for Y (possibly linked to auxiliary variables), the parameters of which are estimated based on the s data. Uncertainity thus arises through model-based prediction errors for units not in the sample. In the design-based case, randomness arises soley from the *random selection* of units which leads to the variation in $\{\hat{\theta}(s), s \in \mathcal{S}\}$, i.e., sampling variance.

7.4.1 *Estimation of model parameters*

We begin this section with a brief treatment of an important class of linear estimators in parametric statistics termed **best linear unbiased (BLU)** estimators. A linear estimator of a model parameter θ is an estimator that can be expressed as $\hat{\theta}_Y = \sum_{i \in s} a_i Y_i$, where Y_i is a random variable associated with unit i and a_i is a fixed coefficient (constant). The particular value (estimate) that a BLU estimator takes on for a sample s depends on the realized values, $\{y_i, i \in s\}$, of the random variables $\{Y_i, i \in s\}$. For a set of realized values associated with sample s, the estimator generates an estimate $\hat{\theta}_y = \sum_{i \in s} a_i y_i$. The usual targets of model-based estimation are fixed *model parameters* (such as the slope of an assumed linear stochastic model) rather than population parameters (such as the mean or total of a target variable in a finite population).

We describe next a number of basic models for the population random variables $\{Y_i\}$ which assume that these variables are *independent*, and derive BLU estimators for the associated model parameters. BLU estimators are (a) linear, (b) *model-unbiased*, and (c) have minimum variance among all model-unbiased linear estimators for these parameters. These estimators are then used in Section 7.4.2 to develop BLU *predictors* of the finite population parameters. We use the notation $E_M(\hat{\theta})$ and $V_M(\hat{\theta})$ to refer, respectively, to the expectation and variance of an estimator or predictor $\hat{\theta}$ *with respect to the model* underlying the $\{Y_i\}$; not with respect to a probability sampling design. For other model-related random quantities, we continue to use the standard notation $E(\cdot)$ and $V(\cdot)$, but want to emphasize that these too refer to expectation and variance with respect to the model. We also emphasize that while all of the models include a quantity σ^2 in their specification, the definition of σ^2 varies among the models (Section 7.4.4), and differs from that for a finite population with *fixed* y values. For the models, σ^2 relates to the variance of a random variable Y, whereas for a finite population, σ^2 measures the actual variation in fixed y values among the population units.

Mean model

The simple mean model, for $i = 1, 2, \ldots, N$, can be expressed as

$$Y_i = \mu + \varepsilon_i, \quad E(\varepsilon_i) = 0, \quad V(\varepsilon_i) = \sigma^2, \quad \text{Cov}(\varepsilon_i, \varepsilon_j) = 0, \text{ for } i \neq j \qquad (7.35)$$

The ε_i term is a random disturbance that leads to Y_i values other than the expectation, μ. The variance of this disturbance term is $V(\varepsilon_i) = V(Y_i) = \sigma^2$, and independence of the $\{Y_i\}$ implies that $\text{Cov}(\varepsilon_i, \varepsilon_j) = 0$, for $i \neq j$. The model parameters μ and σ^2 are common for all of the $\{Y_i\}$.

Suppose that a sample s of size n is selected (unspecified selection process) and the realized values $\{y_i, i \in s\}$ are observed. What is the BLU estimator of the model parameter μ? From fundamental results concerning variances of linear combinations of random variables (Section A.3.5), we know that for a linear estimator $\hat{\mu}_Y = \sum_{i \in s} a_i Y_i$, if the $\{Y_i\}$ are independent random variables having a common variance σ^2, then $V_M(\hat{\mu}_Y) = V_M(\sum_{i \in s} a_i Y_i) = \sum_{i \in s} a_i^2 V_M(Y_i) = \sum_{i \in s} a_i^2 \sigma^2$. We also know that if $\hat{\mu}_Y$ is a BLU estimator then it must, by definition, be model-unbiased for μ: $E_M(\hat{\mu}_Y) = \mu$. Thus, $E_M(\hat{\mu}_Y) = E_M(\sum_{i \in s} a_i Y_i) = \sum_{i \in s} a_i E_M(Y_i) = \sum_{i \in s} a_i \mu = \mu \sum_{i \in s} a_i = \mu$, which implies that $\sum_{i \in s} a_i = 1$ for the mean model.

To find the BLU estimator for μ we need to find $\{a_i, i \in s\}$ such that $V(\hat{\mu}_Y) = \sum_{i \in s} a_i^2 \sigma^2$ is minimized subject to the constraint that $\sum_{i \in s} a_i = 1$. To do this, we form the Lagrange function (Section A.9) $\mathcal{L}(\{a_i\}, \lambda) = V(\hat{\mu}_Y) + \lambda(\sum_{i \in s} a_i - 1) = \sum_{i \in s} a_i^2 \sigma^2 + \lambda(\sum_{i \in s} a_i - 1)$, and partially differentiate $\mathcal{L}$ with respect to the n unknown coefficients $\{a_i\}$ and also λ giving

$$\frac{\partial \mathcal{L}}{\partial a_i} = 2a_i \sigma^2 + \lambda, \quad i \in s$$

$$\frac{\partial \mathcal{L}}{\partial \lambda} = \sum_{i \in s} a_i - 1$$

Setting the first equation equal to zero and solving for a_i gives $a_i = -\lambda/(2\sigma^2)$. Setting the second equation equal to zero reproduces the constraint, $\sum_{i \in s} a_i = 1$. Substituting the solution for a_i into the constraint gives $\sum_{i \in s} a_i = \sum_{i \in s} -\lambda/(2\sigma^2) = 1$, which implies that $\lambda = -2\sigma^2/n$. Substituting this solution for λ back into the solution for a_i gives $a_i = \frac{2\sigma^2/n}{2\sigma^2} = 1/n$. The BLU estimator for μ is thus $\hat{\mu}_Y = \sum_{i \in s} a_i Y_i = \sum_{i \in s} Y_i/n$, the sample mean. For a realized set of values $\{y_i, i \in s\}$, it yields the same result, $\hat{\mu}_y = \sum_{i \in s} y_i/n$, as the design-based estimator $\hat{\mu}_{y,\text{mpu}}$.

Linear models

BLU estimators can also be developed for parameters of more complex models, of course. In this section we develop estimators of parameters of linear (straight line) models relating (fixed) auxiliary and (random) target variables in a finite population of size N. A fairly general family of linear models for the relationship between the target and auxiliary variables for units $i = 1, 2, \ldots, N$, is

$$Y_i = \alpha + \beta x_i + \varepsilon_i, \quad E(\varepsilon_i) = 0, \quad V(\varepsilon_i) = \sigma^2 x_i^\gamma, \quad \text{Cov}(\varepsilon_i, \varepsilon_j) = 0, \text{ for } i \neq j \qquad (7.36)$$

where γ is a specified constant, often ranging from 0 to 2 in ecological settings. How the $\{x_i\}$ themselves came to be realized is left unspecified. The parameters for this model are α (the intercept), β (the slope), and σ^2 (a variance function scalar). The ε_i term is a random disturbance that leads to Y_i values other than its expectation, $\alpha + \beta x_i$. As for the mean model, it is assumed that $E(\varepsilon_i) = 0$ and $\text{Cov}(\varepsilon_i, \varepsilon_j) = 0$, for $i \neq j$.

The variance in Y_i given x_i, $V(\varepsilon_i)$, is a function of x_i and largely determines the form of the BLU estimators within this family of models. Figure 7.4 displays scatterplots of simulated (realized) (x, y) pairs generated from a range of linear models (within the model family) that might be conjectured for finite populations for which $N = 150$ (see Figure 7.4 caption and Section 7.5 for detailed explanation). Note that when $V(\varepsilon_i)$ is constant [$\gamma = 0$, panels (a) and (b)], the scatter of realized y values about the line is very consistent over the range of x, whereas the scatter increases with x for panels (c)–(f).

Zero-intercept model.

For a straight line model passing through the origin (sometimes called a *regression-through-the-origin* model), we set $\alpha = 0$ in Equation (7.36) so that $Y_i = \beta x_i + \varepsilon_i$. Even with this restriction, however, there are a variety of models possible depending on the specification

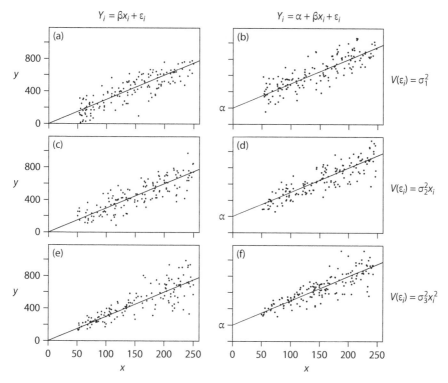

Fig. 7.4 Scatterplots of $N = 150$ simulated (x_i, y_i) pairs that are consistent with linear models of the form $Y_i = a + \beta x_i + \varepsilon_i$, where a and β are fixed parameters (intercept and slope, respectively), and ε_i is a random disturbance. Left column panels reflect *zero-intercept* models ($a = 0, \beta = 3$); right column panels reflect positive intercept models ($a = 200, \beta = 3$). For each panel the $\{x_i\}$ were generated from a continuous uniform(50,250) distribution, and ε_i was generated from a normal(0,$V(\varepsilon_i)$) distribution. For the top row panels $V(\varepsilon_i)$ is a constant, independent of x_i, with $V(\varepsilon_i) = \sigma_1^2 = 100^2$. For the remaining panels $V(\varepsilon_i)$ increases with x_i, and the corresponding σ^2 term was set so that $V(\varepsilon_i) = \sigma_1^2$ at $x_i = 150$ (the midpoint between 50 and 250). Thus, for the middle row panels $V(\varepsilon_i)$ is proportional to x_i, with $V(\varepsilon_i) = \sigma_2^2 x_i$ where $\sigma_2^2 = \sigma_1^2/150$. For the bottom row panels $V(\varepsilon_i)$ is proportional to x_i^2, with $V(\varepsilon_i) = \sigma_3^2 x_i^2$ where $\sigma_3^2 = \sigma_1^2/150^2$. Linear correlations between x and y are 0.90, 0.89, 0.78, 0.83, 0.75, and 0.73 for panels (a)–(f), respectively.

of γ in the variance function (Brewer 1963). Next we derive BLU estimators corresponding to three values of γ for this **zero-intercept model**: $\gamma = 0$, in which case $V(\varepsilon_i) = \sigma^2$ [Figure 7.4 panel (a)]; $\gamma = 1$, in which case $V(\varepsilon_i) = \sigma^2 x_i$ [Figure 7.4 panel (c)]; and $\gamma = 2$, in which case $V(\varepsilon_i) = \sigma^2 x_i^2$ [Figure 7.4 panel (e)].

Derivation of the BLU estimator for model parameter β closely follows the approach used in deriving $\hat{\mu}_Y$ for the mean model. In all cases the BLU estimator is of the form $\hat{\beta}_Y = \sum_{i \in s} a_i Y_i$ so that $V_M(\hat{\beta}_Y) = V_M(\sum_{i \in s} a_i Y_i) = \sum_{i \in s} a_i^2 V_M(Y_i) = \sum_{i \in s} a_i^2 \sigma^2 x_i^\gamma$. And, because a BLU estimator is unbiased, $E_M(\hat{\beta}_Y) = \beta$, so that $E_M(\hat{\beta}_Y) = E_M(\sum_{i \in s} a_i Y_i) = \sum_{i \in s} a_i E_M(Y_i) = \sum_{i \in s} a_i \beta x_i = \beta \sum_{i \in s} a_i x_i = \beta$, which implies that $\sum_{i \in s} a_i x_i = 1$ for the zero-intercept model. Forming the Lagrange function $\mathcal{L}(\{a_i\}, \lambda) = V(\hat{\beta}_Y) + \lambda(\sum_{i \in s} a_i x_i - 1) = \sum_{i \in s} a_i^2 \sigma^2 x_i^\gamma + \lambda(\sum_{i \in s} a_i x_i - 1)$, and partially differentiating $\mathcal{L}$ with respect to the $\{a_i\}$ and λ gives

$$\frac{\partial \mathcal{L}}{\partial a_i} = 2a_i\sigma^2 x_i^{\gamma} + \lambda x_i, \quad i \in s$$

$$\frac{\partial \mathcal{L}}{\partial \lambda} = \sum_{i \in s} a_i x_i - 1$$

Setting the first equation equal to zero and solving for a_i gives $a_i = -\lambda x_i^{1-\gamma}/(2\sigma^2)$. Setting the second equation equal to zero reproduces the constraint, $\sum_{i \in s} a_i x_i = 1$. Substituting the solution for a_i into the constraint gives $\lambda = -2\sigma^2/\sum_{i \in s} x_i^{2-\gamma}$, and substituting this solution for λ back into the solution for a_i gives $a_i = x_i^{1-\gamma}/\sum_{i \in s} x_i^{2-\gamma}$. The BLU estimator for β is thus of general form

$$\hat{\beta}_Y = \sum_{i \in s} a_i Y_i = \sum_{i \in s} \frac{x_i^{1-\gamma}}{\sum_{i \in s} x_i^{2-\gamma}} Y_i = \frac{\sum_{i \in s} x_i^{1-\gamma} Y_i}{\sum_{i \in s} x_i^{2-\gamma}} \qquad (7.37)$$

Case 1: $\gamma = 0$, $V(\varepsilon_i) = \sigma^2$. This is a model of a straight line through the origin with a constant variance in Y_i given x_i. Substituting $\gamma = 0$ into Equation (7.37), the BLU estimator for case 1 is

$$\hat{\beta}_{Y,1} = \frac{\sum_{i \in s} x_i Y_i}{\sum_{i \in s} x_i^2} \qquad (7.38)$$

$\hat{\beta}_{Y,1}$ may be recognized as the ordinary least-squares estimator of the slope of a "straight line through the origin with constant variance".

Case 2: $\gamma = 1$, $V(\varepsilon_i) = \sigma^2 x_i$. This is a model of a straight line through the origin with the variance in Y_i given x_i being proportional to x_i. Substituting $\gamma = 1$ into Equation (7.37), the BLU estimator for case 2 is

$$\hat{\beta}_{Y,2} = \frac{\sum_{i \in s} Y_i}{\sum_{i \in s} x_i} \qquad (7.39)$$

We refer to $\hat{\beta}_{Y,2}$ as the model-based *ratio-of-means estimator*. For a realized set of values $\{y_i, i \in s\}$, it yields the same sample result $\hat{\beta}_{y,2}$, as the design-based ratio estimator of R: $\hat{R}(s) = \sum_{i \in s} y_i / \sum_{i \in s} x_i$.

Case 3: $\gamma = 2$, $V(\varepsilon_i) = \sigma^2 x_i^2$. This is a model of a straight line through the origin with the variance in Y_i given x_i being proportional to x_i^2. Substituting $\gamma = 2$ into Equation (7.37), the BLU estimator for case 3 is

$$\hat{\beta}_{Y,3} = \frac{1}{n} \sum_{i \in s} \frac{Y_i}{x_i} \qquad (7.40)$$

We refer to $\hat{\beta}_{Y,3}$ as the model-based *mean-of-ratios estimator*. For a realized set of values $\{y_i, i \in s\}$, it yields the same sample result, $\hat{\beta}_{y,3}$, as the designed-based mean-of-ratios estimator of R: $\hat{R}_{\mathrm{mrat}}(s) = \frac{1}{n} \sum_{i \in s} y_i/x_i$. This design-based estimator is often considered and applied in forestry contexts (Schreuder et al. 1993 p. 89). Note that the mean-of-ratios estimator is not *finite population consistent* (Section A.7.4). That is, when y is regarded as a fixed variable value and $n = N$, $\hat{R}_{\mathrm{mrat}} = \frac{1}{N} \sum_{i=1}^{N} y_i/x_i$ which is not equal to the population ratio $R = T_y/T_x$.

For linear models of the form of Equation (7.36), an alternative way of deriving the BLU estimators (Valliant et al. 2000) is to minimize the weighted least-squares criterion, with weights inversely proportional to the $\{V(\varepsilon_i)\}$. For the zero-intercept model, the weighted sum of squared errors (SSE) criterion to be minimized is

$$SSE(\hat{\beta}) = \sum_{i \in s} w_i(Y_i - \hat{\beta}x_i)^2$$

where $w_i = 1/V(\varepsilon_i) = 1/(\sigma^2 x_i^{\gamma})$. Differentiating SSE with respect to $\hat{\beta}$, setting the result equal to zero, and solving for $\hat{\beta} = \hat{\beta}_Y$ gives, as we found previously,

$$\hat{\beta}_Y = \frac{\sum\limits_{i \in s} w_i x_i Y_i}{\sum\limits_{i \in s} w_i x_i^2} = \frac{\sum\limits_{i \in s} x_i^{1-\gamma} Y_i}{\sum\limits_{i \in s} x_i^{2-\gamma}} \tag{7.41}$$

Intercept model.

The model of a straight line with intercept is specified in Equation (7.36). Although the method of Lagrange multipliers could be used to derive the BLU estimators of α and β, it is easier in this case to minimize the weighted sum of squared errors criterion

$$SSE(\hat{\alpha}, \hat{\beta}) = \sum_{i \in s} w_i(Y_i - \hat{\alpha} - \hat{\beta}x_i)^2$$

where $w_i = 1/V(\varepsilon_i) = 1/(\sigma^2 x_i^{\gamma})$. The partial derivatives of SSE with respect to $\hat{\beta}$ and $\hat{\alpha}$ are

$$\frac{\partial SSE}{\partial \hat{\beta}} = -2\sum_{i \in s} w_i x_i Y_i + 2\hat{\alpha}\sum_{i \in s} w_i x_i + 2\hat{\beta}\sum_{i \in s} w_i x_i^2$$

$$\frac{\partial SSE}{\partial \hat{\alpha}} = -2\sum_{i \in s} w_i Y_i + 2\hat{\alpha}\sum_{i \in s} w_i + 2\hat{\beta}\sum_{i \in s} w_i x_i$$

Setting these partial derivatives equal to zero and solving for $\hat{\beta} = \hat{\beta}_Y$ and $\hat{\alpha} = \hat{\alpha}_Y$ gives (with a fair bit of algebra!)

$$\hat{\beta}_Y = \frac{\sum\limits_{i \in s} w_i x_i Y_i - \left(\sum\limits_{i \in s} w_i Y_i\right)\left(\sum\limits_{i \in s} w_i x_i\right) \Big/ \sum\limits_{i \in s} w_i}{\sum\limits_{i \in s} w_i x_i^2 - \left(\sum\limits_{i \in s} w_i x_i\right)^2 \Big/ \sum\limits_{i \in s} w_i} = \frac{\sum\limits_{i \in s} w_i(Y_i - \hat{\mu}_{wY})(x_i - \hat{\mu}_{wx})}{\sum\limits_{i \in s} w_i(x_i - \hat{\mu}_{wx})^2} \tag{7.42}$$

$$\hat{\alpha}_Y = \frac{\sum\limits_{i \in s} w_i Y_i - \hat{\beta}_Y \sum\limits_{i \in s} w_i x_i}{\sum\limits_{i \in s} w_i} = \hat{\mu}_{wY} - \hat{\beta}_Y \hat{\mu}_{wx} \tag{7.43}$$

where $\hat{\mu}_{wY} = (\sum_{i \in s} w_i Y_i)/(\sum_{i \in s} w_i)$ and $\hat{\mu}_{wx} = (\sum_{i \in s} w_i x_i)/(\sum_{i \in s} w_i)$ are the weighted sample means. For the Y_i constant variance case, $w_i = 1/V(\varepsilon_i) = 1/\sigma^2$, Equations (7.42) and (7.43) give the specific estimators

$$\hat{\beta}_{Y,4} = \frac{\sum\limits_{i \in s}(Y_i - \hat{\mu}_Y)(x_i - \hat{\mu}_x)}{\sum\limits_{i \in s}(x_i - \hat{\mu}_x)^2} \tag{7.44}$$

$$\hat{\alpha}_{Y,4} = \hat{\mu}_Y - \hat{\beta}_{Y,4}\hat{\mu}_x \tag{7.45}$$

These estimators may be recognized as the ordinary least-squares estimators of the slope and intercept, respectively, of a "straight line with constant variance". For a realized set of values $\{y_i, i \in s\}$, it yields the same sample result, $\hat{\beta}_{y,4}$, as the design-based regression estimator for β [Equation (7.4)].

7.4.2 Prediction of population parameters

A model-based approach may also be used for estimation of finite population parameters, not just for estimation of the assumed model's parameters. Royall and colleagues (e.g., Royall 1992, Royall and Herson 1973) have championed this approach which is described at length in Valliant et al. (2000). Again, with this approach the population values $y_1, y_2, \ldots, y_N$ are regarded as but one possible realization of a set of random variables, $Y_1, Y_2, \ldots, Y_N$. The population total is therefore also regarded as a random variable, $T_Y = \sum_{i=1}^{N} Y_i$. The objective however remains the same: estimation of the *realized* population total, $T_y = \sum_{i=1}^{N} y_i$, from a given sample s. A model is assumed for the joint distribution of $(Y_1, Y_2, \ldots, Y_N)$, and estimation performance (bias, variance) are defined with respect to all possible realizations of $(Y_1, Y_2, \ldots, Y_N)$ under the assumed model, *conditioned* on the set of selected units. Thus, a **model-based predictor** for a population parameter, for example the predictor $\hat{T}_Y$ [Equation (7.46)], is considered to be *model-unbiased* if $E_M(\hat{T}_Y) = E_M(T_Y)$.

An important, potentially controversial, feature of this model-based approach, one that stands in direct contrast to the design-based approach, is that the selection of units for sampling need *not* be at random given that the estimator properties are conditioned on the set of selected units. Indeed, this suggests that the use of purposive sampling might be advantageous (i.e., select a set of sample units for which the predictor properties are expected to be good under the model).

The model-based predictor of the population total has the following general form over the possible realizations of the $Y_1, Y_2, \ldots, Y_N$ for a fixed set of sample units s

$$\hat{T}_Y = \sum_{i \in s} Y_i + \sum_{i \notin s} \hat{Y}_i \tag{7.46}$$

where the $\{\hat{Y}_i, i \notin s\}$ are based on the observable $\{Y_i, i \in s\}$ and possibly some auxiliary information. Thus, the model-based prediction view of the estimation setting is this: given a model outcome, the realized $\{Y_i = y_i, i \in s\}$ will be known so that estimating the total only requires *prediction* of the remaining $\{Y_i, i \notin s\}$ realized values. We denote a realized value of T_Y by T_y, and a prediction of this realized total from a specific sample by $\hat{T}_y$.

The choice of which predictors to use for the $\{\hat{Y}_i, i \notin s\}$, depends on the assumed model, and on the estimator properties judged to be of greatest importance. For the family of linear models previously introduced, a natural choice would be for the predictors to be functions of the BLU estimators of the respective model parameters developed in Section 7.4.1. For the mean, ratio, and regression models, the predictors are based on these BLU estimators, and are in fact the **BLU predictors** for T_Y under their respective models (Valliant et al. 2000).

Mean model

For the mean model,

$$Y_i = \mu + \varepsilon_i, \quad E(\varepsilon_i) = 0, \quad V(\varepsilon_i) = \sigma^2, \quad \text{Cov}(\varepsilon_i, \varepsilon_j) = 0, \text{ for } i \neq j \tag{7.47}$$

Given $\{Y_i, i \in s\}$, the BLU estimator for μ is the sample mean, $\hat{\mu}_Y = \sum_{i \in s} Y_i/n$, and we predict Y_i for $i \notin s$ as $\hat{Y}_i = \hat{\mu}_Y$. The BLU predictor for T_Y is therefore

$$\hat{T}_{Y,m} = \sum_{i \in s} Y_i + \sum_{i \notin s} \hat{Y}_i = n\hat{\mu}_Y + (N - n)\hat{\mu}_Y = N\hat{\mu}_Y \tag{7.48}$$

For a realized set of values $\{y_i, i \in s\}$, it yields the same sample result,

$$\hat{T}_{y,m} = N\hat{\mu}_y \tag{7.49}$$

where $\hat{\mu}_y = \sum_{i \in s} y_i/n$, as the design-based estimator $\hat{T}_{y,\text{mpu}}$, Equation (3.9).

Ratio model

For the model of a straight line through the origin with the variance in Y_i proportional to x_i,

$$Y_i = \beta x_i + \varepsilon_i, \quad E(\varepsilon_i) = 0, \quad V(\varepsilon_i) = \sigma^2 x_i, \quad \text{Cov}(\varepsilon_i, \varepsilon_j) = 0, \text{ for } i \neq j \tag{7.50}$$

Given $\{Y_i, i \in s\}$ and $\{x_i, i = 1, 2, \dots, N\}$, the BLU estimator for β is the model-based ratio estimator, $\hat{\beta}_Y = \sum_{i \in s} Y_i / \sum_{i \in s} x_i = \hat{\mu}_Y/\hat{\mu}_x$, and we predict Y_i for $i \notin s$ as $\hat{Y}_i = \hat{\beta}_Y x_i$. The BLU predictor for T_Y is therefore

$$\hat{T}_{Y,r} = \sum_{i \in s} Y_i + \sum_{i \notin s} \hat{Y}_i = \hat{\beta}_Y \sum_{i \in s} x_i + \hat{\beta}_Y \sum_{i \notin s} x_i = \hat{\beta}_Y T_x \tag{7.51}$$

We refer to $\hat{T}_{Y,r}$ as the model-based *ratio predictor*. For a realized set of values $\{y_i, i \in s\}$, it yields the same sample result,

$$\hat{T}_{y,r} = \hat{\beta}_y T_x \tag{7.52}$$

where $\hat{\beta}_y = \sum_{i \in s} y_i / \sum_{i \in s} x_i = \hat{\mu}_y/\hat{\mu}_x$, as the design-based ratio estimator $\hat{T}_{y,\text{rat}} = \hat{R} T_x$, Equation (7.3).

Regression model

For the model of a straight line with intercept and constant variance in Y_i,

$$Y_i = \alpha + \beta x_i + \varepsilon_i, \quad E(\varepsilon_i) = 0, \quad V(\varepsilon_i) = \sigma^2, \quad \text{Cov}(\varepsilon_i, \varepsilon_j) = 0, \text{ for } i \neq j \tag{7.53}$$

Given $\{Y_i, i \in s\}$ and $\{x_i, i = 1, 2, \dots, N\}$, the BLU estimators for β and α are $\hat{\beta}_Y = \sum_{i \in s}(Y_i - \hat{\mu}_Y)(x_i - \hat{\mu}_x)/\sum_{i \in s}(x_i - \hat{\mu}_x)^2$ and $\hat{\alpha}_Y = \hat{\mu}_Y - \hat{\beta}_Y\hat{\mu}_x$, where $\hat{\mu}_Y = \sum_{i \in s} Y_i/n$, $\hat{\mu}_x = \sum_{i \in s} x_i/n$, and we predict Y_i for $i \notin s$ as $\hat{Y}_i = \hat{\alpha}_Y + \hat{\beta}_Y x_i = \hat{\mu}_Y + \hat{\beta}_Y(x_i - \hat{\mu}_x)$. The BLU predictor for T_Y is therefore

$$\hat{T}_{Y,l} = \sum_{i \in s} Y_i + \sum_{i \notin s} \hat{Y}_i = n\hat{\mu}_Y + \sum_{i \notin s}\left[\hat{\mu}_Y + \hat{\beta}_Y(x_i - \hat{\mu}_x)\right]$$

$$= n\hat{\mu}_Y + (N - n)(\hat{\mu}_Y - \hat{\beta}_Y\hat{\mu}_x) + \hat{\beta}_Y \sum_{i \notin s} x_i = N\left[\hat{\mu}_Y + \hat{\beta}_Y(\mu_x - \hat{\mu}_x)\right] \tag{7.54}$$

with $\mu_x = \sum_{i=1}^{N} x_i/N$, where the last equality follows from $\sum_{i \notin s} x_i = \sum_{i=1}^{N} x_i - \sum_{i \in s} x_i$. We refer to $\hat{T}_{Y,l}$ as the model-based *regression predictor*. For a realized set of values $\{y_i, i \in s\}$, it yields the same sample result,

$$\hat{T}_{y,l} = N\left[\hat{\mu}_y + \hat{\beta}_y(\mu_x - \hat{\mu}_x)\right] \tag{7.55}$$

where $\hat{\mu}_y = \sum_{i \in s} y_i/n$ and $\hat{\beta}_y = \sum_{i \in s}(y_i - \hat{\mu}_y)(x_i - \hat{\mu}_x)/\sum_{i \in s}(x_i - \hat{\mu}_x)^2$, as the design-based regression estimator $\hat{T}_{y,\text{reg}}$, Equation (7.6).

7.4.3 *Prediction error*

Although the (predictor, estimator) pairs $(\hat{T}_{Y,m}, \hat{T}_{y,mpu})$, $(\hat{T}_{Y,r}, \hat{T}_{y,rat})$, and $(\hat{T}_{Y,l}, \hat{T}_{y,reg})$ yield the same results for a realized/selected sample, it would be incorrect to therefore conclude that the paired items are "equivalent to one another". Because the model-based and design-based approaches have fundamentally different logical frameworks, their definition of prediction/estimation *error* and how it is characterized is also fundamentally different.

In the design-based realm, error is characterized in terms of estimation bias, $E(\hat{\theta}) - \theta$, and sampling variance, $E(\hat{\theta} - E(\hat{\theta}))^2$, or mean squared error, $E(\hat{\theta} - \theta)^2$, where θ is a fixed quantity and the expectation is taken over the sample space induced by the sampling design. In the model-based realm, error is characterized in terms of prediction bias, $E_M(\hat{\theta} - \theta)$, and prediction variance, $E_M(\hat{\theta} - E_M(\hat{\theta}))^2$, or prediction mean squared error, $E_M(\hat{\theta} - \theta)^2$, where θ is a random variable and the expectation is taken with respect to repeated outcomes under a presumed model given a fixed set of selected units. In this case, if the bias equals zero, the predictor is said to be *model-unbiased* and the prediction variance and mean squared error are equivalent.

Because the model-based concepts of prediction bias and prediction variance are conceptually so different from the design-based concepts of bias and sampling variance, we believe that it is instructive to elaborate further on the notion of prediction bias and prediction variance. First, it is critical to remember that the concepts of prediction bias and prediction variance are conditioned on a particular sample s of units from the finite population with its associated fixed values of x. Each outcome of the model-based prediction process consists of a realized set of y values associated with the n units in s, a realized set of y values associated with the $(N - n)$ units *not* in the sample, and an associated model-based prediction of the (population) total of these realized y values. Prediction bias is the average difference between the predicted total and the realized (but random) total, where the average is taken over an infinite number of possible realizations (of the presumed model). If this difference is equal to zero, then the predictor is model-unbiased. For prediction variance, we instead calculate the squared difference between the predicted total for a particular realization and the average predicted total (over an infinite number of possible realizations), and we average this squared difference over an infinite number of possible realizations. For a model-unbiased estimator, prediction variance is equivalent to the squared difference between the predicted total and the realized total averaged over an infinite number of possible realizations.

Next we assess prediction error for the BLU predictors of the finite population total associated with the mean, ratio, and regression models that were discussed in Section 7.4.2, namely Equations (7.48), (7.51), and (7.54). In doing this it is helpful to simplify the expression of error, $\hat{T}_Y - T_Y$, by eliminating the term $\sum_{i \in s} Y_i$ that is present in both of its components [Equation (7.46)]

$$\hat{T}_Y - T_Y = \sum_{i \notin s} \hat{Y}_i - \sum_{i \notin s} Y_i \tag{7.56}$$

Because $\hat{Y}_i$ is a function of the sample values $\{Y_i, i \in s\}$, these two sums are independent which means that the variance of the prediction error can be found by summing the variance of the two components if $\hat{T}_Y$ is model-unbiased. Also, the $\{Y_i\}$ are independent, so that the variance of the second sum is equal to the sum of the component variances.

Mean model

For the mean model $E(Y_i) = \mu$, $\hat{Y}_i = \hat{\mu}_Y$, and $E(\hat{\mu}_Y) = \mu$ (BLU estimator). Therefore, the error bias is

$$E_M(\hat{T}_{Y,m} - T) = \sum_{i \notin s} E(\hat{Y}_i) - \sum_{i \notin s} E(Y_i) = \sum_{i \notin s} \mu - \sum_{i \notin s} \mu = 0$$

so that $\hat{T}_{Y,m}$ is model-unbiased. For the error variance, $V(Y_i) = \sigma^2$ and $V(\hat{\mu}_Y) = V(\sum_{i \in s} Y_i/n) = \sum_{i \in s} V(Y_i)/n^2 = \sigma^2/n$. Therefore, the error variance is

$$V_M(\hat{T}_{Y,m} - T) = V\left(\sum_{i \notin s} \hat{Y}_i\right) + \sum_{i \notin s} V(Y_i) = (N-n)^2 V(\hat{\mu}_Y) + \sum_{i \notin s} V(Y_i)$$

$$= (N-n)^2 \frac{\sigma^2}{n} + (N-n)\sigma^2 = N^2\left(\frac{N-n}{N}\right)\frac{\sigma^2}{n} \qquad (7.57)$$

which has the same form as the design-based sampling variance $V(\hat{T}_{y,\text{mpu}})$ [Equation (3.16)], but σ^2 here is the variance of the random variables $\{Y_i\}$, not the finite population variance of the fixed values $\{y_i\}$.

Ratio model

For the ratio model $E(Y_i) = \beta x_i$, $\hat{Y}_i = \hat{\beta}_Y x_i$, and $E(\hat{\beta}_Y) = \beta$ (BLU estimator). Therefore, the error bias is

$$E_M(\hat{T}_{Y,r} - T) = \sum_{i \notin s} E(\hat{Y}_i) - \sum_{i \notin s} E(Y_i) = \sum_{i \notin s} \beta x_i - \sum_{i \notin s} \beta x_i = 0$$

so that $\hat{T}_{Y,r}$ is model-unbiased. For the error variance, $V(Y_i) = \sigma^2 x_i$ and $V(\hat{\beta}_Y) = V(\sum_{i \in s} Y_i/\sum_{i \in s} x_i) = \sum_{i \in s} V(Y_i)/(n\hat{\mu}_x)^2 = \sigma^2/(n\hat{\mu}_x)$, so that

$$V_M(\hat{T}_{Y,r} - T) = V\left(\sum_{i \notin s} \hat{Y}_i\right) + \sum_{i \notin s} V(Y_i) = \left[(N-n)\hat{\mu}_{x'}\right]^2 V(\hat{\beta}_Y) + \sum_{i \notin s} V(Y_i)$$

$$= \left[(N-n)\hat{\mu}_{x'}\right]^2 \frac{\sigma^2}{n\hat{\mu}_x} + \sigma^2(N-n)\hat{\mu}_{x'} = N^2\left(\frac{N-n}{N}\right)\frac{\sigma^2}{n}\left[\frac{\hat{\mu}_{x'}}{\hat{\mu}_x}\mu_x\right] \qquad (7.58)$$

where $\hat{\mu}_{x'} = \sum_{i \notin s} x_i/(N-n) = (N\mu_x - n\hat{\mu}_x)/(N-n)$, the mean value of x for the non-sampled units. This is similar, but not equivalent, to the design-based sampling variance approximation $N^2 V^*(\hat{\mu}_{y,\text{rat}})$ [Equation (7.15)]. Here the formula explicitly indicates that the prediction variance is at its minimum when the sample mean of the auxiliary variable, $\hat{\mu}_x$, is at its maximum. (When $\hat{\mu}_x$ is at its maximum, $\hat{\mu}_{x'}$ will be at its minimum, and their ratio would thus minimize the factor in square brackets.) Therefore, assuming the model were correct, one would minimize the error variance by purposively selecting the n population units that have the largest x values.

Regression model

For the regression model $E(Y_i) = \alpha + \beta x_i$, $\hat{Y}_i = \hat{\alpha}_Y + \hat{\beta}_Y x_i$, and $E(\hat{\alpha}_Y) = \alpha$, $E(\hat{\beta}_Y) = \beta$ (BLU estimators). Therefore, the error bias is

$$E_M(\hat{T}_{Y,l} - T) = \sum_{i \notin s} E(\hat{Y}_i) - \sum_{i \notin s} E(Y_i) = \sum_{i \notin s}(\alpha + \beta x_i) - \sum_{i \notin s}(\alpha + \beta x_i) = 0$$

so that $\hat{T}_{Y,1}$ is model-unbiased. The error variance is given by (Valliant et al. 2000)

$$V_M(\hat{T}_{Y,1} - T) = N^2 \left(\frac{N-n}{N}\right) \frac{\sigma^2}{n} \left[1 + \frac{(\hat{\mu}_x - \mu_x)^2}{\left(\frac{N-n}{N}\right) \frac{\sum_{i \in s}(x_i - \hat{\mu}_x)^2}{n}}\right] \tag{7.59}$$

which is different in form than the sampling variance approximation $N^2 V^*(\hat{\mu}_{y,\text{reg}})$ [Equation (7.16)]. In this case, the formula explicitly indicates that the error variance can be minimized by purposive selection of a sample for which $\hat{\mu}_x = \sum_{i \in s} x_i/n = \mu_x$. Note also that if $\hat{\mu}_x$ were close to but not exactly equal to μ_x, that it would be advantageous to maximize the spread of the sample values of x.

7.4.4 Prediction variance estimators

Estimators for the prediction error variance under the mean, ratio, and regression models can be obtained by substituting into Equations (7.57), (7.58), and (7.59), respectfully, an appropriate estimator for σ^2. All other quantities in these equations are assumed to be known (including, for the ratio and regression predictors, $\{x_i, i \in s\}$ and μ_x).

While the quantity σ^2 is a parameter in all of the models underlying these variance expressions, its definition is different for each of the models. For the mean model, it is defined as the variance of the random variable Y. For the ratio model, it is defined as the proportionality constant between the variance of the random variable Y given x and the value of x. And for the regression model, it is defined as the variance of the random variable Y given x. However, all three of these models are special cases of the model family described in Equation (7.36): for the mean model, $\alpha = \mu$, $\beta = 0$, and $\gamma = 0$; for the ratio model, $\alpha = 0$ and $\gamma = 1$; and for the regression model, $\gamma = 0$. In the following paragraph we motivate an estimator for σ^2 that is applicable to this model family as a whole, and thus to each of these three cases in particular.

From the model family specification, $V(Y_i) = \sigma^2 x_i^\gamma$, $i = 1, 2, \ldots, N$. Therefore, $V(Y_i/x_i^{\gamma/2}) = (1/x_i^\gamma)V(Y_i) = \sigma^2$. By the definition of variance then

$$\sigma^2 = E_M \left[\frac{Y_i}{x_i^{\gamma/2}} - E\left(\frac{Y_i}{x_i^{\gamma/2}}\right)\right]^2 = E_M \left[\frac{Y_i - E_M(Y_i)}{x_i^{\gamma/2}}\right]^2$$

which suggests the following estimator

$$\hat{\sigma}^2 = \hat{E}_M \left[\frac{Y_i - \hat{E}_M(Y_i)}{x_i^{\gamma/2}}\right]^2 = \frac{\sum_{i \in s}(y_i - \hat{y}_i)^2/x_i^\gamma}{v} \tag{7.60}$$

where $\hat{y}_i$ is the predicted value for the sample unit i observation (i.e., the value of the fitted model for unit i), and v is the degrees of freedom associated with the $\{\hat{y}_i\}$. The degrees of freedom for this model family is equal to the sample size n minus the number of estimated model parameters. Thus, for the mean and ratio models, $v = n - 1$, and for the regression model, $v = n - 2$. Therefore, for an estimator of the prediction error variance, substitute the appropriate values of γ and v into Equation (7.60), and then substitute the resulting $\hat{\sigma}^2$ for σ^2 in the corresponding error variance expression [Equation (7.57), (7.58), or (7.59)].

Note that with the mean model, for a realized set of values $\{y_i, i \in s\}$, $\hat{\sigma}^2 = \sum_{i \in s}(y_i - \hat{\mu}_y)^2/(n-1)$, and since $V_M(\hat{T}_{Y,\text{m}})$ is identical in form to $V(\hat{T}_{y,\text{mpu}})$, in this case the prediction- and design-based estimators would generate the same measure of estimator uncertainty from the given sample data. Calculated measures of uncertainty using

$V_M(\hat{T}_{Y,r})$ or $V_M(\hat{T}_{Y,l})$, however, are not equivalent to the design-based estimates of sampling variance. It is also important to recognize that these model-based estimators of prediction variance may be seriously off target is the assumed model is not supported by the observed x and y values.

It is critical, therefore, for model-based prediction to examine various model diagnostics to evaluate whether or not the assumed model is well-supported by the sample data. For example, linearity and the form of the variance function, including the specification of γ, could be addressed by examining the pattern and magnitude of the weighted residuals $[r_i = (y_i - \hat{y}_i)/\sqrt{x_i^{\gamma}}]$ across the range of x_i for $i \in s$. No such diagnostics are required for valid use of design-based estimators of sampling variance, but they may certainly be used to *assist* a design-based practitioner in selection of an estimator which should perform well.

7.5 Monte Carlo performance evaluation

In the preceding presentation of BLU estimators and model-based predictors, we identified some model settings where model-based predictors and design-based ratio and regression estimators would produce identical sample estimates. What are the theoretical implications of these identified correspondences between model-based and design-based estimators? If we return to the design-based setting where x and y are regarded as fixed unit variable values, and we assume that there is a strong positive correlation between x and y in a finite population, then we believe that at least two conclusions are warranted. First, performance of the design-based ratio estimator should be superior to any other linear estimator of T_y when (a) y is proportional to x, (b) the relation between x and y passes through the origin, and (c) the variation in y given x increases in proportion to x. Second, performance of the design-based regression estimator should be superior to any other linear estimator when (a) y is linearly related to x, (b) the relation between x and y has a non-zero intercept, and when (c) the variation in y given x is constant (unaffected by x). These conclusions, if correct, could be used to help us determine when to use ratio or regression estimators in design-based settings. Särndal et al. (1992) have referred to this kind of use of models—to help us understand the circumstances (relations between auxiliary and target variable values) under which certain design-based estimators may be expected to perform well or poorly, to help us choose among competing design-based estimators, or to help us build or imagine new design-based estimators—as *model-assisted estimation*. Inference (estimation and confidence interval construction) remains firmly in the design-based realm when we apply design-based ratio or regression estimators in real world settings, however, even when we may use models for guidance. Design-based estimation and inference make no model assumptions of relationships between auxiliary and target variables.

How well do these model-based theoretical conclusions apply to the actual performance of design-based ratio and regression estimators in finite populations? One approach for assessment of the relative performance of the design-based ratio and regression estimators would be to use Monte Carlo simulations to determine bias, sampling variance and mean square error of these design-based estimators as a function of sample size when they are applied to populations with known variable values that are based on assumed underlying models. For example, R code similar to that displayed immediately below was used to simulate the $N = 150$ population (x, y) values displayed in the Figure 7.4 panels. For each of the panels, $X_i \sim \text{uniform}(50, 250)$ and $(Y_i | X_i = x_i) = \alpha + \beta x_i + \varepsilon_i$, $i = 1, 2, \ldots, N$, with $\beta = 3$ and $\varepsilon_i \sim \text{normal}(0, V(\varepsilon_i))$ where $V(\varepsilon_i) = \sigma^2 x_i^{\gamma}$. The remaining parameters, α, σ^2, and γ varied among the panels. In the case of panel (b), $\alpha = 200$, $\sigma^2 = 100^2$, and $\gamma = 0$ as specified in the following.

```
N <- 150; alpha <- 200; beta <- 3; sigma2 <- 100^2; gamma <- 0
x <- runif(N, min=50, max=250)
V <- sigma2 * (x^gamma)
e <- rnorm(N, mean=0, sd=sqrt(V))
y <- alpha + (beta*x) + e
```

For the remaining panels, the first line of code is revised as necessary to specify the appropriate values of α, σ^2, and γ (see Figure 7.4 caption).

Once the population (x, y) values have been generated, then a large number, K, of independent SRS samples of size n are selected. For each such sample, estimates of T_y are calculated using $\hat{T}_{y,\text{rat}}$ and $\hat{T}_{y,\text{reg}}$, and then (approximate) estimator properties $[B(\hat{T}_y)$, $V(\hat{T}_y)$, and $\text{MSE}(\hat{T}_y)]$ are calculated over the set of K simulated samples. If K is sufficiently large, then the Monte Carlo calculated values will be very close to the actual values.

Results of such Monte Carlo simulations for the (x, y) population data shown in Figure 7.4, panels (b), (c), (e), and (f), are displayed in Figure 7.5. These results generally support the conclusions that emerged from our model-assisted consideration of BLU predictors. That is, the ratio estimator slightly outperformed [smaller $V(\hat{T}_y)$] the regression estimator for the zero intercept model with $V(\varepsilon)$ increasing with x [panel (c)], in which model setting the prediction-based ratio estimator is BLU; whereas the regression estimator outperformed the ratio estimator for the positive intercept model with constant variance [panel (b)] for which the prediction-based regression estimator is BLU. The pattern of

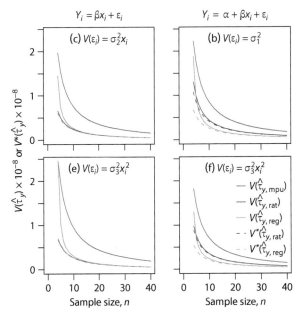

Fig. 7.5 Graphical illustration of the performance of ratio, regression and mean-per-unit estimators of T_y for four simulated populations of (x, y) data displayed in Figure 7.4: panels (c) and (e) (zero-intercept models) and panels (b) and (f) (models with intercept). Results for $V(\hat{T}_y)$ are based on simulated independent selection of 10^6 SRS samples of size $4 \le n \le 40$. Values for $V^*(\hat{T}_y) = N^2 V^*(\hat{\mu}_y)$ are based on Equations (7.15) and (7.16).

sampling variance with n for the ratio and regression estimators exhibited in the top row of panels was very similar to that exhibited in the bottom row of panels [(c) vs (e), (b) vs (f)] indicating that the form of $V(\varepsilon)$ was less influential than the presence of an intercept term in determining their relative performance. For all four populations, sampling variance for the ratio and regression estimators was considerably less than for $\hat{T}_{y,\mathrm{mpu}}$, so it was advantageous to incorporate the auxiliary variable in estimation.

For this simulation setting, one notable advantage of ratio estimation over regression estimation appears to be the more favorable performance at small sample sizes ($n < 10$). For these sample sizes the sampling variance of the regression estimator was unexpectedly high. Indeed, we omitted the $n = 3$ results from the figure panels because $V(\hat{T}_{y,\mathrm{reg}})$ greatly exceeded $V(\hat{T}_{y,\mathrm{mpu}})$ in this case. Moreover, $V^*(\hat{T}_{y,\mathrm{reg}})$ was substantially less than $V(\hat{T}_{y,\mathrm{reg}})$. In contrast, $V^*(\hat{T}_{y,\mathrm{rat}})$ was typically only slightly less than $V(\hat{T}_{y,\mathrm{rat}})$ at small sample sizes. Simulations also showed that the proportional bias, $B(\hat{T}_y)/T_y$, of the ratio and regression estimators was less than 1% at the smallest sample sizes, less than 0.5% for $n > 10$, and in all cases made a negligible contribution to $\mathrm{MSE}(\hat{T}_y)$. Finally, simulations showed that $E[\hat{V}(\hat{T}_{y,\mathrm{rat}})] \approx V^*(\hat{T}_{y,\mathrm{rat}})$, and $E[\hat{V}(\hat{T}_{y,\mathrm{reg}})] \approx V^*(\hat{T}_{y,\mathrm{reg}})$ at all sample sizes. Thus, the V^* expressions appear to provide close approximations to the sampling variance for $n > 10$ for both ratio and regression estimation, and for ratio estimation a good approximation to the sampling variance for even smaller sample sizes. Of course, all of these conclusions are predicated upon the population values considered. A more thorough evaluation would examine, for example, the relevance of the intercept and slope values to these findings.

These same Monte Carlo simulations can also be used to examine the shapes of the sampling distributions of the ratio and regression estimators as a function of sample size and the population (x, y) data generating model. For example, Figure 7.6 displays histograms of the sampling distributions of $\hat{T}_{y,\mathrm{rat}}$ and $\hat{T}_{y,\mathrm{reg}}$ for SRS samples of size $n = 5, 10, 20, 40$ that were selected from the Figure 7.4(b) population, which is the setting most favorable to regression estimation (linear model with intercept term and constant variance). In this case, non-normality (positive skewness) of the sampling distribution

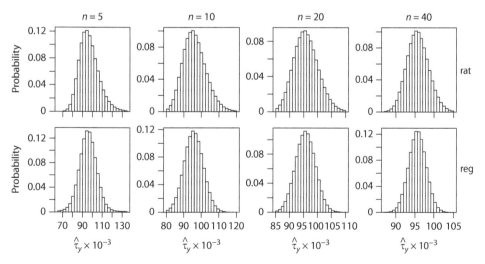

Fig. 7.6 Histograms of simulated sampling distributions of the ratio (top row) and regression (bottom row) estimators of T_y for SRS samples of size $n = 5, 10, 20, 40$ (columns) selected from the Figure 7.4(b) population. Results based on 10^6 independent sample selections for each n value.

of the ratio estimator is very noticeable for $n = 5$ and $n = 10$, slightly evident at $n = 20$, but essentially absent at $n = 40$. In contrast, sampling distributions for the regression estimator are approximately normal at all sample sizes. Thus, the apparent better small sample performance of $\hat{T}_{y,\text{rat}}$ compared to $\hat{T}_{y,\text{reg}}$ must be weighed against the impact of its distribution skewness on confidence interval coverage probabilities.

7.6 Mark-recapture estimation

In wildlife and fisheries work, model-based **mark-recapture estimation** is often used to estimate abundance of a species of interest, especially when the species is highly mobile and the locations of individuals (unlike trees, plants, mussels, etc.) are not fixed to specific locations. There is a very large body of theory and applications of mark-recapture estimation—Amstrup et al. (2005) is a good entry point—and we consider only the most elementary mark-recapture estimator, the Lincoln–Petersen estimator. Our intention here is to illustrate how the design-based perspective may sometimes inform variance estimation in a model-based setting.

We assume that, over the duration of a mark-recapture experiment, the population of interest is closed to recruitment (addition of new individuals). For the Lincoln–Petersen mark-recapture experiment, there is, in principle, a single mark and release sample, and a single subsequent recovery sample. Let M denote the number of individuals marked (e.g., with Passive Integrated Transponder, PIT, tags) in an initial random sample of individuals selected from a population of size N. These marked individuals are released back into the population following marking. Let R_M denote the number of previously marked individuals that are present in a subsequent random recovery sample of size C, where we assume that C is fixed and pre-specified. Given that a number of important assumptions are met (random mixing of the marked individuals within the population, no differential mortality between marked or unmarked individuals[3], no mark loss or errors of mark identification), it is reasonable to suppose that the fraction of marked fish in the recovery sample will, on average, equal M/N

$$\frac{E(R_M)}{C} = \frac{M}{N}$$

Solving this expression for N, and replacing $E(R_M)$ by the observed marked recoveries, R_M, leads to the **Lincoln–Peterson estimator**

$$\hat{N} = \frac{MC}{R_M} \tag{7.61}$$

Given M, and assuming that the recovery sample is an SRS selected from N, and assuming further that there is no mortality between marking and recovery, R_M is distributed as a hypergeometric random variable (Section A.4.6) with expectation $C(M/N)$ and variance

$$V(R_M) = C\left(\frac{M}{N}\right)\left(1 - \frac{M}{N}\right)\left(\frac{N-C}{N-1}\right) \tag{7.62}$$

Because M and C are fixed, the variance of the Lincoln–Peterson estimator of abundance is $V(\hat{N}) = (MC)^2 V(1/R_M)$, and the delta method approximation (Section A.8) for $V(1/R_M)$ is $(1/R_M^4)V(R_M)$, which gives

---

[3] If there is mortality over the course of the mark-recapture experiment, then the mark-recapture approach estimates population size *at the time of marking*, whereas if there is no mortality the population size at time of tagging and recovery will be identical.

$$V(\hat{N}) \approx \frac{(MC)^2}{E(R_M)^4} V(R_M) \tag{7.63}$$

An estimator of $V(\hat{N})$ can thus be formed by replacing $E(R_M)$ by the observed R_M, substituting Equation (7.62) for $V(R_M)$, and replacing N by $\hat{N}$ in Equation (7.62), giving (after some simplification) the expression (Everhart and Youngs 1975 p. 92)

$$\hat{V}(\hat{N}) = \hat{N}^2 \left(\frac{\hat{N}-M}{MC}\right)\left(\frac{\hat{N}-C}{\hat{N}-1}\right) \tag{7.64}$$

In actual practice, few single release mark-recapture experiments are actually carried out in a fashion that is accurately represented by the simple hypergeometric model. In most cases, after the initial mark and release, recovery samples are taken at many different locations and it is common practice to sum the total recoveries and recovery sample sizes (ideally until they reach the total fixed recovery sample size of C) across all locations. Jessen (1978) noted that in such instances the simple Lincoln–Peterson experiment is somewhat analogous to a design-based ratio estimator of a total. Let x denote the number of marked individuals in k independent and randomly selected without replacement recovery samples of size y, assume that $\sum_{i=1}^{k} y_i$ is fixed (and equal to C), and let T_x denote the number of individuals that are marked and released into the population. Using the ratio estimation approach

$$\hat{T}_{y,\text{rat}} = T_x \frac{\sum_{i=1}^{k} y_i}{\sum_{i=1}^{k} x_i} = T_x \hat{R} \tag{7.65}$$

where $\hat{R}$ is the usual ratio estimate, here equal to the reciprocal of the fraction in N that are marked. Equation (7.65) has the identical form as the Lincoln–Peterson estimator ($\hat{N} = \hat{T}_y$, $T_x = M$, $\sum y_i = C$, $\sum x_i = R_M$) and will therefore generate an identical estimate of abundance from sample data. Ignoring the finite population correction, sampling variance for the ratio estimator of abundance could be estimated using

$$\hat{V}(\hat{T}_{y,\text{rat}}) = T_x^2 \hat{V}(\hat{R}) = \frac{T_x^2}{n\hat{\mu}_x^2} \sum_{i=1}^{n} (y_i - \hat{R}x_i)^2/(n-1) \tag{7.66}$$

This estimator would arguably better represent the uncertainty in the mark-recapture estimate of abundance than the usual variance estimator which assumes, falsely, that the recovery sample consists of a single SRS is selected from the population. When the marked individuals are well mixed in the population and the recapture cluster samples are highly variable in size, then the ratio estimator of sampling variance may be less than what might be calculated using Equation (7.64) (which assumes the simple hypergeometric model), but when mixing of marked individuals is poor and the marked fraction varies substantially over the recapture samples, then use of Equation (7.66) may result in a much larger variance estimate than generated by Equation (7.64).

7.7 Chapter comments

When units are selected by SRS, designed-based ratio and regression estimators often have much smaller sampling variance than corresponding mean-per-unit estimators that do not incorporate auxiliary variable information. Though ratio and regression estimators have modest bias, especially for small n, the square of bias is usually small and makes a very small proportional contribution to mean square error. The sampling distributions of ratio and regression estimators are therefore often analogous to the precisely packed but slightly off target set of dart throws [see Figure 2.3 (b)] used in Chapter 2 to illustrate the basic properties of estimators.

Although simple and intuitive in structure, the designed-based ratio estimator of $R = T_y/T_x$ is identical in form to the BLU estimator of the slope of a zero-intercept straight line model relation between Y and x, where the variation in Y given x increases in proportion to x. Thus, in a design-based context, it is hard to beat the performance of the ratio estimator when auxiliary and target variables have this approximate relationship to one another. In our collective experiences in natural resources and environmental settings, we have found it to be quite common for the apparent relation between a target variable and an auxiliary variable to closely match the conditions for which the design-based ratio estimator is also BLU. We provide one such example setting in the next paragraph.

The common guppy (*Poecilia reticulata*) is an *ovoviviparous*, live-bearing fish. Eggs are internally fertilized by male sperm and develop internally, relying on egg yolk for nourishment, until the female gives birth to free-swimming young. Prior to giving birth, females can be dissected to count the number of developing embryos that will be released as young. The potential volume available for eggs to develop is related to the cube of the length of the female. When a female is small, the volume available for eggs to develop is strictly limited and so the average number of young is small and there are narrow limits to the numbers of possible eggs that can be fertilized, develop, and survive to birth. As females grow in length, the volume available for eggs becomes much larger (increasing with the cube of length), the expected number of eggs that are fertilized and survive to birth increases, and the range in the number of embryos that may survive to birth becomes larger as well. Thus, the reproductive biology of this fish species (and of many others species of fish) suggests, a priori, that a zero-intercept linear model linking embryo counts (y) to the cube of body length (x), with variation in y given x increasing with x, might be expected. This expectation seems supported by empirical observations (see Figure 7.7). We also note that in this instance a constant additive error structure seems biologically implausible since it could generate impossible negative embryo counts at small values of x.

The guppy embryo data plotted in Figure 7.7, along with knowledge of the fact that the ratio estimator is BLU for a zero-intercept linear model with variation in y proportional

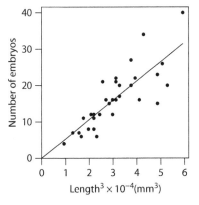

Fig. 7.7 Scatterplot of embryo counts and cube of female standard length (mm^3 x 10^{-4}) for adult female common guppies, *Poecilia reticulata*, reared in aquaria. Based on unpublished data used to generate Figure 11 in Hankin (1980).

to x, provides an appropriate context for model-assisted but design-based estimation: a model-based perspective is used to identify the "best" choice of a design-based sampling strategy. Figure 7.7 also illustrates how simple plots of preliminary data can assist the practitioner in choice of sampling strategy. For example, if a scatterplot of y against x from a preliminary survey indicated an "approximately-linear" relation between y and x but instead suggested a substantial positive intercept, then the regression estimator would be a better choice of design-based estimator than the ratio estimator.

It seems clear that the model-based perspective generates considerable insight into the probable design-based performance of the ratio and regression estimators. In a similar fashion, we believe that the design-based perspective may sometimes inform variance estimation in a model-based setting: the Lincoln–Petersen mark-recapture estimator is a case in point. In most applications of mark-recapture estimation to fish populations, it is clearly false to conjecture that the total recapture sample is a single SRS, thereby justifying a hypergeometric model assumption. Ratio estimation may in this case provide an alternative and probably more realistic perspective for variance estimation. Brewer (2002) is one of very few sampling theorists who has formally explored and promoted combining model-based and design-based inference, but we expect much greater interest in these ideas in the future.

Design-based ratio and regression estimation and linear model-based prediction can both be extended to multivariate auxiliary variables. For example, Cochran (1977 Section 6.20) considers the design-based multivariate ratio estimator, and Särndal et al. (1992 Section 6.4) considers a design-based general multivariate regression estimator that allows for unequal first order inclusion probabilities (see Chapter 8). Valliant et al. (2000) and Chambers and Clark (2012 Chapter 7) consider (model-based) general (multivariate) linear model prediction. It is important to remember, however, that the possibly superior performance (reduced sampling variance) of such multivariate estimators must more than offset the additional cost of obtaining measurements for multiple auxiliary variables, measurements which should ideally be available for all population units, not just for those that appear in a sample. This same caveat applies, of course, to use of a single auxiliary variable in simple ratio and regression estimators. In natural resources and environmental settings, auxiliary variables that are routinely available from remote sensing/mapping programs would seem potentially most cost-effective for inclusion in multivariate estimators. For example, assuming a terrestrial area frame, such auxiliary variables might include slope, aspect, percent cover, elevation, average annual temperature, and so on, all of which might be easily obtained at low cost and might logically be related to a target variable, e.g., productivity of a commercial tree species at a particular unit within an area grid. If productivity were closely tied to a linear function of auxiliary variables, then very accurate and cost-effective estimates of total productivity over an entire area frame might be generated from design-based multivariate estimators or from model-based multivariate prediction.

Problems

Problem 7.1. A habitat survey was executed in the upper reach of a small Oregon coastal stream. A visual estimate (x) of habitat unit area (m^2) was made in all $N = 134$ pools in a preliminary survey. Then a sample of $n = 20$ pool units was selected by SRS and a subsequent accurate measurement (y) of pool area was made in each of these units. Collected data are summarized in the following table.

x	y	x	y	x	y	x	y
58	57.7	92	82.6	137	160.2	44	46.4
29	30.2	31	33.0	39	40.1	160	186.2
78	101.7	181	179.7	22	22.2	33	39.4
45	40.2	35	34.9	72	104.0	19	24.3
159	185.2	169	178.9	174	160.1	161	190.3

Use ratio estimation and regression estimation to estimate the total area of all pools, and to calculate an associated standard error of estimation. Which of these two estimation methods do you feel is more appropriate in this case?

Problem 7.2. Suppose that you are given the following two sets of cross-classification data (collected from two different field locations in Washington State) that were generated for visual (x) and genetic (y) assignments of butterflies to the endangered Fender's Blue Butterfly species (1), *Icaricia icarioides fenderi*, or to the common Silvery Blue butterfly species (0), *Glaucopsyche lygdamus*. The two species co-occur, are difficult to visually distinguish with complete success, and genetic methods currently used to definitively identify species require dissection of a portion of the abdomen, resulting in death.

	Location 1			Location 2	
	y			y	
x	1	0	x	1	0
1	54	1	1	12	1
0	13	31	0	6	89

Assuming that genetic species assignments are 100% correct: (A) Calculate sensitivity and specificity at the two locations (see Section 7.2). (B) Assume that essentially all butterflies present at the two locations on a given day have been visually classified to species and that the visually estimated proportions (π_x) of Fender's Blue at locations 1 and 2 are 0.72 and 0.08, respectively. Calculate ratio estimates of the proportions of Fender's Blue butterfly at each site on this day and determine (at each site) if ratio estimation might provide a more accurate estimate of the species proportions than estimates based on genetic data alone. (Note: A more realistic two-phase setting is described in Problem 10.2.)

Problem 7.3. In R, execute the following code to generate a small population with associated values of y and x that should be generally favorable for application of ratio estimation:

```
set.seed(500); N <- 864
x <- round(runif(N, min=50, max=250))
e <- rnorm(N, mean=0, sd=sqrt(2*var(x)))
y <- round(50 + (3*x) + e)
```

Given these population (x,y) values, construct an R program that will allow you to solve this problem: (A) Construct all possible linear systematic samples, for $k = 72$, for (1) the initial ordering of units, and (2) instead sorting units in increasing order of the values of x. (B) For each possible systematic sample, and for both initial and sorted orderings of units, calculate (1) sample-specific estimates $\hat{T}_{lsys}$ and $\hat{T}_{rat}$, and (2) expected values and sampling variances of the two estimators. (C) For the two alternative orderings of units, compare the (actual) sampling variances generated in Part (B) with the "theoretical sampling variances" that can be calculated for $V(\hat{T}_{y,\text{lsys}})$ [from Equation (4.1)] and $V^*(\hat{T}_{y,\text{rat}})$ [from Equation (7.15)]. (D) Discuss your findings: Is there any apparent reduction in sampling variance of the ratio estimator that results from sorting on x prior to selecting systematic samples? Does the ratio estimator have any apparent substantial advantage over $\hat{\mu}_{\text{lsys}}$ when the population units are first sorted on x and sample selection is linear systematic?

Problem 7.4. Using the Section 7.1.2 example population ($N = 5$) variable values, as well as the entries in Table 7.1, show numerically that $\text{Cor}(\hat{\mu}_x, \hat{\mu}_y) = \text{Cor}(x, y)$.

Problem 7.5. Suppose that the assumed linear model relation between Y and x is of the form $Y = \alpha + \beta x + \varepsilon$, where $E(\varepsilon) = 0$ and $V(\varepsilon) = \sigma^2 x$, and that you are given the following sample data.

x:	47	140	13	150	83	120	136	78	93	71
y:	320	422	124	381	275	280	385	256	302	270

(A) What form do the BLU estimators of α and β [Equations (7.43) and (7.42), respectively] for the general linear model [Equation (7.36)] take when $V(\varepsilon) = \sigma^2 x$? (B) Given these sample data, use your estimators from Part (A) to calculate the BLU estimates of intercept and slope and graphically compare these with the ordinary least-squares (OLS) estimates of intercept and slope (i.e., BLU estimates assuming $V(\varepsilon) = \sigma^2$) by plotting the data along with the fitted lines based on the two sets of estimates.

Problem 7.6. In a reach of a headwater stream on the Olympic Penusula in Washington State, US Forest Service biologists made exhaustive searches of 12 shallow riffle habitat units (selected by SRS, $N = 126$) and counted the number of Pacific giant salamanders (aquatic stage), *Dicamptodon tenebrosus*, that were present in each unit (y). Assume that these counts are equal to the true numbers present in selected units. Biologists also made careful visual estimates (x) of the sizes of all N shallow riffle units, the total of which was 2,520 m^2. Collected sample data are summarized as (x,y) pairs in the following table.

x:	34	13	3	1	33	1	9	35	25	1	110	27
y:	31	4	2	1	9	1	7	6	8	1	54	28

(A) Calculate estimates of the total number of *D. tenebrosus* present in all shallow riffle units and associated estimates of sampling variance for $\hat{T}_{\text{mpu}}$ and $\hat{T}_{\text{rat}}$. (B) For $\hat{T}_{\text{mpu}}$ and $\hat{T}_{\text{rat}}$, respectively, estimate the sample size that would, in expectation, achieve $\text{CV}(\hat{T}) = 0.125$. Verify that your solutions achieve the desired CV and then round the sample sizes up to the nearest integer. (C) The sample sizes calculated in Part (B) are not directly comparable because generally $\hat{T}_{\text{mpu}} \neq \hat{T}_{\text{rat}}$. Construct a weighted estimate $\hat{T}_y = (w_1 \hat{T}_{\text{mpu}} + w_2 \hat{T}_{\text{rat}})/(w_1 + w_2)$, where $w_1 = 1/\hat{V}(\hat{T}_{\text{mpu}})$ and $w_2 = 1/\hat{V}(\hat{T}_{\text{rat}})$. Calculate new estimates of sample sizes required to achieve $\text{CV}(\hat{T}) = 0.125$. (D) Compare/discuss results for Parts (B) and (C).

Problem 7.7. A wildlife biologist wishes to estimate the total population of black-tailed jackrabbits (*Lepus californicus*) that are present on a large fenced military reservation in Arizona. An initial sample, $T_x = M = 1,245$, of these jackrabbits is collected in live-traps set at numerous locations throughout the reserve, marked with PIT tags, and released. Subsequent independent recovery samples, collected using live traps, are taken at $k = 20$ randomly selected locations throughout the reserve. Numbers of marked individuals, $\{R_i\}$, in the k recovery samples of size $\{C_i\}$ are provided in the following table (subscripts omitted)

R	C	R/C	R	C	R/C	R	C	R/C
4	36	0.1111	4	31	0.1290	6	55	0.1091
2	15	0.1333	5	48	0.1042	5	49	0.1020
7	75	0.0933	5	40	0.1250	4	54	0.0741
5	24	0.2083	6	60	0.1000	10	81	0.1235
4	33	0.1212	6	42	0.1429	6	72	0.0833
6	64	0.0938	3	27	0.1111	5	39	0.1282
3	39	0.0769	8	66	0.1212	–	–	–

(A) Calculate estimates of total jackrabbit abundance and the associated (large sample) 95% confidence intervals using (1) the Lincoln–Petersen estimator (assuming a simple hypergeometric model), and (2) the alternative design-based ratio estimation approach (Section 7.6). (B) Compare/discuss your results. (Note: you may find that a scatterplot of live-trap catches and associated marked recoveries is informative.)

Problem 7.8. Suppose that the target variable value *declines* linearly with the value of the auxiliary variable, as it does for the following R code:

```
set.seed(20); N <- 100
x <- runif(N, min=10, max=100)
e <- rnorm(N, mean=0, sd=35)
y <- 500 - (4*x) + e
```

Develop a simulation program in R that allows you to compare the performance (expected value, sampling varance) of the ratio and regression estimators of the mean of the target variable assuming $n = 12$ is selected by SRS from the population generated using the specified R code. Contrast the performance of these two estimators with $\hat{\mu}_{mpu}$ and discuss the relevance of your findings to the setting of the problem.

Problem 7.9. Regard the *y* values in a finite population as realizations of random variables *Y* that are linearly related to the auxiliary variable *x* according to a model which meets the assumptions of the model-based linear regression predictor of the finite population total. Use the following R code to generate a population of *x* values, and to select an SRS sample, *s*, of the population units.

```
set.seed(1500); n <- 20; N <- 200
x <- runif(N, min=40, max=400)
s <- sample(N, size=n, replace=FALSE)
```

For the assessment of model-based predictor performance that follows, recall that the sample units are considered *fixed* and that variability of estimates arises from the random *Y* values associated with the fixed *x* values attached to those units. That is, simulations must capture the range of possible values that the random variables Y can take on given the fixed x values for the fixed set of sample units s. Please assume the following for this problem:

```
a <- 50; b <- 2; sigma <- 10
e <- rnorm(n, mu=0, sd=sigma)
y <- a + (b*x[s]) + e
```

where a $= \alpha$, b $= \beta$, and sigma $= \sqrt{\sigma^2}$. (A) Construct an R program that allows you to compare the expression for the error variance of the model-based linear regression predictor of the population total, Equation (7.59), with the simulated error variance assuming that the linear model applies. (B) Calculate the expected value of the associated model-based estimator of the error variance by substituting into Equation (7.59) an appropriate estimator for σ^2 as discussed in Section 7.4.3. (C) Discuss your findings: (1) Are the BLU estimators of slope, intercept and σ^2 (model-) unbiased? (2) How does the theoretical error variance compare with the simulated error variance? (3) Is the error variance estimator an unbiased estimator of the simulated error variance?

Problem 7.10. Hartley and Ross (1954) proposed the folowing design-unbiased ratio estimator

$$\hat{T}_{y,\text{rat,HR}} = \bar{r}T_x + (\hat{\mu}_y - \bar{r}\hat{\mu}_x)[n(N-1)/(n-1)]$$

where $\bar{r} = (1/n)\sum_{i\in s} y_i/x_i$, the average ratio. (A) Use the Section 7.1.2 example population ($N = 5$) variable values with the $n = 3$ sample space presented in Table 7.1 to numerically verify that this estimator is unbiased for T_y. (B) Calculate the sampling variance of this estimator and compare it with the sampling variance and mean square error of the (biased) ratio estimator, $\hat{T}_{y,\text{rat}}$. (C) Is the Hartley-Ross estimator finite population consistent for the population ratio, $R = T_y/T_x$?

CHAPTER 8

Unequal probability sampling

Fig. 8.1 Young bighorn sheep, *Ovis canadensis*, in Yellowstone National Park. Large-scale programs for estimation of total abundance of this species might rely on unequal probability selection of regions for intensive estimation of local abundance based on region sizes. Photo credit: D. Hankin.

Up to this point in this text, population units or sampling frame units have always been selected with equal probabilities, either with or without replacement. In this chapter we will learn that equal probability sampling is actually a special case of a much more

Sampling Theory: For the Ecological and Natural Resource Sciences. David G. Hankin, Michael S. Mohr, and Ken B. Newman, Oxford University Press (2019). © David G. Hankin, Michael S. Mohr, and Ken B. Newman. DOI: 10.1093/oso/9780198815792.001.0001

general theory of **probability sampling** or **unequal probability sampling**. In this general framework, it is permissible and often desirable to select sample units (individual units or clusters of units) with *unequal* probabilities.

8.1 Unbiased ratio estimator

We begin our consideration of this more general theory not with a fully developed illustration of unequal probability sampling, but instead with a small but very clever "trick" that was apparently first suggested by Lahiri (1951)[1] and allows generation of an unbiased ratio estimator. Let $p_i = \Pr\{$select unit i on first draw$\}$. Suppose that we were to select the *first* sample unit, of a without replacement sample of size n, with unequal selection probabilities, $\{p_i, i = 1, 2, \ldots, N\}$, that are proportional to the size (or some other auxiliary variable) of the population units. That is, let $p_i = x_i / T_x$. Then, select the *remaining* $n - 1$ sample units with equal probability without replacement from the $N - 1$ population units that were not selected on the first draw (i.e., by SRS). If we were to then estimate the total for the target variable using the ratio estimator, $\hat{T}_{y,\text{rat}} = \hat{R} T_x$, we would find that we have an unbiased (ratio) estimator of the total! We'll use a simple numerical example to illustrate this fact based on the Table 8.1 population variables.

Applying Lahiri's method with $n = 2$ to the Table 8.1 population variables results in the Table 8.2 sample space, sample probabilities, and associated ratio estimates of R and T_y, denoted as $\hat{R}_\text{L}$, and $\hat{T}_{y,\text{L,rat}}$, respectively. If instead SRS had been used, then the sample probabilities would have all been equal, $\{p(s) = 1/6, s \in \mathcal{S}\}$, and we would have found that $E(\hat{T}_{y,\text{rat}}) = 27.25$, so that $B(\hat{T}_{y,\text{rat}}) = -0.75$. With the current method of selection, however, the $\{p(s)\}$ are not all equal (Table 8.2), and instead range from 3/30, if units 1 and 2 are selected, to 7/30 if units 3 and 4 are selected. If the first unit is selected with unequal probabilities $\{p_i\}$, we find that

$$E(\hat{T}_{y,\text{L,rat}}) = \sum_{s \in \mathcal{S}} \hat{T}_{y,\text{L,rat}}(s) p(s) = (23.333 \cdot 3/30) + \ldots + (30.0 \cdot 7/30) = 28 = T$$

Thus, the device of selecting the first sample unit with probabilities proportional to the x values causes the ratio estimator to be unbiased.

For $n = 2$, it is easy to calculate the probability of a sample selected using Lahiri's method, for example, $s = \{3, 4\}$. This sample can be selected in one of two ways: (1) unit 3 can be selected first, with probability $p_3 = 30/100$, and unit 4 selected second with probability 1/3 (i.e., with equal probability from the remaining $N - 1 = 3$ units), or (2) unit 4 can be selected first with probability $p_4 = 40/100$, and unit 3 selected second with probability 1/3. Thus, the probability of the *unordered* sample $s = \{3, 4\}$ is the sum of the probabilities

Table 8.1 Population variables and parameters used for illustration in Sections 8.1–8.3.4.

	Units (*i*)				Parameters		
Quantity	1	2	3	4	T	μ	σ^2
Target variable (*y*)	2	5	7	14	28	7	26
Auxiliary variable (*x*)	10	20	30	40	100	25	$166\frac{2}{3}$

[1] Rao (1966) notes that Lahiri (1951) first proved that the ratio estimator would be unbiased if samples were selected with $p(s)$ proportional to $\sum_{i \in s} x_i$. Shortly thereafter, Midzuno (1952) and Sen (1952) showed that this could be achieved by selecting the first unit with probabilities proportional to size and remaining units by SRS.

Table 8.2 Sample space for Lahiri's (1951) method with $n = 2$ selected from the Table 8.1 $N = 4$ population, sample probabilities, $p(s)$, and corresponding ratio estimates of the population ratio, $\hat{R}_L$, and the population total, $\hat{T}_{y,L,\mathrm{rat}}$. First sample unit is selected with probabilities $\{p_i = x_i/T_x\}$ and remaining sample units are selected by SRS. See text for calculation of $p(s)$.

ID	Sample Units (s)	$p(s)$	$\hat{\mu}_y(s)$	$\hat{\mu}_x(s)$	$\hat{R}_L(s)$	$\hat{T}_{y,L,\mathrm{rat}}(s)$
1	1, 2	3/30	3.5	15.0	0.2333	23.333
2	1, 3	4/30	4.5	20.0	0.2250	22.500
3	1, 4	5/30	8.0	25.0	0.3200	32.000
4	2, 3	5/30	6.0	25.0	0.2400	24.000
5	2, 4	6/30	9.5	30.0	0.3166	31.666
6	3, 4	7/30	10.5	35.0	0.3000	30.000
Sum:		1				163.500

of the two ordered selections: $p(\{3,4\}) = p_3(1/3) + p_4(1/3) = (30/100)(1/3) + (40/100)(1/3) = 7/30$. For $n = 2$ selected from $N = 4$

$$p(s) = \frac{\sum\limits_{i \in s} x_i}{T_x} \frac{1}{3}$$

whereas for the more general setting where $n \geq 2$, selected from N using Lahiri's method, the sample probabilities are

$$p(s) = \frac{\sum\limits_{i \in s} x_i}{T_x} \frac{1}{\binom{N-1}{n-1}}$$

The probability that any one of the units in the sample is selected on the first draw is $\sum_{i \in s} x_i/T_x$. Once the first unit has been selected, the probability of selecting the remaining $n - 1$ units in the sample is $1/\binom{N-1}{n-1}$, where $\binom{N-1}{n-1}$ gives the number of equal probability SRS samples of size $n - 1$ that can be selected from the $N - 1$ remaining units.

Unbiasedness of the ratio estimator when Lahiri's method for sample selection is used can be established as follows

$$E(\hat{T}_{y,L,\mathrm{rat}}) = E\left(T_x \frac{\sum\limits_{i \in S} y_i}{\sum\limits_{i \in S} x_i} \right) = \sum_{s \in S} \left[T_x \frac{\sum\limits_{i \in s} y_i}{\sum\limits_{i \in s} x_i} \right] p(s) = \sum_{s \in S} \left[T_x \frac{\sum\limits_{i \in s} y_i}{\sum\limits_{i \in s} x_i} \right] \left[\frac{\sum\limits_{i \in s} x_i}{T_x} \frac{1}{\binom{N-1}{n-1}} \right]$$

$$= \frac{1}{\binom{N-1}{n-1}} \sum_{s \in S} \sum_{i \in s} y_i = \frac{1}{\binom{N-1}{n-1}} \binom{N-1}{n-1} \sum_{i=1}^{N} y_i = \sum_{i=1}^{N} y_i = T_y$$

Convince yourself, using some simple algebraic examples (e.g., $n = 3$, $N = 5$), that $\sum_{s \in S} \sum_{i \in s} y_i = \binom{N-1}{n-1} \sum_{i=1}^{N} y_i$.

Imagine that a large sample is selected according to Lahiri's method. In this case, the estimator remains unbiased, but the sample probabilities will become very similar to one another because unequal probability is used to select only the first of a large number of units. This logic should make it obvious that the bias of the usual ratio estimator, with all units selected by SRS, must be small (near zero) in the case of a large sample. It should also be obvious that there would therefore be no real point to using Lahiri's selection method for ratio estimation unless sample size were quite small. Finally, it should be remembered that unbiasedness sometimes comes at a price. For the Table 8.2 numerical example, how would you calculate $V(\hat{T}_{y,\text{L,rat}})$? How does $V(\hat{T}_{y,\text{L,rat}})$ compare with $\text{MSE}(\hat{T}_{y,\text{rat}})$?

8.2 Sampling with replacement

Suppose that a with replacement sample of size n is selected, where at each draw the probabilities of unit selection are unequal, but fixed across the draws. Let $p_i = \Pr\{\text{select unit } i \text{ on a draw}\}$, $i = 1, 2, \ldots, N$, with $\sum_{i=1}^{N} p_i = 1$. Suppose further that p_i is proportional to an auxiliary variable x, so that $p_i = x_i/T_x$. Typically x is a measure of unit size, and thus this type of selection procedure is often referred to as **probability proportional to size with replacement (PPSWR)** sampling. The generic term **PPS** is used to indicate selection with probability proportional to size, whether selection is with or without replacement.

8.2.1 Hansen–Hurwitz estimator

Hansen and Hurwitz (1943) derived the following unbiased estimator of the target variable total when selection is by PPSWR

$$\hat{T}_{\text{HH}} = \frac{1}{n} \sum_{i \in S} \frac{y_i}{p_i} \tag{8.1}$$

Table 8.3 lists the PPSWR sample space, sample probabilities, and associated $\hat{T}_{\text{HH}}$ for samples of size $n = 2$ drawn from the Table 8.1 $N = 4$ population with $\{p_i = x_i/T_x\} = \{0.1, 0.2, 0.3, 0.4\}$, with $\sum_{i=1}^{N} p_i = 1$. Note that $E(\hat{T}_{\text{HH}}) = \sum_{s \in \mathcal{S}} \hat{T}_{\text{HH}}(s)p(s) = (0.20 \cdot 0.01) + \ldots + (29.166 \cdot 0.24) = 28 = T_y$, verifying numerically that $\hat{T}_{\text{HH}}$ is an unbiased estimator of the total for this example. Note also that the sample probabilities, $\{p(s)\}$, range from 0.01, for sample 1 (that contains unit 1 twice) to 0.24, for sample 10 (that contains the two units

Table 8.3 Sample space for $n = 2$ samples selected by PPSWR from the Table 8.1 $N = 4$ population with probabilities $\{p_i = x_i/T_x\} = \{0.1, 0.2, 0.3, 0.4\}$, sample probabilities, $p(s)$, and corresponding Hansen–Hurwitz estimates of the population total, $\hat{T}_{\text{HH}}$.

ID	Sample Units (s)	$p(s)$	$\hat{T}_{\text{HH}}(s)$	$(\hat{T}_{\text{HH}}(s) - T_y)^2$
1	1, 1	0.01	20.000	64.0000
2	2, 2	0.04	25.000	9.0000
3	3, 3	0.09	23.333	21.7777
4	4, 4	0.16	35.000	49.0000
5	1, 2	0.04	22.500	30.2500
6	1, 3	0.06	21.666	40.1111
7	1, 4	0.08	27.500	0.2500
8	2, 3	0.12	24.166	14.6944
9	2, 4	0.16	30.000	4.0000
10	3, 4	0.24	29.166	1.3611
Sum:		1.00		

with the largest auxiliary variable values), and that $\sum_{s \in \mathcal{S}} p(s) = 0.01 + \ldots + 0.24 = 1.00$. For $n = 2$, the sample probabilities are easily calculated. For a sample that consists of the same unit i selected twice, $p(s) = p_i^2$, as unit i must be selected on the first *and* on the second independent draws. For a sample that consists of distinct units i, j, the unordered sample can be selected in two ways: (1) unit i on the first draw and unit j on the second draw, or (2) unit j on the first draw and unit i on the second draw. Therefore, $p(s) = 2p_i p_j$ for a sample consisting of two distinct units i, j. Thus, for sample 3, $s = \{3, 3\}$, $p(s) = p_3^2 = 0.3^2 = 0.09$, and for sample 9, $s = \{2, 4\}$, $p(s) = 2p_2 p_4 = 2 \cdot 0.2 \cdot 0.4 = 0.16$.

8.2.2 Unbiasedness

Here we prove the unbiasedness of $\hat{T}_{HH}$ for arbitrary sample size, n, selected from a population of arbitrary size, N, with unit selection probabilities at each successive draw of $\{p_i, i = 1, 2, \ldots, N\}$, with $\sum_{i=1}^{N} p_i = 1$. First, we re-express the Hansen–Hurwitz estimator as a sum over the N population units rather than over the n units in the random sample S

$$\hat{T}_{HH} = \frac{1}{n} \sum_{i \in S} \frac{y_i}{p_i} = \frac{1}{n} \sum_{i=1}^{N} \frac{y_i}{p_i} Z_i$$

where Z_i denotes the number of times that unit i appears in the sample. For example, suppose that we had selected a PPSWR sample of size $n = 3$ from $N = 4$ and that $s = \{1, 3, 3\}$. In this case, $Z_1 = 1$, $Z_2 = 0$, $Z_3 = 2$, and $Z_4 = 0$. The Z values thus vary from sample to sample, while y_i and p_i, $i = 1, 2, \ldots, N$ remain fixed. Because the $\{p_i\}$ are fixed values, $Z = (Z_1, Z_2, \ldots, Z_N)$ is a multinomial$(n; p_1, p_2, \ldots, p_N)$ random variable (Section A.4.5) and as a result $E(Z_i) = np_i$ [Equation (A.101)] so that

$$E(\hat{T}_{HH}) = \frac{1}{n} \sum_{i=1}^{N} \frac{y_i}{p_i} E(Z_i) = \frac{1}{n} \sum_{i=1}^{N} \frac{y_i}{p_i} np_i = \sum_{i=1}^{N} y_i = T_y \tag{8.2}$$

The equal probability estimator $\hat{T}_{swr}$ is a special case of the Hansen–Hurwitz estimator, and therefore the unbiasedness of $\hat{T}_{swr}$ follows from the unbiasedness of the Hansen–Hurwitz estimator. Nevertheless, we demonstrate this directly as follows. In SWR, $p_i = 1/N$ for all units so that the Hansen–Hurwitz estimator becomes

$$\hat{T}_{HH,swr} = \frac{1}{n} \sum_{i \in S} \frac{y_i}{1/N} = \frac{N}{n} \sum_{i \in S} y_i = \hat{T}_{swr}$$

and the expected value is

$$E(\hat{T}_{HH,swr}) = \frac{N}{n} \sum_{i=1}^{N} y_i E(Z_i) = \frac{N}{n} \sum_{i=1}^{N} y_i \frac{n}{N} = \sum_{i=1}^{N} y_i = T_y$$

8.2.3 Sampling variance and variance estimation

The sampling variance of $\hat{T}_{HH}$ is

$$V(\hat{T}_{HH}) = \frac{1}{n} \sum_{i=1}^{N} p_i \left(\frac{y_i}{p_i} - T_y \right)^2 \tag{8.3}$$

and can be derived as follows. As in the previous section, the estimator can be re-expressed as a sum over the N population units rather than over the n units in the random sample. In this form, $\hat{T}_{HH}$ is a linear combination of the $\{Z_i\}$ random variables. The variance of a linear combination of random variables is a linear combination of the variances and covariances of the random variables [Equation (A.56)], and for the $\{Z_i\}$ random variables, $V(Z_i) = np_i(1 - p_i)$, and $\text{Cov}(Z_i, Z_j) = -np_ip_j, i \neq j$ [Equations (A.102) and (A.103)], and this leads to Equation (8.3). An unbiased estimator of $V(\hat{T}_{HH})$ that is always positive valued is

$$\hat{V}(\hat{T}_{HH}) = \frac{1}{n} \frac{\sum_{i \in S} \left(\frac{y_i}{p_i} - \hat{T}_{HH} \right)^2}{n - 1} \tag{8.4}$$

In Section 4.3.2, we presented two fundamental theorems that allowed unbiased estimation of sampling variance from interpentrating systematic samples. These same theorems are relevant in the context of $\hat{T}_{HH}$. Theorem 4.1 states that the arithmetic mean of two or more independent unbiased estimators, $\bar{\hat{\theta}}$, is also an unbiased estimator. We have just shown that $\hat{T}_{HH}$ is an unbiased estimator, and it can be viewed as a mean of n independent unbiased estimators, $\hat{\theta}_i = y_i/p_i, i = 1, 2, \ldots, n$. Theorem 4.2 provides a variance estimator for $\bar{\hat{\theta}}$

$$\hat{V}(\bar{\hat{\theta}}) = \frac{\sum_{i \in S}(\hat{\theta}_i - \bar{\hat{\theta}})^2}{n(n - 1)}$$

Making the substitutions $\bar{\hat{\theta}} = \hat{T}_{HH}$ and $\hat{\theta}_i = y_i/p_i$ in this equation gives Equation (8.4).

Use of the Hansen–Hurwitz estimator can sometimes produce sampling variances that are considerably less than what might otherwise be achieved with SRS and mean-per-unit estimation. For example, for the Table 8.1 population, $V(\hat{T}_{HH}) = 17.1666$, considerably less than $V(\hat{T}_{mpu}) = 208$, but slightly more than $V(\hat{T}_{y,\text{rat}}) = 16.3495$. The conditions under which one would expect $\hat{T}_{HH}$ to perform well are similar to those which cause the ratio estimator to perform well. Namely, if the relationship between y and x is linear, passes through the origin, and the linear correlation between the target and auxiliary variables is high and positive, then $\hat{T}_{HH}$ will perform well. In the extreme case where y is perfectly proportional to x, $y = \alpha x$, and in this case $p_i = x_i/\sum x_i = (y_i/\alpha)/\sum(y_i/\alpha) = y_i/\sum y_i = y_i/T_y$. Substituting for p_i would then give

$$\frac{y_i}{p_i} = \frac{y_i}{y_i/T_y} = T_y, \quad i = 1, 2, \ldots, N$$

Thus, for this "ideal" setting, $y_i/p_i = T_y$ for $i = 1, 2, \ldots, N$, so that $\hat{T}_{HH}$ would equal T_y, and the sampling variance of $\hat{T}_{HH}$ would be zero.

8.3 Sampling without replacement

In equal probability sampling, we concluded that it is less efficient to select units with replacement than without replacement unless $n = 1$. The same conclusion applies as a general "rule of thumb" to unequal probability sampling, but, as we will see, there can easily be exceptions to this rule. There are a very large number of published methods that have been developed to select unequal probability without replacement samples,[2] often referred to as **probability proportional to size without replacement (PPSWOR)** sampling.

---

[2] See Hanif and Brewer (1980) and Brewer and Hanif (1983) for reviews of 50 schemes as of 1980, and Sunter (1986) and Tillé (1996) for discussions of the fundamental difficulties in designing these schemes.)

We begin by exploring the performance of this approach when units are selected with **probabilities proportional to the sizes of remaining units** (Yates and Grundy 1953). We first illustrate the simple case of selecting $n = 2$ units without replacement. Let p_i = Pr{select unit i on first draw}, $i = 1, 2, \ldots, N$, with $p_i = x_i/T_x$, just as for unequal probability with replacement selection. After the first sample unit, unit i, has been selected, we then select the second sample unit, unit j, using *conditional* selection probabilities (Section A.2.7), $p_{j|i}$ = Pr{select unit j on second draw | unit i selected on first draw}, $j = 1, 2, \ldots, N, j \neq i$, with $p_{j|i} = x_j/(T_x - x_i)$. For example, for the Table 8.1 $N = 4$ population, suppose that unit 1 has been selected on the first draw with probability $p_1 = x_1/T_x = 10/100 = 1/10$. The conditional selection probabilities for units 2, 3, and 4 on the second draw would then be

$$p_{2|1} = x_2/(T_x - x_1) = 20/(100 - 10) = 2/9$$
$$p_{3|1} = x_3/(T_x - x_1) = 30/(100 - 10) = 3/9$$
$$p_{4|1} = x_4/(T_x - x_1) = 40/(100 - 10) = 4/9$$

There are two distinct ways of obtaining an $n = 2$ sample consisting of units i and j: (1) select unit i first and unit j second, or (2) select unit j first and unit i second. The probability that the first of these occurs is $p_i \cdot p_{j|i}$, and the probability that the second of these occurs is $p_j \cdot p_{i|j}$. Thus, the probability of obtaining the (unordered) sample $s = \{i, j\}$, is $p(s) = (p_i \cdot p_{j|i}) + (p_j \cdot p_{i|j})$. For example, for $s = \{2, 3\}$, $p(s) = (p_2 \cdot p_{3|2}) + (p_3 \cdot p_{2|3}) = (2/10)(3/8) + (3/10)(2/7) = (6/80) + (6/70) = 0.1607$.

8.3.1 *Horvitz–Thompson estimator*

Perhaps the most fundamental estimation equation in sampling theory is the Horvitz and Thompson (1952) estimator of the total for a sample selected with unequal probability without replacement

$$\hat{T}_{HT} = \sum_{i \in S} \frac{y_i}{\pi_i} \tag{8.5}$$

where $\pi_i, i = 1, 2, \ldots, N$, is the **first order inclusion probability** for unit i, defined as the probability that unit i is included in a random sample (Section A.6.3). The values of the $\{\pi_i\}$ depend on the scheme used to select the sample.

For the Table 8.1 $N = 4$ population, with $n = 2$ selected with probabilities proportional to the sizes, x, of remaining units, Table 8.4 lists the sample space, sample probabilities, and associated Horvitz–Thompson estimates of the population total. (Note that for sample 4, $s = \{2, 3\}$, $p(s) = 0.1607$ as was calculated previously.) Table 8.4 reports the sample-specific Horvitz–Thompson estimates of T_y, but these estimates require knowledge of the $\{\pi_i\}$. How are the $\{\pi_i\}$ values calculated? By definition, the probability that unit i is included in a sample of size n selected from N by PPSWOR is equal to the sum of the probabilities of selection of all of those samples that include unit i. Thus

$$\pi_1 = p(\{1, 2\}) + p(\{1, 3\}) + p(\{1, 4\}) = 0.0472 + 0.0762 + 0.1111 = 0.2345$$
$$\pi_2 = p(\{1, 2\}) + p(\{2, 3\}) + p(\{2, 4\}) = 0.0472 + 0.1607 + 0.2333 = 0.4413$$
$$\pi_3 = p(\{1, 3\}) + p(\{2, 3\}) + p(\{3, 4\}) = 0.0762 + 0.1607 + 0.3714 = 0.6083$$
$$\pi_4 = p(\{1, 4\}) + p(\{2, 4\}) + p(\{3, 4\}) = 0.1111 + 0.2333 + 0.3714 = 0.7159$$

Table 8.4 Sample space for PPSWOR $n = 2$ samples selected from the Table 8.1 $N = 4$ population, sample probabilities, $p(s)$, and corresponding Horvitz–Thompson estimates of the population total, $\hat{T}_{HT}$, when successive units are selected without replacement with probabilities proportional to the sizes, x, of remaining units.

ID	Sample Units (s)	$p(s)$	$\hat{T}_{HT}(s)$	$(\hat{T}_{HT}(s) - T_y)^2$
1	1, 2	0.0472	19.859	66.2783
2	1, 3	0.0762	20.035	63.4449
3	1, 4	0.1111	28.084	0.0071
4	2, 3	0.1607	22.838	26.6485
5	2, 4	0.2333	30.887	8.3375
6	3, 4	0.3714	31.063	9.3844
Sum:		1.0000		

Note that $\pi_1 + \pi_2 + \pi_3 + \pi_4 = 2 = n$. In general, $\sum_{i=1}^{N} \pi_i = E(n)$ [Equation (A.129)] so that for fixed sample size designs $\sum_{i=1}^{N} \pi_i = n$ [Equation (A.130)]. For this example, direct calculation of the expected value of $\hat{T}_{HT}$ over the sample space gives

$$E(\hat{T}_{HT}) = \sum_{s \in \mathcal{S}} \hat{T}_{HT}(s)p(s) = (19.859 \cdot 0.0472) + \ldots + (31.063 \cdot 0.3714) = 28 = T_y$$

8.3.2 Unbiasedness

Establishing the unbiasedness of the Horvitz–Thompson estimator is straightforward if we re-express the estimator as a sum over the N population units rather than over the n sample units by introducing the sample membership indicator random variables (Section A.6.4)

$$\hat{T}_{HT} = \sum_{i \in S} \frac{y_i}{\pi_i} = \sum_{i=1}^{N} \frac{y_i}{\pi_i} I_i$$

where $I_i = 1$ if $i \in S$, and $I_i = 0$ if $i \notin S$. The $\{I_i, i = 1, 2, \ldots, N\}$ are Bernoulli random variables (Section A.6.4), with $E(I_i) = \pi_i$ [Equation (A.126)], and as a result the expected value of $\hat{T}_{HT}$ is

$$E(\hat{T}_{HT}) = \sum_{i=1}^{N} \frac{y_i}{\pi_i} E(I_i) = \sum_{i=1}^{N} \frac{y_i}{\pi_i} \pi_i = \sum_{i=1}^{N} y_i = T_y$$

Alternatively, $E(\hat{T}_{HT})$ can be found directly over the sample space

$$E(\hat{T}_{HT}) = E\left(\sum_{i=1}^{N} \frac{y_i I_i}{\pi_i}\right) = \sum_{s \in \mathcal{S}} \sum_{i=1}^{N} \frac{y_i I_i(s)}{\pi_i} p(s) = \sum_{i=1}^{N} \frac{y_i}{\pi_i} \sum_{s \in \mathcal{S}} I_i(s)p(s) = \sum_{i=1}^{N} \frac{y_i}{\pi_i} E(I_i) = T_y$$

Just as we proved the unbiasedness of $\hat{T}_{swr}$ as a special case of Hansen–Hurwitz estimation, we can now do the same for $\hat{T}_{mpu}$. In SRS, $\pi_i = n/N$ for all units so that the Horvitz–Thompson estimator becomes

$$\hat{T}_{HT,srs} = \sum_{i \in S} \frac{y_i}{n/N} = \frac{N}{n} \sum_{i \in S} y_i = \hat{T}_{mpu}$$

and the expected value is

$$E(\hat{T}_{HT,srs}) = \frac{N}{n}\sum_{i=1}^{N} y_i E(I_i) = \frac{N}{n}\sum_{i=1}^{N} y_i \frac{n}{N} = \sum_{i=1}^{N} y_i = T_y$$

Unbiased estimation of the population mean or total is only possible for a **probability sampling design** (Särndal et al. 1992). A probability sampling design is one for which all $\pi_i > 0$. If one or more $\pi_i = 0$ then those units are unobservable, and consequently a design-based estimator of the population mean or total cannot possibly be unbiased.

8.3.3 *Sampling variance and variance estimation*

Continuing with the sample space listing in Table 8.4, we would calculate sampling variance of $\hat{T}_{HT}$ as

$$V(\hat{T}_{HT}) = \sum_{s\in\mathcal{S}} [\hat{T}_{HT}(s) - E(\hat{T}_{HT})]^2 p(s) = \sum_{s\in\mathcal{S}} [\hat{T}_{HT}(s) - T_y]^2 p(s)$$

$$= (66.2783 \cdot 0.0472) + \ldots + (9.3844 \cdot 0.3714) = 17.6783$$

There is a major surprise here, however. If we compare $V(\hat{T}_{HT})$ with $V(\hat{T}_{HH})$ $(= 17.1666)$ we find, for this small example population, that $V(\hat{T}_{HT}) > V(\hat{T}_{HH})$! How can it possibly be that a with replacement estimator outperforms a without replacement estimator with the same sample size? We will explore this matter shortly, when we contrast alternative sample selection schemes for implementing PPSWOR. Before we examine alternative methods for selecting PPSWOR samples, however, we need to examine expressions for the sampling variance of the Horvitz–Thompson estimator.

The **second order inclusion probability**, denoted by π_{ij}, is defined as the probability that two distinct units, $i \neq j$, appear together in a sample (Section A.6.3). For $n = 2$ and $i, j \in s$, $\pi_{ij} = p(s)$, but for larger sample sizes π_{ij} is conceptually equivalent to the sum of the probabilities of all of the samples that contain units i and j. For fixed sample size designs, the $\{\pi_{ij}\}$ satisfy the following relations [Equation (A.130)]

$$\sum_{\substack{j=1 \\ j\neq i}}^{N} \pi_{ij} = (n-1)\pi_i, \quad \text{and} \quad \sum_{i=1}^{N-1}\sum_{j>i}^{N} \pi_{ij} = \frac{1}{2}\sum_{i=1}^{N}\sum_{j\neq i}^{N} \pi_{ij} = n(n-1)/2 \tag{8.6}$$

These relations are useful for evaluating the correctness of often complex calculations of the $\{\pi_i\}$ and $\{\pi_{ij}\}$ for fixed n PPSWOR designs. If they are not satisfied, then there must be an error in the calculation algorithm.

The sampling variance of $\hat{T}_{HT}$ is actually easy to derive. When re-expressed as a sum over the N population units rather than over the n units in the random sample, the estimator is a linear combination of the sample membership indicator random variables, $I_i, i = 1, 2, \ldots, N$, so that its variance is given by [Equation (A.56)].

$$V(\hat{T}_{HT}) = V\left(\sum_{i=1}^{N} \frac{y_i}{\pi_i} I_i\right) = \sum_{i=1}^{N} \left(\frac{y_i}{\pi_i}\right)^2 V(I_i) + \sum_{i=1}^{N}\sum_{j\neq i}^{N} \left(\frac{y_i}{\pi_i}\right)\left(\frac{y_j}{\pi_j}\right)\text{Cov}(I_i, I_j)$$

The $\{I_i\}$ are Bernoulli random variables (Section A.6.4), and as a result $V(I_i) = \pi_i(1 - \pi_i)$ and $\text{Cov}(I_i, I_j) = \pi_{ij} - \pi_i\pi_j, i \neq j$ [Equations (A.127) and (A.128)]. Therefore, as Horvitz and Thompson (1952) found,

$$V_{HT}(\hat{T}_{HT}) = \sum_{i=1}^{N} \left(\frac{1 - \pi_i}{\pi_i} \right) y_i^2 + \sum_{i=1}^{N} \sum_{j \neq i}^{N} \left(\frac{\pi_{ij} - \pi_i \pi_j}{\pi_i \pi_j} \right) y_i y_j \tag{8.7}$$

Horvitz and Thompson (1952) also introduced the following unbiased estimator of $V_{HT}(\hat{T}_{HT})$, assuming that all $\pi_{ij} > 0$

$$\hat{V}_{HT}(\hat{T}_{HT}) = \sum_{i \in S} \left(\frac{1 - \pi_i}{\pi_i^2} \right) y_i^2 + \sum_{i \in S} \sum_{\substack{j \in S \\ j \neq i}} \left(\frac{\pi_{ij} - \pi_i \pi_j}{\pi_{ij} \pi_i \pi_j} \right) y_i y_j \tag{8.8}$$

Equations (8.7) and (8.8) are valid for fixed or random sample size designs (some PPSWOR selection methods do not guarantee a fixed sample size n).

For fixed sample size designs, Sen (1953) and Yates and Grundy (1953) independently derived an alternative expression for $V(\hat{T}_{HT})$, again assuming that all $\pi_{ij} > 0$

$$V_{SYG}(\hat{T}_{HT}) = \frac{1}{2} \sum_{i=1}^{N} \sum_{j \neq i}^{N} (\pi_i \pi_j - \pi_{ij}) \left(\frac{y_i}{\pi_i} - \frac{y_j}{\pi_j} \right)^2 \tag{8.9}$$

An unbiased estimator of $V_{SYG}(\hat{T}_{HT})$ is

$$\hat{V}_{SYG}(\hat{T}_{HT}) = \frac{1}{2} \sum_{i \in S} \sum_{\substack{j \in S \\ j \neq i}} \left(\frac{\pi_i \pi_j - \pi_{ij}}{\pi_{ij}} \right) \left(\frac{y_i}{\pi_i} - \frac{y_j}{\pi_j} \right)^2 \tag{8.10}$$

For fixed n, Equations (8.7) and (8.9) generate identical results. Equations (8.8) and (8.10) have identical expectations, but they generate different sample-specific estimates of sampling variance. If a PPSWOR selection method can guarantee that all $\pi_i \pi_j \geq \pi_{ij}$, then the Sen–Yates–Grundy variance estimator [Equation (8.10)] is guaranteed to be non-negative for all samples, but there is no such similar assurance for $\hat{V}_{HT}(\hat{T}_{HT})$ [Equation (8.8)] which may take on negative values, sometimes with non-negligible probability.

The form of the Sen–Yates–Grundy expression for sampling variance [Equation (8.9)], implies that sampling variance for $\hat{T}_{HT}$ will be zero when, *for all i,j*, (1) $\pi_i \pi_j = \pi_{ij}$, or (2) $y_i/\pi_i = y_j/\pi_j$. The second condition is analogous to the situation in PPSWR, where sampling variance will be zero whenever every y_i/p_i gives an identical result equal to the total, T_y. Thus, assuming that the target variable, y, is directly proportional to the auxiliary variable, x, the theoretically *ideal* sets of first and second order inclusion probabilities would have the following properties:

$\pi_i \pi_j \geq \pi_{ij}$ for all i,j [to ensure non-negativity of Equation (8.10)]

$\pi_{ij} > 0$ for all $i \neq j$ [to allow for unbiased variance estimation (see following paragraph)]

$\pi_i = nx_i/T_x$ for all i (to minimize sampling variance, assuming y is proportional to x).

Unbiased or nearly unbiased estimation of the sampling variance of an estimator of the population mean or total is only possible for a **measureable probability sampling design** (Särndal et al. 1992). A measureable probability sampling design is one for which all $\pi_i > 0$ *and* all $\pi_{ij} > 0$. A non-measureable design is therefore one for which there is at least one pair of population units that can never appear together in the same sample, as in the case of linear systematic sampling (see also Section 8.6). Unbiased or nearly unbiased sampling variance estimators generally lead to *consistent* estimation of the sampling variance, which is necessary to ensure that confidence intervals constructed for a population parameter in fact have, at least approximately, the nominally stated coverage probability (Section A.7.4).

Aronow and Samii (2013) show that the Horvitz–Thompson variance estimator [Equation (8.8)] will have positive bias whenever one or more $\pi_{ij} = 0$ if all y values are positive. (If some y values are non-negative and some are negative, then bias may be positive or negative.) Using methods similar to those used by Aronow and Samii (2013), it can be shown that the Sen–Yates–Grundy variance estimator [Equation (8.10)] will have negative bias whenever one or more $\pi_{ij} = 0$.

8.3.4 *Alternative selection methods*

In this Section we illustrate two very different approaches for selection of PPSWOR samples: (1) **draw-by-draw, all possible samples selection methods** and (2) **list-sequential selection methods**. For the draw-by-draw methods, all possible samples are generated and $\pi_i, i = 1, 2, \ldots, N$, is then calculated by summing the probabilities of all of the samples that contain unit i, and $\pi_{ij}, i \neq j$, is calculated by summing the probabilities of all of the samples that contain units i and j. Selecting units with probabilities proportional to the sizes of remaining units is an example of an all possible samples selection method. It is difficult to extend this kind of method to selection of large samples from a large population because the total number of possible samples, $\binom{N}{n}$, for which sample probabilities would need to be calculated, very quickly gets impractically large for all but very small populations and sample sizes. For example, even for a population of size $N = 40$ and a sample size of $n = 6$, there are $\binom{40}{6} = 3{,}838{,}380$ possible (unordered) samples, and for each of these samples there are $n! = 720$ distinct orders of selection, so that together 2,763,633,600 ordered samples would need to be evaluated (Section A.1.2).

List-sequential schemes can usually be extended to selection of large n from large N and typically rely upon some initial ordering of population units. Beginning at the top of this list, often with the unit having largest auxiliary variable value, some method (usually a very clever one) is used to determine whether or not this unit will be included, temporarily, in the sample. Once this choice has been made, the next unit is evaluated to determine whether or not it should be included, temporarily, in the sample, and, if so, whether or not it should replace one of the units previously included in the sample. These choices are all based on auxiliary variable values. This type of conceptual procedure proceeds through the listing of the N units until a sample of fixed size n has been selected. One important advantage of the list-sequential approach is that calculation of first and second order inclusion probabilities is based on explicit formulas and does not require summation of sample probabilities across all possible samples. The disadvantage of list-sequential schemes is that they sometimes require relatively complex computer alogrithms for implementation and they generally do not allow for calculation of the probability of an individual PPSWOR sample.

We explore two examples of list-sequential procedures that may be used for selecting large size PPSWOR samples from large populations. The first procedure was developed by Chao (1982) and has been the subject of more recent consideration (Sengupta 1989, Tillé 1996). Chao's scheme has nearly ideal properties for most sets of x variable values, but can sometimes produce some $\pi_{ij} = 0$ and therefore does not always generate a measurable design. The second scheme was developed by Sunter (1977) and is recommended by Särndal et al. (1992). Sunter's scheme has less desirable first order inclusion probabilities, but it is guaranteed to produce a measurable design and therefore always allows for unbiased estimation of samplng variance.

Chao's method

For Chao's (1982) procedure, the first order inclusion probabilities are directly proportional to x for all units for which $\pi_i < 1$ (some units may be selected with certainty); $\pi_i \pi_j > \pi_{ij}$ for all $i \neq j$; and usually, but not always, all $\pi_{ij} > 0$, thus usually generating a *measurable* design and nearly achieving the "ideal" $\{\pi_i\}$ according to theory.

For the small population that we have been using for illustration in this chapter (Table 8.1), with $n = 2$, Chao's method gives the ideal $\{\pi_i\}$: $\pi_1 = 0.2, \pi_2 = 0.4, \pi_3 = 0.6$, $\pi_4 = 0.8$. The sample space, sample probabilities, and associated Horvitz–Thompson estimates for PPSWOR when units are selected according to Chao's scheme are summarized in Table 8.5. Note that for $i, j \in s$, $p(s)$ in this example is equivalent to π_{ij}, that $\sum_{s \in S} p(s) = 1$, and that $\sum_{i=1}^{N} \sum_{j \neq i}^{N} \pi_{ij} = n(n-1) = 2$. Unbiasedness of $\hat{T}_{HT}$ is unaffected by the method whereby the PPSWOR sample is selected, but sampling variance is very much less for Chao's procedure (7.0185, calculated immediately below) than for selection with probabilities proportional to the sizes of remaining units (17.6783, calculated previously).

$$E(\hat{T}_{HT}) = \sum_{s \in S} \hat{T}_{HT}(s)p(s) = (22.500 \cdot 0.0444) + \ldots + (29.166 \cdot 0.4444) = 28 = T_y$$

$$V(\hat{T}_{HT}) = \sum_{s \in S} [\hat{T}_{HT}(s) - E(\hat{T}_{HT})]^2 p(s) = \sum_{s \in S} [\hat{T}_{HT}(s) - T_y]^2 p(s)$$

$$= (30.2500 \cdot 0.0444) + \ldots + (1.3611 \cdot 0.4444) = 7.0185$$

When units are selected with probabilities proportional to the sizes of remaining units, there is *distortion* of first order inclusion probabilities in the sense that the $\{\pi_i\}$ may be far from the ideal, direct proportionality with the auxiliary variable. For this example population, $\{\pi_i\} = \{0.2345, 0.4413, 0.6083, 0.7159\}$ when units were selected with probabilities proportional to the sizes of remaining units, as compared to the ideal values of $\{0.2, 0.4, 0.6, 0.8\}$, achieved when units were selected following Chao (1982).

Table 8.5 also lists the Horvitz–Thompson and Sen–Yates–Grundy sampling variance estimates for $\hat{T}_{HT}$ over the sample space using Chao's selection method. Recognizing that the $p(s)$ values in Table 8.5 are rounded, you can convince yourself that both estimators of sampling variance are unbiased for this numerical example and have expected value (approximately) equal to $V(\hat{T}_{HT}) = 7.0185$. The Sen–Yates–Grundy estimates of sampling

Table 8.5 Sample space for $n = 2$ samples selected by PPSWOR using Chao's (1982) method from the Table 8.1 $N = 4$ population, sample probabilities, $p(s)$, and corresponding Horvitz–Thompson estimates of the population total, $\hat{T}_{HT}$, along with sample variance estimates $\hat{V}_{HT}(\hat{T}_{HT})$ and $\hat{V}_{SYG}(\hat{T}_{HT})$.

ID	Sample Units (s)	$p(s)$	$\hat{T}_{HT}(s)$	$(\hat{T}_{HT}(s) - T_y)^2$	$\hat{V}_{HT}(\hat{T}_{HT})(s)$	$\hat{V}_{SYG}(\hat{T}_{HT})(s)$
1	1, 2	0.0444	22.500	30.2500	−26.2501	5.0000
2	1, 3	0.0666	21.666	40.1111	−52.2222	2.2222
3	1, 4	0.0888	27.500	0.2500	−138.7500	45.0000
4	2, 3	0.0888	24.166	14.6944	−347.6389	1.1806
5	2, 4	0.2666	30.000	4.0000	67.5000	5.0000
6	3, 4	0.4444	29.166	1.3611	83.0278	2.7222
Sum:		1.0000				

variance are, however, far more stable (less variable across samples) than the Horvitz–Thompson estimates of sampling variance and, just as important, none of the Sen–Yates–Grundy estimates take on negative values because Chao's selection scheme guarantees that $\pi_i\pi_j \geq \pi_{ij}$ for all $i \neq j$. For these reasons, when PPSWOR sample size is fixed, it is usually best to use to use Equation (8.10) for estimation of sampling variance. Calculating a negative estimate of sampling variance using Equation (8.8) can be an unnerving experience!

Sunter's method

Särndal et al. (1992 Section 3.6.2) recommend use of the list-sequential PPSWOR selection method proposed by Sunter (1977). This scheme guarantees that the $\{\pi_i\}$ are proportional to the $\{x_i\}$ for most units (those with the largest x values). For the remaining units, generally with small x values, the $\{\pi_i\}$ have the same value. Sunter's scheme guarantees that $\pi_i\pi_j > \pi_{ij}$ for all $i \neq j$, and that all $\pi_{ij} > 0$. Thus, use of Sunter's guarantees that the sampling design will be *measurable*, thereby allowing unbiased variance estimation.

Table 8.6 compares first order inclusion probabilities for Sunter (1977) and Chao (1982) PPSWOR selection methods for $N = 12$, $n = 4$, and $\{x_i\} = \{12, 11, 10, 9, 8, 7, 6, 5, 4, 3, 2, 1\}$. The $\{\pi_i\}$ are identical for the largest $\{x_i\}$ (units 1–5). Thereafter, the $\{\pi_i\}$ for Sunter's method are of equal value (0.2051) for all smaller $\{x_i\}$ values (units 6–12), whereas the proportionality of π_i to x_i is maintained for all x_i for Chao's method. For this particular set $\{x_i\}$, all $\pi_{ij} > 0$ for both PPSWOR selection methods, so use of either method would generate a measurable design and would allow unbiased and non-negative variance estimation using Equation (8.10).

8.3.5 *Strategy performance comparisons*

Although sampling strategies relying on unequal probability selection methods have substantial theoretical merit and can deliver a reduced sampling variance when compared to equal probability sampling strategies, it is important to recognize that they will not deliver superior performance in all settings. To illustrate this fact, sampling variances for estimators of the target variable total were compared for several alternative sampling strategies applied to three example populations of size $N = 50$ for sample sizes $4 \leq n \leq 30$. Values of the auxiliary variable, x, were generated as normal$(25, 7^2)$ random variables, and these realized values were identical for all three populations. Associated values of the target variable, y, were simulated as random variables according to three linear model relationships between Y and x, all of the form $Y_i = \alpha + \beta x_i + \varepsilon_i$, $i = 1, 2, \ldots, 50$, with

Table 8.6 Unit auxiliary values $\{x_i\}$ and corresponding first order selection probabilities, $\{\pi_i\}$, for a population of size $N = 12$ from which PPSWOR samples of size $n = 4$ are selected following Sunter (1977) and Chao (1982).

		π_i				π_i	
Unit (i)	x_i	Sunter	Chao	Unit (i)	x_i	Sunter	Chao
1	12	0.6154	0.6154	7	6	0.2051	0.3077
2	11	0.5641	0.5641	8	5	0.2051	0.2564
3	10	0.5128	0.5128	9	4	0.2051	0.2051
4	9	0.4615	0.4615	10	3	0.2051	0.1538
5	8	0.4103	0.4103	11	2	0.2051	0.1026
6	7	0.2051	0.3590	12	1	0.2051	0.0513

the ε_i generated from a normal$[0, V(\varepsilon_i)]$ distribution. The three models that generated Populations (a), (b), and (c) shown in the left column of Figure 8.2 were as follows

(a) $Y_i|x_i \sim \text{normal}(5x_i, 6^2 x_i)$

(b) $Y_i|x_i \sim \text{normal}(2x_i, 10^2)$

(c) $Y_i|x_i \sim \text{normal}(50 + 2x_i, 10^2)$

Thus, the settings for Populations (a) and (c) are most favorable for use of ratio and regression estimation, respectively, whereas the setting for Population (b) is an intermediary one.

Figure 8.2, right column, contrasts the performance (exact or simulation-based sampling variances) of five alternative sampling strategies for estimation of $\mathcal{T}_y$ for each of the

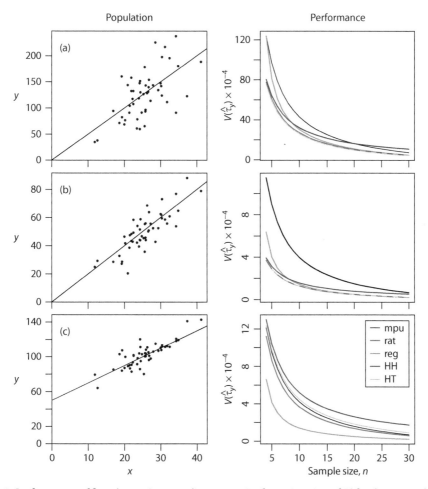

Fig. 8.2 Performance of five alternative sampling strategies for estimation of $\mathcal{T}_y$ for three populations of size $N = 50$. See text for description of the population variables (left column). Population-specific performance is displayed in adjacent panel (right column). $V(\hat{\mathcal{T}}_y)$ for mpu, HH, and HT estimation (using Chao's (1982) method of sample selection), are exact (computed by formula). $V(\hat{\mathcal{T}}_y)$ for ratio (rat) and regression (reg) estimation are approximate (based on 10^6 Monte Carlo simulations for each n).

three populations: SRS/mean-per-unit estimation, SRS/ratio estimation, SRS/regression estimation, PPSWR/HH estimation, and PPSWOR/HT estimation (using Chao's (1982) method of sample selection). For brevity, in the remainder of this section we refer to these five strategies simply as mpu, ratio, regression, HH, and HT estimation, respectively, with the design component left implicit.

For Population (a), the setting most favorable for ratio estimation, ratio and HT estimation deliver nearly identical sampling variances, much lower than for mpu estimation, at all sample sizes. Sampling variance for regression estimation is comparable to that for mpu estimation at $n = 4$, but gradually converges towards that for ratio and HT estimation at the larger sample sizes. Sampling variance for HH estimation is similar to that of ratio and HT estimation at small sample sizes, but its rate of improvement with n is relatively slow (due to with replacement sampling) such that for $n > 20$ its sampling variance actually exceeds that of all other strategies evaluated.

For Population (b), the patterns of relative performance are the same as for Population (a), except that the sampling variance for regression estimation converges much more quickly to that for ratio and HT estimation, and the sampling variance for HH estimation more closely tracks that of ratio and HT estimation over the range of n examined.

For Population (c), the setting most favorable for regression estimation, regression estimation clearly outperforms all other strategies over the range of n evaluated. The sampling variances for ratio, mpu, and HT estimation track each other fairly closely, while the sampling variance for HH estimation is similar for $n \leq 6$, but the rate of improvement is slower for $n > 6$.

Thus, HT estimation performed as well as ratio estimation for Populations (a) and (b), but performed no better than mpu estimation did for Population (c). And, for $n \geq 8$, HH estimation outperformed only mpu estimation for Population (b) and, for $n < 20$, Population (a), otherwise its performance was inferior to the alternative sampling strategies considered.

Visual inspection of Figure 8.2 does suggest some generalizations regarding the use of HH estimation as compared to HT estimation. Generally, HH estimation should only be considered when sample sizes are relatively small compared to the size of the population. When N is large and sample size is small, the performance of HH estimation will be similar to HT estimation, and implementation of HH estimation will be much simpler than implementation of HT estimation.

It is much more difficult to develop general recommendations for use of HH or HT estimation as compared to ratio or regression estimation, however. For the three sets of population (x, y) values used to construct Figure 8.2, HT estimation performed no better than ratio estimation, and for Population (c) was strongly outperformed by regression estimation.[3] However, in other settings, HT estimation may outperform alternative strategies. For example, Figure 8.3 compares performances of the same set of alternative sampling strategies for estimation of T_y on a set of population variable values consisting of the abundance, y, of juvenile coho salmon (*Oncorhynchus kisutch*) residing in $N = 50$ stream pools of various sizes, x, in Knowles Creek on the Oregon coast. HT estimation is the clear "winner" in this setting.

---

[3] The superior performance of the regression estimator over the HT estimator when the underlying model is a regression model is not surprising given that the inclusion probabilities for HT are roughly proportional to x, but y is not proportional to x since the assumed linear model is $E(Y_i) = 50 + 2x_i$.

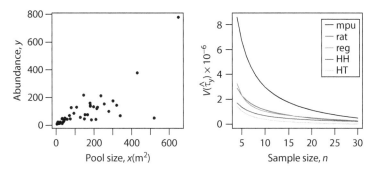

Fig. 8.3 Performance of five alternative sampling strategies for estimation of $\mathcal{T}_y$ for the population variables shown in the left panel, consisting of the abundance, y, of juvenile coho salmon (*Oncorhynchus kisutch*) residing in $N = 50$ stream pools of various sizes, x, in Knowles Creek on the Oregon coast. Variable values sourced from Hankin (1984 Table 2). $V(\hat{\mathcal{T}}_y)$ for mpu, HH, and HT estimation (using Chao's (1982) method of sample selection), are exact (computed by formula). $V(\hat{\mathcal{T}}_y)$ for ratio (rat) and regression (reg) estimation are approximate (based on 10^6 Monte Carlo simulations for each n).

8.3.6 *Survey cost comparisons*

Assessing the relative performance of alternative sampling strategies is further complicated by the possible differential survey costs of strategies that rely on equal or unequal probability sampling. Assuming that x is a measure of unit size, the expected size of a sampled unit is greater for a PPS design than for an equal probability design, because larger units have higher probabilities of selection. Therefore, if the cost of surveying a unit increases with the size of a unit, then the expected total survey costs will be greater for a PPS design than for an equal probability design of the same sample size. In this section we explore these expected costs and cost differentials.

We begin by supposing that the total cost, C, of a realized survey would be equal to the sum of a size-independent component and a size-dependent component

$$C = c_1 n + c_2 \sum_{i \in S} x_i \tag{8.11}$$

where c_1 is the size-independent related cost per sampled unit, and c_2 is the cost per m^2, say, of unit size (e.g., x_i = area of unit i in m^2). Given n, the *expected* total cost of a survey for planning purposes would then be

$$E(C) = c_1 n + c_2 E\left(\sum_{i \in S} x_i\right) \tag{8.12}$$

The expected total size of the selected units can be re-expressed as a sum over the N population units rather than over the n sample units as

$$E\left(\sum_{i \in S} x_i\right) = E\left(\sum_{i=1}^{N} x_i Z_i\right) = \sum_{i=1}^{N} x_i E(Z_i)$$

where the random variable Z_i is the number of times unit i appears in the sample S. Thus

$$E(C) = c_1 n + c_2 \sum_{i=1}^{N} x_i E(Z_i) \tag{8.13}$$

For without replacement designs, an individual unit i may be selected at most one time, $Z_i = 0$ or 1, and $Z_i \sim$ bernoulli(π_i) with $E(Z_i) = \pi_i$ (Sections A.6.4 and 8.3.2). For with replacement designs, an individual unit i may be selected multiple times, $0 \le Z_i \le n$, and $Z_i \sim$ binomial(n, p_i) with $E(Z_i) = np_i$ (Sections A.4.4 and 8.2.2).

For SRS, $\pi_i = n/N$, thus $E(Z_i) = n/N$, while for SWR, $p_i = 1/N$, so that $\pi_i = np_i$, and again $E(Z_i) = n/N$. Thus, the expected total cost of an equal probability SRS or SWR survey (C_e) would be

$$E(C_e) = c_1 n + c_2 \sum_{i=1}^{N} x_i \frac{n}{N} = c_1 n + c_2 n \mu_x$$

For PPSWOR with fixed n, a selection scheme generating the *ideal* π_i would have $E(Z_i) = \pi_i = nx_i/T_x$. For PPSWR, $p_i = x_i/T_x$, and so $E(Z_i) = np_i = nx_i/T_x$, the same as for PPSWOR. Substituting for $E(Z_i)$ in Equation (8.13) thus gives the expected total cost (C_u) of an unequal probability survey

$$E(C_u) = c_1 n + c_2 \sum_{i=1}^{N} x_i \frac{nx_i}{T_x} = c_1 n + c_2 \frac{nN}{T_x}\left(\frac{\sum_{i=1}^{N} x_i^2}{N}\right) = c_1 n + c_2 \frac{n}{\mu_x}\left[\mu_x^2 + V(x)\right]$$

$$= c_1 n + c_2 n \mu_x \left[1 + CV^2(x)\right]$$

where the third equality follows from the finite population relation $V(x) = (\sum_{i=1}^{N} x_i^2)/N - \mu_x^2$. and the fourth equality follows from $CV^2(x) = V(x)/\mu_x^2$. Thus, for these PPS designs the size-related expected costs are increased over those for an equal probability design by the factor $1 + CV^2(x)$.

Denote by λ the proportion of total survey cost of an equal probability survey, $E(C_e)$, that is size-related

$$\lambda = \frac{c_2 n \mu_x}{E(C_e)} = \frac{c_2 n \mu_x}{c_1 n + c_2 n \mu_x} = \frac{c_2 \mu_x}{c_1 + c_2 \mu_x}$$

Then, δ, the expected proportional increase in cost of an unequal probability survey relative to cost of an equal probability survey, would be

$$\delta = \frac{E(C_u) - E(C_e)}{E(C_e)} = \frac{c_2 n \mu_x CV^2(x)}{c_2 n \mu_x / \lambda} = \lambda CV^2(x)$$

Figure 8.4 displays δ as a function of λ for five values (0.0, 0.5, 1.0, 1.5, 2.0) of $CV(x)$, as noted on the five lines. For a given $CV(x)$, δ is proportional to λ and the greater the $CV(x)$ the steeper the slope of the line. For a specific example, consider the Figure 8.2 populations for which $CV(x) \approx 0.22$. For these populations, the expected cost differential for a PPS unequal probability survey relative to the cost of an equal probability survey would be no greater than 5%, but recall that the PPS strategies did not outperform the equal probability strategies for those populations. Contrast this with the Figure 8.3 coho salmon population for which $CV(x) \approx 1.05$. For this population, λ values of 0.2, 0.5, or 0.8, for example, would result in expected costs of a PPS survey that would be approximately 22%, 55%, or 89% greater, respectively, than for an equal probability survey. Note that for this population, however, the PPS strategies did outperform the equal probability surveys and might justify the increased survey costs.

We offer a few comments about the results presented in this section. First, as for ratio and regression estimation, if there are substantial costs associated with the measuring of x for all population units, then those costs must also be considered when evaluating the relative costs and net relative efficiencies of PPS sampling strategies as compared to strategies that

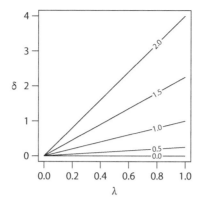

Fig. 8.4 Expected proportional increase in total survey cost of an an unequal probability survey compared to an equal probability survey, $\delta = [E(C_u) - E(C_e)]/E(C_e)$, as a function of the proportion of total survey cost that is size-related (λ), for five values (0.0, 0.5, 1.0, 1.5, 2.0) of $CV(x)$ as noted on the individual lines. Assumes the Equation (8.13) cost function.

rely only on the measurement of y for the sample units. Second, all of these results are in terms of *expected* costs. The actual cost of a survey with a size-related component depends on the realized sample selected [e.g., Equation (8.11)], and may vary considerably across possible samples. This may make it more difficult to keep survey costs under a fixed budget for an unequal probability sampling design than for an equal probability sampling design.

8.4 Sampling distribution

Given a specific set of target variable values $\{y_1, y_2, \ldots, y_N\}$ and associated auxiliary variables, the sampling distribution of an estimator can be approximated based on a Monte Carlo simulation of a large number of independent samples of size n using the sampling design of interest. Alternatively, when computationally feasible, the exact sampling distribution can be directly determined over the sample space consisting of all possible samples. In either case, the resulting distribution can then be visualized by constructing a histogram of the sample estimates. For equal probability selection methods (e.g., SRS), for which all sample selections are equally likely, an unweighted histogram of the estimates is appropriate, but for unequal probability selection methods (e.g., PPSWOR) a weighted histogram is required if based on all possible estimates (an unweighted histogram is appropriate if based on Monte Carlo simulation, as explained next). While the actual sampling distribution is discrete, a histogram is useful for assessing its general shape which, for the purpose of confidence interval construction, we assume is approximately normal.

For example, panel (a) of Figure 8.5 displays a population of $N = 40$ (x, y) variable values, with a target variable total of $T_y = 4{,}622.8$. For SRS with $n = 4$, there are a total of $\binom{N}{n} = \binom{40}{4} = 91{,}390$ possible samples. For each of these samples, $\hat{T}_{y,\mathrm{mpu}}(s) = N \sum_{i \in s} y_i / n$ was calculated, and a frequency distribution of these estimates is displayed in panel (b) of Figure 8.5. What is desired, however, is a histogram of the sampling distribution, with the vertical axis representing probability, not frequency. That is, the height of each bin should be equal to the sum of the $\{p(s)\}$ that produce estimates falling within that bin interval. For SRS all possible samples are equally likely, $p(s) = \binom{N}{n}^{-1}$ for all $s \in \mathcal{S}$, and therefore the sum of the $\{p(s)\}$ for a bin interval is equal to the bin frequency multiplied by $p(s) = 1/91{,}390$; i.e., the proportion of estimates falling within the bin interval. This results in the probability scale displayed on the right margin of panel (b) of Figure 8.5, and the sum of the bin probabilities being equal to one, as for the probability mass function (pmf, Section A.3.2) of a discrete random variable. In this particular case, the shape of the frequency histogram does not change because the bin frequencies are simply scaled by a constant. If distribution shape is all that is of interest (e.g., to assess normality), then rescaling is unnecessary.

Continuing with the example, for PPSWOR sampling with probabilities proportional to sizes of the remaining units (namely using the Yates and Grundy (1953) method discussed at the beginning of Section 8.3) and $n = 4$, there are also a total of $\binom{40}{4} = 91{,}390$ possible unordered samples, but they are not all equally likely. For a given sample s, the number of possible unit orderings (Section A.1.1) is $n! = 4! = 24$, each having its own probability (Section 8.3), and $p(s)$ was computed as the sum of these probabilities for each $s \in \mathcal{S}$. Given the $p(s)$, the $\{\pi_i\}$ could then be determined and this allowed for the calculation of $\hat{T}_{y,\mathrm{HT}}(s) = \sum_{i \in s} y_i/\pi_i$ for each sample. A histogram of these estimates on the proportion scale is shown in panel (c) of Figure 8.5. However, because the samples are not equally likely this generates a distorted view of the sampling distribution of $\hat{T}_{y,\mathrm{HT}}$. For an undistorted view, what is needed is the sum of the $p(s)$ that produce the estimates falling within each interval; i.e., a weighted histogram where the occurence of each estimate is weighted according to its $p(s)$. Panel (c) of Figure 8.5 displays both histograms, showing that in this case the histogram of observed proportions overweights the $\hat{T}_{y,\mathrm{HT}}$ that are further from T_y, and underweights the $\hat{T}_{y,\mathrm{HT}}$ that are closer to T_y.

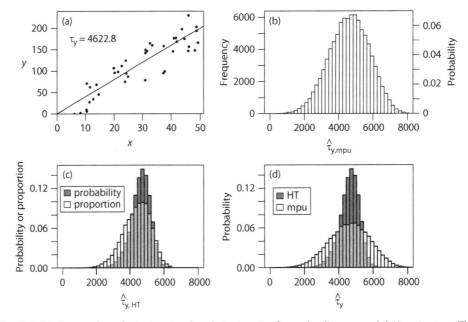

Fig. 8.5 (a) Scatterplot of $N = 40$ simulated (x, y) pairs from the linear model $Y_i = 4x_i + \varepsilon_i$. The $\{x_i\}$ were generated from a continuous uniform(5, 50) distribution, and the $\{\varepsilon_i\}$ were generated from a normal(0, 30^2) distribution. (b)–(d) Associated histograms of sample-specific estimates for all possible (unordered) samples of size $n = 4$ for $\hat{T}_{y,\mathrm{mpu}}$ (SRS selection) and $\hat{T}_{y,\mathrm{HT}}$ (PPSWOR selection, with probabilities proportional to x of remaining units). For panels (c) and (d), light grey color indicates overlap of two distributions. Histograms were drawn using R package weights function wtd.hist() (Pasek 2018). See text for additional details.

Finally, the sampling distributions of the two estimators, $\hat{T}_{y,\text{mpu}}$ and $\hat{T}_{y,\text{HT}}$, for this example can now be visually compared on the (same) probability scale [panel (d) of Figure 8.5]. The $\hat{T}_{y,\text{HT}}$ distribution is more heavily concentrated around the true value of $T_y = 4{,}622.8$, i.e., has smaller sampling variance, but is moderately skewed.

As noted previously, confidence interval construction using $\hat{T}_{y,\text{HT}} \pm 2\sqrt{\hat{V}(\hat{T}_{y,\text{HT}})}$ assumes that the sampling distribution of $\hat{T}_{y,\text{HT}}$ is approximately normal when stating that its coverage probability is 95%. For with replacement selection methods, equal or unequal probability, the y_i/p_i are independent and the central limit theorem can be readily applied, so that the sampling distribution of $\hat{T}_{y,\text{HH}}$ (or of the estimated mean, $\hat{T}_{y,\text{HH}}/N$) is asymptotically normal (Chaudhary and Sen 2002, Prášková and Sen 2009). For without replacement sampling, several classes of PPSWOR selection methods have been shown to result in asymptotic normality of the $\hat{T}_{y,\text{HT}}$ sampling distribution, including successive-type methods (Berger 1998) and rejective-type methods (Prášková and Sen 2009), as well as Chao's (1982) list sequential method (Berger 2005).

For those PPSWOR methods that have not yet been shown to result in an asymptotically normal sampling distribution for $\hat{T}_{y,\text{HT}}$, directly determining the distribution via the all possible samples approach would not appear feasible for larger sample sizes. For example, suppose $N = 40$, $n = 20$, and sample units are selected by PPSWOR with probabilities proportional to the sizes of remaining units. The number of possible (unordered) samples would be equal to $\binom{40}{20} \approx 10^{11}$, and for each of these samples there are $20! \approx 10^{18}$ distinct orders of selection, so that together approximately 10^{29} ordered samples would need to be evaluated.

With list sequential methods, however, for which larger n can often be easily accomodated, it is possible to approximate the sampling distribution of $\hat{T}_{y,\text{HT}}$ based on Monte Carlo simulation of a large number of samples selected independently according to the list-sequential method of interest. In this case the probability of a particular sample $S = s$ being selected during the simulation will be equal to $p(s)$, and s may be selected on multiple occasions (independent trials), so that the proportion of trials that sample s is expected to be selected is $p(s)$. Therefore, an unweighted histogram showing the proportion of the simulation's resulting estimates in each bin interval will provide an undistorted visual representation of the estimator sampling distribution.

Figure 8.6 demonstrates the results of such a Monte Carlo simulation for the sampling distribution of $\hat{T}_{y,\text{HT}}$ based on applying PPSWOR using Chao's (1982) selection method to the population variable values displayed in panel (a) of Figure 8.5. For $n < 10$, histograms of the sampling distributions have modest skew to the left, but by $n = 20$ the distribution is fairly close to being normal, in agreement with the asymptotically normal theoretical result.

8.5 Systematic sampling

If values of an auxiliary variable are known for all units in a population, then a systematic sample may be selected in a fashion that ensures that first order inclusion probabilities, $\{\pi_i\}$, are directly proportional to x. Murthy (1967 pp. 215–218) nicely illustrates the original application of this procedure, termed PPS systematic sampling, with integer random start, r. Its virtues lie in its simplicity of application (as for other methods of systematic sampling), its achievement of the desirable proportionality of π_i to x_i, and its substantial reduction of sampling variance (relative to SRS with mean-per-unit estimation)

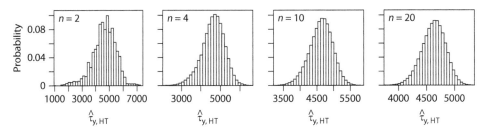

Fig. 8.6 Changes in shapes of histograms of sampling distributions of $\hat{T}_{y,\mathrm{HT}}$ based on 10^6 independently selected samples of size $n = 2, 4, 10$, and 20 selected from population variable values shown in panel (a) of Figure 8.5 ($N = 40$) using Chao's (1982) list-sequential method for sample selection.

whenever y is roughly proportional to x. With integer random start, however, sample size is not fixed and so, analogous to equal probability systematic sampling, unbiased estimation generally does not result.

If, however, the initial random start is instead drawn from a continuous uniform$(0, k)$ distribution (analogous to the fractional interval approach in linear systematic sampling), then sample size is fixed and the PPS systematic estimator becomes unbiased. This scheme can be implemeted as follows.

1. Form a vector of the cumulative sums of the auxiliary variable, beginning at zero:
   $X = (0, x_1, x_1 + x_2, x_1 + x_2 + x_3, \ldots, T_x)$.

2. Set $k = T_x/n$, where k is (typically) non-integer, and n is desired sample size.

3. Generate a random start, r, from the continuous uniform$(0, k)$ distribution, and create the *selection vector* $R = (r, r+k, r+2k, r+3k, \ldots, r+(n-1)k)$.

4. Compare the values in this selection vector with the vector of cumulative sums to determine which units will be included in the systematic sample. The hth selected unit will be population unit i, where $X_i < R_h \le X_{i+1}$.

If all $x_i < k$, then $\pi_i = nx_i/T_x$ and the Horvitz–Thompson estimator can be used to estimate the total of the target variable. [Note that if one or more of the $x_i \ge k$, then those units will be selected with certainty. If there are m such sample units, one could set those units aside as sure-to-sample, then take a random sample from the remaining $n - m$ units using the fractional interval scheme with $k' = (T_x - \sum_{j=1}^{m} x_i)/(n-m)$.]

For example, suppose we wish to select a PPS systematic sample of fixed size $n = 2$ from $N = 4$ units with $x_1 = 11, x_2 = 18, x_3 = 32, x_4 = 46$. We first form the cumulative sums vector, $X = (0, 11, 29, 61, 107)$. We then set $k = \sum_i x/n = 107/2 = 53.5$ and generate a value at random from the continuous uniform$(0, 53.5)$ distribution. If the random value of r is, say, 7.02, we form the selection vector $R = (r, r+k) = (7.02, 60.52)$. We then compare the values in R with X. Because $X_1 < 7.02 \le X_2$, the first selection is unit 1, and because $X_3 < 60.52 \le X_4$, the second selection is unit 3. If instead the random start r had been, say, 48.96, then the selection vector would have been $r = (48.96, 102.46)$, and the selected units would have been units 3 and 4.

As for equal probability systematic sampling, unbiased estimation will result (because n is fixed) and sampling variance for this strategy may be much less than for SRS if y is positively correlated with x. Also as for equal probability systematic sampling, however, unbiased variance estimation is impossible because π_{ij} may be 0 for many pairs of units. In this example, $\pi_{12} = 0$ because it is impossible to generate a random start, r, such that the selection vector would result in selection of units 1 and 2. The very smallest possible value of r would be 0, and the next value in the selection vector would then be 53.5, resulting in selection of units 1 and 3. Finally, as for equal probability systematic sampling, the order of the population units may have an important influence on sampling variance if there is a trend in the values of y_i with i and/or in the relations between y_i and x_i.

If it is possible to randomize the order of population units prior to generating the vector of cumulative sums for the auxiliary variable, then the expected sampling variance of the randomized PPS systematic strategy should be the same as for a PPWSOR strategy which has $\pi_i = nx_i/T_x$ and which is usually much more difficult to implement. For this randomized order case, Hartley and Rao (1962) proposed various approximations to the $\{\pi_{ij}\}$ that would allow use of Equation (8.10) for estimation of approximate sampling variance. One of the Hartley and Rao (1962) expressions for estimation of sampling variance is directly analogous to the use of SRS formulas for estimating sampling variance in a random order linear systematic sampling setting.[4]

8.6 Generality of Horvitz–Thompson estimation

Overton and Stehman (1995) note that the Horvitz–Thompson estimator, and its associated definitions of first and second order inclusion probabilities, provides a unifying framework for probability sampling from a finite population, as was clearly understood by Horvitz and Thompson (1952 pp. 670–673). A specific sampling design can be viewed as inducing a specific sample space (a specific set of possible samples) and a specific set of first and second order inclusion probabilities associated with the finite population units. We illustrate the generality of Horvitz–Thompson estimation with three examples, in each case using the labels i and j, $i \neq j$ to denote particular population units.

In simple random sampling, n units are selected from N units with equal probability without replacement. For this method of selecting samples, the sample space consists of $\binom{N}{n}$ equally likely samples of size n. The first and second order inclusion probabilities are

$$\pi_i = \frac{n}{N}$$

$$\pi_{ij} = \frac{n(n-1)}{N(N-1)}$$

For stratified sampling, let h denote stratum. Suppose that we select n_h from N_h by SRS within each stratum. Then

$$\pi_i = \frac{n_h}{N_h}, \quad \text{for } i \text{ in stratum } h$$

$$\pi_{ij} = \begin{cases} \dfrac{n_h(n_h-1)}{N_h(N_h-1)}, & \text{for } i,j \text{ in stratum } h \\[2ex] \dfrac{n_h n_{h'}}{N_h N_{h'}}, & \text{for } i,j \text{ in different strata } (h,h') \end{cases}$$

---

[4] Stehman and Overton (1994) carried out extensive simulation studies exploring the performance of Equations (8.8) and (8.10) for estimating sampling variance of the randomized PPS systematic Horvitz–Thompson estimator, relying on two alternative approximations to π_{ij}, including one from Hartley and Rao (1962).

For equal probability linear systematic sampling, assuming $N/k = n$ is integer-valued, denote the possible systematic samples by $h = 1, 2, \ldots, k$. Then

$$\pi_i = \frac{1}{k}, \quad \text{for } i \text{ in sample } h$$

$$\pi_{ij} = \begin{cases} \frac{1}{k}, & \text{for } i, j \text{ in sample } h \\ 0, & \text{for } i, j \text{ in different samples} \end{cases}$$

Unbiased variance estimation is not possible in linear systematic sampling because many $\pi_{ij} = 0$, i.e. this is not a *measurable* sampling design.

8.7 Generalized Horvitz–Thompson estimation

The Horvitz–Thompson estimator can be used to estimate the total of population variables other than x and y, including functions of the x and y variables themselves. For an arbitrary variable, z, the total is defined in the same way, $T_z = \sum_{i=1}^N z_i$, and its Horvitz–Thompson estimator is $\hat{T}_{z,\text{HT}} = \sum_{i \in S} z_i / \pi_i$. For example, the population total for x^2 is $T_{x^2} = \sum_{i=1}^N x_i^2$ and its Horvitz–Thompson estimator is $\hat{T}_{x^2,\text{HT}} = \sum_{i \in S} x_i^2 / \pi_i$. Other totals of general interest in addition to T_x and T_y are T_{xy}, T_{x^2}, T_{y^2}, and, the seemingly trivial, $T_{z=1} = \sum_{i=1}^N 1 = N$, which leads to the useful estimator $\hat{N}_{\text{HT}} = \sum_{i \in S} 1/\pi_i$. (This estimator sees frequent application).

Often some other population parameter of interest, θ, can be expressed as a function of several such totals, $\theta = f(T_{z_1}, T_{z_2}, \ldots, T_{z_r})$, and in this case a general principle (Särndal et al. 1992 Chapter 5) to use in forming an estimator of θ, what we'll call a *generalized Horvitz–Thompson estimator*, is to substitute for each unknown total in f its Horvitz–Thompson estimator: $\hat{\theta}_{\text{gen-HT}} = f(\hat{T}_{z_1,\text{HT}}, \hat{T}_{z_2,\text{HT}}, \ldots, \hat{T}_{z_r,\text{HT}})$. For example, for the generalized Horvitz–Thompson estimator of the ratio, $R = \mu_y / \mu_x$, re-express R in terms of the respective totals, $R = T_y / T_x$, and then substitute in the respective Horvitz–Thompson estimators to give $\hat{R}_{\text{gen-HT}} = \hat{T}_{y,\text{HT}} / \hat{T}_{x,\text{HT}}$.

If f is a linear function of these estimated totals, then $\hat{\theta}_{\text{gen-HT}}$ is unbiased for θ, and its sampling variance can be readily developed based on the known variance and covariance relationships for linear combinations of random variables (Sections A.3.5 and A.3.6). If f is a non-linear function of the estimated totals, then $\hat{\theta}_{\text{gen-HT}}$ is not unbiased, and its sampling variance can be approximated using the Delta method (Section A.8). Though biased, the estimator may be *asymptotically unbiased* (bias approaches 0 with increasing n), and may be *consistent* (converges in probability to θ with increasing n) and/or *finite population consistent* ($\hat{\theta}_{\text{gen-HT}} = \theta$ for $n = N$)[5] (Section A.7.4). Fisher (1959 pp. 143–147) considered consistency to be the *fundamental criterion of estimation*. In general, given a set of consistent estimators $\{\hat{\theta}_i\}$, an estimator $\hat{\theta} = f(\hat{\theta}_1, \hat{\theta}_2, \ldots, \hat{\theta}_r)$ will very often be a consistent estimator for $\theta = f(\theta_1, \theta_2, \ldots, \theta_r)$. For example, if $\hat{T}_x$ and $\hat{T}_y$ are consistent estimators of the population means T_x and T_y, then the ratio estimator, $\hat{T}_y / \hat{T}_x$, is consistent for $T_y / T_x = R$. The Horvitz–Thompson estimator for an arbitrary total, $\hat{T}_{z,\text{HT}}$, is finite population consistent for all fixed-n without replacement sampling designs, including PPSWOR.

---

[5] Not all estimators are finite population consistent. An example is the *mean of ratios* estimator, $\hat{R} = (1/n) \sum_{i \in S} y_i / x_i$, of $R = T_y / T_x$ that was considered briefly in Chapter 7. When $n = N$, this estimator takes on the value $(1/N) \sum_{i=1}^N y_i / x_i \neq R$.

8.7.1 *Variance, covariance, and correlation estimators*

When a sample of size n is selected from N by PPSWOR, the usual estimators of σ_y^2, σ_{xy}, and $\mathrm{Cor}(x,y)$ are not necessarily the best ones. In this section we develop alternative, generalized Horvitz–Thompson estimators for these quantities which may be preferred, by appealing to their basic definitions (Section A.5.2).

By definition, $\sigma_y^2 = \sum_{i=1}^{N}(y_i - \mu_y)^2/(N-1) = \left(T_{y^2} - T_y^2/N\right)/(N-1)$, and, by analogy, $\sigma_x^2 = \sum_{i=1}^{N}(x_i - \mu_x)^2/(N-1) = \left(T_{x^2} - T_x^2/N\right)/(N-1)$. Substituting the Horvitz–Thompson estimators for each of the component totals thus gives

$$\hat{\sigma}_{y,\text{gen-HT}}^2 = \left(\hat{T}_{y^2,\text{HT}} - \hat{T}_{y,\text{HT}}^2/\hat{N}_{\text{HT}}\right)\bigg/\left(\hat{N}_{\text{HT}} - 1\right) \tag{8.14}$$

Similarly, $\sigma_{xy} = \sum_{i=1}^{N}(x_i - \mu_x)(y_i - \mu_y)^2/(N-1) = (T_{xy} - T_x T_y/N)/(N-1)$. Therefore

$$\hat{\sigma}_{xy,\text{gen-HT}} = \left(\hat{T}_{xy,\text{HT}} - \hat{T}_{x,\text{HT}}\hat{T}_{y,\text{HT}}/\hat{N}_{\text{HT}}\right)\bigg/\left(\hat{N}_{\text{HT}} - 1\right) \tag{8.15}$$

where $\hat{T}_{xy,\text{HT}} = \sum_{i \in S} x_i y_i/\pi_i$. Finally, $\mathrm{Cor}(x,y) = \sigma_{xy}/\sqrt{\sigma_x^2 \sigma_y^2}$. Substituting the generalized Horvitz–Thompson estimators just developed for σ_{xy}, σ_x^2, and σ_y^2 thus gives

$$\widehat{\mathrm{Cor}(x,y)}_{\text{gen-HT}} = \hat{\sigma}_{xy,\text{gen-HT}}\bigg/\sqrt{\hat{\sigma}_{x,\text{gen-HT}}^2 \hat{\sigma}_{y,\text{gen-HT}}^2} \tag{8.16}$$

8.7.2 *Mean-per-unit, ratio, and regression estimators*

Generalized Horvitz–Thompson estimators for the total T_y in the contexts of mean-per-unit, ratio, and regression estimation are developed in the following paragraphs as possible alternatives to the standard equal probability estimators.

Re-expressing the mean in terms of population totals, $\mu_y = T_y/N$, and substituting in the Horvitz–Thompson estimators of the two totals gives

$$\hat{\mu}_{y,\text{gen-HT}} = \hat{T}_{y,\text{HT}}/\hat{N}_{\text{HT}}, \quad \text{and} \quad \hat{T}_{y,\text{gen-HT,mpu}} = N\hat{\mu}_{y,\text{gen-HT}} \tag{8.17}$$

For the ratio $R = T_y/T_x$,

$$\hat{R}_{\text{gen-HT}} = \hat{T}_{y,\text{HT}}/\hat{T}_{x,\text{HT}}, \quad \text{and} \quad \hat{T}_{y,\text{gen-HT,rat}} = T_x \hat{R}_{\text{gen-HT}} \tag{8.18}$$

For regression estimation, the linear regression estimator of the total is $\hat{T}_{y,\text{reg}} = \hat{T}_{y,\text{mpu}} + \hat{\beta}(T_x - \hat{T}_{x,\text{mpu}})$, where $\hat{\beta} = \hat{\sigma}_{xy}/\hat{\sigma}_x^2$. Substituting in the respective Horvitz–Thompson estimators of the totals gives

$$\hat{\beta}_{\text{gen-HT}} = \hat{\sigma}_{xy,\text{gen-HT}}/\hat{\sigma}_{x^2,\text{gen-HT}}, \quad \text{and} \quad \hat{T}_{y,\text{gen-HT,reg}} = \hat{T}_{y,\text{HT}} + \hat{\beta}_{\text{gen-HT}}(T_x - \hat{T}_{x,\text{HT}}) \tag{8.19}$$

8.7.3 *Performance of generalized Horvitz–Thompson estimators*

It is tough to generalize about the performance of these generalized Horvitz–Thompson estimators! For example, sometimes the estimator of finite population variance that would be used in the conventional equal probability setting will outperform the generalized Horvitz–Thompson estimator of finite population variance, even when sampling is with unequal probability. Courbois and Urquhart (2004) concluded that the estimator of finite population variance that is recommended is the usual one, $\hat{\sigma}_y^2 = \sum_{i \in S}(y_i - \hat{\mu}_y)^2/(n-1)$, when $\mathrm{Cor}(x,y)$ is low (< 0.40), but that the generalized Horvitz–Thompson estimator of σ_y^2 [Equation (8.14)] is best for "high" correlations. We suspect that a similar finding might emerge for estimation of the linear correlation. The generalized Horvitz–Thompson estimator of the mean, $\hat{\mu}_{y,\text{gen-HT}} = \hat{T}_{y,\text{HT}}/\hat{N}_{\text{HT}}$, is believed to be (usually) superior to the estimator $\hat{T}_{y,\text{HT}}/N$ due to the positive correlation between $\hat{T}_{y,\text{HT}}$ and $\hat{N}_{\text{HT}}$ over the sample space (Särndal et al. 1992 Result 5.7.1, S. Overton personal communication). (Thompson (2012 Section 6.3) provides expressions for sampling variance and for estimation of sampling variance for $\hat{\mu}_{y,\text{gen-HT}}$.) $\hat{\mu}_{y,\text{gen-HT}}$ also seems clearly superior to $\hat{T}_{y,\text{HT}}/N$ for selection methods that result in variable sample size (see Problems 8.3 and 8.4 concerning Poisson sampling). In Section 8.9.2 we provide evidence that $\hat{\mu}_{y,\text{gen-HT}}$ may be especially valuable when there are nonresponse units in a PPSWOR setting. In our own (unpublished) explorations of the performance of $\hat{T}_{y,\text{gen-HT,reg}}$, we have found that this generalized regression estimator may be a very good choice of estimator when the sample is selected by PPSWOR, there is a large positive intercept for an approximately straight line relation between y and x, and there is a strong positive correlation between y and x.

8.8 Poisson sampling

Poisson sampling is the unequal probability analog of Bernoulli sampling which was considered in Section 3.1.4. As for Bernoulli sampling, this sample selection method may be useful when there is no pre-existing list of population units and sample selection must be done "on the fly". Suppose, for example, that population units consist of pool habitat units in a well-defined reach of a small stream and that the target of estimation is the total number of some aquatic salamander species in all pools in this reach. An accurate assessment of the number of salamanders in a pool is a time-consuming process and so only a sample of pool habitat units can be selected for measurement. Suppose that, as field personnel move upstream, individual pools are identified and are selected for inclusion in the sample with probabilities $\{\pi_i = \alpha x_i\}$, where x_i is an ocular (visual) estimate of pool area for unit i, and $\alpha > 0$ is set so that $\pi_i < 1$ for the largest pool area imagined pre-survey. This sampling protocol can be achieved in the field by selecting, for each unit i, a continuous uniform$(0,1)$ random variate, u_i, using a pocket calculator, and then comparing this value to π_i. If $u_i \leq \pi_i$, then unit i is selected and protocols for counting salamanders are executed; if not, then field personnel move up to the next pool habitat unit and repeat this process. As for Bernoulli sampling, sample size is a random variable. For Poisson sampling, $E(n) = \sum_{i=1}^{N} \pi_i$ [Equation (A.129)].

Because units are selected independently of one another, the second order inclusion probabilities are of the form $\pi_{ij} = \pi_i \pi_j$. Problem 8.3 asks the reader to construct a simulation program to numerically support the facts that with Poisson sampling: (1) the Horvitz–Thompson estimator, $\hat{T}_{y,\text{P,HT}} = \sum_{i \in S} y_i/\pi_i$, calculated over the sample S of random size n, gives an (unconditionally) unbiased estimator of the population total (e.g., total number of salamanders in all pool units), and (2) the (unconditional) sampling variance of this estimator is given by the Horvitz-Thompson expression for sampling variance, Equation (8.7) (valid for random n), recognizing that all terms involving $\pi_{ij} - \pi_i \pi_j$ have value zero so that

$$V(\hat{T}_{y,\text{P,HT}}) = \sum_{i}^{N} \left(\frac{1 - \pi_i}{\pi_i} \right) y_i^2 \tag{8.20}$$

As Problem 8.4 numerically demonstrates, however, the *conditional* expectations of the Horvitz–Thompson estimator under Poisson selection are not unbiased but instead increase with the size of the realized sample, n_s. For that reason, an alternative generalized mean-per-unit estimator of the total is often recommended for Poisson sampling because it should have more stable conditional expectation over the random sample size and, as a consequence, reduced sampling variance (Särndal et al. 1992 Remark 3.5.1)

$$\hat{T}_{y,\mathrm{P,gen\text{-}HT,mpu}} = N\hat{T}_{y,\mathrm{P,HT}}/\hat{N}_{\mathrm{P,HT}} \tag{8.21}$$

where $\hat{N}_{\mathrm{P,HT}}$ is the Horvitz–Thompson estimator of N under Poisson sampling. The approximate sampling variance of $\hat{T}_{y,\mathrm{P,gen\text{-}HT,mpu}}$ is (Särndal et al. 1992)

$$V(\hat{T}_{y,\mathrm{P,gen\text{-}HT,mpu}}) \approx \sum_{i=1}^{N} \frac{(y_i - \mu_y)^2}{\pi_i} - N\sigma_y^2 \tag{8.22}$$

Gregoire and Valentine (2008 Section 3.3.2) note that, if an auxiliary variable is available for all population units, then a generalized ratio estimator may have even better performance

$$\hat{T}_{y,\mathrm{P,gen\text{-}HT,rat}} = T_x \frac{\hat{T}_{y,\mathrm{P,HT}}}{\hat{T}_{x,\mathrm{P,HT}}} \tag{8.23}$$

where $\hat{T}_{x,\mathrm{P,HT}}$ is the Horvitz–Thompson estimator of T_x under Poisson sampling. The approximate sampling variance of $\hat{T}_{y,\mathrm{P,gen\text{-}HT,rat}}$ is (Gregoire and Valentine 2008)

$$V(\hat{T}_{y,\mathrm{P,gen\text{-}HT,rat}}) \approx \sum_{i=1}^{N} \left(\frac{1-\pi_i}{\pi_i}\right)(y_i - Rx_i)^2 \tag{8.24}$$

8.9 Nonresponse and oversampling

As noted in Section 3.5, whenever nonresponse in a unit is due to some factor which depends on the value of y, then equal probability estimators based on the realized sample which excludes the nonresponse units will be biased to an unknown degree. The same result applies to unequal probability estimators. When nonresponse is due to *pure chance* (i.e., unrelated to the value of y), however, unbiased or nearly unbiased estimation is possible in the unequal probability selection setting.

8.9.1 *Hansen–Hurwitz estimator*

As in Section 3.5, let S_r denote the set of units in the random sample S for which observations or measurements of y have been made (i.e., the set of response units), and let n_r denote the number of units in S_r. In the event of pure chance nonresponse, the Hansen–Hurwitz estimator of the total, calculated over the S_r response units remains unbiased for T_y. That is

$$E(\hat{T}_{y,\mathrm{HH}}) = E\left[\left(\frac{1}{n_r}\right)\sum_{i\in S_r}\frac{y_i}{p_i}\right] = T_y$$

and Equation (8.4), calculated over the units in S_r, is an unbiased estimator of $V(\hat{T}_{y,\mathrm{HH}})$. Thus, for pure chance nonresponse, the usual Hansen–Hurwitz estimator, calculated over the units providing response, remains unbiased when faced with nonresponse, as for the equal probability setting considered in Section 3.5.

8.9.2 *Horvitz–Thompson estimator*

For the Horvitz-Thompson estimator, the usual estimator of the total, calculated over the units in S_r, will be negatively biased (assuming $y > 0$) whenever there is nonresponse, regardless of the reason for nonresponse. Assuming that $n_r < n$ and that $y > 0$, it must always be the case that $\sum_{i \in S_r} y_i/\pi_i < \sum_{i \in S} y_i/\pi_i$ because the sum over all units in S would contain at least one more term than the corresponding sum calculated over the reduced set of units in S_r. As the sum over all n units in S would be unbiased, it cannot also be true that the sum over the reduced set of units is unbiased.

In the case of nonresponse due to pure chance, however, the generalized mean-per-unit estimator of T_y, should provide a nearly unbiased estimator of the total. That is

$$E(\hat{T}_{y,\text{gen-HT,mpu}}) = E\left(N\hat{\mu}_{y,\text{gen-HT}}\right) = NE\left(\hat{\mu}_{y,\text{gen-HT}}\right) \approx T_y$$

Consider the following simulation results, based on 5×10^4 independent samples generated using Chao's (1982) method for selection of PPSWOR samples. Values for target and auxiliary variables for a finite population of size $N = 50$ (favorable for regression estimation) were generated based on the following R code:

```
set.seed(100); N <- 50
x <- rnorm(N, mean=25, sd=7)
x <- sort(x, decreasing=TRUE)
y <- rnorm(N, mean=50+(2*x), sd=10)
```

Samples of size $n = 5$ were selected, and then $k = 1$, 2, or 3 of the units in S were assigned with equal probability to be (pure chance) nonresponse units. The reduced samples were therefore of sizes 4, 3, or 2. For each realized sample of response units, $S_r = s_r$, estimates of μ_y were calculated using $\hat{\mu}_{y,\text{HT}}$ and $\hat{\mu}_{y,\text{gen-HT}}$. Estimator simulated expectations, sampling variances and mean square errors are shown in Table 8.7.

Table 8.7 Simulated expected value, sampling variance, and mean square error (5×10^4 independent samples) for the Horvitz–Thompson estimator of μ_y ($\hat{\mu}_{y,\text{HT}}$) and the generalized Horvitz–Thompson estimator of μ_y ($\hat{\mu}_{y,\text{gen-HT}}$) when $k = 1, 2$, or 3 units are pure chance nonresponse units for PPSWOR $n = 5$ samples selected using Chao's (1982) method from a population for which $\mu_y = 102.36$.

k	$E(\hat{\mu}_{y,\text{HT}})$	$E(\hat{\mu}_{y,\text{gen-HT}})$	$V(\hat{\mu}_{y,\text{HT}})$	$V(\hat{\mu}_{y,\text{gen-HT}})$	$\text{MSE}(\hat{\mu}_{y,\text{HT}})$	$\text{MSE}(\hat{\mu}_{y,\text{gen-HT}})$
1	81.91	103.05	31.68	51.59	449.88	52.07
2	61.42	103.27	24.17	68.38	1700.25	69.21
3	40.93	103.85	16.24	103.67	3789.88	105.89

Table 8.7 shows that $\hat{\mu}_{y,\text{gen-HT}}$ remains approximately unbiased even when three of five selected units are nonresponse units, whereas the magnitude of the negative bias of $\hat{\mu}_{y,\text{HT}}$ increases rapidly with increasing numbers of nonresponse units. The slight bias of $\hat{\mu}_{y,\text{gen-HT}}$ is a reflection of the fact that this estimator is the ratio of two estimators (random variables). Sampling variance of $\hat{\mu}_{y,\text{gen-HT}}$ is considerably greater than for $\hat{\mu}_{y,\text{HT}}$, but the mean square error of $\hat{\mu}_{y,\text{gen-HT}}$ is much less than for $\hat{\mu}_{y,\text{HT}}$ due to the large squared bias contribution.

If some modest level of pure chance nonresponse is anticipated in a survey or if it is critical that a survey achieve its desired sample size, n, then *oversampling* can be used in the unequal probability setting, at least for those PPSWOR selection methods and sample sizes for which it is true that the $\{\pi_i\}$ are proportional to x (e.g., Chao's (1982) method) for the desired sample size of n as well as for the oversample of size $n' = \gamma n$. As for the equal probability setting described in Section 3.5, first select a PPSWOR sample of size $n' = \gamma n$, where n is desired sample size and γ is large enough to ensure, with near certainty, that $n' - n_{nr} \geq n$, where n_{nr} is the number of nonresponse units. Second, randomize the order of the selected units to be sure that they are in random order with respect to values of x in selected units. Third, use the first n response units in the oversample to achieve the desired sample size n. Finally, calculate $\hat{T}_{y,HT}$ using the PPSWOR scheme's $\{\pi_i\}$ *for the fixed sample size n*. This procedure will result in unbiased estimation of the population total and will have sampling variance equal to $V(\hat{T}_{y,HT})$ for fixed n for the adopted PPSWOR selection method.

8.10 Chapter comments

As previously discussed in Section 8.6, Overton and Stehman (1995) noted that Horvitz–Thompson estimation provides a unifying framework for design-based sampling theory and establishes the fundamental precepts of *probability sampling*. A sampling scheme applied to a sampling frame (the sampling design) generates the sample space and establishes the probabilities associated with specific samples. These sample probabilities in turn determine the values of the first and second order inclusion probabilities, $\{\pi_i\}$ and $\{\pi_{ij}\}$. Any valid probability sampling design must have all $\pi_i > 0$, and unbiased estimation of sampling variance requires that all $\pi_{ij} > 0$, which is true only for a measurable sampling design. Systematic sampling designs, whether equal or unequal probability, have numerous $\pi_{ij} = 0$ and therefore do not allow for unbiased estimation of sampling variance. Equal probability sampling designs, for which all π_i are equal, can be viewed as special cases of the more general notion of probability sampling in which units may be selected with unequal probabilities. Proofs of the unbiasedness of the Hansen–Hurwitz and Horvitz–Thompson estimators allow related simple proofs of the unbiasedness of the equal probability estimators $\hat{\mu}_{swr}$ and $\hat{\mu}_{mpu}$ as special cases.

The feasibility of selecting large samples using PPSWOR selection methods and of calculating first and second order inclusion probabilities for these methods have changed dramatically over the past 30 years as a result of advancements in computational capacities. The two list-sequential schemes presented in this chapter (Sunter 1977, Chao 1982) today allow for rapid calculation of the $\{\pi_i\}$ and selection of large samples from large populations (say, $n = 10^3, N = 10^4$). Calculation of the $\{\pi_{ij}\}$ is far less rapid, but seems quite feasible for relatively large n and N (say, $n = 10^2, N = 10^3$) for both of these schemes, though remains time-consuming for Chao's (1982) scheme. With continuing advances in computational capacities, we suspect that feasible calculation of the $\{\pi_{ij}\}$ for large n and N for these two schemes (and many others for which there are explicit equations for the $\{\pi_{ij}\}$), and estimation of sampling variance using Equation (8.10), will prove routine within the next decade.

The past and continuing difficulties of calculating second order inclusion probabilities for large n and N for many PPSWOR selection methods have, however, prompted development and exploration of the performance of approximate estimators of sampling variance that rely only on the first order inclusion probabilities, $\{\pi_i\}$, which can usually be rapidly calculated. For readers interested in the peformance of these approximate expressions, see Berger and Tillé (2009 Section 4), Henderson (2006), Haziza et al. (2008)

and Hassan et al. (2009). Note that the performance of these approximations depends on the particular PPSWOR selection method that is used and that no approximation method has clearly superior performance characteristics for all selection schemes.

For PPSWOR selection methods that produce some $\pi_{ij} = 0$, unbiased variance estimation is impossible. If only a *few* $\pi_{ij} = 0$ out of many thousands of $\{\pi_{ij}\}$, then the consequence for estimation of sampling variance is probably ignorable.[6] But if a substantial fraction of the $\pi_{ij} = 0$, as for unequal probability systematic sampling, then it is probably wise to use the conservative estimator of sampling variance proposed by Aronow and Samii (2013). A conservative estimator of sampling variance will have expected value at least as large as the sampling variance.

PPSWR selection of samples and estimation of the population total and sampling variance using the Hansen–Hurwitz estimators does not require complex computer code. PPSWR selection of individual units can easily be accomplished in the following manner. First, form a vector of the cumulative sums of the $\{p_i\}$ for all N population units, beginning at 0: $P = (0, p_1, p_1 + p_2, \ldots, 1)$, where $p_i = x_i / T_x$. Then, generate a vector, R, of n independent continuous uniform(0, 1) random variates, and compare R to P to determine which units to select for the sample. The hth selected unit will be population unit i, where $P_i < R_h \leq P_{i+1}$. (See Section 8.5 for a description of a similar procedure for selecting a PPSWOR systematic sample.)

Selection of units by PPSWOR using any one of the three procedures described in this chapter, or many of the other published methods for selecting PPSWOR samples, generally requires construction and use of computer programs for all but the smallest sample sizes (say, $n = 2$ or $n = 3$). And even when selection of a PPSWOR sample may not be too difficult, calculation of the associated $\{\pi_i\}$ and $\{\pi_{ij}\}$ almost always requires implementation of a complex algorithm. For that reason, we provide readers with a set of R programs for selection of samples and calculation of associated first and second order inclusion probabilities for (a) selection with probabilities proportional to the sizes of remaining units (an all possible samples selection method), and for the list-sequential selection methods of (b) Sunter (1977) and (c) Chao (1982). These programs can be found at www.oup.co.uk/companion/hankin.

Problems

Problem 8.1. You are given the following variable values for a small population

Population unit (i):	1	2	3	4	5
Auxiliary variable (x_i):	5	10	15	30	40
Target variable (y_i):	2	5	7	14	17

(A) Construct all possible samples of size $n = 3$ that can be selected with unequal probability without replacement with probabilities proportional to the sizes, x, of remaining units. For each sample, calculate the corresponding sample probability, $p(s)$, and verify that $\sum_{s \in S} p(s) = 1$. Use the calculated $\{p(s)\}$ to calculate the $\{\pi_i\}$ and $\{\pi_{ij}\}$. For each sample, calculate $\hat{T}_{\text{HT}}$ and the Horvitz–Thompson and Sen–Yates–Grundy estimates of sampling variance (using Equations (8.8) and (8.10), respectively). (B) Calculate sampling variance

---

[6] Sengupta (1989) provides the conditions which guarantee that all $\pi_{ij} > 0$ for Chao's (1982) PPSWOR scheme which otherwise sometimes produces a few $\pi_{ij} = 0$.

over the sample space and numerically show that Equations (8.8) and (8.10) are unbiased for $V(\hat{T}_{HT})$. Is the performance of the Sen–Yates–Grundy variance estimator superior to the performance of the Horvitz–Thompson variance estimator?

Problem 8.2. You are given the following small population with associated values of auxiliary and target variables

Population unit (i):	1	2	3	4
Auxiliary variable (x_i):	10	20	30	40
Target variable (y_i):	6	10	13	15

For samples of size $n = 3$ selected by PPSWR, with $\{p_i = x_i/T_x\}$: (A) Construct all possible with replacement samples of size $n = 3$. (B) For each sample, calculate: $p(s)$, $\hat{T}_{HH}$, and $\hat{V}(\hat{T}_{HH})$. (C) Show numerically that the sampling variance of $\hat{T}_{HH}$, calculated over the sample space, gives the same result as Equation (8.3). (D) Show numerically that $E(\hat{T}_{HH}) = T_y$ and that $E[\hat{V}(\hat{T}_{HH})] = V(\hat{T}_{HH})$.

Problem 8.3. In R, use the following code to construct values for auxiliary and target variables.

```
set.seed(100); N <- 40
x <- 1:N
y <- rnorm(N, mean=15+(2*x), sd=8)
```

Define the inclusion probabilities $\{\pi_i = np_i = nx_i/T_x\}$, where n is a fixed and pre-specified value, and all $\pi_i \leq 1$. (A) Construct a simulation program that sets $n = 15$ and selects 2×10^5 independent Poisson samples from the population. (B) Use this program to provide numerical support for the following theoretical results: $E(n_S) = \sum_{i=1}^{N} \pi_i$; $V(n_S) = \sum_{i=1}^{N} \pi_i(1 - \pi_i)$; and $E(\hat{T}_{y,P,HT}) = T_y$, where $\hat{T}_{y,P,HT} = \sum_{i \in S} y_i/\pi_i$, and n_S is the size of the random sample S. (C) Show that the simulated sampling variance of $\hat{T}_{y,P,HT}$ is consistent with what would be expected for Poisson sampling given the original Horvitz–Thompson expression for sampling variance of $\hat{T}_{HT}$, Equation (8.7), which is valid when sample size is random. The following R code may be helpful for identifying the unit labels for any particular independently selected Poisson sample (where pi.i = $\{nx_i/T_x\}$).

```
r <- runif(N, min=0, max=1)
s <- (1:N)[r < pi.i]
```

Problem 8.4. This problem follows up on Problem 8.3. The generalized mean-per-unit estimator of T_y under Poisson sampling is $\hat{T}_{y,P,\text{gen-HT,mpu}} = N\hat{T}_{y,P,HT}/\hat{N}_{P,HT}$, where $\hat{N}_{P,HT} = \sum_{i \in S} 1/\pi_i$ is the Horvitz–Thompson estimator of N, calculated over the random sample size under Poisson sampling. Modify the R code developed for Problem 8.3 to allow calculation of the unconditional sampling variance of $\hat{T}_{y,P,\text{gen-HT,mpu}}$ given the same $\{\pi_i\}$, $\{x_i\}$ and $\{y_i\}$. Does this estimator have reduced sampling variance compared to $\hat{T}_{y,P,HT}$? If so, why do you think this might be the case? Is the conditional expectation of this estimator fairly stable over the various realized sample sizes?

Problem 8.5. S. Overton (personal communication) noted that one of the potential pitfalls of using unequal probability selection in U.S. government sample surveys is that interested parties may sue (under the Freedom of Information Act) for access to collected data and, having obtained data, may engage in misleading analyses that fail to account for the underlying probability structure of the collected data. Suppose that a small population has the following values of auxiliary and target variables

Population unit (i):	1	2	3	4	5	6
Auxiliary variable (x_i):	300	120	80	60	20	10
Target variable (y_i):	400	320	240	160	85	40

(A) Use program PPSWOR.remaining.units.R (at www.oup.co.uk/companion/hankin) to construct a listing of all possible samples of size $n = 3$, associated sample probabilities, $\{p(s)\}$, and first order inclusion probabilities, $\{\pi_i\}$, when units are selected with probabilities proportional to the sizes, $\{x_i\}$, of remaining units. Consider the following estimators:

$$\hat{T}_{y,\text{HT}} = \sum_{i \in s} y_i/\pi_i; \qquad \hat{T}_{y,\text{rat}} = T_x \sum_{i \in s} y_i / \sum_{i \in s} x_i; \qquad \hat{T}_{y,\text{mpu}} = N \sum_{i \in s} y_i/n.$$

(B) Calculate expected value, bias, sampling variance, and mean square error for each of these estimators *assuming that samples had been selected with probabilities proportional to the sizes of remaining units*. That is, assume that the $\{p(s)\}$ are the same for all estimators and calculate, e.g., $E(\hat{\theta}) = \sum_{s \in S} \hat{\theta}(s)p(s)$. (C) Discuss the relevance of your findings to the concerns that collected data may be analyzed without taking proper account of the underlying probability structure of the collected data.

Problem 8.6. From the Horvitz–Thompson perspective (Section 8.6), a sampling design applied to a finite population generates a specific set of $\{\pi_i\}$ and $\{\pi_{ij}\}$ that can be summarized in the form of an $N \times N$ matrix (S. Overton, personal communication), as for a population of $N = 6$ units illustrated in the following matrix.

$$\begin{bmatrix} i/j & 1 & 2 & 3 & 4 & 5 & 6 \\ 1 & \pi_1 & \pi_{12} & \pi_{13} & \pi_{14} & \pi_{15} & \pi_{16} \\ 2 & \cdots & \pi_2 & \pi_{23} & \pi_{24} & \pi_{25} & \pi_{26} \\ 3 & \cdots & \cdots & \pi_3 & \pi_{34} & \pi_{35} & \pi_{36} \\ 4 & \cdots & \cdots & \cdots & \pi_4 & \pi_{45} & \pi_{46} \\ 5 & \cdots & \cdots & \cdots & \cdots & \pi_5 & \pi_{56} \\ 6 & \cdots & \cdots & \cdots & \cdots & \cdots & \pi_6 \end{bmatrix}$$

Note that the lower left entries of this matrix reproduce the upper right entries of the matrix because $\pi_{ij} = \pi_{ji}$, and that the diagonal entries are $\{\pi_{ii} = \pi_i\}$. Specify the $N \times N$ matrix entries for the following sampling designs: (A) Suppose that $n = 4$ are selected by circular systematic sampling from $N = 7$, with $k = 3$ ($n = 3$). (B) Suppose that stratified SRS is used to select $n_1 = 2$ from $N_1 = 3$, and $n_2 = 3$ from $N_2 = 4$. Let the units in Stratum 1 be denoted by units 1–3; let the units in Stratum 2 be denoted by units 4–7. (C) Suppose a two-stage design (SRS at both stages) is used to select $n = 2$ clusters and $m = 2$ subunits within each cluster from a population consisting of three clusters of size $M = 3$. Let units 1–3 denote the subunits within the first cluster; units 4–6 denote the subunits within the second cluster; and units 7–9 denote the subunits within the third cluster. For each sampling design, verify that $\sum_i \pi_i = r$, and $\sum_i \sum_{j>i} \pi_{ij} = r(r-1)/2$, where r is the total sample size (number of selected subunits or population units).

Problem 8.7. Two herpetologists have designed a sampling strategy for estimating the abundance of coastal tailed frogs (*Ascaphus truei*) in individual riffle habitat units in small streams when the frogs are juveniles (tadpoles) which attach to rocks with an oral sucker. All rocks within 1 m² square metal frames are temporarily removed and numbers of clinging tadpoles are recorded. The metal frames are deployed at "random" locations within an identified riffle habitat unit. For a given habitat unit, n sample locations (i.e., locations for placing the 1 m² metal frames) are selected randomly with replacement according to the following scheme. First, the length (L) of the habitat unit is measured to the nearest m and a random choice, l, is made with equal probability on the integers 1 through L. Second, the width of the habitat unit (W) is measured to the nearest m at the selected random choice l and a random selection, w, is made with equal probability on the integers 1 through W. The set of random choices (l, w) determines the location for placement of each metal frame, and the number of tailed frog tadpoles are counted within each of n such frame locations. (A) The herpetologists think that they have designed a "random" with replacement method of selection and that density of tailed frog tadpoles (tadpoles/m²) within a given habitat unit can therefore be unbiasedly estimated using $\hat{\mu}_{swr} = \sum_{i \in S} y_i / n$, where y_i is the unit i count. Are they correct? (B) If not, how would you characterize their selection method and what alternative estimator would you propose? (C) Is a habitat unit "map" required to implement their original idea or your alternative proposal?

Problem 8.8. Suppose that the variable values for a small population are

Population unit (i):	1	2	3	4	5	6
Auxiliary variable (x_i):	2	5	7	4	10	3
Target variable (y_i):	6	12	16	8	20	7

Contrast the performance $[E(\hat{\theta}), B(\hat{\theta}), V(\hat{\theta}), \text{MSE}(\hat{\theta})]$ of the following estimators for without replacement samples of size $n = 4$ when applied to this population: $\hat{T}_{y,mpu}, \hat{T}_{y,rat}, \hat{T}_{y,reg}$, and $\hat{T}_{y,rat,L}$ (Lahiri's ratio estimator when the first sample unit is selected with probabilities $\{p_i = x_i / T_x\}$ and remaining units are selected by SRS).

Problem 8.9. Rao (1966) proposed the following unbiased estimator of sampling variance of $\hat{R}$ for Lahiri's unbiased ratio estimation strategy

$$\hat{V}(\hat{R}) = \hat{R}^2 - \frac{\sum_{i \in S} y_i^2 + \frac{N-1}{n-1} \sum_{i \in S} \sum_{\substack{j \in S \\ j \neq i}} y_i y_j}{T_x \sum_{i \in S} x_i} \qquad (8.25)$$

For the same set of population variables values listed in Problem 8.8: (A) Numerically determine if $\hat{V}(\hat{R})$ can be used to produce an unbiased estimator for $V(\hat{T}_{y,rat,L})$. (B) Is there anything odd about the sample-to-sample variation in these estimates of sampling variance? Is the structure of Rao's (1966) variance estimator consistent with the estimator's behavior?

Problem 8.10. (A) Using the Table 8.1 population variables which led to the Hansen–Hurwitz estimator sample space displayed in Table 8.3, confirm numerically that under PPSWR sampling, $\pi_i = 1 - (1 - p_i)^n, i = 1, 2, \ldots, N$. Algebraically, justify this expression for π_i. (B) Confirm numerically that under PPSWR sampling, $\pi_{ij} = \pi_i + \pi_j + [1 - (1 - p_i - p_j)^n]$, $i, j = 1, 2, \ldots, N, i \neq j$ for the same sample space. Algebraically, justify this expression by appealing to Equation (A.14). (Note that in this case the $\{\pi_{ij}\}$ are the same as the sample probabilities.) (C) Write a generalized R program that simulates selection of a large number of independent PPSWR samples of size n from a population of size N with specified auxiliary and target variable values. Use this program to calculate simulated $\{\pi_i\}$ and $\{\pi_{ij}\}$ and compare these simulated values to the exact values that can be calculated from the analytic expressions for $\{\pi_i\}$ and $\{\pi_{ij}\}$.

Problem 8.11. Show numerically that, for the $N = 7$, $k = 2$, $n = 4$ circular systematic sample space illustrated in Chapter 4, Table 4.2: (A) All $\pi_{ij} > 0$. (B) $\sum_i \sum_{j>i} \pi_{ij} = n(n-1)/2$. (C) Use of the Sen–Yates–Grundy (SYG) expression [Equation (8.9)] for sampling variance of $\hat{\mu}_{csys}$ gives exactly the same sampling variance as that calculated over the sample space. Would the related SYG estimator [Equation (8.10)] provide an unbiased estimator of sampling variance for this specific situation? Would it be possible that some estimates of sampling variance might be negative? Are all $\pi_i \pi_j > \pi_{ij}$ for this sample space?

CHAPTER 9

Multi-stage sampling

Fig. 9.1 Two adult male coho salmon, *Oncorhynchus kisutch* exhibiting competitive spawning behaviors in a small northern California stream. Red coloration and extended mouth (kype) develop at full maturity in freshwater, but are absent in the marine environment. Judgment selection of survey locations led to substantial over-estimation of abundance of this species in Oregon coastal streams (Section 2.6). Photo credit: Thomas Dunklin.

In many practical contexts, it is natural to carry out sample selection in two or more *stages* using what is termed **multi-stage sampling**. For example, suppose one wished to estimate the mean length of a particular species of fish that was landed at some coastal port over a specific three day period. The **first stage of sampling** might involve selection of a random sample of vessels that unloaded catch of that species. Suppose that fish are unloaded from vessels in large bins. Then, at the **second stage of sampling** one might select a random sample of bins from each of the vessels selected at the first stage. Finally, at the **third stage** of sampling, one might select a random sample of fish from each of the bins selected from a particular vessel and take length measurements of these fish. Errors of estimation arise at each of these stages of sampling. **First-stage variance** measures variation in average fish lengths among the first-stage units, in this

Sampling Theory: For the Ecological and Natural Resource Sciences. David G. Hankin, Michael S. Mohr, and Ken B. Newman, Oxford University Press (2019). © David G. Hankin, Michael S. Mohr, and Ken B. Newman. DOI: 10.1093/oso/9780198815792.001.0001

case vessels. Different vessels may fish at different depths or locations, causing average fish age and size to vary, perhaps substantially. **Second-stage variance** reflects variation in mean fish lengths among second-stage units (bins) within first-stage units, which would here measure variation of mean fish size among bins on a particular vessel. If a vessel had a long trip and fished at many locations and depths, then it is possible that there might be substantial variation in average fish length across bins within a particular vessel. Finally, **third-stage variance** reflects variation of fish length within selected bins. This last source of variation would reflect the interaction of the length distribution of fish subject to harvest, the size-selective properties of the gear used to harvest the fish, and any size limits that might be in place.

Effective implementation of multi-stage designs requires that one utilize all of the sampling theory that we have thus far developed. For example, how should the first-stage sample of vessels be selected? Would it make sense to select vessels by PPSWOR because larger vessels typically land greater catch and therefore have greater impact on the average length of fish landed? Would it be possible to select a first-stage sample by PPSWOR if the number and sizes of vessels landing catch were not known in advance? Could or should vessel size be incorporated in estimation schemes in some other fashion, perhaps as an auxiliary variable via ratio estimation? Although bins from a vessel might be feasibly selected by SRS, it is unlikely that an SRS sample of fish from within bins could feasibly be selected. Instead, systematic sampling with $\hat{\mu}_{\mathrm{lsys}}$ might prove an excellent strategy for estimating mean fish length within selected bins. Once selection methods and estimators have been adopted, there are many other questions remaining. For example, what is the optimal allocation of a multi-stage sample? What fraction of vessels should be selected at the first stage relative to the fraction of bins selected at the second stage relative to the fraction of fish selected from individual bins?

The multi-stage context presents an opportunity for substantial creativity in development of a survey design. The selection methods used at each stage of sampling will influence the choice of estimators of the mean at that particular stage of sampling (i.e., estimating mean length within a bin on a vessel, across all bins on a particular vessel, across all vessels) and will lead to choice of a specific set of alternative three-stage estimators of the mean (and of the associated sampling variance) that might be considered. As might be anticipated, these estimators and associated expressions for sampling variance are of considerably more complex form than those that we have thus far considered.

We begin our consideration of multi-stage sampling with the simplest two-stage design: **two-stage cluster sampling** with equal size clusters. In single-stage cluster sampling, measurements of the target variable were taken for all subunits within a selected cluster (primary unit) so that sample cluster means were known exactly. In two-stage cluster sampling, only a fraction of the subunits within clusters are selected for measurement. Why completely enumerate each and every cluster that is selected, when one might generate accurate estimates of cluster means based on just a fraction of the subunits within each selected cluster? If fewer subunits were measured within selected clusters, then a larger number of clusters might be selected at the same survey cost, and sampling variance for an optimally designed two-stage survey might be considerably less than for an equal cost single-stage survey.

9.1 Two-stage sampling: Clusters of equal size

We begin by considering the simplest possible two-stage cluster sampling setting. Suppose that a population is partitioned into three clusters, with each cluster consisting of three subunits. At the first stage of sampling, we select $n = 2$ from $N = 3$ clusters by SRS. At the second stage of sampling, within each of the selected clusters, we select $m = 2$ units from the $M = 3$ units by SRS. In the two-stage setting, we refer to clusters as **primary units**, and to units within clusters as **subunits** or **secondary units**. The total number of subunits in the population is here equal to 9, but in general equals NM. We'll use the following notation:

$y_{ij} =$ variable value of jth subunit in i^{th} cluster

$T_i = i^{\text{th}}$ cluster total

$\mu_i = i^{\text{th}}$ cluster mean

$\sigma_{2i}^2 = i^{\text{th}}$ cluster finite population variance

$T =$ population total

$\mu =$ population mean per subunit

and define these quantities as

$$T_i = \sum_{j=1}^{M} y_{ij}, \quad \mu_i = \sum_{j=1}^{M} y_{ij}/M = T_i/M, \quad \sigma_{2i}^2 = \sum_{j=1}^{M} (y_{ij} - \mu_i)^2/(M-1)$$

$$T = \sum_{i=1}^{N} \sum_{j=1}^{M} y_{ij} = \sum_{i=1}^{N} T_i, \quad \mu = T/(NM) = \sum_{i=1}^{N} T_i/(NM) = \sum_{i=1}^{N} \mu_i/N$$

We denote the random sample of n primary units selected from $\{i = 1, 2, \ldots, N\}$ as S_1, and the random sample of m secondary units (subunits) selected from $\{j = 1, 2, \ldots, M\}$ within a selected primary unit i as S_{2i}. The overall sample, S, consists of the primary/secondary label pairs $\{ij, i \in S_1, j \in S_{2i}\}$ for all of the selected subunits (i.e., the paired ij subscript values that index the sample y values, $\{y_{ij}\}$). The corresponding sample spaces for S_1 and S_{2i} are denoted as $\mathcal{S}_1$ and $\mathcal{S}_{2i}$, respectively, and the overall sample space is denoted by $\mathcal{S}$ as in single-stage sampling. Also, recall that $p(s)$ is shorthand for $\Pr\{S = s\}$, where s is a particular sample, and we now introduce $p(s|s_1)$ as shorthand for the conditional probability $\Pr\{S = s|S_1 = s_1\}$, where s_1 is a particular first-stage sample of primary units.

9.1.1 *Estimation of the population mean*

In two-stage sampling, we build estimates "from the bottom up". We thus begin with estimation of cluster means (primary unit means) at the second stage of sampling. For SRS sampling at the second stage, the use of mean-per-unit estimators is appropriate

$$\hat{\mu}_i = \frac{\sum\limits_{j \in S_{2i}} y_{ij}}{m}, \quad i \in S_1 \tag{9.1}$$

We then combine these estimators to form a two-stage estimator of μ, the population mean per subunit

$$\hat{\mu}_{2s} = \frac{\sum\limits_{i \in S_1} \hat{\mu}_i}{n} \tag{9.2}$$

where the subscript "2s" denotes "two-stage" estimation. The estimator $\hat{\mu}_{2s}$ is the two-stage analog to the single stage, equal size cluster sampling estimator $\hat{\mu}_c$ [Equation (6.1)], and is unbiased for μ.

Consider a population of $N = 3$ clusters, each consisting of $M = 3$ subunits, with variable values and parameters as shown in Table 9.1. The population total is $\mathcal{T} = \sum \mathcal{T}_i = 81$, the population mean per subunit is $\mu = \mathcal{T}/(NM) = 81/(3 \cdot 3) = 9$, and suppose that μ is the target of estimation. First- and second-stage samples will be taken of sizes $n = 2$ and $m = 2$, respectively. The first-stage sample will thus consist of two clusters, say i and i', so that $S_1 = \{i, i'\}$. There are $\binom{N}{n} = \binom{3}{2} = 3$ such samples possible, the S_1 sample space being $\mathcal{S}_1 = \{\{1,2\}, \{1,3\}, \{2,3\}\}$. For a selected cluster i, the second stage will consist of selecting $m = 2$ subunits, say j and j', so that $S_{2i} = \{j, j'\}$. There are $\binom{M}{m} = \binom{3}{2} = 3$ such samples possible. For example, if $i = 1$ (Cluster 1) is selected at the first stage, the three possible sets of $m = 2$ subunit y values that could result from this sampling of Cluster 1 would be $\{5, 10\}$, $\{5, 12\}$, $\{10, 12\}$. Because the sampling of subunits at the second stage is independent between the n clusters selected at the first stage, there are thus a total of $\binom{M}{m}^n = \binom{3}{2}^2 = 9$ possible second-stage samples *given a specific first-stage sample*, $S_1 = s_1$. The total number of possible two-stage samples is thus $\binom{N}{n}\binom{M}{m}^n = 3 \cdot 9 = 27$.

In the two-stage setting, the realized sample $S = s$ thus consists of a particular selection of primary units, $S_1 = s_1$, from which particular sets of subunits are selected: $S_{2i} = s_{2i}, i \in s_1$. With SRS used at both stages, all of the possible samples are equally likely, so that $p(s) = 1/[\binom{N}{n}\binom{M}{m}^n] = 1/27$. And, given a particular first-stage sample $S_1 = s_1$, all of the samples that include s_1 are also equally likely, so that $p(s|s_1) = 1/\binom{M}{m}^n = 1/9$. The sample space for this two-stage sampling design as applied to the Table 9.1 population is listed in Table 9.2, along with the selected subunit y values, sample probabilities, and two-stage estimates $(\hat{\mu}_{2s})$ of the population mean per subunit.

9.1.2 Expectation

Two approaches can be used to determine the (unconditional) expectation of a two-stage estimator $\hat{\theta}_{2s}$. The first approach is to use the usual, *unconditional*, definition of expectation

$$E(\hat{\theta}_{2s}) = \sum_{s \in \mathcal{S}} \hat{\theta}_{2s}(s)p(s)$$

All of the possible samples listed in Table 9.2 are equally likely ($p(s) = 1/27, s \in \mathcal{S}$) so that

$$E(\hat{\mu}_{2s}) = [(8.25 + \ldots + 10.75) + (8.50 + \ldots + 9.75) + (9.25 + \ldots + 9.50)] \cdot \frac{1}{27}$$

$$= [76.5 + 85.5 + 81.0](1/27) = 9 = \mu$$

numerically verifying that $\hat{\mu}_{2s}$ is unbiased when clusters are of equal size.

The second approach, however, is a more insightful approach which captures the underlying nature of expectation in a two-stage setting. This approach invokes the

Table 9.1 Clustered population variable values and parameters used for illustration in two-stage equal size cluster sampling.

Cluster (i)	(y_{i1}, y_{i2}, y_{i3})	$\mathcal{T}_i$	μ_i	σ_{2i}^2
1	5, 10, 12	27	9	13
2	3, 15, 6	24	8	39
3	13, 6, 11	30	10	13

Table 9.2 Listing of all possible two-stage SRS cluster samples of size $n = 2$ clusters selected from $N = 3$, and $m = 2$ subunits selected from $M = 3$ subunits per cluster for the Table 9.1 population. Selected clusters are labeled generically as A and B and are identified by cluster number ($i = 1, 2, 3$).

					Sample						
		Units (s)				Variable values					
		Cluster A		Cluster B		Cluster A	Cluster B				
ID	i	$j \in s_{2i}$	i	$j \in s_{2i}$		$y_{ij}, j \in s_{2i}$	$y_{ij}, j \in s_{2i}$	$p(s)$	$p(s\|s_1)$	$\hat{\mu}_A$	$\hat{\mu}_B$	$\hat{\mu}_{2s}$
1	1	1, 2	2	1, 2		5, 10	3, 15	1/27	1/9	7.5	9.0	8.25
2	1	1, 2	2	1, 3		5, 10	3, 6	1/27	1/9	7.5	4.5	6.00
3	1	1, 2	2	2, 3		5, 10	15, 6	1/27	1/9	7.5	10.5	9.00
4	1	1, 3	2	1, 2		5, 12	3, 15	1/27	1/9	8.5	9.0	8.75
5	1	1, 3	2	1, 3		5, 12	3, 6	1/27	1/9	8.5	4.5	6.50
6	1	1, 3	2	2, 3		5, 12	15, 6	1/27	1/9	8.5	10.5	9.50
7	1	2, 3	2	1, 2		10, 12	3, 15	1/27	1/9	11.0	9.0	10.00
8	1	2, 3	2	1, 3		10, 12	3, 6	1/27	1/9	11.0	4.5	7.75
9	1	2, 3	2	2, 3		10, 12	15, 6	1/27	1/9	11.0	10.5	10.75
							Sum:	1/3	1			76.50
10	1	1, 2	3	1, 2		5, 10	13, 6	1/27	1/9	7.5	9.5	8.50
11	1	1, 2	3	1, 3		5, 10	13, 11	1/27	1/9	7.5	12.0	9.75
12	1	1, 2	3	2, 3		5, 10	6, 11	1/27	1/9	7.5	8.5	8.00
13	1	1, 3	3	1, 2		5, 12	13, 6	1/27	1/9	8.5	9.5	9.00
14	1	1, 3	3	1, 3		5, 12	13, 11	1/27	1/9	8.5	12.0	10.25
15	1	1, 3	3	2, 3		5, 12	6, 11	1/27	1/9	8.5	8.5	8.50
16	1	2, 3	3	1, 2		10, 12	13, 6	1/27	1/9	11.0	9.5	10.25
17	1	2, 3	3	1, 3		10, 12	13, 11	1/27	1/9	11.0	12.0	11.50
18	1	2, 3	3	2, 3		10, 12	6, 11	1/27	1/9	11.0	8.5	9.75
							Sum:	1/3	1			85.50
19	2	1, 2	3	1, 2		3, 15	13, 6	1/27	1/9	9.0	9.5	9.25
20	2	1, 2	3	1, 3		3, 15	13, 11	1/27	1/9	9.0	12.0	10.50
21	2	1, 2	3	2, 3		3, 15	6, 11	1/27	1/9	9.0	8.5	8.75
22	2	1, 3	3	1, 2		3, 6	13, 6	1/27	1/9	4.5	9.5	7.00
23	2	1, 3	3	1, 3		3, 6	13, 11	1/27	1/9	4.5	12.0	8.25
24	2	1, 3	3	2, 3		3, 6	6, 11	1/27	1/9	4.5	8.5	6.50
25	2	2, 3	3	1, 2		15, 6	13, 6	1/27	1/9	10.5	9.5	10.00
26	2	2, 3	3	1, 3		15, 6	13, 11	1/27	1/9	10.5	12.0	11.25
27	2	2, 3	3	2, 3		15, 6	6, 11	1/27	1/9	10.5	8.5	9.50
							Sum:	1/3	1			81.00

concepts of **total expectation** and **conditional expectation** formally considered in Section A.3.7. The law of total expectation provides that the unconditional expectation of $\hat{\theta}_{2s}$ may be expressed as

$$E(\hat{\theta}_{2s}) = E_1[E_2(\hat{\theta}_{2s}|S_1)] \tag{9.3}$$

where E_1 denotes **first-stage expectation**—expectation over all possible first-stage samples (i.e., over the first-stage sample space, S_1), and E_2 denotes **second-stage expectation**—the conditional expectation of the estimator over all possible second-stage sample selections *given a particular first-stage selection*. $E_2(\hat{\theta}_{2s}|S_1)$ is thus a *conditional* expectation (i.e., it depends on which first-stage sample is selected) and will have as many values as there are possible first-stage samples. $E(\hat{\theta}_{2s}) = E_1[E_2(\hat{\theta}_{2s}|S_1)]$ is an *unconditional* expectation—over the entire two-stage sample space—and has just a single value. (A proof of Equation (9.3) is provided in the expanded online version of Appendix A at www.oup.co.uk/companion/hankin.)

For the second-stage (conditional) expectation, applying the general definition of expectation provides

$$E_2(\hat{\theta}_{2s}|S_1 = s_1) = \sum_{\substack{s \in S \\ S_1 = s_1}} \hat{\theta}_{2s}(s)p(s|s_1), \quad s_1 \in S_1 \tag{9.4}$$

That is, the second-stage (conditional) expectation of an estimator is equal to the weighted average of the values that the estimator takes on over all possible samples *given a particular first-stage selection*, with weights equal to $\{p(s|s_1)\}$. For each of the possible first-stage selections of clusters listed in Table 9.2, there are nine possible two-stage samples, all of which are equally likely ($p(s|s_1) = 1/9, s \in S, s_1 \in S_1$) so that, for $\hat{\mu}_{2s} = \hat{\theta}_{2s}$

$$E_2(\hat{\mu}_{2s}|S_1 = \{1,2\}) = (8.25 + 6.00 + \ldots + 10.75)(1/9) = 76.50/9 = 8.5$$

$$E_2(\hat{\mu}_{2s}|S_1 = \{1,3\}) = (8.50 + 9.75 + \ldots + 9.75)(1/9) = 85.50/9 = 9.5$$

$$E_2(\hat{\mu}_{2s}|S_1 = \{2,3\}) = (9.25 + 10.50 + \ldots + 9.50)(1/9) = 81.00/9 = 9.0$$

For the overall expection

$$E(\hat{\theta}_{2s}) = E_1[E_2(\hat{\theta}_{2s}|S_1)] = \sum_{s_1 \in S_1} E_2(\hat{\theta}_{2s}|S_1 = s_1)p(s_1) \tag{9.5}$$

Each of the three possible first-stage selections of primary units are equally likely ($p(s_1) = 1/3, s_1 \in S_1$) so that

$$E(\hat{\mu}_{2s}) = (8.5 + 9.5 + 9.0)(1/3) = 27/3 = 9 = \mu$$

The same logic applies to evaluation of expectation at more than two stages of sampling. For example, for the expectation of a three-stage estimator ($E(\hat{\theta}_{3s}) = E_1\{E_2[E_3(\hat{\theta}_{3s}|S_1, S_2)]\}$): (1) First evaluate the (conditional) third-stage expectations of the estimator by averaging estimates over all possible third-stage samples given a particular set of first- and second-stage selections; (2) Second, evaluate the (conditional) second-stage expectations of the estimator by averaging third-stage expectations over all possible second-stage samples given a particular first-stage selection; (3) Finally, evaluate the (unconditional) first-stage expectation of the estimator by averaging the second-stage expectations over all possible first-stage samples.

9.1.3 *Sampling variance and its estimation*

As for estimator expectation, estimator sampling variance can also be evaluated using an unconditional or conditional approach. For a two-stage estimator $\hat{\theta}_{2s}$, the unconditional formula is

$$V(\hat{\theta}_{2s}) = E([\hat{\theta}_{2s} - E(\hat{\theta}_{2s})]^2) = \sum_{s \in \mathcal{S}} [\hat{\theta}_{2s}(s) - E(\hat{\theta}_{2s})]^2 p(s)$$

All of the possible samples listed in Table 9.2 are equally likely ($p(s) = 1/27, s \in \mathcal{S}$) so that, for $\hat{\mu}_{2s} = \hat{\theta}_{2s}$

$$V(\hat{\mu}_{2s}) = \left[([8.25 - 9]^2 + \ldots) + ([8.50 - 9]^2 + \ldots) + ([9.25 - 9]^2 + \ldots) \right](1/27)$$

$$= (21.75 + 12 + 19.5)(1/27) = 1.972$$

As for two-stage expectation, however, the unconditional approach does not reveal the nature of sampling variance in a two-stage setting. The **law of total variance** (Section A.3.7, with a proof in the expanded online version of Appendix A at www.oup.co.uk/companion/hankin), expressed in terms relevant to two-stage estimators, provides that

$$V(\hat{\theta}_{2s}) = V_1[E_2(\hat{\theta}_{2s}|S_1)] + E_1[V_2(\hat{\theta}_{2s}|S_1)] \tag{9.6}$$

The first term in Equation (9.6) reflects uncertainty at the first stage of sampling (variation of conditional second-stage expectations over all possible first-stage samples)—first-stage variance, whereas the second term reflects uncertainty at the second-stage of sampling (average conditional second-stage sampling variance)—second-stage variance. If clusters are of equal size and SRS is used to select n from N at the first stage and m from M at the second stage, Equation (9.6) for the sampling variance of $\hat{\mu}_{2s}$ takes the form (Cochran 1977)

$$V(\hat{\mu}_{2s}) = \left(\frac{N-n}{N} \right) \frac{\sigma_1^2}{n} + \left(\frac{M-m}{M} \right) \frac{\sigma_2^2}{nm} \tag{9.7}$$

where

$$\sigma_1^2 = \frac{\sum\limits_{i=1}^{N}(\mu_i - \mu)^2}{N-1}; \quad \sigma_2^2 = \frac{\sum\limits_{i=1}^{N}\sum\limits_{j=1}^{M}(y_{ij} - \mu_i)^2}{N(M-1)} = \frac{\sum\limits_{i=1}^{N}\sigma_{2i}^2}{N}$$

The first term in Equation (9.7) is equal to the first-stage variance, whereas the second term is equal to the second-stage variance.[1] Note that an increase in n will reduce both the first- and second-stage variance, whereas an increase in m will reduce only the second-stage variance. Continuing with the clustered population variable values and parameters presented in Table 9.1

$$\sigma_1^2 = \frac{(9-9)^2 + (8-9)^2 + (10-9)^2}{3-1} = 1, \quad \text{and} \quad \sigma_2^2 = \frac{13 + 39 + 13}{3} = 21.667$$

If we substitute for n, N, m, M, σ_1^2, and σ_2^2 in Equation (9.7), we get

$$V(\hat{\mu}_{2s}) = \left(\frac{3-2}{3} \right) \frac{1}{2} + \left(\frac{3-2}{3} \right) \frac{21.667}{2 \cdot 2} = 0.167 + 1.805 = 1.972$$

---

[1] Note that, from the ANOVA approach to sampling variance in single-stage equal size cluster sampling (Section 6.2.1), $\sigma_1^2 = MS(B)/M$, and $\sigma_2^2 = MS(W)$.

as previously calculated. For this numerical example, nearly all of the sampling variance of $\hat{\mu}_{2s}$ is due to second-stage variance because average within cluster variance ($\sigma_2^2 = 21.667$) is so much larger than between cluster variation of cluster means ($\sigma_1^2 = 1$). In most practical applications of two-stage sampling, however, first-stage variance typically dominates second-stage variance—variation between primary unit means is usually large compared to variation in variable values within primary units.

Further exploration of the form of Equation (9.7) is informative. First, suppose that $m = M$, i.e., that all subunits in each selected cluster are selected at the second stage. Then, the second term in Equation (9.7) has value zero and the two-stage variance formula reduces to that for single-stage equal size cluster sampling [i.e., substituting $\sigma_1^2 = MS(B)/M$ gives Equation (6.3)]. Second, suppose instead that all clusters are sampled so that $n = N$. In that case, the first term in Equation (9.7) has value zero and the two-stage variance formula reduces to that for a special case of stratified sampling, namely the case where (a) stratum sizes are equal (all $N_h = M$), and (b) stratum sample sizes are equal (all $n_h = m$). For this special case, the stratum weights are all $W_h = 1/N$, where N is the number of strata (clusters), and the stratum fpc all equal $(M - m)/M$.

An unbiased estimator of $V(\hat{\mu}_{2s})$ is

$$\hat{V}(\hat{\mu}_{2s}) = \left(\frac{N-n}{N}\right)\frac{\hat{\sigma}_1^2}{n} + \frac{n}{N}\left(\frac{M-m}{M}\right)\frac{\hat{\sigma}_2^2}{nm} \tag{9.8}$$

where

$$\hat{\sigma}_1^2 = \frac{\sum\limits_{i\in S_1}(\hat{\mu}_i - \hat{\mu}_{2s})^2}{n-1}, \quad \hat{\sigma}_2^2 = \frac{\sum\limits_{i\in S_1}\sum\limits_{j\in S_{2i}}(y_{ij} - \hat{\mu}_i)^2}{n(m-1)} = \frac{\sum\limits_{i\in S_1}\hat{\sigma}_{2i}^2}{n} \tag{9.9}$$

and

$$\hat{\sigma}_{2i}^2 = \frac{\sum\limits_{j\in S_{2i}}(y_{ij} - \hat{\mu}_i)^2}{(m-1)}, \quad i\in S_1 \tag{9.10}$$

Let $f_1 = n/N$ denote the **first-stage sampling fraction**, and let $f_2 = m/M$ denote the **second-stage sampling fraction**. Then, we may re-express Equation (9.8) as

$$\hat{V}(\hat{\mu}_{2s}) = (1 - f_1)\frac{\hat{\sigma}_1^2}{n} + f_1(1 - f_2)\frac{\hat{\sigma}_2^2}{nm}$$

Some special cases of variance estimation are of interest. First, if $f_2 = 1$ (i.e., $m = M$), then $\hat{V}(\hat{\mu}_{2s}) = (1 - f_1)\hat{\sigma}_1^2/n$, the unbiased estimator of sampling variance in single-stage cluster sampling [Equation (6.14)]. Second, if $f_1 = 1$, then $\hat{V}(\hat{\mu}_{2s}) = (1 - f_2)\hat{\sigma}_2^2/(nm)$, an unbiased estimator for the special case of stratified sampling previously noted. Finally, if $f_1 = n/N \approx 0$, then

$$\hat{V}(\hat{\mu}_{2s}) \approx \frac{\hat{\sigma}_1^2}{n} = \frac{\sum\limits_{i=1}^{n}(\hat{\mu}_i - \hat{\mu}_{2s})^2/(n-1)}{n}$$

This last result implies that, when $f_1 = n/N$ is quite small, an approximately unbiased estimate of two-stage sampling variance can be made simply from the variation among the estimated primary unit means. Thus, for example, when f_1 is small, systematic sampling could be used for the second stage of sampling because it could be used to generate unbiased estimates (e.g., using $\hat{\mu}_{lsys,u}$ or $\hat{\mu}_{csys}$) of primary unit means, but unbiased estimates of within primary unit variances, σ_{2i}^2, would not be needed for the purpose of estimating $V(\hat{\mu}_{2s})$.

Estimation of first- and second-stage contributions

Although the expression for sampling variance of $\hat{\mu}_{2s}$ [Equation (9.7)] has two strictly additive terms that account separately for first-stage and second-stage variance contributions, it is incorrect to interpret the estimator [Equation (9.8)] in this same way. The two terms in Equation (9.8) are *not* unbiased estimates of variance contributions from the first and second stages of sampling, respectively. The first term in Equation (9.8), on average, overestimates the first-stage variance, sometimes greatly so, because $\hat{\sigma}_1^2$ is not an unbiased estimator of σ_1^2. This is because $\hat{\sigma}_1^2$ includes variation beyond that of the variation between cluster means due to its use of $\{\hat{\mu}_i, i \in S_1\}$ in place of $\{\mu_i, i \in S_1\}$. $\hat{\sigma}_2^2$ is, however, an unbiased estimator of σ_2^2, and an unbiased estimator of σ_1^2 is (Cochran 1977)

$$\hat{\sigma}_{1,un}^2 = \hat{\sigma}_1^2 - \frac{\hat{\sigma}_2^2(1-f_2)}{m} \tag{9.11}$$

Continuing with the clustered population variable values and parameters presented in Table 9.1 and the associated Table 9.2 sample space, Table 9.3 lists for all possible samples the estimates $\hat{\sigma}_1^2$, $\hat{\sigma}_{1,un}^2$, $\hat{\sigma}_2^2$, $\hat{V}(\hat{\mu}_{2s})$, as well as the first and second terms in Equation (9.8) (denoted as $\hat{V}_a$ and $\hat{V}_b$, respectively). The calculated second-stage expectations of these estimators, $E_2(\hat{\theta}|S_1)$, are also presented, as are the unconditional expectations, $E(\hat{\theta}) = E_1[E_2(\hat{\theta}|S_1)]$, which are the averages of the second-stage expectations because the first-stage samples are equally likely. Note that the expected value of $\hat{\sigma}_2^2 = 21.667 = \sigma_2^2$ (i.e., is unbiased), but the expected value of $\hat{\sigma}_1^2$ (4.6111) is much greater than σ_1^2 (1.00). As a consequence, the expected value of $\hat{V}_a$ (0.7685) is much larger than the actual first-stage variance contribution (0.166). Thus, if one used $\hat{V}_a$ as a proxy for the first-stage variance, on average one would infer that the first stage of sampling makes a much larger contribution to sampling variance than is actually the case. Finally, although the calculated expected values show that $\hat{\sigma}_{1,un}^2$ is indeed an unbiased estimator of σ_1^2, more than half of the samples resulted in negative estimates. (Note that the form of Equation (9.11) does not guarantee a positive result.)

9.1.4 *Optimal allocation*

For a given total survey cost, what is the optimal number of clusters to select and what is the optimal number of subunits to sample from within selected clusters? Intuitively, one would expect that first-stage sample size (n) should be large when between cluster variation is relatively large and that second-stage sample size (m) should be small when within cluster variation is relatively small. If the reverse were true, then perhaps a small number of clusters might be selected at the first stage, but second-stage sample size within selected clusters might be large.

As for stratified sampling, we address these questions by minimizing the sampling variance of the estimator subject to a fixed total survey cost through choice of n and m. We assume that the cost per subunit within clusters is the same for all clusters, and define the total cost function as $C = c_1 n + c_2 nm$, where c_1 is the cost per primary unit, and c_2 is the cost per subunit within primary units. One may interpret c_1 as the cost of travel and setup to take measurements within a given primary unit, whereas c_2 may be viewed as the cost per subunit of measuring y_{ij}. A two-stage sample requires that measurements be made in a total of nm subunits. We assume SRS selection of primary units and subunits at first and second stages. Optimal allocation results are identical for estimation of the mean or total, so we present results only for estimation of the mean of the target variable.

Table 9.3 Estimates of σ_1^2 ($\hat{\sigma}_1^2$, biased, and $\hat{\sigma}_{1,un}^2$, unbiased), σ_2^2 ($\hat{\sigma}_2^2$), and two-stage sampling variance, $V(\hat{\mu}_{2s})$ [$\hat{V}(\hat{\mu}_{2s})$], for the Table 9.2 two-stage sample space based on the Table 9.1 population. The estimated two-stage sampling variance, $\hat{V}(\hat{\mu}_{2s})$, is equal to the sum of the first term ($\hat{V}_a$) and second term ($\hat{V}_b$) in Equation (9.8). $E_2(\cdot|S_1 = \{i,i'\})$ denotes conditional second-stage expectation for indicated first-stage selection of two clusters.

Sample s_1	ID	$\hat{\sigma}_1^2$	$\hat{\sigma}_{1,un}^2$	$\hat{\sigma}_2^2$	$\hat{V}_a$	$\hat{V}_b$	$\hat{V}(\hat{\mu}_{2s})$
clusters 1, 2	1	1.125	−5.917	42.25	0.1875	2.3472	2.5347
	2	4.500	3.083	8.50	0.7500	0.4722	1.2222
	3	4.500	0.083	26.50	0.7500	1.4722	2.2222
	4	0.125	−7.917	48.25	0.0208	2.6806	2.7014
	5	8.000	5.583	14.50	1.3333	0.8056	2.1389
	6	2.000	−3.417	32.50	0.3333	1.8056	2.1389
	7	2.000	−4.167	37.00	0.3333	2.0556	2.3889
	8	21.125	20.583	3.25	3.5208	0.1806	3.7014
	9	0.125	−3.417	21.25	0.0208	1.1806	1.2014
$E_2(\cdot|S_1 = \{1,2\})$:		4.833	0.500	26.00	0.8056	1.4444	2.2500
clusters 1, 3	10	2.000	−1.083	18.50	0.3333	1.0278	1.3611
	11	10.125	8.917	7.25	1.6875	0.4028	2.0903
	12	0.500	−1.583	12.50	0.0833	0.6944	0.7778
	13	0.500	−3.583	24.50	0.0833	1.3611	1.4444
	14	6.125	3.917	13.25	1.0208	0.7361	1.7569
	15	0.000	−3.083	18.50	0.0000	1.0278	1.0278
	16	1.125	−1.083	13.25	0.1875	0.7361	0.9236
	17	0.500	0.167	2.00	0.0833	0.1111	0.1944
	18	3.125	1.917	7.25	0.5208	0.4028	0.9236
$E_2(\cdot|S_1 = \{1,3\})$:		2.666	0.500	13.00	0.4444	0.7222	1.1660
clusters 2, 3	19	0.125	−7.917	48.25	0.0208	2.6806	2.7014
	20	4.500	−1.667	37.00	0.7500	2.0556	2.8056
	21	0.125	−6.917	42.25	0.0208	2.3472	2.3681
	22	12.500	10.083	14.50	2.0833	0.8056	2.8889
	23	28.125	27.583	3.25	4.6875	0.1806	4.8681
	24	8.000	6.583	8.50	1.3333	0.4722	1.8056
	25	0.500	−4.917	32.50	0.0833	1.8056	1.8889
	26	1.125	−2.417	21.25	0.1875	1.1806	1.3681
	27	2.000	−2.417	26.50	0.3333	1.4722	1.8056
$E_2(\cdot|S_1 = \{2,3\})$:		6.333	2.000	26.00	1.0556	1.4444	2.5000
Unconditional $E(\cdot)$:		4.611	1.000	21.67	0.7685	1.2037	1.9722

Initially, we formulate the minimization problem in terms of n and the second-stage sampling fraction, $f_2 = m/M$, rather than in terms of n and m. Expressing $V(\hat{\mu}_{2s})$ [Equation (9.7)] in terms of f_2 gives

$$V(\hat{\mu}_{2s}) = \left(\frac{N-n}{N}\right)\frac{\sigma_1^2}{n} + \frac{(1-f_2)\sigma_2^2}{nf_2M}$$

and the cost function becomes $C = c_1n + c_2nf_2M$. To minimize $V(\hat{\mu}_{2s})$ subject to the fixed total cost C, we form the Lagrange function (Section A.9)

$$\mathcal{L}(\lambda, n, f_2) = V(\hat{\mu}_{2s}) + \lambda(C - c_1n - c_2nf_2M)$$

and partially differentiate $\mathcal{L}(\lambda, n, f_2)$ with respect to n and f_2

$$\frac{\partial \mathcal{L}}{\partial n} = -\frac{1}{n^2}\left[\sigma_1^2 + \frac{(1-f_2)\sigma_2^2}{f_2M}\right] - \lambda(c_1 + c_2f_2M)$$

$$\frac{\partial \mathcal{L}}{\partial f_2} = -\frac{\sigma_2^2}{f_2^2nM} - \lambda c_2nM$$

Setting these partial derivatives equal to zero, solving each resulting equation for the unknown multiplier λ, and then equating the two solutions for λ gives

$$\frac{1}{n^2}\left[\sigma_1^2 - \frac{\sigma_2^2}{M}\right] = \frac{1}{n^2}\left[\frac{c_1\sigma_2^2}{c_2f_2^2M^2}\right]$$

Cancelling out $1/n^2$ from both sides of the equation allows us to solve for the optimal second-stage sampling fraction, $f_{2,\mathrm{opt}}$. The solution, in a form accentuating the ratios c_1/c_2 and σ_2^2/σ_1^2, is

$$f_{2,\mathrm{opt}} = \sqrt{\frac{c_1}{c_2} \cdot \frac{(\sigma_2^2/\sigma_1^2)(1/M^2)}{1 - (\sigma_2^2/\sigma_1^2)(1/M)}} \tag{9.12}$$

and

$$m_{\mathrm{opt}} = f_{2,\mathrm{opt}}M = \sqrt{\frac{c_1}{c_2} \cdot \frac{\sigma_2^2/\sigma_1^2}{1 - (\sigma_2^2/\sigma_1^2)(1/M)}} \tag{9.13}$$

for $\sigma_2^2/M < \sigma_1^2$.[2] The solution indicates that m_{opt} increases with the relative costs of sampling at the cluster and subunit levels, and increases with the relative variation within and between clusters, as one would expect.

Note that $f_{2,\mathrm{opt}}$ and m_{opt} are independent of the first-stage sample size, n, which implies that there is an optimal fraction (number) of subunits to select from within each cluster, irrespective of the number of clusters that are selected or how much money is available to carry out a two-stage cluster sample survey. Once m_{opt} has been calculated, then the corresponding number of clusters to select, for a fixed cost C, is

$$n_{\mathrm{opt}} = C/(c_1 + c_2m_{\mathrm{opt}}) \tag{9.14}$$

Example 9.1. Suppose $M = 49$, $\sigma_1^2 = 5$, $\sigma_2^2 = 49$, $c_1 = 12$, $c_2 = 3$, and that $C = \$1,000$. From Equation (9.13)

$$m_{\mathrm{opt}} = \sqrt{\frac{12}{3} \cdot \frac{49/5}{1 - (49/5)(1/49)}} = 7$$

---

[2] If $\sigma_2^2/M \geq \sigma_1^2$, then $m = M$ and there is no benefit to be gained from multi-stage sampling.

Given the solution for m_{opt}, from Equation (9.14) $n_{opt} = 1{,}000 / (12 + 3 \cdot 7) = 1{,}000 / 33 \approx 30$ primary units.[3]

9.1.5 *Net relative efficiency*

To evaluate the net relative efficiency of an optimally allocated two-stage strategy using $\hat{\mu}_{2s}$ compared to a single-stage SRS survey strategy using $\hat{\mu}_{mpu}$ requires that a cost function be specified for the single-stage survey. We assume that $C = c_2 n_{srs}$, that is, no cluster structure in the population and no fixed costs,[4] where n_{srs} is the number of units selected from a total of NM units. In this case

$$\text{NRE}(\hat{\mu}_{2s,opt}, \hat{\mu}_{mpu}) = \frac{V(\hat{\mu}_{mpu}|C)}{V(\hat{\mu}_{2s,opt}|C)} = \frac{\left(\frac{NM-n_{srs}}{NM}\right)\frac{\sigma^2}{n_{srs}}}{\left(\frac{N-n_{opt}}{N}\right)\frac{\sigma_1^2}{n_{opt}} + \left(\frac{M-m_{opt}}{M}\right)\frac{\sigma_2^2}{n_{opt}m_{opt}}} \quad (9.15)$$

The dependence of $\text{NRE}(\hat{\mu}_{2s,opt}, \hat{\mu}_{mpu})$ on the various parameters can be clarified somewhat by assuming that N is large relative to n_{srs} and n_{opt} (i.e., that $n_{opt}/N \approx 0$ and $n_{srs}/(NM) \approx 0$). Making these simplifications and expressing the NRE in terms of the ratios c_1/c_2, σ_2^2/σ_1^2, and σ^2/σ_1^2 leads to

$$\text{NRE}(\hat{\mu}_{2s,opt}, \hat{\mu}_{mpu}) \approx \frac{\sigma^2/\sigma_1^2}{\left[\frac{c_1}{c_2} + m_{opt}\right]\left[1 + \left(\frac{M-m_{opt}}{M}\right)\frac{\sigma_2^2/\sigma_1^2}{m_{opt}}\right]} \quad (9.16)$$

where m_{opt} itself also depends on c_1/c_2, σ_2^2/σ_1^2, and M [Equation (9.13)]. With this equation graphical methods could be used to explore the dependence of the NRE on the four variables (the three ratios and M), as in Section 10.2.

If a pilot or previous two-stage survey had been conducted, estimates of the NRE evaluation parameters would be available: $\hat{\sigma}_{1,un}^2$ for σ_1^2 and $\hat{\sigma}_2^2$ for σ_2^2, along with M, c_1, and c_2. An estimator for σ^2 based on a two-stage survey is provided in the following paragraph. If C was also known, or particular values of C were being contemplated, there would be no need to use the NRE approximation [Equation (9.16)]. Instead, n_{opt} and n_{srs} could be calculated from the value of C, and substituted directly into Equation (9.15) to obtain the NRE.

Estimation of finite population variance

By analogy to the estimator for finite population variance in single-stage cluster sampling, Equation (6.18), an unbiased estimator of finite population variance in two-stage equal size cluster sampling, with SRS used at both stages, is

$$\hat{\sigma}_{2s}^2 = \frac{(N-1)M\hat{\sigma}_{1,un}^2 + N(M-1)\hat{\sigma}_2^2}{NM-1} \quad (9.17)$$

where $\hat{\sigma}_2^2$ and $\hat{\sigma}_{1,un}^2$ are given by Equations (9.9) and (9.11), respectively.

---

[3] Note that if C were large enough, it would be theoretically possible that $n_{opt} > N$. In such a situation, one might set $n_{opt} = N$, and then solve for an adjusted $m_{opt}^* = [(C/N) - c_1]/c_2$, assuming that m_{opt}^* remained less than M.

[4] Assuming that $C = c_2 n_{srs}$ results in a *minimum* value of $\text{NRE}(\hat{\mu}_{2s,opt}, \hat{\mu}_{mpu})$ because it ignores possible survey costs related to traveling between units selected in an SRS sample, and thus inflates the number of units that could be surveyed using single-stage SRS for an equivalent cost. A similar ambiguity in calculation of relative efficiency arises in adaptive cluster sampling (Section 11.1.4).

9.2 Two-stage sampling: Clusters of unequal size

Although the previous treatment of two-stage equal size cluster sampling is informative, it is far less useful than a more general treatment that would allow cluster size to vary. When cluster sizes vary, there are a large number of sampling strategies that might be considered for estimating the population mean per subunit or population total and the complexity of estimators is considerably greater than for the simpler case of equal size clusters. When clusters are of unequal size, these alternative two-stage sampling strategies are basically "multi-stage versions" of strategies that might otherwise be used in a simpler, single-stage setting. For that reason, we begin with a consideration of alternative estimators that might be considered in a single-stage setting when clusters are of unequal size.

9.2.1 *Single-stage cluster sampling*

When clusters are of unequal sizes, there are many alternative sampling strategies for estimation of the population mean-per-subunit and population total, and the relative performance of these strategies will depend, generally, on the correlation between the cluster totals and cluster sizes. When clusters are of unequal size, the total number of subunits in the population equals $\sum_{i=1}^{N} M_i = N\overline{M}$, where $\overline{M} = \sum_{i=1}^{N} M_i/N$, the mean cluster size. The mean-per-subunit for the ith cluster is $\mu_i = T_i/M_i$, and the population mean-per-subunit is $\mu = T/(N\overline{M})$.

Estimation of the population mean and total

For SRS selection of n clusters from N clusters of unequal size $\{M_i\}$, there are two obvious alternative estimators of the population mean-per-subunit and population total that we might consider—the "mean-per-unit" estimator, and the ratio estimator (for which cluster size is treated as an auxiliary variable). If clusters are instead selected with unequal probability, then we might consider the Hansen–Hurwitz estimator for PPSWR selection (with $p_i = M_i/\sum_{i=1}^{N} M_i$) or the Horvitz–Thompson estimator for PPSWOR selection (with π_i based on the $\{M_i\}$ but dependent on the method of selection).

Each of these estimators for the population total is displayed below, with S denoting the single-stage random sample of cluster units. These estimators for T are the standard ones (Equations (3.9), (7.3), (8.1), and (8.5), respectively), but given that the sampling unit is the cluster rather than the subunit, we replace y_i and x_i with T_i and M_i, respectively

$$\hat{T}_{\text{c,mpu}} = N\frac{\sum_{i \in S} T_i}{n}, \qquad \hat{T}_{\text{c,rat}} = N\overline{M}\frac{\sum_{i \in S} T_i}{\sum_{i \in S} M_i}, \qquad \hat{T}_{\text{c,HH}} = \frac{1}{n}\sum_{i \in S}\frac{T_i}{p_i}, \qquad \hat{T}_{\text{c,HT}} = \sum_{i \in S}\frac{T_i}{\pi_i}$$

For the corresponding population mean-per-subunit estimator, $\hat{\mu}_{\text{c}}$, divide the respective T estimator by $N\overline{M}$, the number of population subunits.

All of these estimators, except for the ratio estimator, are unbiased. For the ratio estimator, the bias becomes smaller as the number of selected clusters increases. If the cluster totals have substantial positive correlation with cluster sizes, then the ratio and unequal probability estimators should have smaller sampling variance than the mean-per-unit estimator.

Sample space illustration

We illustrate application of these estimators for the population mean-per-subunit to the clustered population presented in Table 9.4. For the population overall, $N = 4$, $\sum M_i = N\overline{M} = 12$, $\overline{M} = 3$, $\mu = 4.75$, $\sigma^2 = 9.2955$.

Table 9.4 Clustered population variable values and parameters used for illustration of single-stage, unequal size cluster sampling alternative estimators of the population mean-per-subunit, μ.

Cluster (i)	M_i	($y_{ij}, j = 1, 2, \ldots, M_i$)	T_i	μ_i	σ_{2i}^2
1	2	1, 4	5	2.50	2
2	4	7, 3, 10, 5	25	6.25	4
3	2	2, 8	10	5.00	2
4	4	2, 4, 9, 2	17	4.25	4

For the mean-per-unit and ratio estimators, $\hat{\mu}_{c,mpu}$ and $\hat{\mu}_{c,rat}$, $n = 2$ clusters were selected from the $N = 4$ clusters by SRS. For the Hansen–Hurvitz estimator, $\hat{\mu}_{c,HH}$, $n = 2$ clusters were selected from the $N = 4$ clusters by PPSWR with $p_i = M_i / \sum_{i=1}^{N} M_i$. In this case, $p_1 = p_3 = 1/6$ and $p_2 = p_4 = 1/3$. For the Horvitz–Thompson estimator, $\hat{\mu}_{c,HT}$, $n = 2$ clusters were selected from the $N = 4$ clusters by PPSWOR with π_i, π_{ij} based on the population $\{M_i\}$ following Chao (1982). For this PPSWOR selection scheme, $\pi_1 = \pi_3 = 1/3$ and $\pi_2 = \pi_4 = 2/3$. The sample space for these sampling designs, corresponding estimates, and estimator performance measures are shown in Table 9.5.

The results of this example numerically verify that $\hat{\mu}_{c,mpu}$, $\hat{\mu}_{c,HH}$, and $\hat{\mu}_{c,HT}$ are all unbiased estimators for the population mean-per-subunit, $\mu = 4.75$, whereas $\hat{\mu}_{c,rat}$ is not. The bias of the ratio estimator is small, however, and mean square error of the ratio estimator (0.620) is less than the alternative estimators except for $\hat{\mu}_{c,HT}$ ($V(\hat{\mu}_{c,HT}) = 0.542$). Most striking is the relatively poor performance of $\hat{\mu}_{c,mpu}$ ($V(\hat{\mu}_{mpu}) = 2.10$). When cluster totals ($\{T_i\}$)

Table 9.5 Sample space for single-stage, unequal size cluster sampling with $n = 2$ clusters selected from the $N = 4$ cluster population (Table 9.4) by SRS ($\hat{\mu}_{c,mpu}$ and $\hat{\mu}_{c,rat}$ estimators), PPSWR ($\hat{\mu}_{c,HH}$ estimator), and PPSWOR ($\hat{\mu}_{c,HT}$ estimator) using Chao's (1982) method based on the population $\{M_i\}$. Corresponding sample probabilities labeled as p_{srs}, p_{HH}, and p_{HT}, respectively.

Sample		Probability			Estimate			
ID	Clusters (s)	$p_{srs}(s)$	$p_{HH}(s)$	$p_{HT}(s)$	$\hat{\mu}_{c,mpu}$	$\hat{\mu}_{c,rat}$	$\hat{\mu}_{c,HH}$	$\hat{\mu}_{c,HT}$
1	1, 2	0.1666	0.1111	0.1333	5.000	5.000	4.375	4.375
2	1, 3	0.1666	0.0555	0.0666	2.500	3.750	3.750	3.750
3	1, 4	0.1666	0.1111	0.1333	3.666	3.666	3.375	3.375
4	2, 3	0.1666	0.1111	0.1333	5.833	5.833	5.625	5.625
5	2, 4	0.1666	0.2222	0.4000	7.000	5.250	5.250	5.250
6	3, 4	0.1666	0.1111	0.1333	4.500	4.500	4.625	4.625
7	1, 1		0.0277				2.500	
8	2, 2		0.1111				6.250	
9	3, 3		0.0277				5.000	
10	4, 4		0.1111				4.250	
			$E(\hat{\mu})$:		4.750	4.667	4.750	4.750
			$V(\hat{\mu})$:		2.100	0.613	0.844	0.542
			$MSE(\hat{\mu})$:		2.100	0.620	0.844	0.542

have high variation and there is a substantial positive correlation between cluster totals and cluster sizes ($\{M_i\}$), then one can expect the mean-per-unit estimator to have large sampling variance compared to estimators that take advantage of the correlation. For our simple numerical example, the two small cluster totals ($T_1 = 5$, $T_3 = 10$) were associated with the two small clusters ($M_1 = M_3 = 2$), whereas the two large cluster totals ($T_2 = 25$, $T_4 = 17$) were associated with the two large clusters ($M_2 = M_4 = 4$), and the correlation was equal to 0.897.

9.2.2 *Two-stage estimation of the population mean and total*

When cluster sizes vary and only a sample of the subunits within selected clusters are surveyed, then there are two stages of sampling and there are a large number of alternative sampling strategies that might be considered. To simplify initial conceptual presentation of these alternative strategies, we temporarily ignore details of selection of subunits within clusters and of estimation of cluster totals. Instead, we assume that some unspecified method has been used to generate unbiased estimates of T_i for selected clusters that appear in a first-stage sample. This simplification allows us to very quickly form two-stage estimators of the population mean-per-subunit and population total by simply substituting these estimated primary unit totals $\{\hat{T}_i\}$ for the primary unit totals $\{T_i\}$ in the single-stage estimators described in the previous section

$$\hat{T}_{2s,mpu} = N \frac{\sum_{i \in S_1} \hat{T}_i}{n}, \qquad \hat{T}_{2s,rat} = N\bar{M} \frac{\sum_{i \in S_1} \hat{T}_i}{\sum_{i \in S_1} M_i}, \qquad \hat{T}_{2s,HH} = \frac{1}{n} \sum_{i \in S_1} \frac{\hat{T}_i}{p_i}, \qquad \hat{T}_{2s,HT} = \sum_{i \in S_1} \frac{\hat{T}_i}{\pi_i}$$

For the corresponding population mean-per-subunit estimator, $\hat{\mu}_{2s}$, divide the respective T estimator by $N\bar{M}$, the number of population subunits.

9.2.3 *Sampling variance and its estimation*

General expressions

Exact or approximate expressions for sampling variance and for estimation of sampling variance for these two-stage estimators of the population mean-per-subunit and population total depend not just on the selection method used to select primary units (clusters), but also on the method used to select subunits within clusters and the estimator used to estimate individual cluster totals (i.e., the within cluster sampling strategy). For example, subunits within clusters might be of unequal size, some PPSWOR selection method might be used to select subunits within clusters, and the Horvitz–Thompson estimator might be used for estimation of cluster totals. In that case, if the $\{m_i\}$ were fixed, then the Sen–Yates–Grundy variance expressions [Equations (8.9) and (8.10)] could be used to calculate $V(\hat{T}_i)$ and $\hat{V}(\hat{T}_i)$, respectively. Alternatively, subunits within clusters might be selected by SRS and the usual mean-per-unit estimator or perhaps a ratio estimator might be used for estimation of cluster totals, and appropriate associated expressions used for $V(\hat{T}_i)$ and $\hat{V}(\hat{T}_i)$. Therefore, a sampling theory text may contain a rather bewildering array of two-stage variance formulas and it can be a daunting and often impossible task to locate an expression for sampling variance that exactly matches the selection methods and estimators used at different stages of sampling in a complex multi-stage design.

Raj (1968) derived generalized expressions for sampling variance of the four two-stage estimators of a population total. These expressions are valid for any probability sampling strategy that might be used at the second stage of sampling. As we will see in the following, specification of the particular selection methods and estimators used at the second stage of sampling will lead to *particular* expressions that are special cases of these generalized expressions.

$$V(\hat{T}_{2s,\text{mpu}}) = N^2 \left(\frac{N-n}{N}\right) \frac{\sigma_{1,\text{mpu}}^{*2}}{n} + \frac{N}{n} \sum_{i=1}^{N} V(\hat{T}_i) \tag{9.18}$$

$$V(\hat{T}_{2s,\text{rat}}) \approx N^2 \left(\frac{N-n}{N}\right) \frac{\sigma_{1,\text{rat}}^{*2}}{n} + \frac{N}{n} \sum_{i=1}^{N} V(\hat{T}_i) \tag{9.19}$$

$$V(\hat{T}_{2s,\text{HH}}) = \frac{1}{n} \sum_{i=1}^{N} p_i \left(\frac{T_i}{p_i} - T\right)^2 + \frac{1}{n} \sum_{i=1}^{N} \frac{V(\hat{T}_i)}{p_i} \tag{9.20}$$

$$V(\hat{T}_{2s,\text{HT}}) = \frac{1}{2} \sum_{i=1}^{N} \sum_{j \neq i}^{N} (\pi_i \pi_j - \pi_{ij}) \left(\frac{T_i}{\pi_i} - \frac{T_j}{\pi_j}\right)^2 + \sum_{i=1}^{N} \frac{V(\hat{T}_i)}{\pi_i} \tag{9.21}$$

where $\sigma_{1,\text{mpu}}^{*2} = \sum_{i=1}^{N}(T_i - \bar{M}\mu)^2/(N-1)$, the between cluster variation in the cluster totals, and $\sigma_{1,\text{rat}}^{*2} = \sum_{i=1}^{N}(T_i - M_i\mu)^2/(N-1) = \sum_{i=1}^{N} M_i^2(\mu_i - \mu)^2/(N-1)$, the between cluster (weighted) variation in the mean-per-subunit. Notice that the first-stage variance component in each of these formulas has the same form as the single-stage version of these strategies (no subsampling) that we have seen in previous chapters, with y_i and x_i being replaced by T_i and M_i, respectively.

Sampling variance estimators for the strategies using without replacement sampling at the first stage are, assuming n is fixed,

$$\hat{V}(\hat{T}_{2s,\text{mpu}}) = N^2 \left(\frac{N-n}{N}\right) \frac{\hat{\sigma}_{1,\text{mpu}}^{*2}}{n} + \frac{N}{n} \sum_{i \in S_1} \hat{V}(\hat{T}_i) \tag{9.22}$$

$$\hat{V}(\hat{T}_{2s,\text{rat}}) = N^2 \left(\frac{N-n}{N}\right) \frac{\hat{\sigma}_{1,\text{rat}}^{*2}}{n} + \frac{N}{n} \sum_{i \in S_1} \hat{V}(\hat{T}_i) \tag{9.23}$$

$$\hat{V}(\hat{T}_{2s,\text{HT}}) = \frac{1}{2} \sum_{i \in S_1} \sum_{\substack{j \in S_1 \\ j \neq i}} \left(\frac{\pi_i \pi_j - \pi_{ij}}{\pi_{ij}}\right) \left(\frac{\hat{T}_i}{\pi_i} - \frac{\hat{T}_j}{\pi_j}\right)^2 + \sum_{i \in S_1} \frac{\hat{V}(\hat{T}_i)}{\pi_i} \tag{9.24}$$

where $\hat{\sigma}_{1,\text{mpu}}^{*2} = \sum_{i \in S_1}(\hat{T}_i - \bar{M}\hat{\mu}_{2s,\text{mpu}})^2/(n-1)$, and $\hat{\sigma}_{1,\text{rat}}^{*2} = \sum_{i \in S_1}(\hat{T}_i - M_i\hat{\mu}_{2s,\text{rat}})^2/(n-1)$. Equations (9.22) and (9.24) are unbiased if $\hat{T}_i$ and $\hat{V}(\hat{T}_i)$ are unbiased estimators.

If with replacement sampling is used at the first stage, then an unbiased estimator of sampling variance is (Raj 1968 Section 6.5)

$$\hat{V}(\hat{T}_{2s,\text{HH}}) = \frac{1}{n} \frac{\sum_{i \in S_1} [(\hat{T}_i/p_i) - \hat{T}_{2s,\text{HH}}]^2}{n-1} \tag{9.25}$$

Thus, when clusters are selected by PPSWR, unbiased estimation of sampling variance does not require estimation of $V(\hat{T}_i)$. Therefore, systematic sampling (e.g., using $\hat{\mu}_{\text{csys}}$ or $\hat{\mu}_{\text{lsys,u}}$) could be used for the second stage of sampling if the primary units were selected by PPSWR, even though systematic sampling would not allow for unbiased variance estimation at the second stage.

Sampling variance and estimators of sampling variance for the population mean-per-subunit estimators, $\hat{\mu}_{2s}$, can be obtained from those for the corresponding T estimator by dividing the respective $V(\hat{T}_{2s})$ or $\hat{V}(\hat{T}_{2s})$ formula by $(N\overline{M})^2$, the number of population subunits squared.

As noted previously, the particular forms that $\hat{T}_i$, $V(\hat{T}_i)$ and $\hat{V}(\hat{T}_i)$ take on will depend on the sampling strategy used to estimate cluster totals. For example, suppose that subunits within clusters were of variable size and were selected by PPSWR or PPSWOR. In that case $\hat{T}_i$, $V(\hat{T}_i)$ and $\hat{V}(\hat{T}_i)$ would have forms that correspond to the single-stage Hansen–Hurwitz or Horvitz–Thompson estimators of a total that are considered in Chapter 8 (Equations (8.1), (8.3), and (8.4) for PPSWR or Equations (8.5), (8.7), and (8.8), or (8.10) for PPSWOR). If SRS were instead used to select subunits from clusters, then the usual expressions for the mean-per-unit estimator of a total would be appropriate, and Equations (3.16) and (3.20) would be the appropriate expressions for $V(\hat{T}_i)$ and $\hat{V}(\hat{T}_i)$.

SRS within clusters

Because SRS is often used for selection of subunits at the second stage of sampling, we provide the particular forms that Raj's (1968) generalized formulas take for estimation of the population total for this common setting (and by extension for the population mean-per-subunit). If for selected primary unit i, m_i from M_i secondary units are selected by SRS, and the primary unit i total is estimated using the mean-per-unit estimator

$$\hat{T}_i = M_i \frac{\sum\limits_{j \in S_{2i}} y_{ij}}{m_i}, \quad i \in S_1$$

then Raj's (1968) formulas take on the following specific forms.

Mean-per-unit estimator.

$$V(\hat{T}_{2s,mpu}) = N^2 \left(\frac{N-n}{N}\right) \frac{\sigma^{*2}_{1,mpu}}{n} + \frac{N}{n} \sum_{i=1}^{N} M_i^2 \left(\frac{M_i - m_i}{M_i}\right) \frac{\sigma^2_{2i}}{m_i} \tag{9.26}$$

for which an unbiased estimator is

$$\hat{V}(\hat{T}_{2s,mpu}) = N^2 \left(\frac{N-n}{N}\right) \frac{\hat{\sigma}^{*2}_{1,mpu}}{n} + \frac{N}{n} \sum_{i \in S_1} M_i^2 \left(\frac{M_i - m_i}{M_i}\right) \frac{\hat{\sigma}^2_{2i}}{m_i} \tag{9.27}$$

Ratio estimator.

$$V(\hat{T}_{2s,rat}) \approx N^2 \left(\frac{N-n}{N}\right) \frac{\sigma^{*2}_{1,rat}}{n} + \frac{N}{n} \sum_{i=1}^{N} M_i^2 \left(\frac{M_i - m_i}{M_i}\right) \frac{\sigma^2_{2i}}{m_i} \tag{9.28}$$

and

$$\hat{V}(\hat{T}_{2s,rat}) = N^2 \left(\frac{N-n}{N}\right) \frac{\hat{\sigma}^{*2}_{1,rat}}{n} + \frac{N}{n} \sum_{i \in S_1} M_i^2 \left(\frac{M_i - m_i}{M_i}\right) \frac{\hat{\sigma}^2_{2i}}{m_i} \tag{9.29}$$

Note that the second-stage contributions to the sampling variance are identical for mean-per-unit and ratio estimators [compare Equations (9.26) and (9.28)]. The first-stage variance terms may, however, be quite different depending on the value of $\sigma^{*2}_{1,mpu}$ compared to $\sigma^{*2}_{1,rat}$.

Hansen–Hurwitz estimator.

Assuming that the first-stage sample of primary units is selected by PPSWR with $p_i = M_i/\sum_{i=1}^{N} M_i$

$$V(\hat{T}_{2s,HH}) = \frac{1}{n}\sum_{i=1}^{N} p_i \left(\frac{T_i}{p_i} - T\right)^2 + \frac{1}{n}\sum_{i=1}^{N} M_i^2 \left(\frac{M_i - m_i}{m_i}\right)\frac{\sigma_{2i}^2}{m_i}\frac{1}{p_i} \tag{9.30}$$

for which an unbiased estimator is given by Equation (9.25).

Horvitz–Thompson estimator.

Assuming that the first-stage (fixed size) sample of n primary units is selected by PPSWOR with π_i based on the $\{M_i\}$

$$V(\hat{T}_{2s,HT}) = \frac{1}{2}\sum_{i=1}^{N}\sum_{j\neq i}^{N} (\pi_i\pi_j - \pi_{ij})\left(\frac{T_i}{\pi_i} - \frac{T_j}{\pi_j}\right)^2 + \sum_{i=1}^{N} M_i^2 \left(\frac{M_i - m_i}{M_i}\right)\frac{\sigma_{2i}^2}{m_i}\frac{1}{\pi_i} \tag{9.31}$$

for which an unbiased estimator is

$$\hat{V}(\hat{T}_{2s,HT}) = \frac{1}{2}\sum_{i\in S_1}\sum_{\substack{j\in S_1 \\ j\neq i}} \left(\frac{\pi_i\pi_j - \pi_{ij}}{\pi_{ij}}\right)\left(\frac{\hat{T}_i}{\pi_i} - \frac{\hat{T}_j}{\pi_j}\right)^2 + \sum_{i\in S_1} M_i^2 \left(\frac{M_i - m_i}{M_i}\right)\frac{\hat{\sigma}_{2i}^2}{m_i}\frac{1}{\pi_i} \tag{9.32}$$

Sample space illustration: Horvitz-Thompson estimation

Consider the population of primary and secondary units, variable values, and associated primary unit parameters presented in Table 9.6. Overall population parameters are $T = 129$, $\mu = T/(N\overline{M}) = 129/(3\cdot4) = 43$, and the target of estimation is T. We provide an illustration of the two-stage sample space for $\hat{T}_{2s,HT}$ when primary units are selected by PPSWOR ($n = 2$), and secondary units are selected by SRS ($m_1 = 3$, $m_2 = 2$, $m_3 = 4$).

Chao's (1982) method was used for the PPSWOR selection of primary units for which the first order inclusion probabilities were $\pi_1 = 2/3$, $\pi_2 = 1/2$, and $\pi_3 = 5/6$. Second order inclusion probabilities, here denoted by $\pi_{ii'}$, were $\pi_{12} = 1/6$, $\pi_{13} = 1/2$, and $\pi_{23} = 1/3$. What is the probability then of selecting a particular sample? For $n = 2$, $\pi_{ii'}$ is equivalent to the probability of selecting the first-stage sample $S_1 = \{i, i'\}$. Given this particular first-stage sample, the number of possible second stage without replacement samples is equal to $\binom{M_i}{m_i}\binom{M_{i'}}{m_{i'}}$ and, because SRS was used at the second stage, all of these samples are equally likely so that $p(s|s_1) = 1/[\binom{M_i}{m_i}\binom{M_{i'}}{m_{i'}}]$. Therefore, the probability of any particular sample (first- and second-stages) being selected is

$$p(s) = p(s_1)\cdot p(s|s_1) = \frac{\pi_{ii'}}{\binom{M_i}{m_i}\binom{M_{i'}}{m_{i'}}}, \quad \text{for } s_1 = \{i, i'\}; i, i' = 1, 2, 3; i \neq i'$$

Table 9.6 Clustered population variable values and parameters used in Section 9.2.3.

Primary unit (*i*)	M_i	$(y_{ij}, j = 1, 2, \ldots, M_i)$	T_i	μ_i	σ_{2i}^2
1	4	8, 10, 12, 10	40	10.0	2.666
2	3	7, 9, 11	27	9.0	4.000
3	5	12, 14, 10, 12, 14	62	12.4	2.800

For example, with primary unit sample $s_1 = \{1, 2\}$, $\binom{M_1}{m_1} = \binom{4}{3} = 4$, and $\binom{M_2}{m_2} = \binom{3}{2} = 3$. Thus, $p(s_1) = \pi_{12} = 1/6$, $p(s|s_1) = 1/(4 \cdot 3) = 1/12$, and $p(s) = (1/6) \cdot (1/12) = 1/72$ for samples that include primary units 1 and 2. For each of the possible samples, primary unit totals were estimated as $\hat{T}_i = M_i \hat{\mu}_{i,\text{mpu}}$, where $\hat{\mu}_{i,\text{mpu}} = \sum_{j \in S_{2i}} y_{ij}/m_i$, and T was then estimated as $\hat{T}_{2s,\text{HT}} = \sum_{i \in S_1} \hat{T}_i/\pi_i$. $\hat{V}(\hat{T}_{2s,\text{HT}})$ was calculated using Equation (9.32).

The resulting sample space, sample probabilities, and associated Horvitz–Thompson estimates for the two-stage design where primary units are selected by PPSWOR using Chao's (1982) method and secondary units are selected by SRS is presented in Table 9.7. The table is organized according to the three possible primary unit samples, $\mathcal{S}_1 = \{\{1, 2\}, \{1, 3\}, \{2, 3\}\}$. Note that $\sum_{s \in \mathcal{S}, S_1 = s_1} p(s) = p(s_1)$. (Note also that this listing of all possible samples would be identical for two-stage mean-per-unit or ratio estimators with SRS selection at both stages—see Problems 9.8 and 9.9—but of course the corresponding sample probabilities and estimates would differ.)

The expectation and variance of the $\hat{T}_{2s,\text{HT}}$ estimator can now be calculated. We do this first using the usual, unconditional formulas.

$$E(\hat{T}_{2s,\text{HT}}) = \sum_{s \in \mathcal{S}} \hat{T}_{2s,\text{HT}}(s) p(s)$$

$$= (108 + \ldots + 124)(1/72) + (132 + \ldots + 139)(1/40) + (120 + \ldots + 135)(1/45)$$

$$= (1368 \cdot 1/72) + (2688 \cdot 1/40) + (1926 \cdot 1/45) = 129 = T$$

$$V(\hat{T}_{2s,\text{HT}}) = E([\hat{T}_{2s,\text{HT}} - E(\hat{T}_{2s,\text{HT}})]^2) = \sum_{s \in \mathcal{S}} [\hat{T}_{2s,\text{HT}}(s) - 129]^2 p(s)$$

$$= ([108 - 129]^2 + \ldots)(1/72) + ([132 - 129]^2 + \ldots)(1/40) + ([120 - 129]^2 + \ldots)(1/45)$$

$$= (3084 \cdot 1/72) + (844 \cdot 1/40) + (441 \cdot 1/45) = 73.73$$

Now we calculate the unconditional expectation and sampling variance of $\hat{T}_{2s,\text{HT}}$ using the laws of total expectation and variance. At the heart of these formulations are the second-stage *conditional* expectations, $E_2(\hat{T}_{2s,\text{HT}}|S_1)$, and sampling variances, $V_2(\hat{T}_{2s,\text{HT}}|S_1)$, which vary with the particular first-stage selection of primary units, $S_1 = s_1; s_1 \in \mathcal{S}_1$. Given a particular primary unit sample, $S_1 = s_1$, these conditional expectations and sampling variances are weighted averages of $\hat{T}_{2s,\text{HT}}$ and $[\hat{T}_{2s,\text{HT}} - E_2(\hat{T}_{2s,\text{HT}}|S_1 = s_1)]^2$ over all of the samples that can be selected (given $S_1 = s_1$), where the weights are the sample probabilities $\{p(s|s_1)\}$. With SRS at the second stage of sampling, the $\{p(s|s_1)\}$ are equal for all of these samples, so that the weights are equal, and the weighted averages reduce to simple averages. From Table 9.7, the second-stage conditional expectations and sampling variances are

$$E_2(\hat{T}_{2s,\text{HT}}|S_1 = s_1) = \begin{cases} 114.0, & s_1 = \{1, 2\} \\ 134.4, & s_1 = \{1, 3\}, \\ 128.4, & s_1 = \{2, 3\} \end{cases} \quad \text{and} \quad V_2(\hat{T}_{2s,\text{HT}}|S_1 = s_1) = \begin{cases} 32.00, & s_1 = \{1, 2\} \\ 13.04, & s_1 = \{1, 3\} \\ 29.04, & s_1 = \{2, 3\} \end{cases}$$

The law of total expectation gives the unconditional expectation

$$E(\hat{T}_{2s,\text{HT}}) = E_1[E_2(\hat{T}_{2s,\text{HT}}|S_1)] = \sum_{s_1 \in \mathcal{S}_1} E_2(\hat{T}_{2s,\text{HT}}|S_1 = s_1) p(s_1)$$

$$= (114 \cdot 1/6) + (134.4 \cdot 1/2) + (128.4 \cdot 1/3) = 129 = T$$

Table 9.7 Sample space, sample probabilities, and associated Horvitz–Thompson estimates for two-stage design with unequal size primary units. Primary units selected by PPSWOR ($n = 2$) using Chao's (1982) method and secondary units selected by SRS ($m_1 = 3, m_2 = 2, m_3 = 4$) from the clustered population ($M_1 = 4, M_2 = 3, M_3 = 5$) presented in Table 9.6. $E_2(\cdot|S_1 = \{i, i'\})$ denotes conditional second-stage expectation for indicated first-stage selection of two clusters.

	Sample units (s)								
	Cluster A		Cluster B						
ID	i	$j \in s_{2i}$	i	$j \in s_{2i}$	$p(s)$	$p(s|s_1)$	$\hat{T}_{2s,HT}$	$[\hat{T}_{2s,HT} - E_2(\hat{T}_{2s,HT}|S_1 = s_1)]^2$	$\hat{V}(\hat{T}_{2s,HT})$
1	1	1, 2, 3	2	1, 2	1/72	1/12	108.0	36.00	158.00
2	1	1, 2, 3	2	1, 3	1/72	1/12	114.0	0.00	68.00
3	1	1, 2, 3	2	2, 3	1/72	1/12	120.0	36.00	14.00
4	1	1, 2, 4	2	1, 2	1/72	1/12	104.0	100.00	72.67
5	1	1, 2, 4	2	1, 3	1/72	1/12	110.0	16.00	30.67
6	1	1, 2, 4	2	2, 3	1/72	1/12	116.0	4.00	24.67
7	1	1, 3, 4	2	1, 2	1/72	1/12	108.0	36.00	158.00
8	1	1, 3, 4	2	1, 3	1/72	1/12	114.0	0.00	68.00
9	1	1, 3, 4	2	2, 3	1/72	1/12	120.0	36.00	14.00
10	1	2, 3, 4	2	1, 2	1/72	1/12	112.0	4.00	264.67
11	1	2, 3, 4	2	1, 3	1/72	1/12	118.0	16.00	126.67
12	1	2, 3, 4	2	2, 3	1/72	1/12	124.0	100.00	24.67
				Sum:	1/6	1	$E_2(\cdot|S_1 = \{1,2\})$: 114.0	32.00	85.33
13	1	1, 2, 3	3	1, 2, 3, 4	1/40	1/20	132.0	5.76	28.00
14	1	1, 2, 3	3	1, 2, 3, 5	1/40	1/20	135.0	0.36	38.50
15	1	1, 2, 3	3	1, 2, 4, 5	1/40	1/20	138.0	12.96	46.00
16	1	1, 2, 3	3	1, 3, 4, 5	1/40	1/20	132.0	5.76	28.00
17	1	1, 2, 3	3	2, 3, 4, 5	1/40	1/20	135.0	0.36	38.50
18	1	1, 2, 4	3	1, 2, 3, 4	1/40	1/20	128.0	40.96	35.11
19	1	1, 2, 4	3	1, 2, 3, 5	1/40	1/20	131.0	11.56	48.28
20	1	1, 2, 4	3	1, 2, 4, 5	1/40	1/20	134.0	0.16	58.44
21	1	1, 2, 4	3	1, 3, 4, 5	1/40	1/20	128.0	40.96	35.11
22	1	1, 2, 4	3	2, 3, 4, 5	1/40	1/20	131.0	11.56	48.28
23	1	1, 3, 4	3	1, 2, 3, 4	1/40	1/20	132.0	5.76	28.00
24	1	1, 3, 4	3	1, 2, 3, 5	1/40	1/20	135.0	0.36	38.50
25	1	1, 3, 4	3	1, 2, 4, 5	1/40	1/20	138.0	12.96	46.00
26	1	1, 3, 4	3	1, 3, 4, 5	1/40	1/20	132.0	5.76	28.00
27	1	1, 3, 4	3	2, 3, 4, 5	1/40	1/20	135.0	0.36	38.50
28	1	2, 3, 4	3	1, 2, 3, 4	1/40	1/20	136.0	2.56	13.78
29	1	2, 3, 4	3	1, 2, 3, 5	1/40	1/20	139.0	21.16	21.61
30	1	2, 3, 4	3	1, 2, 4, 5	1/40	1/20	142.0	57.76	26.44
31	1	2, 3, 4	3	1, 3, 4, 5	1/40	1/20	136.0	2.56	13.78
32	1	2, 3, 4	3	2, 3, 4, 5	1/40	1/20	139.0	21.16	21.61
				Sum:	1/2	1	$E_2(\cdot|S_1 = \{1,3\})$: 134.4	13.04	33.97

Table 9.7 Continued

	Sample units (s)							
	Cluster A		Cluster B					
ID	i $j \in s_{2i}$		i	$j \in s_{2i}$	$p(s)$	$p(s|s_1)$	$\hat{T}_{2s,HT}$ $[\hat{T}_{2s,HT} - E_2(\hat{T}_{2s,HT}|S_1 = s_1)]^2$	$\hat{V}(\hat{T}_{2s,HT})$
33	2 1, 2	3	1, 2, 3, 4		1/45	1/15	120.0 70.56	154.00
34	2 1, 2	3	1, 2, 3, 5		1/45	1/15	123.0 29.16	193.75
35	2 1, 2	3	1, 2, 4, 5		1/45	1/15	126.0 5.76	233.00
36	2 1, 2	3	1, 3, 4, 5		1/45	1/15	120.0 70.56	154.00
37	2 1, 2	3	2, 3, 4, 5		1/45	1/15	123.0 29.16	193.75
38	2 1, 3	3	1, 2, 3, 4		1/45	1/15	126.0 5.76	109.00
39	2 1, 3	3	1, 2, 3, 5		1/45	1/15	129.0 0.36	139.75
40	2 1, 3	3	1, 2, 4, 5		1/45	1/15	132.0 12.96	170.00
41	2 1, 3	3	1, 3, 4, 5		1/45	1/15	126.0 5.76	109.00
42	2 1, 3	3	2, 3, 4, 5		1/45	1/15	129.0 0.36	139.75
43	2 2, 3	3	1, 2, 3, 4		1/45	1/15	132.0 12.96	46.00
44	2 2, 3	3	1, 2, 3, 5		1/45	1/15	135.0 43.56	67.75
45	2 2, 3	3	1, 2, 4, 5		1/45	1/15	138.0 92.16	89.00
46	2 2, 3	3	1, 3, 4, 5		1/45	1/15	132.0 12.96	46.00
47	2 2, 3	3	2, 3, 4, 5		1/45	1/15	135.0 43.56	67.75
				Sum:	1/3	1	$E_2(\cdot|S_1 = \{2,3\})$: 128.4 29.04	127.50

The law of total variance gives the unconditional sampling variance

$$V(\hat{T}_{2s,HT}) = V_1[E_2(\hat{T}_{2s,HT}|S_1)] + E_1[V_2(\hat{T}_{2s,HT}|S_1)]$$

$$= \sum_{s_1 \in \mathcal{S}_1} [E_2(\hat{T}_{2s,HT}|S_1 = s_1) - E(\hat{T}_{2s,HT})]^2 p(s_1) + \sum_{s_1 \in \mathcal{S}_1} V_2(\hat{T}_{2s,HT}|S_1 = s_1)p(s_1)$$

$$= \left[(114.0 - 129)^2(1/6) + (134.4 - 129)^2(1/2) + (128.4 - 129)^2(1/3)\right]$$

$$+ [32 \cdot (1/6) + 13.04 \cdot (1/2) + (29.04 \cdot (1/3)]$$

$$= 52.2 + 21.53 = 73.73$$

Alternatively, $V(\hat{T}_{2s,HT})$ could be calculated directly using the previously presented two-stage (SRS at the second stage) Horvitz–Thompson variance expression [Equation (9.31)]. Use of this equation gives the same result, $V(\hat{T}_{2s,HT}) = 52.2 + 21.53 = 73.73$.

Finally, the expected value of $\hat{V}(\hat{T}_{2s,HT})$ can also be calculated. From Table 9.7

$$E[\hat{V}(\hat{T}_{2s,HT})] = E_1(E_2[\hat{V}(\hat{T}_{2s,HT}|S_1)]) = \sum_{s_1 \in \mathcal{S}_1} E_2[\hat{V}(\hat{T}_{2s,HT}|S_1 = s_1)]p(s_1)$$

$$= (85.33 \cdot 1/6) + (33.97 \cdot 1/2) + (127.50 \cdot 1/3) = 73.73$$

numerically confirming that $\hat{V}(\hat{T}_{2s,HT})$ is unbiased for $V(\hat{T}_{2s,HT})$.

More than two stages

Raj's (1968) generalized expressions for sampling variance and for estimation of sampling variance for two-stage sampling have the identical form of Equations (9.18) and (9.22) (for

$\hat{T}_{2s,mpu}$), Equations (9.19) and (9.23) (for $\hat{T}_{2s,rat}$), Equations (9.20) and (9.25) (for $\hat{T}_{2s,HH}$), and Equations (9.21) and (9.24) (for $\hat{T}_{2s,HT}$) if there are three or more stages of sampling. The $V(\hat{T}_i)$ [or $\hat{V}(\hat{T}_i)$] are then sampling variances (or estimates of sampling variance) of the estimated primary unit totals based on sampling at the *second and subsequent stages of sampling*, and $\hat{T}_i$ is an estimate of the ith primary unit total based on sampling at the second and subsequent stages. For example, in a three-stage setting, a two-stage estimation process would be used to generate an estimate of the total for the ith cluster. For each cluster selected at the first stage of sampling, one would use appropriate two-stage formulas for calculating $V(\hat{T}_i)$ [or $\hat{V}(\hat{T}_i)$] and then these results would be substituted into Raj's (1968) general expressions for multi-stage sampling variance. For example, if SRS mean-per-unit estimators were used at the second and third stages of sampling, then Equations (9.18) and (9.22) would be used for $V(\hat{T}_i)$ and $\hat{V}(\hat{T}_i)$, respectively. When primary units are selected by PPSWR, Equation (9.25) remains an unbiased estimator of sampling variance even when there are three or more stages of sampling.

9.2.4 *Optimal allocation*

Given a particular selection of n primary units, one might argue, from optimal allocation results in stratified sampling [Jessen 1978 Equation (9.33)], that m_i should be proportional to $M_i\sigma_{2i}/\sqrt{c_{2i}}$, where c_{2i} is the (second-stage) cost per subunit in the ith selected cluster.[5] The form of the sampling variance for such a *conditional allocation scheme* would be complicated, however, and this idea is not used in practice. Instead, we derive here simpler rules that are approximately optimal. First, we assume that the cost per subunit is the same for all clusters (i.e., $c_{2i} = c_2$). Second, we note that in stratified sampling, simple proportional allocation is often *near optimal*. Simple proportional allocation in stratified sampling would be analogous to setting a single second-stage sampling fraction common to all selected primary units (i.e., setting $f_{2i} = f_2$). We therefore assume a constant second-stage sampling fraction, $f_2 = m_i/M_i$, for all selected primary units. We evaluate the case for which sampling is SRS at first and second stages, and a two-stage mean-per-unit estimator is used to estimate the population total. Our objective is to find the optimal second-stage sampling fraction, $f_{2,opt}$, subject to a specified cost function and cost constraint, and thereafter to determine the corresponding n_{opt}.

We begin by assuming a cost function of the form $C = c_1 n + c_2 \sum_{i \in S_1} m_i = c_1 n + c_2 f_2 \sum_{i \in S_1} M_i$. With a constant second-stage sampling fraction f_2, the total number of secondary units that will be sampled is a random variable (it depends on the selected $\{M_i\}$). For a given n and f_2, the *expected* cost would be $E(C) = c_1 n + c_2 f_2 E(\sum_{i \in S_1} M_i)$. In Section 8.3.6, we showed that an expectation such as this can be re-expressed as a sum over the population units rather than as a sum over the sample units as $E(\sum_{i=1}^N M_i Z_i) = \sum_{i=1}^N M_i E(Z_i)$, where Z_i is a binary (0, 1) random variable indicating whether or not primary unit i is selected, and that for SRS $E(Z_i) = \pi_i = n/N$. Therefore, $E(C) = c_1 n + c_2 f_2 \sum_{i=1}^N M_i n/N = c_1 n + c_2 n f_2 \bar{M}$. The following optimal allocation analysis is subject to this expected cost function and expected cost constraint.

Expressing $V(\hat{T}_{2s,mpu})$ [Equation (9.26)] in terms of f_2 gives

$$V(\hat{T}_{2s,mpu}) = N^2 \left(\frac{N-n}{N}\right)\frac{\sigma_{1,mpu}^{*2}}{n} + \frac{N}{n}\left(\frac{1-f_2}{f_2}\right)\sum_{i=1}^N M_i\sigma_{2i}^2 \qquad (9.33)$$

---

[5] The same logic applies to the equal size cluster sampling context (Section 9.1.4), because within cluster variances ordinarily vary across clusters. Thus, the optimal allocation results in equal size cluster sampling are also only approximately optimal.

To minimize $V(\hat{T}_{2s,mpu})$ subject to the expected cost constraint, we form the Lagrange function

$$\mathcal{L}(\lambda, n, f_2) = V(\hat{T}_{2s,mpu}) + \lambda[E(C) - c_1 n - c_2 n f_2 \overline{M}] \tag{9.34}$$

and partially differentiate $\mathcal{L}(\lambda, n, f_2)$ with respect to n and f_2

$$\frac{\partial \mathcal{L}}{\partial n} = -\frac{N}{n^2}\left[N\sigma^{*2}_{1,mpu} + \left(\frac{1-f_2}{f_2}\right)\sum M_i \sigma^2_{2i}\right] - \lambda(c_1 + c_2 f_2 \overline{M}) \tag{9.35}$$

$$\frac{\partial \mathcal{L}}{\partial f_2} = -\frac{N\sum M_i \sigma^2_{2i}}{n f_2^2} - \lambda c_2 n \overline{M} \tag{9.36}$$

Setting these partial derivatives equal to zero, solving each resulting equation for the unknown multiplier λ, and then equating the two solutions for λ gives

$$\frac{N}{n^2}\left[\frac{N\sigma^{*2}_{1,mpu} + \left(\frac{1-f_2}{f_2}\right)\sum M_i \sigma^2_{2i}}{c_1 + c_2 f_2 \overline{M}}\right] = \frac{N}{n^2}\left[\frac{\sum M_i \sigma^2_{2i}}{c_2 f_2^2 \overline{M}}\right]$$

Cancelling out N/n^2 from both sides of the equation allows us to solve for the optimal second-stage sampling fraction, $f_{2,opt}$[6]

$$f_{2,opt} = \sqrt{\frac{c_1}{c_2} \cdot \frac{\sum M_i \sigma^2_{2i}}{\overline{M}\left[N\sigma^{*2}_{1,mpu} - \sum M_i \sigma^2_{2i}\right]}} \tag{9.37}$$

and

$$m_{i,opt} = f_{2,opt} M_i \tag{9.38}$$

If primary units are of equal sizes, Equation (9.38) can be shown to be equivalent to Equation (9.13), the result previously generated for m_{opt} when primary units are of equal sizes (i.e., $\overline{M} = M_i = M$)

$$m_{opt} = f_{2,opt} M = \sqrt{\frac{c_1}{c_2} \cdot \frac{\sigma^2_2/\sigma^2_1}{1 - (\sigma^2_2/\sigma^2_1)(1/M)}}$$

where $\sigma^2_1 = \sum_i^N (\mu_i - \mu)^2/(N-1)$ and $\sigma^2_2 = \sum_i^N \sigma^2_{2i}/N$. (Note that $\sigma^{*2}_1 = M^2 \sigma^2_1$ when primary units are of equal size.)

9.3 Chapter comments

9.3.1 *Generality of the multi-stage framework*

While Horvitz–Thompson estimation may be said to establish a unifying framework for the conceptual basis of design-based sampling theory (see Chapter 8 comments), multi-stage sampling may be said to establish a general framework for specification of a wide range of sampling designs and sampling strategies. For example, as noted in Section 9.1.3, a two-stage design with equal probability sampling at both stages represents (a) stratified sampling when the first-stage sampling fraction $f_1 = n/N = 1$, (b) (single-stage) cluster sampling when $n/N < 1$ and the second-stage sample fractions $f_{2i} = m_i/M_i = 1$, and (c)

---

[6] An analogous solution for $f_{2,opt}$ exists for the two-stage ratio estimator [Equation (9.28)]. Just substitute $\sigma^{*2}_{1,rat}$ for $\sigma^{*2}_{1,mpu}$ in the solution for $f_{2,opt}$, Equation (9.37).

a full two-stage design whenever $f_1 < 1$ and $f_2 < 1$. In the two-stage setting, whatever sampling strategy appears most appropriate and cost-effective for estimation of primary unit totals may be employed. Indeed, these strategies may vary across primary units so long as they allow (ideally) unbiased or nearly unbiased estimation of primary unit totals and (usually but not always required) unbiased estimation of sampling variance of the estimated primary unit totals. Sampling strategies used at the first stage of sampling can also be similarly varied. The two-stage sampling context therefore incorporates all of the ideas and concepts that have been covered thus far in this text, and can be extended to three or more stages of sampling using Raj's (1968) general expressions for sampling variance and for estimation of sampling variance in the multi-stage setting (Equations (9.18)–(9.21) and Equations (9.22)–(9.24), respectively).

9.3.2 *Taking advantage of ecological understanding*

When considering implementation of a multi-stage sampling strategy, one can and should take advantage of everything that is known about the sampling setting to guide selection and adoption of such a strategy. For example, in surveys of small streams, where estimation of the abundance of a particular species of fish or stream amphibian is the target of estimation, it makes good biological and statistical sense to first stratify the stream units according to habitat type. For most species of fish and many aquatic amphibians, habitat preferences lead to substantial differences in densities of organisms per unit of habitat area across habitat types. Given that areas of habitat units within strata typically vary substantially, it makes good statistical and biological sense to expect that there will be a substantial positive correlation, within a habitat type stratum, between unit abundance (T) and habitat unit area (x). Therefore, two obvious candidates for two-stage sampling strategies for estimation of the total number of fish or, say, aquatic salamanders, across all primary units (within a given habitat type stratum) would be (a) SRS selection of primary units and use of a two-stage ratio estimator, or (b) PPSWOR selection of primary units and use of a Horvitz–Thompson estimator. Within each selected primary unit, conventional design-based methods might be used to count slow-moving salamander numbers within m_i subunits (e.g., strip transects) selected by SRS from the M_i subunits within each primary unit and $\hat{T}_{\mathrm{mpu}}$ might be used to estimate primary unit totals (Welsh et al. 1997). If the target of estimation were instead fish, which are a *rapidly moving target* with respect to space, then various model-based approaches such as removal method estimation based on electro-fishing might be used to estimate primary unit totals (Hankin 1984). Raj's (1968) general multi-stage sampling formulas remain approximately correct when a model-based estimator rather than a design-based estimator is used to estimate primary unit totals, assuming that model-based estimators provide approximately (model-) unbiased estimates of primary unit totals. Indeed, we believe that it is quite common in estimation of abundance of organisms over large areas to embed model-based estimates of unit-specific abundances (e.g., number of an endangered owl species on forest holding i) within the general design-based framework of multi-stage sampling.

9.3.3 *Implications for large-scale natural resource surveys*

In most applications of multi-stage sampling, errors of estimation arising from first-stage variance (a reflection of variation in primary unit totals or means, depending on sampling strategy) are large compared to errors of estimation that arise at the second and subsequent stages of sampling. This is especially true when there is large variation in primary unit sizes or habitat qualities. Such variation would ordinarily generate large variation in primary unit totals or means, respectively. If clusters are of equal size but there is a high degree of spatial structure underlying the distribution of target variable values over the population (Chapter 12), then between cluster variation will again be large if clusters consist of geographically proximate areal units. Also, it is important to remember that an increase in first-stage sample size will reduce errors of estimation at first and second stages of sampling, whereas an increase in sample size (or sampling effort) at the second stage will reduce errors of estimation at the second stage only. Together, these basic observations have important implications for large-scale abundance surveys in natural resource contexts. If the primary target of estimation is total abundance over a large number of geographic areas (primary units), then it is extremely important to have as large a first-stage sample size as can be afforded and, in many settings, it is far more important to have a large first-stage sample size than it is to have extremely accurate estimates of primary unit totals. Most ecologists and biologists receive extensive training in model-based estimation of abundance using removal method estimation, elaborate mark-recapture methods, or other approaches (Seber 1982), but they rarely receive training in how to develop coordinated population estimation programs across multiple geographic areas or how to combine estimates across these areas. The multi-stage sampling framework provides exactly these capabilities.

Problems

Problem 9.1. Consider the following population consisting of $N = 5$ clusters:

Cluster (i)	$(y_{i1}, y_{i2}, y_{i3}, y_{i4}, y_{i5})$
1	7, 15, 24, 6, 10
2	8, 20, 30, 10, 14
3	9, 12, 5, 4, 12
4	40, 50, 16, 20, 18
5	6, 7, 10, 14, 18

Suppose that you have $5,800 to sample the population and that the costs of sampling are (a) $1,050 per cluster for travel and set up, and (b) $105 per subunit within each cluster (once travel and set-up have been accomplished). Assuming SRS selection and mean-per-unit estimation at both stages of sampling, and using Section 9.2.4 as a guide: (A) Find the optimal number of subunits to select within each cluster, and the associated optimal number of clusters to select. (Round optimal sample sizes to the nearest integer.) (B) Calculate the sampling variance associated with this optimal two-stage allocation. (C) Suppose instead that a single SRS sample with the same number of subunits could be selected from the population's $N\overline{M}$ subunits. What would sampling variance be for this single-stage SRS sample? (D) Is it *fair* or meaningful to compare the sampling variances calculated in parts (B) and (C)? Why or why not?

Problem 9.2. Suppose that n unequal size clusters are selected by SRS from N, and that m_i subunits are selected by PPSWOR from the M_i subunits within selected cluster i. Let y_{ij} and x_{ij} denote the y and x variable values for subunit j in cluster i, respectively; let $\pi_{i(j)}$ (or $\pi_{i(j')}$) denote Pr{subunit j (or j') is included in a second-stage sample selected from cluster i}, and let $\pi_{i(jj')}$ denote Pr{subunits j and j' are included in a second-stage sample from cluster i}. (A) Using Equation (9.18) as a guide, develop an expression for sampling variance of the two-stage mean-per-unit estimator of T_y for this sampling design. (B) Using Equation (9.19) as a guide, develop an expression for sampling variance of the two-stage ratio estimator of T_y for this sampling design.

Problem 9.3. Two herpetologists have proposed a two-stage design for estimation of the total number of tailed frog (*Ascaphus truei*) tadpoles in all N shallow riffle habitat units in a specified reach of a small stream. To do this, they propose to first identify and flag all such riffle habitat units, measure the area (m^2) of each unit, and then to select $n = 10$ from N by SRS. Within selected riffles, they propose to create M_i 1 m cross-stream *strips* and to select $m = 4$ of these strips by SRS. Within each strip, they will remove all rocks and count the number of attached tadpoles. (A) Propose at least two alternative two-stage estimators of total tadpole abundance that would be consistent with the proposed sampling design (and which ignore possible variation in strip lengths within primary units). (B) What are the expressions for sampling variance for these alternative two-stage estimators?

Problem 9.4. Show algebraically that the solution for the optimal number of subunits to select from each selected primary unit when primary units are of unequal sizes [Equations (9.37) and (9.38)] simplifies to the solution for m_{opt} [Equation (9.13)] when primary units are of equal size.

Problem 9.5. Suppose that unequal size first-stage units are selected by SRS and that $\sigma^{*2}_{1,\text{rat}} < \sigma^{*2}_{1,\text{mpu}}$. How would $f_{2,\text{opt,rat}}$ compare with $f_{2,\text{opt,mpu}}$?

Problem 9.6. Suppose that n primary units are selected from N by PPSWOR. At the second stage, suppose that subunit sizes vary within primary units and that a sample of m_i subunits is selected from the M_i subunits within selected primary unit i using the Sen-Midzuno procedure (first unit selected with $p_{ij} = x_{ij}/T_{x,i}$, where x_{ij} is the size of the subunit j in primary unit i and $T_{x,i}$ is the known total of all subunit sizes in primary unit i, and the remaining units are selected by SRS). Primary unit totals are estimated using Lahiri's unbiased ratio estimator. (A) Construct the appropriate two-stage expressions for (1) estimation of the total of the target variable, and (2) sampling variance of this estimator, letting $V(\hat{R}_{L,i})$ denote sampling variance of Lahiri's ratio estimator in primary unit i. Would the estimator of the population total be unbiased? (B) Suppose that you collected a particular two-stage sample using this design. What expression would you use to estimate the sampling variance? Would you have any concerns about using this expression? (See Problem 8.9 and Equation (8.25).)

Problem 9.7. For the equal size cluster example population (Table 9.1), verify numerically that Equation (9.17) is unbiased for σ^2, the finite population variance (i.e., assuming no cluster structure).

Problem 9.8. For the unequal size cluster example population presented in Table 9.6 and corresponding sample space listed in Table 9.7: (A) Calculate all two-stage sample mean-per-unit estimates, $\hat{T}_{2s,\text{mpu}}$, and associated estimates of sampling variance, $\hat{V}(\hat{T}_{2s,\text{mpu}})$. (B) Calculate the second-stage conditional expectation of the two estimators, $E_2(\hat{T}_{2s,\text{mpu}})$ and $E_2[\hat{V}(\hat{T}_{2s,\text{mpu}})]$, and also the second-stage conditional sampling variance,

$V_2(\hat{T}_{2s,mpu})$, for each of the three possible first-stage selections of two primary units. (C) Using Equation (9.6), calculate the (unconditional) sampling variance of $\hat{T}_{2s,mpu}$ over the sample space, calculate the (unconditional) expectations of $\hat{T}_{2s,mpu}$ and $\hat{V}(\hat{T}_{2s,mpu})$, and verify numerically that these are both unbiased estimators. (D) Verify numerically that Equation (9.26) gives the same result that you obtained for $V(\hat{T}_{2s,mpu})$ from your direct calculations over the sample space.

Problem 9.9. For the unequal size cluster example population presented in Table 9.6 and corresponding sample space listed in Table 9.7: (A) Calculate all two-stage sample ratio estimates, $\hat{T}_{2s,rat}$, and associated estimates of sampling variance, $\hat{V}(\hat{T}_{2s,rat})$. (B) Calculate the second-stage conditional expectation of the two estimators, $E_2(\hat{T}_{2s,rat})$ and $E_2[\hat{V}(\hat{T}_{2s,rat})]$, and also the second-stage conditional sampling variance, $V_2(\hat{T}_{2s,rat})$, for each of the three possible first-stage selections of two primary units. (C) Using Equation (9.6), calculate the (unconditional) sampling variance of $\hat{T}_{2s,rat}$ over the sample space, calculate the (unconditional) expectations of $\hat{T}_{2s,rat}$ and $\hat{V}(\hat{T}_{2s,rat})$. Are these estimators unbiased? (D) Verify numerically that Equation (9.28) gives approximately the same result that you obtained for $V(\hat{T}_{2s,rat})$ from your direct calculations over the sample space. (E) Compare your results to those obtained in Problem 9.8. What is the relative efficiency of $\hat{T}_{2s,rat}$ compared to $\hat{T}_{2s,mpu}$ for this population?

CHAPTER 10

Multi-phase sampling

Fig. 10.1 Adult male Galapagos land iguana, *Conolophus subcristatus*, from Isabella Island. A double sampling with stratification strategy could be used to estimate mean adult weight. Photo credit: D. Hankin.

In **multi-phase sampling**, there are two or more levels or *phases* of sample selection and different sets of observations are made at each phase of sampling. The simplest setting, which we consider in this chapter, has just two phases of sampling and is often termed **double sampling**. (As noted in Chapter 1, Neyman (1938) developed the initial theory for double sampling.) In double sampling, a large first-phase sample of units, of size n_1, is selected according to some scheme and values of an easily available and inexpensive to measure auxiliary variable, x, are recorded for every unit in n_1. Then, a smaller second-phase sample, of size n_2, is selected, usually from the first-phase sample of n_1 units; alternatively n_2 could be independently selected from N. If the second-phase sample is selected from the first-phase sample, then only values of the target variable, y, need to be recorded or measured in the second-phase sample because the auxiliary variable values

Sampling Theory: For the Ecological and Natural Resource Sciences. David G. Hankin, Michael S. Mohr, and Ken B. Newman, Oxford University Press (2019). © David G. Hankin, Michael S. Mohr, and Ken B. Newman. DOI: 10.1093/oso/9780198815792.001.0001

are known from the first-phase sample. If the second-phase sample has been selected independently of the first-phase sample, however, then values of both the auxiliary and target variables need to be recorded (or measured) at the second phase of sampling because there would generally be little or no overlap between the two samples.

In this chapter, we illustrate two quite different applications of double sampling. In the first application, we show how (single-phase) ratio and regression estimation (considered in Chapter 7) can be extended to a two-phase setting. In the two-phase setting, a large first-phase sample is used to get an accurate estimate of the mean or total of the auxiliary variable, whereas these values are assumed known in single-phase ratio or regression estimation. In the second application, we show how the double sampling approach can be used to develop two-phase stratified estimators, similar to the (single-phase) stratified estimators considered in Chapter 5. Here, the large first-phase sample is used to estimate unknown stratum weights (stratum sizes), when the number of strata and stratum membership criteria are known or specified prior to sampling. In each of these applications, measurements of target variable values are only made in the second-phase sample of size n_2 and, since auxiliary variable population parameters or stratum sizes are estimated rather than known without error, errors of estimation will be larger for a two-phase survey than for a single-phase survey with sample size n_2 where these quantities are known. In some settings, however, values of auxiliary variables are not known for all units (or auxiliary variable population parameters or stratum sizes are not known), but the cost of obtaining auxiliary variable values may be considerably less than for target variables. In such instances, the double sampling approach may be quite cost-effective.

In most double sampling application settings, the key to success is finding an inexpensively recorded or measured auxiliary variable that is highly correlated with the target variable which is typically much more expensive to record (or measure). For example, in forestry work, a *timber cruiser* can walk through a large first-phase random sample of n_1 stands of timber, selected from a population of N stands. In this first-phase sample, the timber cruiser makes inexpensive and rapid visual assessments of timber volume that might be generated from each stand in n_1. In a small random subsample, n_2, of these n_1 stands, field technicians make detailed tree-specific measurements of timber volume which, when added together, provide nearly error-free measurements of actual timber volumes in second-phase units. If the timber cruiser's visual estimates of timber volume are highly correlated with actual measured volumes, then the two-phase survey may prove a more cost-effective approach for estimating total timber volume than a single-phase survey of equivalent total cost in which only expensive measurements of actual timber volume are made. That is, for the same total survey cost, the two-phase survey may have smaller sampling variance than a single-phase survey. In surveys of fish and habitat in small streams, first-phase measurements might be visual estimates of habitat unit areas or visual counts of fish made by divers while snorkeling; corresponding second-phase measurements would be very accurate measurements of habitat areas or very accurate estimates of fish abundance. In the treatment that follows, we assume that the second-phase measurements of target variable values are error-free.

10.1 Two-phase estimation of the population mean and total

We begin our treatment of two-phase or double sampling with ratio and regression estimation of the mean and total. We assume that first-phase units, n_1, are selected by SRS from N, and that second-phase units, n_2, are selected by SRS from the n_1 units selected at the first phase. As for the multi-stage strategies explored in Chapter 9, however, the basic framework of two-phase sampling allows for alternative methods of selecting units at the first- and/or second-phases of sampling (Section 10.4).

10.1.1 *Estimators*

Let y_i and x_i denote the target and auxiliary variable values, respectively, for unit i, $i = 1, 2, \ldots, N$, and denote the corresponding population totals as $T_y = \sum_{i=1}^{N} y_i$ and $T_x = \sum_{i=1}^{N} x_i$, and population means as $\mu_y = T_y/N$ and $\mu_x = T_x/N$. Define the following first- and second-phase estimators of μ_x and μ_y

$$\hat{\mu}_{1x} = \sum_{i \in S_1} x_i/n_1, \qquad \hat{\mu}_{2x} = \sum_{i \in S_2} x_i/n_2, \qquad \hat{\mu}_{2y} = \sum_{i \in S_2} y_i/n_2$$

where S_1 and S_2 denote the random sets of population units selected at the first and second phases of sampling, respectively. The two-phase ratio estimator of μ_y is

$$\hat{\mu}_{y,2p,\mathrm{rat}} = \hat{\mu}_{1x}\hat{R}, \quad \text{where } \hat{R} = \frac{\hat{\mu}_{2y}}{\hat{\mu}_{2x}} \tag{10.1}$$

and the two-phase (linear) regression estimator of μ_y is

$$\hat{\mu}_{y,2p,\mathrm{reg}} = \hat{\mu}_{2y} + \hat{\beta}(\hat{\mu}_{1x} - \hat{\mu}_{2x}) \tag{10.2}$$

where

$$\hat{\beta} = \frac{\sum\limits_{i \in S_2} (y_i - \hat{\mu}_{2y})(x_i - \hat{\mu}_{2x})}{\sum\limits_{i \in S_2} (x_i - \hat{\mu}_{2x})^2}$$

For the corresponding estimators of the population total, multiply the respective μ_y estimators by N, the number of population units.

Note the obvious analogies between these two-phase estimators of μ_y and the single-phase ratio and regression estimators considered in Chapter 7

$$\hat{\mu}_{y,\mathrm{rat}} = \mu_x\hat{R}$$
$$\hat{\mu}_{y,\mathrm{reg}} = \hat{\mu}_y + \hat{\beta}(\mu_x - \hat{\mu}_x)$$

In two-phase sampling, the large first-phase sample is intended to give an accurate estimate, $\hat{\mu}_{1x}$, of the mean of the auxiliary variable, μ_x. That accurate estimate replaces the value for μ_x that is assumed to be known in single-phase ratio and regression estimation.

10.1.2 *Sampling variance and its estimation*

As for single-phase ratio and regression estimators, the two-phase estimators of the mean are not unbiased and there are no exact expressions for sampling variance. The two-phase expressions for sampling variance must account for error of estimation of the mean, μ_x, of the auxiliary variable. Assuming that the first-phase sample is selected by SRS and that the second-phase sample is selected by SRS from the first-phase sample, a good approximation to the sampling variances of $\hat{\mu}_{y,2p,\mathrm{rat}}$ and $\hat{\mu}_{y,2p,\mathrm{reg}}$ is (Cochran 1977)

$$V(\hat{\mu}_{y,2p,r}) \approx \left(\frac{N - n_1}{N}\right)\frac{\sigma_y^2}{n_1} + \left(\frac{n_1 - n_2}{n_1}\right)\frac{\sigma_{y,r}^2}{n_2} \tag{10.3}$$

where the subscript "r" is shorthand for "rat" or "reg", as appropriate, and

$$\sigma_{y,r}^2 = \begin{cases} R^2\sigma_x^2 + \sigma_y^2 - 2R\,\text{Cor}(x,y)\sigma_x\sigma_y, & \text{for r = rat} \\ \sigma_y^2(1 - \text{Cor}(x,y)^2), & \text{for r = reg} \end{cases} \tag{10.4}$$

with $R = \mu_y/\mu_x$, $\text{Cor}(x,y) = \sigma_{x,y}/(\sigma_x\sigma_y)$, and the finite population variances (σ_x^2, σ_y^2) and covariance $(\sigma_{x,y})$ defined as in previous chapters. An equivalent expression for $\sigma_{y,r}^2$ is $\sum_{i=1}^{N}(y_i - y_{i,r}^*)^2/(N-1)$, where $y_{i,\text{rat}}^* = Rx_i$ and $y_{i,\text{reg}}^* = \mu_y + \beta(x_i - \mu_x)$ with $\beta = \sigma_{x,y}/\sigma_x^2$. For the sampling variances of the corresponding estimators of the population total, $V(\hat{T}_{y,2p,r}) = N^2 V(\hat{\mu}_{y,2p,r})$.

The form of $V(\hat{\mu}_{y,2p,r})$ is readily interpretted. It consists of a first-phase component and a second-phase component, each having the same general form as the sampling variance of a single-phase mean-per-unit estimator under SRS, except that the second-phase component is based on $\sigma_{y,r}^2$ rather than σ_y^2. This second-phase component, including Equation (10.4), is analogous to the sampling variance approximation formulas we have previously seen for the single-phase ratio and regression estimators (Equations (7.13) and (7.16), respectively), except that in this case an SRS of size n_2 is selected from n_1 units, rather than an SRS of size n selected from N units.

The special cases of Equation (10.3) are also noteworthy. If $n_1 = N$, then μ_x is known and the two-phase estimator is equivalent to the single-phase estimator, and the first-phase sampling variance component is equal to zero. If $n_2 = n_1$, then $\hat{\mu}_{2x} = \hat{\mu}_{1x}$ and the ratio and regression estimators reduce to a mean-per-unit estimator, $\hat{\mu}_{1y}$, and the sampling variance second-phase component is equal to zero. Finally, we note that for the regression estimator, if $\text{Cor}(x,y) = 1$, then $\sigma_{y,\text{reg}}^2 = 0$ and again the sampling variance second-phase component is equal to zero. Thus, under the most favorable of circumstances, the sampling variance of the two-phase regression estimator should be similar to that of a single-phase mean-per-unit estimator of the mean with a sample size of n_1, even though observations of the target variable have been made only on the smaller second-phase sample of size n_2.

The sampling variance can be estimated as

$$\hat{V}(\hat{\mu}_{y,2p,r}) = \left(\frac{N - n_1}{N}\right)\frac{\hat{\sigma}_y^2}{n_1} + \left(\frac{n_1 - n_2}{n_1}\right)\frac{\hat{\sigma}_{y,r}^2}{n_2} \tag{10.5}$$

where the estimators $\hat{\sigma}_y^2$ and $\hat{\sigma}_{y,r}^2$ are based entirely on the second-phase sample x and y values. For $\hat{\sigma}_{y,r}^2$, estimates of individual population parameters could be substituted in Equation (10.4) (as in Chapter 7 for single-phase ratio or regression estimators). Alternatively, the following simpler expressions could be used. For the ratio estimator, $\hat{\sigma}_{y,\text{rat}}^2 = \sum_{i \in S_2}(y_i - \hat{y}_{i,\text{rat}}^*)^2/(n_2 - 1)$ with $\hat{y}_{i,\text{rat}}^* = \hat{R}x_i$. For the regression estimator, $\hat{\sigma}_{y,\text{reg}}^2 = \sum_{i \in S_2}(y_i - \hat{y}_{i,\text{reg}}^*)^2/(n_2 - 2)$ with $\hat{y}_{i,\text{reg}}^* = \hat{\mu}_{2y} + \hat{\beta}(x_i - \hat{\mu}_{2x})$.

10.1.3 Sample space illustration

We illustrate the performance of two-phase ratio and regression estimators of the mean when applied to a small population for which, in principle, both estimators should perform moderately well. The population x and y values are presented in the following table. The target of estimation is $\mu_y = 225.2$. Other parameter values are $\mu_x = 19.8$, $\sigma_y^2 = 6{,}493.2$, $\sigma_x^2 = 61.2$, and $\text{Cor}(x,y) = 0.8087$.

Population unit (i):	1	2	3	4	5
Auxiliary variable (x_i):	28	13	10	24	24
Target variable (y_i):	341	184	145	275	181

For the first phase, $n_1 = 4$ units are selected from the $N = 5$ units by SRS, and at the second phase $n_2 = 3$ units are selected by SRS from the first-phase sample of n_1 units. There are thus a total of $\binom{5}{4} = 5$ possible first-phase samples, and for each of these first-phase samples, there are a total of $\binom{4}{3} = 4$ possible second-phase samples, so that the total number of possible samples overall is $\binom{5}{4}\binom{4}{3} = 5 \cdot 4 = 20$. Table 10.1 displays each of these possible samples, along with the corresponding ratio and regression estimates of μ_y, and associated estimates of sampling variance. The expected values of estimators can be calculated as simple averages over the twenty equally likely two-phase samples and are 225.997 and 216.985, respectively, for the ratio and regression estimators. Thus, bias

Table 10.1 Two-phase sample space for ratio and regression estimators for the $N = 5$ example population for which $\mu_y = 225.2$ and $Cor(x,y) = 0.8087$. Note that for each of the five possible first-phase samples S_1 ($n_1 = 4$), there are four possible second-phase samples S_2 ($n_2 = 3$). For each such two-phase sample, two-phase ratio and regression estimates of μ_y are calculated using Equations (10.1) and (10.2), respectively, and estimates of sampling variance are calculated using Equation (10.5). $E(\cdot)$ denotes expected values of estimators.

	Sample units						
ID	$i \in s_1$	$i \in s_2$	$p(s)$	$\hat{\mu}_{y,2p,\text{rat}}$	$\hat{\mu}_{y,2p,\text{reg}}$	$\hat{V}(\hat{\mu}_{y,2p,\text{rat}})$	$\hat{V}(\hat{\mu}_{y,2p,\text{reg}})$
1	1, 2, 3, 4	1, 2, 3	0.05	246.324	242.151	583.25	540.16
2	1, 2, 3, 4	1, 2, 4	0.05	230.769	237.500	352.58	340.88
3	1, 2, 3, 4	1, 3, 4	0.05	230.141	233.599	534.18	522.64
4	1, 2, 3, 4	2, 3, 4	0.05	240.957	229.121	292.35	228.74
5	1, 2, 3, 5	1, 2, 3	0.05	246.324	242.151	583.25	540.16
6	1, 2, 3, 5	1, 2, 5	0.05	203.654	211.468	816.32	1141.65
7	1, 2, 3, 5	1, 3, 5	0.05	201.714	205.627	916.81	1227.29
8	1, 2, 3, 5	2, 3, 5	0.05	203.457	175.561	418.69	72.59
9	1, 2, 4, 5	1, 2, 4	0.05	273.846	272.500	352.58	340.88
10	1, 2, 4, 5	1, 2, 5	0.05	241.670	240.106	816.32	1141.65
11	1, 2, 4, 5	1, 4, 5	0.05	233.332	178.563	647.59	691.43
12	1, 2, 4, 5	2, 4, 5	0.05	233.443	221.000	468.45	510.88
13	1, 3, 4, 5	1, 3, 4	0.05	263.895	262.392	534.18	522.64
14	1, 3, 4, 5	1, 3, 5	0.05	231.298	229.597	916.81	1227.29
15	1, 3, 4, 5	1, 4, 5	0.05	225.467	157.375	647.59	691.43
16	1, 3, 4, 5	3, 4, 5	0.05	222.784	213.179	516.37	593.43
17	2, 3, 4, 5	2, 3, 4	0.05	228.106	220.109	292.35	228.74
18	2, 3, 4, 5	2, 3, 5	0.05	192.606	173.758	418.69	72.59
19	2, 3, 4, 5	2, 4, 5	0.05	186.230	203.000	468.45	510.88
20	2, 3, 4, 5	3, 4, 5	0.05	183.927	190.946	516.37	593.43
			$E(\cdot)$:	225.997	216.985	554.66	586.97

(0.797 and -8.215, respectively) and relative bias (+0.35% and -3.65%, respectively) are small for both estimators. Sampling variance, calculated over the set of 20 equally likely two-phase samples, of the ratio estimator (563.86) is less than the sampling variance of the regression estimator (881.78), and the mean square errors of the two estimators are 564.49 and 949.27, respectively. Note that the stability of the two-phase regression estimator of sampling variance seems considerably less than for the two-phase ratio estimator; estimates of sampling variance range from 72.59–1,227.29 for the regression estimator as compared to 292.35–916.81 for the ratio estimator. The expected value of $\hat{V}(\hat{\mu}_{y,2p,rat})$ (554.66) is relatively close to the actual sampling variance (563.86) and mean square error (564.49), whereas the expected value of $\hat{V}(\hat{\mu}_{y,2p,reg})$ (586.97) is substantially less than the actual sampling variance (881.78) and mean square error (949.27).

If instead a single-phase SRS of size $n = 3$ had been selected, then the sampling variance would have been $V(\hat{\mu}_{y,mpu}) = [(5-3)/5]\sigma_y^2/3 = 1{,}082.2$, considerably larger than the sampling variance of the two-phase ratio estimator and slightly larger than the sampling variance of the two-phase regression estimator. This comparison may be misleading, however, because the cost of the two-phase survey may be greater than the cost of a single-phase survey of size n_2, as some cost must be attached to the first-phase measurement or recording of auxiliary variable values. This issue is further explored in Section 10.1.5 which considers the net relative efficiencies of these two-phase estimators.

10.1.4 Optimal allocation

We begin by assuming a simple cost function: $C = c_1 n_1 + c_2 n_2$, where c_1 is the cost per unit of obtaining the first-phase measurements of the auxiliary variable, and c_2 is the cost per unit of measuring the target variable in second-phase units. Typically, c_2 will be much greater than c_1. The objective is to minimize the sampling variance subject to a fixed cost (or to achieve a specified sampling variance at minimum cost). Optimal allocation results are identical for estimation of the mean or total, so we present results only for estimation of the mean of the target variable.

Initially, we formulate the minimization problem in terms of n_1 and the second-phase sampling fraction, $f_2 = n_2/n_1$, rather than in terms of n_1 and n_2. Re-expressing $V(\hat{\mu}_{2p,r})$ [Equation (10.3)] in terms of f_2 gives

$$V(\hat{\mu}_{2p,r}) = \left(\frac{N-n_1}{N}\right)\frac{\sigma_y^2}{n_1} + (1-f_2)\frac{\sigma_{y,r}^2}{n_1 f_2}$$

and the cost function becomes $C - c_1 n_1 - c_2 n_1 f_2$. To minimize $V(\hat{\mu}_{2p,r})$ subject to the fixed total cost C, we form the Lagrange function (Section A.9)

$$\mathcal{L}(\lambda, n_1, f_2) = V(\hat{\mu}_{2p,r}) + \lambda(C - c_1 n_1 - c_2 n_1 f_2)$$

and partially differentiate $\mathcal{L}(\lambda, n_1, f_2)$ with respect to n_1 and f_2

$$\frac{\partial \mathcal{L}}{\partial n_1} = -\frac{1}{n_1^2}\left[\sigma_y^2 + (1-f_2)\frac{\sigma_{y,r}^2}{f_2}\right] - \lambda(c_1 + c_2 f_2)$$

$$\frac{\partial \mathcal{L}}{\partial f_2} = -\frac{\sigma_{y,r}^2}{n_1 f_2^2} - \lambda c_2 n_1$$

Setting these partial derivatives equal to zero, solving each resulting equation for the unknown multiplier λ, and then equating the two solutions for λ gives

$$\frac{1}{n_1^2}\left[\sigma_y^2 - \sigma_{y,\mathrm{r}}^2\right] = \frac{1}{n_1^2}\left[\frac{c_1 \sigma_{y,\mathrm{r}}^2}{c_2 f_2^2}\right]$$

Cancelling out $1/n_1^2$ from both sides of the equation allows us to solve for the optimal second-phase sampling fraction

$$f_{2,\mathrm{opt}} = \sqrt{\frac{c_1}{c_2} \cdot \frac{\sigma_{y,\mathrm{r}}^2}{\sigma_y^2 - \sigma_{y,\mathrm{r}}^2}} \tag{10.6}$$

The associated optimal first-phase sample size can be found by substituting $f_{2,\mathrm{opt}}$ into the cost function and solving for $n_{1,\mathrm{opt}}$

$$n_{1,\mathrm{opt}} = \frac{C}{c_1 + c_2 f_{2,\mathrm{opt}}}$$

and $n_{2,\mathrm{opt}} = f_{2,\mathrm{opt}} n_{1,\mathrm{opt}}$.

Two-phase ratio estimation

To make the $f_{2,\mathrm{opt}}$ formula explicit for the ratio estimator, referring to Equation (10.4) and substituting for $\sigma_{y,\mathrm{rat}}^2$ we find

$$f_{2,\mathrm{opt}} = \sqrt{\frac{c_1}{c_2}\left(\frac{\sigma_y^2}{2R\mathrm{Cor}(x,y)\sigma_x\sigma_y - R^2\sigma_x^2} - 1\right)} = \sqrt{\frac{c_1}{c_2}\left(\frac{1}{\frac{\mathrm{CV}(x)}{\mathrm{CV}(y)}\left[2\mathrm{Cor}(x,y) - \frac{\mathrm{CV}(x)}{\mathrm{CV}(y)}\right]} - 1\right)} \tag{10.7}$$

noting that $R\sigma_x/\sigma_y = \mathrm{CV}(x)/\mathrm{CV}(y)$. For ratio estimation, $f_{2,\mathrm{opt}}$ depends only on the cost ratio of obtaining auxiliary as compared to target variable values (c_1/c_2), the correlation between the auxiliary and target variables [$\mathrm{Cor}(x,y)$], and the CV ratio of the auxiliary and target variables, [$\mathrm{CV}(x)/\mathrm{CV}(y)$]. Larger $\mathrm{Cor}(x,y)$ and smaller ratios (c_1/c_2) and [$\mathrm{CV}(x)/\mathrm{CV}(y)$] make for more favorable conditions (smaller $f_{2,\mathrm{opt}}$) for two-phase ratio estimation. For ratio estimation, it is evident from the denominator of Equation (10.7) that $\mathrm{Cor}(x,y)$ must exceed $0.5 \cdot \mathrm{CV}(x)/\mathrm{CV}(y)$ in order for a positive real solution for $f_{2,\mathrm{opt}}$ to exist.

For the illustrative population presented at the beginning of Section 10.1.3, $\mathrm{Cor}(x,y) = 0.8087$ and $\mathrm{CV}(x)/\mathrm{CV}(y) = 1.1042$. Suppose $C = \$1,000$, $c_1 = 2$, $c_2 = 5$. Solving Equation (10.7) gives $f_{2,\mathrm{opt}} = 0.5530$, so that $n_{1,\mathrm{opt}} = 209.9$ and $n_{2,\mathrm{opt}} = 116.1$. In practice, these values would need to be rounded up or down depending on the flexibility with C.

Two-phase regression estimation

Making the $f_{2,\mathrm{opt}}$ formula explicit for the regression estimator is more straigtforward. Referring to Equation (10.4) and substituting for $\sigma_{y,\mathrm{reg}}^2$ gives

$$f_{2,\mathrm{opt}} = \sqrt{\frac{c_1}{c_2}\left(\frac{1 - \mathrm{Cor}(x,y)^2}{\mathrm{Cor}(x,y)^2}\right)} = \sqrt{\frac{c_1}{c_2}\left(\frac{1}{\mathrm{Cor}(x,y)^2} - 1\right)} \tag{10.8}$$

For regression estimation, the optimal allocation of the two-phase survey depends on the correlation between auxiliary and target variables [$\mathrm{Cor}(x,y)$], and on the cost ratio of obtaining auxiliary as compared to target variable values (c_1/c_2). As for ratio estimation, larger correlations and smaller cost ratios lead to more favorable settings (smaller $f_{2,\mathrm{opt}}$) for application of two-phase regression estimation.

For the illustrative population presented at the beginning of Section 10.1.3, $\text{Cor}(x,y) = 0.8087$. Suppose $C = \$1{,}000$, $c_1 = 2$, $c_2 = 5$. Solving Equation (10.8) gives $f_{2,\text{opt}} = 0.4600$, so that $n_{1,\text{opt}} = 232.6$ and $n_{2,\text{opt}} = 107.0$. In practice, these values would need to be rounded up or down depending on the flexibility with C.

10.1.5 *Net relative efficiency*

The solutions for $f_{2,\text{opt}}$ [Equations (10.7) and (10.8)] allow one to determine the optimal allocations of two-phase ratio or regression estimation surveys. The optimally allocated two-phase survey will not necessarily have smaller sampling variance than an equal cost single-phase SRS survey, however, because the second-phase sample size must be reduced (when compared to an equal cost single-phase survey) to accommodate the cost of measurement or recording of auxiliary values in the large first-phase sample.

A comparable single-phase SRS survey with mean-per-unit estimation, for which only values of the target variable would be measured, would have sample size $n_{\text{srs}} = C/c_2$. The net relative efficiency of an optimally allocated two-phase strategy using $\hat{\mu}_{2\text{p}}$ compared to this single-phase strategy would be

$$\text{NRE}(\hat{\mu}_{y,2\text{p,r}}, \hat{\mu}_{y,\text{mpu}}) = \frac{V(\hat{\mu}_{y,\text{mpu}})|C}{V(\hat{\mu}_{y,2\text{p,r}})|C} = \frac{\left(\frac{N-n_{\text{srs}}}{N}\right)\frac{\sigma_y^2}{n_{\text{srs}}}}{\left(\frac{N-n_{1,\text{opt}}}{N}\right)\frac{\sigma_y^2}{n_{1,\text{opt}}} + (1-f_{2,\text{opt}})\frac{\sigma_{y,\text{r}}^2}{n_{1,\text{opt}}f_{2,\text{opt}}}}$$

To simplify this equation, we assume that N is large relative to $n_{1,\text{opt}}$ (i.e., that $n_{1,\text{opt}}/N \approx 0$ and $n_{\text{srs}}/N \approx 0$). Making these simplifications gives

$$\text{NRE}(\hat{\mu}_{y,2\text{p,r}}, \hat{\mu}_{y,\text{mpu}}) \approx \frac{\frac{\sigma_y^2}{n_{\text{srs}}}}{\frac{\sigma_y^2}{n_{1,\text{opt}}} + (1-f_{2,\text{opt}})\frac{\sigma_{y,\text{r}}^2}{n_{1,\text{opt}}f_{2,\text{opt}}}} = \frac{f_{2,\text{opt}}\left(\frac{n_{1,\text{opt}}}{n_{\text{srs}}}\right)}{f_{2,\text{opt}} + (1-f_{2,\text{opt}})\frac{\sigma_{y,\text{r}}^2}{\sigma_y^2}}$$

For the sample size ratio $n_{1,\text{opt}}/n_{\text{srs}}$, the cost equations provide

$$\frac{n_{1,\text{opt}}}{n_{\text{srs}}} = \frac{C/(c_1 + c_2 f_{2,\text{opt}})}{C/c_2} = \frac{c_2}{c_1 + c_2 f_{2,\text{opt}}} = \frac{1}{\frac{c_1}{c_2} + f_{2,\text{opt}}}$$

and for $\sigma_{y,\text{r}}^2/\sigma_y^2$, rearranging the $f_{2,\text{opt}}$ equation [Equation (10.6)] provides $\sigma_{y,\text{r}}^2/\sigma_y^2 = \frac{f_{2,\text{opt}}^2}{\frac{c_1}{c_2} + f_{2,\text{opt}}^2}$.

Making these substitutions and simplifying gives

$$\text{NRE}(\hat{\mu}_{y,2\text{p,r}}, \hat{\mu}_{y,\text{mpu}}) \approx \frac{1}{\left[\frac{c_1}{c_2} + f_{2,\text{opt}}\right]\left[1 + \frac{f_{2,\text{opt}}(1-f_{2,\text{opt}})}{\frac{c_1}{c_2} + f_{2,\text{opt}}^2}\right]} \tag{10.9}$$

This simplification of the NRE has not produced a readily interpretable equation, but it does make clear that the NRE depends on the cost ratio (c_1/c_2) and on $f_{2,\text{opt}}$, which in turn depends on the cost ratio, the correlation [$\text{Cor}(x,y)$], and for ratio estimation, on the CV ratio [$\text{CV}(x)/\text{CV}(y)$], as we discovered in the previous section. We now show how this formulation of the NRE can be used in a graphical analysis to understand the dependence of the NRE on these factors.

Graphical analysis

The dependence of $f_{2,\text{opt}}$ and $\text{NRE}(\hat{\mu}_{y,2\text{p,r}}, \hat{\mu}_{y,\text{mpu}})$ on the cost ratio (c_2/c_1) and $\text{Cor}(x,y)$ for a CV ratio of $\text{CV}(x)/\text{CV}(y) = 1$ is shown in Figure 10.2 for the two-phase regression estimator

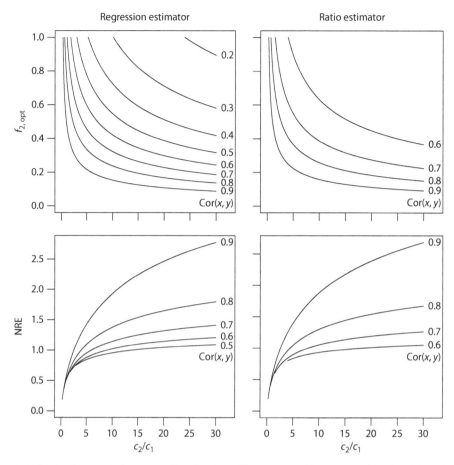

Fig. 10.2 Optimal second-phase sampling fractions ($f_{2,\text{opt}}$, top row) [Equations (10.8) and (10.7)] and net relative efficiencies given optimal allocation NRE($\hat{\mu}_{y,2p,r}, \hat{\mu}_{y,\text{mpu}}$), bottom row) [Equation (10.9)] for the two-phase regression and ratio estimators as a function of cost ratio (c_2/c_1, horizontal axis) and Cor(x,y) (individual lines drawn on graphs). The cost ratio is the cost per unit of sampling at the second phase compared to the first phase. NRE curves not shown for Cor(x,y) values for which NRE($\hat{\mu}_{y,2p,r}, \hat{\mu}_{y,\text{mpu}}$) < 1 for all $n \leq 30$.

(left column) and ratio estimator (right column). Comparison of the $f_{2,\text{opt}}$ panels (top row) shows that, in principle, the two-phase regression estimator may be applied over a much broader range of Cor(x,y) than the two-phase ratio estimator. However, once Cor(x,y) exceeds about 0.8, the curves for $f_{2,\text{opt}}$ as a function of the cost ratio are very similar to one another. For both estimators, the rate of change in $f_{2,\text{opt}}$ is rapid as the cost ratio increases from 0 to about 7, but thereafter $f_{2,\text{opt}}$ changes very slowly as the cost ratio increases to 30 (maximum level on graphs). Comparison of the NRE panels (bottom row) shows that patterns of increase in the NRE as a function of the cost ratio and Cor(x,y) are also very similar for the regression and ratio estimators. For Cor(x,y) ≤ 0.7, there is not much advantage in using double sampling for either approach, even at very favorable cost ratios (values of the NRE at Cor(x,y) = 0.7 and c_2/c_1 = 30 are about 1.4 and 1.25 for regression and ratio estimation, respectively). For Cor(x,y) ≥ 0.8, however, the NRE can be quite favorable for the two estimators (1.5–3.0), and the curves relating the NRE to the

Table 10.2 Illustrative calculations for optimally allocated two-phase regression and ratio strategies and corresponding $\text{NRE}(\hat{\mu}_{y,2p,r}, \hat{\mu}_{y,mpu})$ assuming $N = 2{,}000$, $\text{Cor}(x,y) = 0.75$, $\text{CV}(x)/\text{CV}(y) = 1$, total survey cost = \$10,000, cost per unit at the first phase of sampling, c_1, is \$10, and several alternative costs per unit at the second phase of sampling ($c_2 = \$20$, \$50, \$100, or \$200). n_{srs} is sample size for a single-phase mean-per-unit estimator.

c_2/c_1	$n_{1,opt}$ reg	$n_{1,opt}$ rat	$n_{2,opt}$ reg	$n_{2,opt}$ rat	n_{srs}	$f_{2,opt}$ reg	$f_{2,opt}$ rat	NRE reg	NRE rat
2	445	414	278	293	500	0.624	0.707	0.704	0.686
5	336	309	133	138	200	0.394	0.447	1.006	0.955
10	264	240	74	76	100	0.279	0.316	1.238	1.154
20	202	183	40	41	50	0.197	0.224	1.455	1.336

cost ratio are very similar for the two estimators. For ratio estimation, values of the CV ratio $[\text{CV}(x)/\text{CV}(y)]$ less than or greater than 1 would result in a greater or lesser value, respectively, of the NRE compared to that displayed in the figure.

Example 10.1. Suppose that $N = 2{,}000$, $\text{Cor}(x,y) = 0.75$, $\text{CV}(x)/\text{CV}(y) = 1$, $C = \$10{,}000$, $c_1 = \$10$, and that $c_2 = \$20$, \$50, \$100, or \$200, so that there are four alternative cost ratios ($c_2/c_1 = 2$, 5, 10, or 20). Equations (10.7) and (10.8) are used to calculate $f_{2,opt}$ for the two-phase ratio and regression estimators, respectively, and the corresponding optimal sample sizes are calculated as $n_{1,opt} = C/(c_1 + c_2 f_{2,opt})$ and $n_{2,opt} = f_{2,opt} n_{1,opt}$. If a single-phase SRS sample were instead selected, then the sample size would be $n_{srs} = C/c_2$. The $\text{NRE}(\hat{\mu}_{y,2p,r}, \hat{\mu}_{y,mpu})$ of the optimally allocated two-phase ratio and regression estimators for each of the alternative cost ratios are calculated using Equation (10.9).

   The results of these calculations, summarized in Table 10.2, show that the NRE of the optimally allocated two-phase regression estimator of the mean is modestly larger than for the optimally allocated two-phase ratio estimator of the mean, but that neither approach would be more efficient than a single-phase SRS estimator of the mean of the target variable, unless $c_2/c_1 > 5$ (for $\hat{\mu}_{y,2p,reg}$) or $c_2/c_1 > 10$ (for $\hat{\mu}_{y,2p,rat}$). At the highest cost ratio ($c_2/c_1 = 20$), however, both two-phase estimators are more efficient than single-phase SRS. For $c_2/c_1 = 20$, a modest reduction in second phase sample size—from 50 (for an equal cost single-phase SRS survey) to 41 or 40 for the two-phase surveys—is more than offset by the reduction in sampling variance achieved from double sampling and taking advantage of the large first-phase sample of about 200 units for which only the auxiliary variable is measured.

10.2 Two-phase ratio estimation of a proportion

In this section we extend single-phase ratio estimation of a proportion (Section 7.2) to the two-phase setting, referring the reader to Section 7.2 for additional background and motivation for this estimation approach in an ecological context. In this case, the auxiliary and target variables are both binary (0, 1) variables. A first-phase, larger sample is used to obtain an estimate of the population proportion π_x, and the second-phase, smaller sample is used to obtain an estimate of the ratio of the proportions $R = \pi_y/\pi_x$. The population proportion π_y is then estimated as

$$\hat{\pi}_{y,2p,rat} = \hat{\pi}_{1x}\hat{R}, \quad \text{where } \hat{R} = \frac{\hat{\pi}_{2y}}{\hat{\pi}_{2x}} \tag{10.10}$$

All of the two-phase formulas presented in Section 10.1 applicable to the population mean and its estimation can be directly applied in this setting for the population proportion, $\mu_y = \pi_y$, where the x and y population values for individual units are equal to 0 or 1. However, greater insight into the NRE($\hat{\pi}_{y,2p,rat}, \hat{\pi}_{y,mpu}$) can be achieved by re-expressing it in terms of the population proportion π_y and the following two conditional proportions[1]

$$Sensitivity = P_{sens} = \mathcal{P}(x = 1|y = 1) = 1 - \mathcal{P}(\text{false negative})$$

$$Specificity = P_{spec} = \mathcal{P}(x = 0|y = 0) = 1 - \mathcal{P}(\text{false positive})$$

where $\mathcal{P}(\cdot)$ denotes the proportion of units meeting the parenthetical criteria stated. Thus, sensitivity is the proportion of units for which the auxiliary variable equals 1 given that the target variable equals 1, whereas specificity is the proportion of units for which the auxiliary variable equals 0 given that the target variable equals 0. For sensitivity and specificity to be high, the proportions of false negatives, $\mathcal{P}(x = 0|y = 1)$, and false positives, $\mathcal{P}(x = 1|y = 0)$, respectively, must be low. The proportion π_x can also be expressed in terms of π_y, P_{sens}, and P_{spec} as follows. The proportion of $x = 1$ units is the proportion of units that either (a) have $y = 1$ and are correctly classified by x, or (b) have $y = 0$ but are incorrectly classified by x. Thus, [Equation (A.18)]

$$\pi_x = \pi_y P_{sens} + (1 - \pi_y)(1 - P_{spec}) \tag{10.11}$$

To evaluate the dependence of the NRE($\hat{\pi}_{y,2p,rat}, \hat{\pi}_{y,mpu}$) [Equation (10.9)] on π_y, P_{sens}, and P_{spec}, we could express $f_{2,opt}$ in terms of these quantities by either (1) expressing Cor(x,y) and CV(x)/CV(y) in terms of π_y, P_{sens}, and P_{spec}, and substituting these expressions into Equation (10.7), or (2) expressing σ_y^2 and $\sigma_{y,rat}^2$ in terms of π_y, P_{sens}, and P_{spec}, and substituting these expressions into Equation (10.6). We take the latter approach here. Expanding the $\sigma_y^2 = \sum_{i=1}^{N}(y_i - \mu)^2/(N-1)$ and $\sigma_{y,rat}^2 = \sum_{i=1}^{N}(y_i - Rx_i)^2/(N-1)$ terms, noting for a binary (0, 1) variable y that $y_i^2 = y_i$, and substituting proportions for means we find that

$$\sigma_y^2 = \left(\frac{N}{N-1}\right)\pi_y(1 - \pi_y) \tag{10.12}$$

$$\sigma_{y,rat}^2 = \left(\frac{N}{N-1}\right)\frac{\pi_y}{\pi_x}(1 - \pi_y P_{sens} - (1 - \pi_y)P_{spec}) \tag{10.13}$$

It proves more convenient to work with the ratio $\sigma_y^2/\sigma_{y,rat}^2$ than the difference $\sigma_y^2 - \sigma_{y,rat}^2$, so we re-express Equation (10.7) for $f_{2,opt}$ as

$$f_{2,opt} = \sqrt{\frac{c_1}{c_2} \cdot \frac{1}{(\sigma_y^2/\sigma_{y,rat}^2) - 1}} \tag{10.14}$$

where the ratio $\sigma_y^2/\sigma_{y,rat}^2$, on substitution of Equations (10.12), (10.13), and (10.11), is

$$\frac{\sigma_y^2}{\sigma_{y,rat}^2} = \frac{(1 - \pi_y)\pi_x}{1 - \pi_y P_{sens} - (1 - \pi_y)P_{spec}} = \frac{(1 - \pi_y)[\pi_y P_{sens} + (1 - \pi_y)(1 - P_{spec})]}{1 - \pi_y P_{sens} - (1 - \pi_y)P_{spec}} \tag{10.15}$$

---

[1] Recall from Section 7.2 that sensitivity and specificity can be estimated from sample data as: $\hat{P}_{sens} = \sum_{i \in S_2} x_i y_i / \sum_{i \in S_2} y_i$, $\hat{P}_{spec} = \sum_{i \in S_2}(1 - x_i)(1 - y_i)/\sum_{i \in S_2}(1 - y_i)$.

Note then that $f_{2,\text{opt}}$ depends on c_1/c_2, π_y, P_{sens}, and P_{spec}. Finally, substituting this expression for $f_{2,\text{opt}}$ into the $\text{NRE}(\hat{\pi}_{y,2\text{p,rat}}, \hat{\pi}_{y,\text{mpu}})$ formula [Equation (10.9)], shows that the NRE (approximation) is soley a function of c_1/c_2, π_y, P_{sens}, and P_{spec}.

This NRE function can now be used to determine the graphical *boundaries* that illustrate the conditions under which (assuming SRS at both phases, and n_2 selected from n_1) an optimally allocated two-phase ratio estimation strategy can be recommended over an equal cost single-phase mean-per-unit estimation strategy (Hankin et al. 2009). Analysis of these conditions is more complex than for the two-phase ratio estimation setting when x and y are continuous variables because we have expressed the NRE here in terms of four quantities (c_1/c_2, π_y, P_{sens}, and P_{spec}) rather than three quantities [c_1/c_2, $\text{Cor}(x,y)$, and $\text{CV}(x)/\text{CV}(y)$], but it is also more informative for this binary (0, 1) variable setting. The analysis indicates that overall, given a particular value of P_{spec}, the setting for application of a two-phase ratio estimator of a proportion becomes more favorable as c_2/c_1 gets larger and as P_{sens} approaches 1 (i.e., as $\mathcal{P}$(false negative) approaches 0).

Figure 10.3 graphically illustrates the $\text{NRE}(\hat{\pi}_{y,2\text{p,rat}}, \hat{\pi}_{y,\text{mpu}}) = 1$ boundaries assuming the second-phase sampling fraction is optimally allocated. The four panels correspond to four distinct P_{spec} values (0.7, 0.8, 0.9, 1.0). For each panel, the vertical axis gives the cost ratio c_2/c_1, and the horizontal axis gives the target variable proportion π_y. The four curves in each panel are lines along which the NRE = 1 for four distinct P_{sens} values (0.7, 0.8, 0.9, 1.0). Given a particular ($P_{\text{spec}}, P_{\text{sens}}, \pi_y$) combination, NRE > 1 for cost ratios above the curve, and NRE < 1 for cost ratios below the curve.

Careful inspection of the four panels in Figure 10.3 suggests a complex pattern of relative performance of the two-phase ratio estimator of a proportion. For example, for $P_{\text{spec}} = 0.7$, $P_{\text{sens}} = 0.8$, and $c_2/c_1 \leq 100$, the (optimally allocated) $\hat{\pi}_{y,2\text{p,rat}}$ estimator can only be recommended over the relatively narrow (approximate) range $0.05 \leq \pi_y \leq 0.56$; over this restricted range of π_y the cost ratio, c_2/c_1, must be a minimum of about 26 for $\pi_y = 0.3$, but increases to a minimum value close to 100 for $\pi_y = 0.05$ or $\pi_y = 0.56$. In more favorable settings, the region over which the $\hat{\pi}_{y,2\text{p,rat}}$ estimator can be recommended may be much larger and the required cost ratios may be much smaller. For example, for $P_{\text{spec}} = 0.9$, $P_{\text{sens}} = 0.9$, and $c_2/c_1 \leq 100$, the optimally allocated $\hat{\pi}_{y,2\text{p,rat}}$ estimator can outperform the $\hat{\pi}_{y,\text{mpu}}$ estimator over the approximate range $0.01 \leq \pi_y \leq 0.87$; over this range, minimum cost ratios vary from about 4.4 at the middle of the π_y range (i.e., $\pi_y \approx 0.45$) to about 10 for $\pi_y = 0.08$ and $\pi_y = 0.8$. At the extreme values of the boundary region for π_y, however, the cost ratio must again be very large (close to 100 at $\pi_y = 0.01$ and $\pi_y = 0.87$).

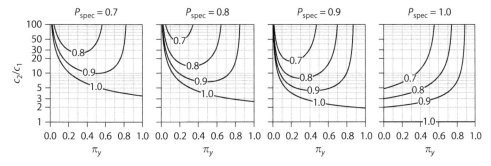

Fig. 10.3 Net relative efficiency boundaries [NRE = 1] and their dependence on P_{spec}, P_{sens} (curve-specific), the cost ratio c_2/c_1, and π_y. Given a particular ($P_{\text{spec}}, P_{\text{sens}}, \pi_y$) combination, the optimally allocated $\hat{\pi}_{y,2\text{p,rat}}$ estimator will be more efficient than the $\hat{\pi}_{y,\text{mpu}}$ estimator (NRE > 1) for cost ratios above the curve, and less efficient (NRE < 1) for cost ratios below the curve.

Overall, Figure 10.3 makes clear that the two-phase ratio estimator, even when optimally allocated, cannot be generally recommended over a single-phase estimator whenever the proportion π_y is very low (< 0.05), even with P_{sens} close to 1, unless the probability of false positives ($1 - P_{\text{spec}}$) is close to zero. Similarly, the optimally allocated two-phase ratio estimator cannot be generally recommended for very high proportions ($\pi_y > 0.90$), even with P_{spec} close to 1, unless the probability of false negatives ($1 - P_{\text{sens}}$) is close to zero. These conclusions are similar to those arrived at in considering the single-phase ratio estimator of a proportion (Section 7.2).

10.3 Two-phase sampling with stratification

In some settings, the number of population units belonging to individual strata (N_h) are unknown but the number of strata (L) may be known. For example, suppose that interest lies in the average length (or age, or survival rate) of brook trout in a well-identified set of Adirondack lakes in upstate New York and that pre-specified criteria for strata membership have been identified according to the pH ranges of lake water (e.g., < 5.5, 5.5–6.0, 6.0–6.5, 6.5–7.0, > 7.0). The total number of lakes (N) is well-identified, but stratum membership may vary interannually and is not known with certainity.

10.3.1 Estimation of the population mean and total

Suppose that in such a setting a larger, first-phase sample, S_1, of size n_1 is selected from N by SRS. For this first-phase sample, let $x_{1ih} = 1$ if unit i belongs to stratum h, and $x_{1ih} = 0$ if unit i does not belong to stratum h, $h = 1, 2, \ldots, L$. Assuming that n_1 is large enough so that all L strata are represented in n_1, then the stratum h weight, $W_h = N_h/N$, can be un unbiasedly estimated as $\widehat{W}_h = \sum_{i \in S_1} x_{1ih}/n_1 = n_{1h}/n_1$, where n_{1h} is the number of units in S_1 that belong to stratum h. Note that the $\{n_{1h}\}$ are random variables, analogous to the situation of post-stratified estimation following selection of a simple random sample (Section 5.9). Now, suppose that the second-phase sampling fractions, $f_{2h} = n_{2h}/n_{1h}$, $h = 1, 2, \ldots, L$, are *fixed* (pre-specified) for all strata and that the $\{y_{ih}\}$ are measured for all second-phase units. The stratum h mean, μ_{yh}, can be unbiasedly estimated as $\hat{\mu}_{2yh} = \sum_{i \in S_{2h}} y_{ih}/(f_{2h}n_{1h}) = \sum_{i \in S_{2h}} y_{ih}/n_{2h}$, where S_{2h} is the random set of selected phase-two units that belong to stratum h, n_{2h} is the number of such units, and the total second-phase sample size is $n_2 = \sum_{h=1}^{L} n_{2h}$. An unbiased two-phase stratified estimator of the population mean, μ_y, is

$$\hat{\mu}_{y,2p,\text{st}} = \sum_{h=1}^{L} \frac{n_{1h}}{n_1} \hat{\mu}_{2yh} = \sum_{h=1}^{L} \widehat{W}_h \hat{\mu}_{2yh} \tag{10.16}$$

The corresponding estimator of the population total, T_y, is $\hat{T}_{y,2p,\text{st}} = N\hat{\mu}_{y,2p,\text{st}}$. Recall that with stratified sampling, $\hat{\mu}_{y,\text{st}} = \sum_{h=1}^{L} \frac{N_h}{N} \hat{\mu}_{yh} = \sum_{h=1}^{L} W_h \hat{\mu}_{yh}$. Thus, in two-phase stratified sampling, the $W_h = N_h/N$ are replaced by $\widehat{W}_h = n_{1h}/n_1$, similar to the replacement of μ_x by $\hat{\mu}_{1x}$ in two-phase ratio estimation.

10.3.2 Sampling variance and its estimation

The sampling variance of $\hat{\mu}_{y,2p,\text{st}}$ is given by (Rao 1973)

$$V(\hat{\mu}_{y,2p,\text{st}}) = \left(\frac{N - n_1}{N}\right) \frac{\sigma^2}{n_1} + \sum_{h=1}^{L} W_h \left(\frac{1}{f_{2h}} - 1\right) \frac{\sigma_h^2}{n_1} \tag{10.17}$$

For the corresponding estimator of the total, $V(\hat{T}_{y,2p,st}) = N^2 V(\hat{\mu}_{y,2p,st})$. The first component of this equation is the same as that for two-phase ratio and regression estimation [Equation (10.3)], while the second component accounts for errors associated with stratification and mean-per-unit estimation. Raj (1968 Section 7.8) showed that $V(\hat{\mu}_{y,2p,st})$ can also be expressed as

$$\sum_{h=1}^{L} W_h^2 \left(\frac{N_h - n_{2h}}{N_h} \right) \frac{\sigma_h^2}{n_{2h}} + b_1 \sum_{h=1}^{L} W_h (1 - W_h) \left(\frac{N_h - n_{2h}}{N_h} \right) \frac{\sigma_h^2}{n_{2h}} + b_1 \sum_{h=1}^{L} W_h (\mu_{yh} - \mu_y)^2$$

where $b_1 = [(N - n_1)/(N - 1)](1/n_1)$. The first term in this equation *is equivalent to* the sampling variance for the single-phase stratified estimator [Equation (5.4)], where the stratum sizes (weights) are known, and the stratum sample sizes are pre-specified. Thus, the second and third terms in this equation can be viewed as the *penalty* for having to estimate the stratum weights.

A non-negative, unbiased estimator for $V(\hat{\mu}_{y,2p,st})$ is (Rao 1973)

$$\hat{V}(\hat{\mu}_{y,2p,st}) = \frac{N-1}{N} \sum_{h=1}^{L} \left(\frac{n_{1h} - 1}{n_1 - 1} - \frac{n_{2h} - 1}{N - 1} \right) \frac{\widehat{W_h \hat{\sigma}_h^2}}{n_{2h}} + \frac{N - n_1}{N(n_1 - 1)} \sum_{h=1}^{L} \widehat{W}_h (\hat{\mu}_{2yh} - \hat{\mu}_{y,2p,st})^2$$

(10.18)

provided that n_1 is sufficiently large that $n_{2h} \geq 2$, $h = 1, 2, \ldots, L$, so that the $\{\sigma_h^2\}$ can be unbiasedly estimated. For the corresponding estimator of the total, $\hat{V}(\hat{T}_{y,2p,st}) = N^2 \hat{V}(\hat{\mu}_{y,2p,st})$.

10.3.3 *Optimal allocation*

We consider optimal *proportional* allocation, where $f_{2h,opt} = f_{2,opt}$, $h = 1, 2, \ldots, L$. As noted in Chapter 5, the performance (sampling variance) of a proportionally allocated stratified sample is often close to the performance of an optimally allocated stratified sample.

We begin by assuming a cost function of the form $C = c_1 n_1 + \sum c_{2h} n_{2h} = c_1 n_1 + \sum c_{2h} f_2 n_{1h}$, where c_1 is the cost per unit at the first phase (used to stratify units and estimate the W_h), $\{c_{2h}\}$ are the second-phase stratum-specific costs per unit (for measurement or recording of the $\{y_{ih}\}$), $f_2 = n_{2h}/n_{1h}$ is the constant second-phase sampling fraction for all strata, and the $\{n_{1h}\}$ are the numbers of first-phase units belonging to each of the strata. The survey cost, C, is therefore a random variable because the $\{n_{1h}\}$ are random variables (collectively multivariate hypergeometric, Section A.4.7). Given n_1, $E(n_{1h}) = n_1 W_h$, so that given n_1 and f_2, the *expected* cost of the survey would be $E(C) = c_1 n_1 + \sum c_{2h} f_2 E(n_{1h}) = n_1 (c_1 + f_2 \sum W_h c_{2h})$. The following optimal allocation analysis is subject to this expected cost function and expected cost constraint.

With a constant second-phase sampling fraction, $f_{2h} = f_2$, $V(\hat{\mu}_{y,2p,st})$ [Equation (10.17)] becomes

$$V(\hat{\mu}_{y,2p,st}) = \left(\frac{N - n_1}{N} \right) \frac{\sigma^2}{n_1} + \left(\frac{1}{f_2} - 1 \right) \sum_{h=1}^{L} W_h \frac{\sigma_h^2}{n_1}$$

To minimize $V(\hat{\mu}_{y,2p,st})$ subject to the fixed, expected total cost, $E(C)$, we form the Lagrange function (Section A.9)

$$\mathcal{L}(\lambda, n_1, f_2) = V(\hat{\mu}_{y,2p,st}) + \lambda[E(C) - n_1 (c_1 + f_2 \sum W_h c_{2h})]$$

and partially differentiate $\mathcal{L}(\lambda, n_1, f_2)$ with respect to n_1 and f_2

$$\frac{\partial \mathcal{L}}{\partial n_1} = -\frac{1}{n^2}\left[\sigma^2 + \left(\frac{1}{f_2} - 1\right)\sum W_h \sigma_h^2\right] - \lambda(c_1 + f_2 \sum W_h c_{2h})$$

$$\frac{\partial \mathcal{L}}{\partial f_2} = -\frac{\sum W_h \sigma_h^2}{n_1 f_2^2} - \lambda n_1 \sum W_h c_{2h}$$

Setting these partial derivatives equal to zero, solving each resulting equation for the unknown multiplier λ, and then equating the two solutions for λ gives

$$\frac{1}{n_1^2}\left[\sigma_2 - \sum W_h \sigma_h^2\right] = \frac{1}{n_1^2}\left[\frac{c_1 \sum W_h \sigma_h^2}{f_2^2 \sum W_h c_{2h}}\right]$$

Cancelling out $1/n_1^2$ from both sides of the equation allows us to solve for the optimal second-stage sampling fraction, $f_{2,\text{opt}}$. The solution[2] in a form accentuating the cost and variance ratios is

$$f_{2,\text{opt}} = \sqrt{\frac{c_1}{\sum W_h c_{2h}} \cdot \frac{\sum W_h \sigma_h^2/\sigma^2}{1 - \sum W_h \sigma_h^2/\sigma^2}} \tag{10.19}$$

and

$$n_{1,\text{opt}} = \frac{E(C)}{c_1 + f_{2,\text{opt}} \sum W_h c_{2h}}$$

$\sum W_h \sigma_h^2$ is the average within stratum variance and, with proportional allocation at the second phase, $\sum W_h c_{2h}$ gives the average cost per unit at the second phase. Thus, $f_{2,\text{opt}}$ is a function of (a) the ratio of cost per unit at the first phase compared to average cost per unit at the second phase, and (b) the ratio of average within stratum variance compared to the finite population variance. Figure 10.4(a) displays $f_{2,\text{opt}}$ as a function of the cost

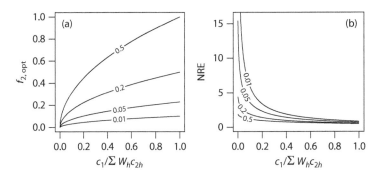

Fig. 10.4 Optimal second-phase sampling fraction [$f_{2,\text{opt}}$, panel (a)] [Equation (10.19)] and net relative efficiency given optimal allocation [NRE, panel (b)] [Equation (10.20)] for two-phase proportional allocation stratified sampling as a function of the cost ratio ($c_1/\sum W_h c_{2h}$), for four values (0.01, 0.05, 0.2, 0.5) of the variance ratio ($\sum W_h \sigma_h^2/\sigma^2$) as noted on the individual curves.

---

[2] Cochran (1977 Equation 12.23) arrives at a solution similar to Equation (10.19) by taking the general solution for $f_{2h,\text{opt}}$ and then noting that proportional allocation is optimal when all $c_{2h} = c_2$ and all $\sigma_h^2 = \sigma_w^2$, giving

$$f_{2,\text{opt}} = \sqrt{\frac{c_1}{c_2} \cdot \frac{\sigma_w^2/\sigma^2}{1 - \sigma_w^2/\sigma^2}}$$

ratio, $c_1/\sum W_h c_{2h}$, for four values (0.01, 0.05, 0.2, 0.5) of the variance ratio, $\sum W_h \sigma_h^2/\sigma^2$, as noted on the four curves. The smallest $f_{2,\mathrm{opt}}$ are achieved at the (most favorable) lowest values for both ratios.

10.3.4 *Net relative efficiency*

We explore the net relative efficiency of optimal two-phase proportional allocation stratified SRS sampling compared to an equal cost single-phase SRS survey. For a single-phase SRS survey of sample size n_{srs}, the number of selected units belonging to the various strata (for which the $\{y_{ij}\}$ measurement or recording costs may differ), $\{n_{\mathrm{srs},h}\}$, is a random variable, just as it is for the first-phase SRS strata membership results, $\{n_{1h}\}$. Thus, a comparable single-phase SRS survey of sample size n_{srs} with mean-per-unit estimation, for which only values of the target variable would be measured, would have an *expected* cost of $E(C) = \sum c_{2h} E(n_{\mathrm{srs},h}) = \sum c_{2h} n_{\mathrm{srs}} W_h = n_{\mathrm{srs}} \sum W_h c_{2h}$. The net relative efficiency would be

$$\mathrm{NRE}(\hat{\mu}_{y,2p,\mathrm{st}}, \hat{\mu}_{\mathrm{mpu}}) = \frac{\left(\frac{N-n_{\mathrm{srs}}}{N}\right)\frac{\sigma^2}{n_{\mathrm{srs}}}}{\left(\frac{N-n_{1,\mathrm{opt}}}{N}\right)\frac{\sigma^2}{n_{1,\mathrm{opt}}} + \left(\frac{1}{f_{2,\mathrm{opt}}} - 1\right)\sum W_h \frac{\sigma_h^2}{n_{1,\mathrm{opt}}}}$$

To simplify this equation, we assume that N is large relative to $n_{1,\mathrm{opt}}$ (i.e., that $n_{1,\mathrm{opt}}/N \approx 0$ and $n_{\mathrm{srs}}/N \approx 0$). Making these simplifications gives

$$\mathrm{NRE}(\hat{\mu}_{y,2p,\mathrm{st}}, \hat{\mu}_{y,\mathrm{mpu}}) \approx \frac{\frac{\sigma^2}{n_{\mathrm{srs}}}}{\frac{\sigma^2}{n_{1,\mathrm{opt}}} + \left(\frac{1}{f_{2,\mathrm{opt}}} - 1\right)\sum W_h \frac{\sigma_h^2}{n_{1,\mathrm{opt}}}} = \frac{\left(\frac{n_{1,\mathrm{opt}}}{n_{\mathrm{srs}}}\right)}{1 + \left(\frac{1}{f_{2,\mathrm{opt}}} - 1\right)\sum W_h \sigma_h^2/\sigma^2}$$

For the sample size ratio $n_{1,\mathrm{opt}}/n_{\mathrm{srs}}$, the cost equations provide

$$\frac{n_{1,\mathrm{opt}}}{n_{\mathrm{srs}}} = \frac{E(C)/(c_1 + f_{2,\mathrm{opt}}\sum W_h c_{2h})}{E(C)/(\sum W_h c_{2h})} = \frac{\sum W_h c_{2h}}{c_1 + f_{2,\mathrm{opt}}\sum W_h c_{2h}} = \frac{1}{\frac{c_1}{\sum W_h c_{2h}} + f_{2,\mathrm{opt}}}$$

and on making this substitution the NRE becomes

$$\mathrm{NRE}(\hat{\mu}_{y,2p,\mathrm{st}}, \hat{\mu}_{\mathrm{mpu}}) \approx \frac{1}{\left[\frac{c_1}{\sum W_h c_{2h}} + f_{2,\mathrm{opt}}\right]\left[1 + \left(\frac{1}{f_{2,\mathrm{opt}}} - 1\right)\sum W_h \sigma_h^2/\sigma^2\right]} \tag{10.20}$$

If cost per unit at the first phase is small compared to average cost per unit at the second phase, and if average within stratum variance is also small compared to the finite population variance, then the NRE may be considerably greater than 1. For example, with $c_1/\sum W_h c_{2h} = 0.1$ and $\sum W_h^2 \sigma_h^2/\sigma^2 = 0.05$, the NRE is about 3.5 and more favorable settings generate even larger NRE [Figure 10.4(b)].

10.4 Chapter comments

Just as multi-stage sampling can be shown to be a generalized framework within which stratified sampling and single-stage cluster sampling can be viewed as special cases, multi-phase sampling can be shown to provide a generalized framework for stratified sampling (with unknown stratum sizes) and ratio and regression estimation (with μ_x unknown). In many practical sampling contexts, auxiliary variables are not available or *on file* for all units in a population, but they may be relatively inexpensive to measure compared to the target variable. In such cases, assuming that target and auxiliary variables are highly correlated, two-phase ratio or regression estimators might be efficiently employed.

As for the multi-stage strategies explored in Chapter 9, the basic framework of double sampling allows selection of units at the first and/or second phases of sampling to be by other than the SRS method used in this Chapter. Raj (1968 pp. 139–152) and Jessen (1978 Chapter 10) present additional illustrations of double sampling strategies, including independent selection of first and second-phase samples, multivariate regression estimators, and settings where the first or second-phase samples are selected with unequal probability. Särndal and Swensson (1987) present a generalized framework for estimation in double sampling, with arbitrary first and second order inclusion probabilities at each phase of sampling and also link the generalized theory of double sampling to the problem of nonresponse in sample surveys. (The analogy is that a single large (first-phase) sample, n_1, is selected, but that responses are obtained only for n_2 of the n_1 units.) Särndal and Swensson's (1987) design-unbiased two-phase generalized Horvitz–Thompson-like estimator for the total of the target variable is

$$\hat{T}_{2p,\text{gen}} = \sum_{i \in S_2} y_i / \pi_i^*$$

where S_2 is the random second-phase sample, selected with conditional probabilities $\{\pi_{i|S_1}\}$, $\{\pi_i\}$ are the first order inclusion probabilities for units included in the random first-phase sample outcome, S_1, and $\pi_i^* = \pi_i \cdot \pi_{i|S_1}$.

Finally, as for multi-stage sampling, multi-phase sampling can be extended far beyond the simple two-phase or double sampling structure considered in this chapter. For example, Pradhan (2013) describes a three-phase survey where the first phase is used for stratification, the second is used to get a good estimate of auxiliary variable means, and the third phase is used for observation of the target variable. Jessen (1978 Section 10.12, Table 10.2) describes a six-phase sampling strategy, with one or more stages of sampling within each phase, devised to allow estimation of the total tonnage of oranges hanging on trees in Florida. Crudely summarizing the nature of this strategy, the first three phases of sampling (including use of aerial photos) were used to generate an estimate of the total number of orange trees, whereas the following three phases of sampling were used to generate an estimate of the mean weight of fruits on each tree. The overall estimator is the product of six estimators of population parameters, with each phase producing an estimate of one of the parameters.

Problems

Problem 10.1. Suppose that in a habitat survey of a reach of small stream, a total of $N = 146$ shallow pool units have been identified, unit boundaries have been flagged with bright tape, and units have been labeled $(1, 2, \ldots, N)$. In a first-phase sample of $n_1 = 60$ units selected by SRS, single visual counts (x) of juvenile coho salmon (*Oncorhynchus kisutch*) are made by snorkeling. The mean snorkel count for this first-phase sample is 9.808. In a smaller second-phase sample of size $n_2 = 13$ selected by SRS from n_1, "exhaustive" electrofishing removals (a maximum of 5 successive removals or until two successive removals generate no catches) are used to generate unit-specific estimates (y) of juvenile coho abundance (total count of removed fish) that we will assume are free of error. (Seber (1982 Section 7.2.1) provides further details on removal method estimation.) The following table lists the removal and snorkel counts for the n_2 second-phase units.

x	y	x	y	x	y	x	y	x	y
7	12	5	6	31	27	10	14	23	30
18	27	8	13	9	14	15	22	17	13
13	19	4	5	10	11				

(A) Calculate two-phase ratio and regression estimates of the total abundance of juvenile coho salmon in all N shallow pools and also calculate associated standard errors of estimation.

(B) Which estimator do you think is more appropriate in this context?

Problem 10.2. This problem builds on Problem 7.2 which concerns single-phase ratio estimation of the proportion of an endangered species of butterfly that co-occurs with a common species that is similar in appearance. Visual classification may be good, but imperfect; genetic classification is assumed 100% correct. Problem 7.2 provides cross-classifications of visual (x) and genetic (y) species identifications made at each of two field locations. Suppose that: (a) the cross-classification data presented in Problem 7.2 were based on second-phase SRS samples of butterflies selected from first-phase samples of $n_1 = 1,000$ butterflies at each location, (b) first-phase visual estimates at the two locations are 0.72 and 0.08, respectively, as in Problem 7.2, and (c) the cost per butterfly of collecting and genetically determining species identity is \$50 as compared to a cost of \$2.50 for collecting and visually determining probable species identity, and (d) $n_1/N \approx 0$. For each of the two field locations: (A) Find the optimal two-phase allocation strategy assuming use of a two-phase ratio estimator of the proportion of Fender's blue butterflies. (B) Find an estimate of the net relative efficiency of the optimally allocated two-phase ratio estimation strategy compared to the single-phase estimator $\hat{\pi}_{mpu} = \hat{\mu}_{mpu}$ based only on second-phase genetic data. (C) Discuss/explain your findings.

Problem 10.3. Construct a simulation program in R that allows you to contrast the (simulated) sampling variances (use 5×10^4 independent trials) of the double sampling regression estimator of μ_y [Equation (10.2)] when n_2 is selected by SRS from n_1 (also selected by SRS) as compared to when n_2 is selected by SRS independently of n_1 (selected by SRS). Begin your R code with the following lines

```
set.seed(1000); N <- 500; n1 <- 100; n2 <- 20
x <- rnbinom(N, mu=10, size=5)
e <- rnorm(N, mean=0, sd=4*sqrt(x))
y <- 10 + (6*x) + e
```

Use the program to evaluate several values of n_1 and n_2, so that you can better discuss the relative performance of $\hat{\mu}_{y,2p,reg}$ when n_2 is selected from n_1 as compared to when n_1 and n_2 are selected independently by SRS. Summarize your simulation results and discuss your findings.

Problem 10.4. In multi-phase sampling, as well as in single-phase sampling, one may wonder why the known or relatively accurate first-phase estimate of σ_x^2 is not used in the estimation of sampling variance. Instead, the second-phase estimate of σ_x^2 is used. (A) Write a simulation program in R, using 5×10^5 independent trials, that allows: (1) accurate determination of the sampling variance of $\hat{\mu}_{2p,rat}$, and (2) comparison of the performances of estimators of sampling variance using Equation (10.5) where $\hat{\sigma}_{y,\text{rat}}^2$ is obtained by substituting estimates of individual population parameters into Equation (10.4) (as in Chapter 7 for the single-phase ratio estimator), but where $\hat{\sigma}_x^2$ is taken as (i) the actual value σ_x^2, (ii) the first-phase estimate of σ_x^2, and (iii) the second-phase estimate of σ_x^2. Begin your program with the same four lines of R code provided for Problem 10.3. (B) In your characterization of the performance of these three alternatives, calculate expected values, sampling variances, and the probabilities that $\hat{V}(\hat{\mu}_{y,\text{2p,rat}}) < 0$. (C) Summarize and discuss your simulation results.

Problem 10.5. Suppose that a population study of Galapagos land iguanas (*Conolophus subcristatus*, Figure 10.1) is being carried out on Isabella Island. Study objectives are to estimate adult mean weight and total adult biomass. An approximately unbiased mark-recapture estimate of abundance is available from recent tagging: 2,785 adults, with an estimated standard error of 416 adults. A two-phase stratified estimator (Section 10.3) has been used to estimate mean weight. A total of 350 adults were sexed and 250 were found to be females. Twenty percent of each sex in this initial sample were weighed with mean weights being 3.4 kg for females and 6.2 kg for males. Corresponding sample variances ($\hat{\sigma}^2$) of weights were 0.38 and 1.62 for females and males, respectively. (A) Construct a 95% confidence interval for mean adult weight using the two-phase stratified estimator, with Equation (10.18) used to estimate sampling variance and assuming $N = 2,785$. (B) Construct an approximate 95% confidence interval for total population biomass, B, using the product estimator $\hat{B} = \hat{N}\hat{\mu}_{y,\text{2p,st}}$, where $\hat{N}$ is estimated total adult population size.

CHAPTER 11

Adaptive sampling

Fig. 11.1 A cluster of black morel mushrooms, genus *Morchella*, a contagiously distributed and highly prized edible mushroom for which adaptive cluster sampling might be effectively used for estimation of abundance (and for harvesting mushrooms!). Photo credit: D. Hankin.

The lead publication associated with the theory of **adaptive sampling** is Thompson's (1990) paper on **adaptive cluster sampling**. Thompson later followed up with an extensive overview of the theory (Thompson 1992), and then devoted an entire text to the topic of adaptive sampling (Thompson and Seber 1996). The theory is directed toward applications for which the distribution of y values over the population's spatial domain tends to be highly clumped or contagious, with many units having very low values of y and isolated clusters of units having relatively high values of y. This kind of distribution is often characteristic of rare items (e.g., rare or endangered plants, amphibians, birds, and fish).

Sampling Theory: For the Ecological and Natural Resource Sciences. David G. Hankin, Michael S. Mohr, and Ken B. Newman, Oxford University Press (2019). © David G. Hankin, Michael S. Mohr, and Ken B. Newman. DOI: 10.1093/oso/9780198815792.001.0001

While out on a mushroom hunt a few years past, the "obvious" logic of adaptive sampling struck home to one of us in a fashion that had not happened previously. Many, though not all, species of mushrooms tend to be distributed in a highly aggregated or contagious fashion rather than being more uniformly distributed over the forest floor. One such mushroom that is highly prized in the Pacific Northwest is the matsutake (*Tricholoma magnivelare*). This mushroom is often quite rare, but when one is found several others are likely to be found nearby. Therefore, the successful mushroom hunter proceeds as follows. First, an extensive and often long (and often completely unsuccessful!) search is made to locate the first matsutake. When a matsutake is finally discovered, a very thorough search is made in the neighborhood of this initial success, often resulting in the location of additional matsutakes. This local search is ended when no more matsutake can be found in the local area. Then, another extensive and often long (and, again, often completely unsuccessful!) search is made to locate a "new" matsutake. Once one is found, then a very thorough search for more matsutake is made around the location of this new find and the search area is increased until matsutake can no longer be found. This "sampling scheme" often makes for a successful matsutake hunt, although it sometimes proves almost impossible to locate that initial mushroom that leads to others, and is usually a more successful scheme than "random" searching not followed by the extensive searches in the neighborhoods of known matsutake locations. This mushroom hunting strategy, identical to the successfull strategy for finding morel mushrooms (Figure 11.1) when they are not abundant, is similar to *adaptive cluster sampling*, the topic of this chapter.

In single-stage adaptive cluster sampling, an initial SRS sample, S_1, of fixed size n_1, is selected from N population units. Then, for the subset of n_1 units for which the y values meet or exceed some minimum criterion value, c (e.g., $c = 1$), additional follow-up sampling is carried out in the *neighborhood* (defined in the following paragraph) of these units. This follow-up or *adaptive* sampling is carried out in successive iterative steps until one encounters units for which y no longer meets or exceeds c. Sounds like a successful matsutake hunting strategy, doesn't it?

11.1 Adaptive cluster sampling

For adaptive cluster sampling, the population units are represented as a grid with N equal size cells. The selection scheme is as follows: (a) select n_1 units from N by SRS, generating an *initial* random sample, S_1, of fixed size, and (b) for $i \in S_1$, select its *neighboring* units as well if $y_i \geq c$ (e.g., $c = 1$, as in the matsutake example) and continue this process until $y < c$ for the final iteration of units selected in this way. (Units that are in the **neighborhood** of a selected unit are those that share cell edges with the selected unit.) This kind of sampling procedure will lead to clusters of units being selected around some, but probably not all (depending on the value of c), of the units in the initial sample, i.e., an **adaptive cluster sample**. The total number of units in the realized adaptive sample, n^*, may be highly variable depending on the number of units in the initial sample for which $y \geq c$, and on the distribution of the neighboring y values.

11.1.1 *Basic scheme*

The formal specification of adaptive sampling is perhaps most easily understood using an example of a 3×3 grid of spatially arranged units ($N = 9$) with associated target variable values, as depicted in Table 11.1.

Table 11.1 Example 3×3 grid of spatially arranged population units and associated y values used for illustration throughout this chapter.

$y_1 = 1$	$y_4 = 0$	$y_7 = 0$
$y_2 = 0$	$y_5 = 0$	$y_8 = 1$
$y_3 = 0$	$y_6 = 0$	$y_9 = 3$

Suppose that an initial SRS realized sample, s_1, of size $n_1 = 2$, consisting of units 1 and 7 has been selected (light shade).

$y_1 = 1$	$y_4 = 0$	$y_7 = 0$
$y_2 = 0$	$y_5 = 0$	$y_8 = 1$
$y_3 = 0$	$y_6 = 0$	$y_9 = 3$

Let $c = 1$. Then, adaptive sampling would be conducted in the neighborhood of unit 1 (because $y_1 \geq c$), but not in the neighborhood of unit 7 (because $y_7 < c$). The adaptive sampling in the neighborhood of unit 1 would lead to selection of units 2 and 4 (next grid, dark shade) that share edges with unit 1, but then the adaptive sampling would stop because $y_2 < c$ and $y_4 < c$.

$y_1 = 1$	$y_4 = 0$	$y_7 = 0$
$y_2 = 0$	$y_5 = 0$	$y_8 = 1$
$y_3 = 0$	$y_6 = 0$	$y_9 = 3$

Thus, the size of the realized sample would be $n^* = 4$ units; two selected in the initial sample (light shade) and two selected in the adaptive sampling (dark shade).

Suppose instead that the initial sample consisted of units 2 and 8 (next grid, light shade). In this case, unit 8 (for which $y_8 \geq c$) would lead to a first iteration of adaptive sampling in its neighborhood (units 5, 7, 9; dark shade). Unit 9 (for which $y_9 \geq c$) would then lead to a second iteration of adaptive sampling in its neighborhood (unit 6; medium shade), but then the adaptive search would stop because $y_6 < c$. The size of the realized sample would in this case be $n^* = 6$ units.

$y_1 = 1$	$y_4 = 0$	$y_7 = 0$
$y_2 = 0$	$y_5 = 0$	$y_8 = 1$
$y_3 = 0$	$y_6 = 0$	$y_9 = 3$

11.1.2 *Definitions*

The following definitions are fundamental to adaptive cluster sampling.

network: a collection of units such that initial selection of any unit within the network would lead, through the adaptive sampling protocol, to selection of all other units in the network (i.e., to selection of all other neighboring units for which $y \geq c$).

edge unit: a unit such that $y < c$, but which is in the neighborhood of a unit for which $y \geq c$.

b_i = number of units in the network to which unit i belongs (i.e., its network *size*).

a_i = total number of units in the networks for which unit i is an edge unit.

It is also important to note the following:

1. Each unit belongs to one, and only one, network.
2. All units have a network size of at least 1 (i.e., contains at least itself).
3. Two kinds of units may appear in the initial sample that do not lead to further sampling: (a) a network of size 1, but not an edge unit, for which $y < c$, or (b) an edge unit.
4. If an initially selected unit falls in a network and $y \geq c$, then (a) all other units in that network will be selected, and (b) all edge units of that network will be included in the realized adaptive sample.

For example, with our previous 3×3 grid of spatially arranged units, each of the units for which $y < c$, $(i = 2, 3, \ldots, 7)$, form a network unto themselves ($b_i = 1$). Unit 1 also forms a network unto itself ($b_1 = 1$) with two edge units (units 2 and 4). Units 8 and 9 form a network ($b_8 = b_9 = 2$) with three edge units (units 5, 6, and 7). Therefore

Unit, i:	1	2	3	4	5	6	7	8	9
b_i:	1	1	1	1	1	1	1	2	2
a_i:	0	1	0	1	2	2	2	0	0

11.1.3 Inclusion probabilities and expected sample size

The initial sample, S_1, of size n_1 can be selected by SRS or SWR, but we examine only the case for which SRS is used. For n_1 selected from N by SRS, the first order inclusion probabilities for adaptive cluster sampling are

$$\pi_i = 1 - \frac{\binom{N-b_i-a_i}{n_1}}{\binom{N}{n_1}}, \quad i = 1, 2, \ldots, N \tag{11.1}$$

In Equation (11.1), $\binom{N-b_i-a_i}{n_1}$ gives the total number of samples of initial size n_1 that could be selected from the population which would not, through an adaptive search, eventually lead to selection of unit i. Thus, the fraction $\binom{N-b_i-a_i}{n_1} / \binom{N}{n_1}$ gives the proportion of initial samples that would not result in selection of unit i, and the complement of this proportion gives the probability that unit i will be included in an adaptive cluster sample. For example, with our previous 3×3 grid of spatially arranged units, for unit 1, $\pi_1 = 1 - \binom{9-1-0}{2} / \binom{9}{2} = 1 - \binom{8}{2} / \binom{9}{2} = 1 - 7/9 = 0.2222$, and, for unit 5, $\pi_5 = 1 - \binom{9-1-2}{2} / \binom{9}{2} = 1 - \binom{6}{2} / \binom{9}{2} = 1 - 5/12 = 0.5833$.

Table 11.2 lists the sample space containing the 36 possible adaptive samples for $n_1 = 2$ selected from the Table 11.1 population of $N = 9$ units. Let S^* denote the set of units selected in the adaptive sample (including the initial units selected), and let n^* denote the adaptive sample size, which varies from 2 to 8. For a sampling design with fixed sample size, n, including SRS, $\sum_{i=1}^{N} \pi_i = n$. For a sampling design with random sample size, n^*, including adaptive sampling, however, $\sum_{i=1}^{N} \pi_i = E(n^*)$. For our example, $n_1 = 2$, but $E(n^*) = \frac{1}{36} \sum_{s_1 \in S_1} n^*(s_1) = 139/36 \approx 3.86$, where S_1 denotes the sample space for the initial selection of units, and $n^*(s_1)$ denotes the realized sample size given a particular initial sample, $S_1 = s_1$. Note also that, because each of the possible adaptive samples is equally likely due to SRS selection of the initial sample, π_i can be directly calculated by summing the number of samples that include unit i (Table 11.2, "Sum" row) and dividing

Table 11.2 Listing of the 36 possible initial samples, the resulting adaptive samples, and the adaptive sample sizes (n^*) given an initial SRS selection of $n_1 = 2$ from the Section 11.1.1 3×3 grid of spatially arranged units ($N = 9$). Entries of 1 and 0 denote inclusion and exclusion, respectively, of individual units from the adaptive samples.

Initial sample		Adaptive sample units (s^*)									
ID	units (s_1)	1	2	3	4	5	6	7	8	9	$n^*(s_1)$
1	1, 2	1	1	0	1	0	0	0	0	0	3
2	1, 3	1	1	1	1	0	0	0	0	0	4
3	1, 4	1	1	0	1	0	0	0	0	0	3
4	1, 5	1	1	0	1	1	0	0	0	0	4
5	1, 6	1	1	0	1	0	1	0	0	0	4
6	1, 7	1	1	0	1	0	0	1	0	0	4
7	1, 8	1	1	0	1	1	1	1	1	1	8
8	1, 9	1	1	0	1	1	1	1	1	1	8
9	2, 3	0	1	1	0	0	0	0	0	0	2
10	2, 4	0	1	0	1	0	0	0	0	0	2
11	2, 5	0	1	0	0	1	0	0	0	0	2
12	2, 6	0	1	0	0	0	1	0	0	0	2
13	2, 7	0	1	0	0	0	0	1	0	0	2
14	2, 8	0	1	0	0	1	1	1	1	1	6
15	2, 9	0	1	0	0	1	1	1	1	1	6
16	3, 4	0	0	1	1	0	0	0	0	0	2
17	3, 5	0	0	1	0	1	0	0	0	0	2
18	3, 6	0	0	1	0	0	1	0	0	0	2
19	3, 7	0	0	1	0	0	0	1	0	0	2
20	3, 8	0	0	1	0	1	1	1	1	1	6
21	3, 9	0	0	1	0	1	1	1	1	1	6
22	4, 5	0	0	0	1	1	0	0	0	0	2
23	4, 6	0	0	0	1	0	1	0	0	0	2
24	4, 7	0	0	0	1	0	0	1	0	0	2
25	4, 8	0	0	0	1	1	1	1	1	1	6
26	4, 9	0	0	0	1	1	1	1	1	1	6
27	5, 6	0	0	0	0	1	1	0	0	0	2
28	5, 7	0	0	0	0	1	0	1	0	0	2
29	5, 8	0	0	0	0	1	1	1	1	1	5
30	5, 9	0	0	0	0	1	1	1	1	1	5
31	6, 7	0	0	0	0	0	1	1	0	0	2
32	6, 8	0	0	0	0	1	1	1	1	1	5
33	6, 9	0	0	0	0	1	1	1	1	1	5
34	7, 8	0	0	0	0	1	1	1	1	1	5
35	7, 9	0	0	0	0	1	1	1	1	1	5
36	8, 9	0	0	0	0	1	1	1	1	1	5
Sum:		8	15	8	15	21	21	21	15	15	139
Avg:		0.22	0.42	0.22	0.42	0.58	0.58	0.58	0.42	0.42	3.86

by 36, (Table 11.2, "Avg" row). These directly calculated inclusion probabilities agree with application of Equation (11.1). Note that the largest first order inclusion probabilities are for units 5, 6, and 7, which are all edge units for the largest network (that contains units 8 and 9). This property of adaptive cluster sampling seems undesirable since edge units, by definition, have $y < c$.

The pattern of y values across the population units, combined with a specific choice of c, leads to a unique specification of the π_i and $\sum_{i=1}^{N} \pi_i = E(n^*)$. If the π_i could be calculated for all units in the population, then it would in principle be possible to use a Horvitz–Thompson estimator to calculate an unbiased estimate of $\mathcal{T}$. Unfortunately, no auxiliary variable, readily available for all units, has been used to establish the π_i. Instead, the π_i depend directly on the magnitude of the y_i and the network configurations, none of which are known in advance of sample selection. It is not even possible to calculate the π_i for the particular units that may appear in an adaptive sample because many of the a_i will be unknown whenever $n^* < N$. A particular edge unit and one of its associated networks may appear in a sample, but it is possible that this same unit is an edge unit for another (unknown) network which does not appear in the sample. Therefore, the a_i cannot be known with certainty even for those units included in the adaptive sample.

Consider the dilemma posed by a second 3×3 grid of spatially arranged units depicted immediately below this paragraph, where units 4 and 8 are selected for the initial sample (light shade), leading to a first iteration adaptive sample selection of units 5, 7, and 9 (dark shade), and a second iteration adaptive sample selection of unit 6 (medium shade). To calculate π_i for all units in this sample, the value of a_i would need to be known for edge unit 5. However, from the sample information, it would only be known that edge unit 5 is associated with a single network of size 2 (units 8 and 9), not that it is also an edge unit for another network of size 2 (units 1 and 2). Therefore, the value of a_i for unit 5 would be unknown, ruling out standard application of the Horvitz–Thompson estimator. This same problem would occur for unit 4, which would not be known to be an edge unit for the same network of size 2 (units 1 and 2).

$y_1 = 1$	$y_4 = 0$	$y_7 = 0$
$y_2 = 1$	$y_5 = 0$	$y_8 = 1$
$y_3 = 0$	$y_6 = 0$	$y_9 = 3$

11.1.4 *Estimators and relative efficiency*

To circumvent the issues for estimation just described, Thompson (1990) proposed two estimators that (1) utilize the data for the sampled networks, and (2) dismiss the edge unit data collected from the sampled networks. Both estimators are unbiased for the population mean. Salehi (2003) provides evidence that it will usually be the case that the second estimator ($\hat{\mu}_{\text{ad,HT}}$) is more efficient than the first estimator ($\hat{\mu}_{\text{ad,mpu}}$).

Adaptive mean-per-unit estimation

Recall that each population unit belongs to one, and only one, network. The collection of networks thus form a partition of the N population units. Label the networks $k = 1, 2, \ldots, K$, let U_k denote the set of population unit labels for network k, and let N_k denote the number of these units (network size). Now define a new population variable $z = \{\mu_1, \mu_2, \ldots, \mu_N\}$, where μ_i is the mean of the y values for the network to which unit i belongs. For example, for our Table 11.1 population of $N = 9$ units

Unit, i:	1	2	3	4	5	6	7	8	9
y_i:	1	0	0	0	0	0	0	1	3
μ_i:	1	0	0	0	0	0	0	2	2

The mean for network k is thus represented in the z values a total of N_k times. Letting μ_k denote the mean of the y values for network k, $\mu_k = \sum_{i \in U_k} y_i / N_k$, we now show that the population mean of z, μ_z, is in fact equal to μ

$$\mu_z = \frac{\sum_{i=1}^{N} \mu_i}{N} = \frac{\sum_{k=1}^{K} N_k \mu_k}{N} = \frac{\sum_{k=1}^{K} \sum_{i \in U_k} y_i}{N} = \frac{\sum_{i=1}^{N} y_i}{N} = \mu \qquad (11.2)$$

For the above listing of y_i and μ_i, $\mu_z = \mu = 5/9 = 0.5556$.

This suggests an option for estimating μ: because the μ_i values are known for all networks encountered in the initial sample (due to the adaptive protocol), use the initial SRS sample of units, S_1, with a mean-per-unit estimator for μ_z

$$\hat{\mu}_{\text{ad,mpu}} = \frac{\sum_{i \in S_1} z_i}{n_1} = \frac{\sum_{i \in S_1} \mu_i}{n_1} \qquad (11.3)$$

This estimator, being a mean-per-unit type of estimator, is unbiased for $\mu_z = \mu$, with sampling variance

$$V(\hat{\mu}_{\text{ad,mpu}}) = \left(\frac{N - n_1}{N}\right) \frac{\sigma_z^2}{n_1}, \quad \sigma_z^2 = \frac{\sum_{i=1}^{N} (\mu_i - \mu)^2}{N - 1} \qquad (11.4)$$

which is unbiasedly estimated by

$$\hat{V}(\hat{\mu}_{\text{ad,mpu}}) = \left(\frac{N - n_1}{N}\right) \frac{\hat{\sigma}_z^2}{n_1}, \quad \hat{\sigma}_z^2 = \frac{\sum_{i \in S_1} (\mu_i - \hat{\mu}_{\text{ad,mpu}})^2}{n_1 - 1} \qquad (11.5)$$

Note that $\hat{\mu}_{\text{ad,mpu}}$ implicitly incorporates all of the $y \geq c$ values from the adaptive sample, but includes $y < c$ values only if these units were selected in the initial sample, S_1. In addition, the initial sample size n_1 is fixed for estimation purposes, even if the adaptive sample size n^* is not.

Following up on our Table 11.2 adaptive sample space example, Table 11.3 lists for each of the 36 possible initial samples the associated $\hat{\mu}_{\text{ad,mpu}}$ and $\hat{V}(\hat{\mu}_{\text{ad,mpu}})$. Because we have assumed the initial sample is selected by SRS, each of the possible without replacement samples are equally likely, $p(s_1) = 1/36, s_1 \in S_1$. Thus, the expected value and sampling variance of $\hat{\mu}_{\text{ad,mpu}}$ can be calculated directly over the sample space (S_1) of the initial samples of size $n_1 = 2$ as

$$E(\hat{\mu}_{\text{ad,mpu}}) = \frac{1}{36} \sum_{s_1 \in S_1} \hat{\mu}_{\text{ad,mpu}}(s_1) = 0.5556 = \mu$$

$$V(\hat{\mu}_{\text{ad,mpu}}) = \frac{1}{36} \sum_{s_1 \in S_1} (\hat{\mu}_{\text{ad,mpu}}(s_1) - \mu)^2 = 0.3025$$

The first result demonstrates the unbiasedness of the estimator, and the second result demonstrates agreement with Equation (11.4): $V(\hat{\mu}_{\text{ad,mpu}}) = ((9 - 2)/9)(0.7778/2) = 0.3025$. Also, the expectation of the sampling variance estimator, $\hat{V}(\hat{\mu}_{\text{ad,mpu}})$, over the sample space is equal to 0.3025, demonstrating its unbiasedness for $V(\hat{\mu}_{\text{ad,mpu}})$. All of these results are consistent with what one would expect for an SRS/mpu setting, even though this is an adaptive cluster sampling setting.

Table 11.3 Listing of the 36 possible initial samples for the Table 11.2 example with correspond-ing $\hat{\mu}_{ad,mpu}$ and $\hat{\mu}_{ad,HT}$ estimates and associated estimates of sampling variance. Note that sample ID = 36 consists of the two population units $s_1 = \{8,9\}$, which belong to the same network, $k = 8$. In this case, the estimator $\hat{V}(\hat{\mu}_{ad,HT})$ fails, and as a result the sum and average of $\hat{V}(\hat{\mu}_{ad,HT})$ over the sample space do not exist.

Initial sample		Mean-per-unit			Horvitz–Thompson		
ID	units (s_1)	$\mu_i, i \in s_1$	$\hat{\mu}_{ad,mpu}(s_1)$	$\hat{V}(\hat{\mu}_{ad,mpu})(s_1)$	Networks (s_1)	$\hat{\mu}_{ad,HT}(s_1)$	$\hat{V}(\hat{\mu}_{ad,HT})(s_1)$
1	1, 2	1, 0	0.5000	0.1944	1, 2	0.5000	0.1944
2	1, 3	1, 0	0.5000	0.1944	1, 3	0.5000	0.1944
3	1, 4	1, 0	0.5000	0.1944	1, 4	0.5000	0.1944
4	1, 5	1, 0	0.5000	0.1944	1, 5	0.5000	0.1944
5	1, 6	1, 0	0.5000	0.1944	1, 6	0.5000	0.1944
6	1, 7	1, 0	0.5000	0.1944	1, 7	0.5000	0.1944
7	1, 8	1, 2	1.5000	0.1944	1, 8	1.5667	0.5026
8	1, 9	1, 2	1.5000	0.1944	1, 8	1.5667	0.5026
9	2, 3	0, 0	0.0000	0.0000	2, 3	0.0000	0.0000
10	2, 4	0, 0	0.0000	0.0000	2, 4	0.0000	0.0000
11	2, 5	0, 0	0.0000	0.0000	2, 5	0.0000	0.0000
12	2, 6	0, 0	0.0000	0.0000	2, 6	0.0000	0.0000
13	2, 7	0, 0	0.0000	0.0000	2, 7	0.0000	0.0000
14	2, 8	0, 2	1.0000	0.7778	2, 8	1.0667	0.6637
15	2, 9	0, 2	1.0000	0.7778	2, 8	1.0667	0.6637
16	3, 4	0, 0	0.0000	0.0000	3, 4	0.0000	0.0000
17	3, 5	0, 0	0.0000	0.0000	3, 5	0.0000	0.0000
18	3, 6	0, 0	0.0000	0.0000	3, 6	0.0000	0.0000
19	3, 7	0, 0	0.0000	0.0000	3, 7	0.0000	0.0000
20	3, 8	0, 2	1.0000	0.7778	3, 8	1.0667	0.6637
21	3, 9	0, 2	1.0000	0.7778	3, 8	1.0667	0.6637
22	4, 5	0, 0	0.0000	0.0000	4, 5	0.0000	0.0000
23	4, 6	0, 0	0.0000	0.0000	4, 6	0.0000	0.0000
24	4, 7	0, 0	0.0000	0.0000	4, 7	0.0000	0.0000
25	4, 8	0, 2	1.0000	0.7778	4, 8	1.0667	0.6637
26	4, 9	0, 2	1.0000	0.7778	4, 8	1.0667	0.6637
27	5, 6	0, 0	0.0000	0.0000	5, 6	0.0000	0.0000
28	5, 7	0, 0	0.0000	0.0000	5, 7	0.0000	0.0000
29	5, 8	0, 2	1.0000	0.7778	5, 8	1.0667	0.6637
30	5, 9	0, 2	1.0000	0.7778	5, 8	1.0667	0.6637
31	6, 7	0, 0	0.0000	0.0000	6, 7	0.0000	0.0000
32	6, 8	0, 2	1.0000	0.7778	6, 8	1.0667	0.6637
33	6, 9	0, 2	1.0000	0.7778	6, 8	1.0667	0.6637
34	7, 8	0, 2	1.0000	0.7778	7, 8	1.0667	0.6637
35	7, 9	0, 2	1.0000	0.7778	7, 8	1.0667	0.6637
36	8, 9	2, 2	2.0000	0.0000	8	1.0667	—
Sum:			20.0000	10.8889		20.0000	—
Avg:			0.5556	0.3025		0.5556	—

We now contrast the performance of adaptive cluster sampling using $\hat{\mu}_{ad,mpu}$ with that of an SRS/mpu strategy ($\hat{\mu}_{mpu}$). Clearly $\hat{\mu}_{ad,mpu}$ will have a smaller sampling variance than an SRS of fixed sample size n_1 mpu estimator whenever $\sigma_z^2 < \sigma^2$ [compare Equations (11.4) and (3.15)]. This relative precision comparison would not be fair, however, because the realized sample size of the adaptive sample, n^*, will presumably be larger than the initial SRS sample size n_1, and sometimes much larger. Therefore, a more appropriate measure of comparison would be the relative precision of the two strategies based on a sample size of $E(n^*)$ for the non-adaptive SRS strategy ($\hat{\mu}_{mpu}$). The relative efficiency (RE) of the two strategies is thus

$$\text{RE}(\hat{\mu}_{ad,mpu}, \hat{\mu}_{mpu}) = \frac{V(\hat{\mu}_{mpu}|n = E(n^*))}{V(\hat{\mu}_{ad,mpu}|n_1)} = \frac{N - E(n^*)}{N - n_1} \cdot \frac{\sigma^2/E(n^*)}{\sigma_z^2/n_1} \qquad (11.6)$$

For large N relative to n_1 and $E(n^*)$, the RE will exceed 1 (adaptive strategy more efficient than non-adaptive strategy) if $\sigma_z^2/n_1 < \sigma^2/E(n^*)$.

Net relative efficiency [NRE, Equation (3.35)] might offer a more appropriate comparison, however, in that it would take into account the costs of the two designs, which might be cheaper for adaptive sampling. This is because the average travel cost between units in the neighborhood of those that trigger adaptive searches may be less than the average travel cost between units for SRS. If this were the case, then $\text{NRE}(\hat{\mu}_{ad,mpu}, \hat{\mu}_{mpu})$ would exceed $\text{RE}(\hat{\mu}_{ad,mpu}, \hat{\mu}_{mpu})$. This is not unlike the situation previously described for multi-stage sampling (Chapter 9).

Example 11.1. *Performance of adaptive cluster sampling using $\hat{\mu}_{ad,mpu}$.* For our Table 11.1 population of $N = 9$ units, $\sigma^2 = 1.0278$ and $\sigma_z^2 = 0.7778$. Thus, for an initial sample of size $n_1 = 2$, $V(\hat{\mu}_{ad,mpu}) = [(9 - 2)/9](0.7778/2) = 0.3025$, while for a non-adaptive SRS sample of the same size, $V(\hat{\mu}_{mpu}) = [(9 - 2)/9](1.0278/2) = 0.3997$, so that $V(\hat{\mu}_{ad,mpu}) < V(\hat{\mu}_{mpu})$ because $\sigma_z^2 < \sigma^2$, and $\text{RP}(\hat{\mu}_{ad,mpu}, \hat{\mu}_{mpu}) = 0.3997/0.3025 \approx 1.32$. However, for the relative efficiency of the two strategies, an initial sample size of $n_1 = 2$ for the adaptive design would be paired with an SRS sample size of $n = E(n^*) \approx 4$ (see Table 11.2, $E(n^*) = 139/36 \approx 3.86$). For $n = 4$, $V(\hat{\mu}_{mpu}) = [(9 - 4)/9](1.0278/4) = 0.1427$, so that $\text{RE}(\hat{\mu}_{ad,mpu}, \hat{\mu}_{mpu}) = 0.1427/0.3025 \approx 0.47$.

Adaptive Horvitz–Thompson estimation

As was previously discussed, with adaptive cluster sampling designs the sample inclusion probability, π_i, will not necessarily be known for a sampled edge unit because it depends on a_i [Equation (11.1)], the total number of network units for which unit i is an edge unit, and this may not be known. However, Thompson (1990) showed that a Horvitz–Thompson type estimator can be formed by explicitly excluding the adaptively selected network edge units from the estimator but including those same population units if selected as part of the initial sample, where they are considered networks of size 1. The inclusion probability for this estimator is the probability that a unit is included in the *estimator* rather than included in the *sample*. The probability that unit i is *not included* in the estimator is the probability that no members of its network are included in the initial sample, because inclusion of any one of these members would trigger inclusion of the remaining network members (Section 11.1.2, note 4). This probability is equal to the fraction of all possible initial samples that do not include members of the unit i network, $\binom{N-b_i}{n_1}/\binom{N}{n_1}$, where b_i is the unit i network size. Thus, the probability that unit i is included in the estimator, π_i', is the complement of this probability, $\pi_i' = 1 - \binom{N-b_i}{n_1}/\binom{N}{n_1}$, $i = 1, 2, \ldots, N$ and the estimator of μ is

$$\hat{\mu}_{ad,HT} = \frac{1}{N} \sum_{i \in S^*} \frac{y_i J_i}{\pi'_i} \qquad (11.7)$$

where J_i is an indicator random variable equal to 0 if unit i is not selected in the initial sample and $y_i < c$, and is equal to 1 otherwise. S^* is the adaptive random sample of units (including the initial units selected), of variable size n^*. Note that Equation (11.7) is analogous in form to the Horvitz–Thompson estimator of the total [Equation (8.5)] scaled by $1/N$.

Equation (11.7) can be alternatively expressed in a more convenient form as a scaled summation of weighted network totals. Recall that each population unit belongs to one, and only one, network, and that the collection of networks thus form a partition of the N population units. Recall also the network notation from the previous section: for network k, $k = 1, 2, \ldots, K$, U_k denotes the set of population units, N_k denotes the number of these units (network size), and let $T_k = \sum_{i \in U_k} y_i$ denote its y total. Because of the adaptive protocol, for any particular network k, either no network members are sampled or all network members are sampled, and thus all network members have the same probability of being included in the estimator, which we'll denote as α_k

$$\alpha_k = 1 - \frac{\binom{N-N_k}{n_1}}{\binom{N}{n_1}}, \quad k = 1, 2, \ldots, K \qquad (11.8)$$

Therefore, $\hat{\mu}_{ad,HT}$ can be re-expressed as

$$\hat{\mu}_{ad,HT} = \frac{1}{N} \sum_{k=1}^{K} \frac{T_k I_k}{\alpha_k} \qquad (11.9)$$

where I_k is an indicator random variable equal to 1 if any unit of network k is selected in the initial sample, and equal to 0 otherwise. With this formulation, the networks can be viewed as the estimator "units", the T_k as the variable values, and the α_k as the network inclusion probabilities.

Note that $\hat{\mu}_{ad,HT}$ is unbiased for μ

$$E\left(\hat{\mu}_{ad,HT}\right) = E\left(\frac{1}{N} \sum_{k=1}^{K} \frac{T_k I_k}{\alpha_k}\right) = \frac{1}{N} \sum_{k=1}^{K} \frac{T_k E(I_k)}{\alpha_k} = \frac{1}{N} \sum_{k=1}^{K} T_k = \frac{1}{N} T = \mu \qquad (11.10)$$

because $E(I_k) = \Pr\{I_k = 1\} = \alpha_k$.

The sampling variance of $\hat{\mu}_{ad,HT}$ follows per the standard Horvitz–Thompson estimator, once the second order network inclusion probabilities can be established. The probability that two particular networks k and h are both represented in the estimator, α_{kh}, is equal to the complement of the probability that network k or h are not represented. That is, $\alpha_{kh} = 1 - \Pr\{I_k = 0 \text{ or } I_h = 0\} = 1 - [\Pr\{I_k = 0\} + \Pr\{I_h = 0\} - \Pr\{I_k = 0, I_h = 0\}]$ [see Equation (A.28)]. $\Pr\{I_k = 0\}$ is equal to the fraction of all possible initial samples that do not include members of the network, $\binom{N-N_k}{n_1}/\binom{N}{n_1}$, and $\Pr\{I_k = 0, I_h = 0\}$ is equal to the fraction that does not include members from either network, $\binom{N-N_k-N_h}{n_1}/\binom{N}{n_1}$. Therefore

$$\alpha_{kh} = 1 - \frac{\binom{N-N_k}{n_1} + \binom{N-N_h}{n_1} - \binom{N-N_k-N_h}{n_1}}{\binom{N}{n_1}}, \quad h \neq k \qquad (11.11)$$

with $\alpha_{kh} = \alpha_k$ for $h = k$. The sampling variance of $\hat{\mu}_{ad,HT}$ in Horvitz–Thompson form [Equation (8.7)] is

$$V(\hat{\mu}_{ad,HT}) = \frac{1}{N^2}\left[\sum_{k=1}^{K} T_k^2 \frac{1-\alpha_k}{\alpha_k} + \sum_{k=1}^{K}\sum_{\substack{h=1\\h\neq k}}^{K} T_k T_h \frac{\alpha_{kh} - \alpha_k \alpha_h}{\alpha_k \alpha_h}\right] \quad (11.12)$$

with unbiased Horvitz–Thompson estimator (assuming that the number of networks observed in the initial sample is greater than one, i.e., $\sum_{k=1}^{K} I_k > 1$)

$$\hat{V}(\hat{\mu}_{ad,HT}) = \frac{1}{N^2}\left[\sum_{k=1}^{K} T_k^2 I_k \frac{1-\alpha_k}{\alpha_k^2} + \sum_{k=1}^{K}\sum_{\substack{h=1\\h\neq k}}^{K} T_k T_h I_k I_h \frac{\alpha_{kh} - \alpha_k \alpha_h}{\alpha_{kh}\alpha_k \alpha_h}\right] \quad (11.13)$$

Note that this approach explicitly incorporates all of the $y \geq c$ values from the adaptive sample, but includes $y < c$ values only if these units were selected in the initial sample, S_1. Note also that while the π_i values may be unknown for some of the sampled units (edge units), the α_k and α_{kh} values will be known for the sampled networks and thus permit application of the modified Horvitz–Thompson estimators. Finally, note that while the initial sample size n_1 is fixed, the number of networks sampled is a random variable for estimation purposes.

Continuing with the Table 11.2 adaptive sample space example, the estimators $\hat{\mu}_{ad,HT}$ and $\hat{V}(\hat{\mu}_{ad,HT})$ require specification of the associated α_k and α_{kh} for the respective networks. For this particular example, there are only two unique values of α_k ($k = 1, 2, \ldots, 8$). For the seven networks of size one: $\alpha_k = 1 - \binom{9-1}{2}/\binom{9}{2} = 8/36$ ($k = 1, 2, \ldots, 7$). For the remaining network of size two: $\alpha_8 = 1 - \binom{9-2}{2}/\binom{9}{2} = 15/36$. Similarly, for this particular example, there are only two values of α_{kh} ($k = 1, 2, \ldots, 8; h = 1, 2, \ldots, 8; h \neq k$). For the joint inclusion probabilities involving networks of size one: $\alpha_{kh} = 1 - \left[\binom{9-1}{2} + \binom{9-1}{2} - \binom{9-1-1}{2}\right]/\binom{9}{2} = 1/36$ ($k \neq 8, h \neq 8$). For the joint inclusion probabilities involving the size two network with one of the size one networks: $\alpha_{kh} = 1 - \left[\binom{9-1}{2} + \binom{9-2}{2} - \binom{9-1-2}{2}\right]/\binom{9}{2} = 2/36$ ($k = 8$ or $h = 8$). All of these inclusion probabilities can be directly verfied by counting the number of initial samples in Table 11.3 that include the respective network pairs and dividing by the number of possible samples. Table 11.3 lists for each of the 36 possible initial samples the associated $\hat{\mu}_{ad,HT}$ and $\hat{V}(\hat{\mu}_{ad,HT})$ estimates. Because the initial sample is selected by SRS, each of the possible without replacement samples are equally likely, $p(s_1) = 1/36$, for $s_1 \in S_1$. The expected value and sampling variance of $\hat{\mu}_{ad,HT}$ can be calculated directly over the sample space (S_1) of the initial samples of size $n_1 = 2$ as

$$E(\hat{\mu}_{ad,HT}) = \frac{1}{36}\sum_{s_1 \in S_1} \hat{\mu}_{ad,HT}(s_1) = 0.5556 = \mu$$

$$V(\hat{\mu}_{ad,HT}) = \frac{1}{36}\sum_{s_1 \in S_1} (\hat{\mu}_{ad,HT}(s_1) - \mu)^2 = 0.2802$$

The first result demonstrates the unbiasedness of the estimator, and the second result demonstrates agreement with Equation (11.12) as follows. For this particular example, Equation (11.12) is simple to apply because the $T_k = 0$ for all networks except networks 1 (unit 1) and 8 (units 8 and 9). Whenever $T_k = 0$, the corresponding term in Equation (11.12) makes no contribution to the sampling variance. Thus, Equation (11.12) reduces to

$$V(\hat{\mu}_{ad,HT}) = \frac{1}{N^2} \left[T_1^2 \frac{1-\alpha_1}{\alpha_1} + T_8^2 \frac{1-\alpha_8}{\alpha_8} \right] + 2\frac{1}{N^2} \left[T_1 T_8 \frac{\alpha_{1,8} - \alpha_1 \alpha_8}{\alpha_1 \alpha_8} \right]$$

$$= \frac{1}{9^2} \left[1^2 \frac{1-(8/36)}{8/36} + 4^2 \frac{1-(15/36)}{15/36} \right] + 2\frac{1}{9^2} \left[1 \cdot 4 \frac{(2/36) - (8/36)(15/36)}{(8/36)(15/36)} \right]$$

$$= 0.2802$$

Interestingly, in this particular example, the expectation of the sampling variance estimator, $\hat{V}(\hat{\mu}_{ad,HT})$, does not exist because the last of the possible samples listed (ID = 36) consists of a single network (units 8 and 9 both belong to network 8), and for a single network the sampling variance estimator is undefined. This illustrates that while $n_1 = 2$ is fixed, the number of networks sampled is not fixed (for this particular example, the initial sample may consist of either one or two networks).

Consider now the performance of adaptive cluster sampling using $\hat{\mu}_{ad,HT}$ with that of an SRS/mpu strategy ($\hat{\mu}_{mpu}$). The relative precision, $RP(\hat{\mu}_{ad,HT}, \hat{\mu}_{mpu})$, would not be a fair comparison of the two strategies because the realized sample size of the adaptive sample, n^*, will presumably be larger than the initial SRS sample size n_1, and sometimes much larger. Therefore, a more appropriate measure of comparison would be the relative efficiency of the two strategies, based on a sample size of $E(n^*)$ for the non-adaptive design. The relative efficiency (RE) of the two strategies would be

$$\text{RE}(\hat{\mu}_{ad,HT}, \hat{\mu}_{mpu}) = \frac{V(\hat{\mu}_{mpu}|n = E(n^*))}{V(\hat{\mu}_{ad,HT}|n_1)} \tag{11.14}$$

where the numerator is based on Equation (3.15) and the denominator is based on Equation (11.12). Substituting these equations into Equation (11.14) leads to a complicated expression which we do not show here. Again, however, net relative efficiency [NRE, Equation (3.35)] might offer the more appropriate comparison because it would take into account the costs of the two designs, which might be cheaper for adaptive sampling. If this were the case, $\text{NRE}(\hat{\mu}_{ad,HT}, \hat{\mu}_{mpu})$ would exceed $\text{RE}(\hat{\mu}_{ad,HT}, \hat{\mu}_{mpu})$.

Example 11.2. *Performance of adaptive cluster sampling using $\hat{\mu}_{ad,HT}$.* For our example of population units given in Table 11.1 with $\sigma^2 = 1.0278$, and following our example of the performance of the mean-per-unit approach for $n_1 = 2$, it was shown previously that $V(\hat{\mu}_{ad,HT}) = 0.2802$. For an SRS sample of equivalent size $n = 2$, $V(\hat{\mu}_{mpu}) = [(9-2)/9](1.0278/2) = 0.3997$, so that the relative precision of the two strategies would be $RP(\hat{\mu}_{ad,mpu}, \hat{\mu}_{mpu}) = 0.3997/0.2802 \approx 1.43$. However, for the relative efficiency of the two strategies, an initial sample size of $n_1 = 2$ for the adaptive design would be paired with an SRS sample size of $n = E(n^*) \approx 4$ (see Table 11.2, $E(n^*) = 139/36 \approx 3.86$). For $n = 4$, $V(\hat{\mu}_{mpu}) = [(9-4)/9](1.0278/4) = 0.1427$, so that $\text{RE}(\hat{\mu}_{ad,mpu}, \hat{\mu}_{mpu}) = 0.1427/0.2802 \approx 0.51$. Finally, we compare the relative performance of the two adaptive cluster sampling estimation approaches. In this case, the sampling designs are equivalent, so that for a given initial sample size n_1, the relative precision, relative efficiency, and net relative efficiency are equivalent to each other. For this example,

$\text{NRE}(\hat{\mu}_{ad,HT}, \hat{\mu}_{ad,mpu}) = V(\hat{\mu}_{ad,mpu}|n_1)/V(\hat{\mu}_{ad,HT}|n_1) = 0.3025/0.2802 \approx 1.08$, so that $\hat{\mu}_{ad,HT}$ is the more efficient estimator of the two, as Salehi (2003) showed will usually be the case.

11.2 Other adaptive sampling designs

11.2.1 Single-stage strip and systematic designs

The adaptive cluster sampling procedures described previously, with the initial sample of units selected by SRS, seem the most obvious way to implement an adaptive sampling scheme, but Thompson (1991) extended the approach to accommodate other selection schemes for the initial sample. With these other selection schemes, the population units are regarded as *secondary units*, the collection of which is partitioned into a smaller set of *primary units* from which the initial sample is selected. Given an initial sample of primary units, the adaptive sampling procedure is carried out as before at the secondary unit level.

For a rectangular grid of spatially arranged population units, one such option is to treat the rows or columns (*strips*) of secondary units as the primary units. For example, with a 5×5 grid there would be five vertical (or horizontal) strips (primary units), each consisting of five secondary units. Table 11.4 depicts a realized adaptive cluster sample for this configuration, in which $n_1 = 2$ vertical strips were selected by SRS for the initial sample (primary units 1 and 3, light shade). With $c = 1$, a first iteration of adaptive sampling was then made in the neighborhood of secondary units 2 and 3 in the first strip (leading to selection of secondary units 7 and 8, dark shade), and in the neighborhood of secondary unit 11 in the third strip (leading to selection of secondary units 6 and 16, dark shade). A second iteration of adaptive sampling was then made in the neighborhood of secondary unit 8 (leading to selection of secondary unit 9, medium shade), and the adaptive search then stopped because $y_9 < c$.

Another option, one that has substantial appeal from a spatial perspective, would be to treat the possible systematic samples of secondary units as the primary units, and to select one or more of these for the initial sample. For example, with the 5×5 grid of secondary units, if the systematic interval was $k = 4$ with possible random starts r equal to units 1, 2, 3, 4, there would be four primary units consisting of 7, 6, 6, and 6 secondary units, respectively. Table 11.5 depicts a realized adaptive cluster sample for this configuration, in which a single systematic sample ($n_1 = 1$) of 6 secondary units was selected for the initial

Table 11.4 Realization of an iterative adaptive cluster sampling scheme with primary units defined by vertical strips of secondary units, and an initial SRS sample size of $n_1 = 2$ primary units, each consisting of five secondary units, with $c = 1$. Initial primary unit selections in light shade; first iteration of adaptive sampling secondary unit selections in dark shade; second iteration in medium shade.

$y_1 = 0$	$y_6 = 0$	$y_{11} = 1$	$y_{16} = 0$	$y_{21} = 0$
$y_2 = 2$	$y_7 = 0$	$y_{12} = 0$	$y_{17} = 1$	$y_{22} = 0$
$y_3 = 4$	$y_8 = 2$	$y_{13} = 0$	$y_{18} = 3$	$y_{23} = 0$
$y_4 = 0$	$y_9 = 0$	$y_{14} = 0$	$y_{19} = 0$	$y_{24} = 0$
$y_5 = 0$	$y_{10} = 0$	$y_{15} = 0$	$y_{20} = 0$	$y_{25} = 0$

Table 11.5 Realization of an iterative adaptive cluster sampling scheme with primary units defined by possible systematic samples of secondary units ($k = 4$, $r = 1, 2, 3, 4$), and an initial SRS sample size of $n_1 = 1$ primary unit containing 6 secondary units, with $c = 1$. Initial primary unit selection (for $r = 3$) in light shade; first iteration of adaptive sampling secondary unit selections in dark shade; second iteration in medium shade.

$y_1 = 0$	$y_6 = 0$	$y_{11} = 1$	$y_{16} = 0$	$y_{21} = 0$
$y_2 = 2$	$y_7 = 0$	$y_{12} = 0$	$y_{17} = 1$	$y_{22} = 0$
$y_3 = 4$	$y_8 = 2$	$y_{13} = 0$	$y_{18} = 3$	$y_{23} = 0$
$y_4 = 0$	$y_9 = 0$	$y_{14} = 0$	$y_{19} = 0$	$y_{24} = 0$
$y_5 = 0$	$y_{10} = 0$	$y_{15} = 0$	$y_{20} = 0$	$y_{25} = 0$

sample ($r = 3$, light shade), followed by an adaptive search ($c = 1$) in the neighborhood of the initial and subsequently selected secondary units.

11.2.2 Two-stage complete allocation cluster sampling

Salehi and Brown (2010) proposed a **complete allocation** stratified adaptive sampling strategy. Salehi and Seber (2017) [see also Salehi (2017)] later generalized this approach for **two-stage complete allocation (adaptive cluster) sampling** (TSCAS). (Recall that when the first-stage number of selected clusters, n, equals the total number of clusters, N, then stratified sampling can be considered a special case of two-stage cluster sampling.) For the original stratified sampling setting (Salehi and Brown 2010), initial samples of size n_h are selected from the N_h units from all strata, as for conventional stratified sampling. Then, *all units* in stratum h are (adaptively) sampled (hence the term *complete allocation*) whenever any unit in the initial sample n_h has $y_{hj} > 0$. A similar procedure may be used for two-stage adaptive cluster sampling. First, select a random sample of clusters according to some selection scheme. Second, select independent samples of subunits within each cluster selected at the first stage. Finally, adaptively select all subunits within selected cluster i whenever any of the sampled subunits has $y_{ij} > c$.[1] For the special case of $c = 0$ and stratified sampling, Salehi and Brown (2010) show that a single-stage Horvitz–Thompson estimator can be used for unbiased estimation of the population total over all strata. For the case $c > 0$, in the stratified or two-stage settings, Murthy's (1957) estimator (described in the following section) is used to unbiasedly estimate individual stratum or cluster totals, and a two-stage Horvitz–Thompson estimator is used to unbiasedly estimate the total over all clusters or strata (Salehi and Seber 2017).

For TSCAS, primary units may be of unequal size and the first-stage sample, S_1, may be selected, if desired, with unequal probability without replacement. For $i \in S_1$, a second-stage SRS sample, S_{2i}, of m_i subunits is selected from the M_i subunits. If the value of the target variable for any one of the m_i subunits exceeds the adaptive search criterion, c, then *all* of the subunits within the primary unit are selected. One advantage of this scheme is that, given knowledge of the initial primary unit selections, $S_1 = s_1$, the total size of the realized adaptive sample across the selected primary units has a known upper limit, $\sum_{i \in s_1} M_i$. In contrast, in (single-stage) adaptive cluster sampling (Section 11.1), the size

---

[1] Note that the Salehi and Brown (2010) and Salehi and Seber (2017) adaptive search trigger is specified as $y_{ij} > c$ rather than $y_{ij} \geq c$. This allows $c = 0$ for the special case of stratified sampling proposed by Salehi and Brown (2010).

of the adaptive cluster sample may have an upper limit of $\sum_{i=1}^{N} M_i$, the total number of subunits in the population, depending on the y values and c.

We begin with the usual two-stage arrangement, with N primary units of sizes M_i, $i = 1, 2, \ldots, N$. At the first stage of sampling, select a random first-stage sample, S_1, of n primary units from N without replacement, with equal or unequal probability, with π_i denoting the first-stage primary unit inclusion probabilities. At the second stage of sampling, select second-stage SRS samples, S_{2i}, of m_i subunits from M_i within the selected primary units. At this point, the adaptive sampling begins. Compare the observed values of the target variable in selected subunits (y_{ij}, for $j \in S_{2i}$) to the adaptive search criterion c. If $y_{ij} \le c$ for all of the subunits in S_{2i}, then no further subunits in primary unit i are selected. If $y_{ij} > c$ for any one of the subunits in m_i, then *all* M_i subunits in primary unit i are selected. Thus, for a primary unit included in the first-stage sample, the random set of subunits adaptively selected from primary unit i, S_{2i}^*, is either of size m_i or M_i.

For this sampling design, Horvitz–Thompson estimators cannot be used to estimate the primary unit totals, $\mathcal{T}_i$, $i = 1, 2, \ldots, N$, because inclusion probabilities for the subunits in selected primary units are unknown when none of the subunits in S_{2i} trigger selection of all remaining $M_i - m_i$ subunits. Salehi and Seber (2017) show that Murthy's (1957) estimator can instead be used to derive an unbiased estimator of $\mathcal{T}_i$ for TSCAS.

Murthy's estimator for $\mathcal{T}_i$

Let $U_i = \{1, 2, \ldots, M_i\}$ denote the set of subunit labels for primary unit i. If primary unit i is selected at the first stage of sampling, then following the initial second-stage sample of subunits, $S_{2i} = s_{2i}$, either no further sampling will be conducted ($y_{ij} \le c$ for all s_{2i} subunits), in which case $S_{2i}^* = s_{2i}$, or all of the remaining subunits will be sampled ($y_{ij} > c$ for one or more s_{2i} subunits), in which case $S_{2i}^* = U_i$. Thus, let s_{2i}^* denote the realized set of secondary units in the adaptive sample within primary unit i, equal to s_{2i} (adaptive sampling not triggered) or U_i (adaptive complete allocation triggered). Murthy's (1957) unbiased estimator for $\mathcal{T}_i$ depends on which of these two events occurs

$$\hat{\mathcal{T}}_i = \sum_{j \in S_{2i}^*} \left[\frac{\Pr\{S_{2i}^* = s_{2i}^* | \text{subunit } j \text{ selected first}\}}{\Pr\{S_{2i}^* = s_{2i}^*\}} \right] y_{ij}, \quad s_{2i}^* = U_i, s_{2i} \tag{11.15}$$

where the conditional probability is given that a particular subunit, j, is the first one selected in the s_{2i} sample. Note that this conditional probability is meaningful only for a draw-by-draw or sequential sample selection method like SRS or, in unequal probability sampling, a method like selection of successive units with probabilities proportional to the sizes of remaining units (see Section 8.3).

Case 1: $S_{2i}^* = U_i$.

Following Salehi and Seber (2017), partition U_i into two subsets: U_{ic}, those subunits for which $y_{ij} > c$, and $U_{ic'}$, those subunits for which $y_{ij} \le c$. Denote the number of subunits in U_{ic} by k_i, so that the number of subunits in $U_{ic'}$ is $M_i - k_i$. Define

$$F_i(x) = 1 - \frac{\binom{M_i - k_i - x}{m_i - x}}{\binom{M_i - x}{m_i - x}}, \quad x = 0, 1, 2$$

The probabilities in Equation (11.15) for $\hat{\mathcal{T}}_i$ can be expressed in terms of $F_i(x)$ as follows. $\Pr\{S_{2i}^* = U_i\}$ is equal to the probability that $y_{ij} > c$ for at least one subunit in S_{2i}, which is equal to 1 minus the probability that $y_{ij} \le c$ for all subunits in S_{2i}. The number of possible samples of size m_i from $U_{ic'}$ is $\binom{M_i - k_i}{m_i}$, and the number of possible samples of size m_i from

U_i is $\binom{M_i}{m_i}$. Therefore,

$$\Pr\{S_{2i}^* = U_i\} = 1 - \frac{\binom{M_i - k_i}{m_i}}{\binom{M_i}{m_i}} = F_i(0)$$

The conditional probability in Equation (11.15), $\Pr\{S_{2i}^* = U_i | \text{subunit } j \text{ selected first}\}$, depends on whether $y_{ij} > c$ or $y_{ij} \le c$ for subunit j. That is, whether subunit j is a member of U_{ic} or $U_{ic'}$. If $y_{ij} > c$ for subunit j, then all of the unit i subunits are sampled so that $S_{2i}^* = U_i$ with probability equal to 1. If instead $y_{ij} \le c$ for subunit j, then $\Pr\{S_{2i}^* = U_i | \text{subunit } j \text{ selected first}\}$ is equal to the probability that $y_{ij} > c$ for at least one of the $m_i - 1$ remaining subunits in S_{2i}, which is equal to 1 minus the probability that $y_{ij} \le c$ for all $m_i - 1$ remaining subunits in S_{2i}. The number of possible samples of size $m_i - 1$ from the $M_i - k_i - 1$ remaining subunits in $U_{ic'}$ is $\binom{M_i - k_i - 1}{m_i - 1}$, and the number of possible samples of size $m_i - 1$ from the $M_i - 1$ remaining subunits in U_i is $\binom{M_i - 1}{m_i - 1}$. Therefore,

$$\Pr\{S_{2i}^* = U_i | \text{subunit } j \text{ selected first}, y_{ij} \le c\} = 1 - \frac{\binom{M_i - k_i - 1}{m_i - 1}}{\binom{M_i - 1}{m_i - 1}} = F_i(1)$$

Partitioning Equation (11.15) into its U_{ic} and $U_{ic'}$ components thus gives

$$\hat{T}_i = \frac{1}{F_i(0)} \sum_{j \in U_{ic}} y_{ij} + \frac{F_i(1)}{F_i(0)} \sum_{j \in U_{ic'}} y_{ij} \qquad (11.16)$$

An unbiased estimator of sampling variance of $\hat{T}_i$ for this case is (Salehi 2017)

$$\hat{V}(\hat{T}_i) = \frac{k_i}{F_i(0)} \left[k_i \left(1 - \frac{1}{F_i(0)} \right) + (M_i - k_i) \left(1 - \frac{F_i(1)}{F_i(0)} \right) \right] V(y)_{ic}$$

$$+ \frac{(M_i - k_i) k_i}{F_i(0)} \left(1 - \frac{F_i(1)}{F_i(0)} \right) (\mu_{ic} - \mu_{ic'})^2 \qquad (11.17)$$

$$+ \frac{(M_i - k_i)}{F_i(0)} \left[(M_i - k_i) \left(F_i(2) - \frac{F_i^2(1)}{F_i(0)} \right) + k_i \left(1 - \frac{F_i(1)}{F_i(0)} \right) \right] V(y)_{ic'}$$

where $\mu_{ic} = \sum_{j \in U_{ic}} y_{ij} / k_i$, $\mu_{ic'} = \sum_{j \in U_{ic'}} y_{ij} / (M_i - k_i)$, $V(y)_{ic} = \sum_{j \in U_{ic}} (y_{ij} - \mu_{ic})^2 / k_i$, and $V(y)_{ic'} = \sum_{j \in U_{ic'}} (y_{ij} - \mu_{ic'})^2 / (M_i - k_i)$. Note that μ_{ic}, $\mu_{ic'}$, $V(y)_{ic}$ and $V(y)_{ic'}$ are known because all subunits in primary unit i have been selected in the adaptive search in this case.

Case 2: $S_{2i}^* = s_{2i}$.

Following Salehi and Seber (2017), Equation (11.15) may be re-expressed using the conditional probability relations outlined in Section A.3.3 as follows. Let A denote the event $S_{2i}^* = s_{2i}$, and B denote the event that subunit j is selected first. Then, $\hat{T}_i = \Pr\{A|B\} / \Pr\{A\} = \Pr\{B|A\} / \Pr\{B\}$. From first principles, $\Pr\{B\} = 1/M_i$, the probability that subunit j is the first one selected at the second stage of sampling, and $\Pr\{B|A\} = 1/m_i$, the probability that subunit j is the first one selected given the set of subunits selected at the second stage, $S_{2i}^* = s_{2i}$. Thus, in this case, Equation (11.15) reduces to

$$\hat{T}_i = \sum_{j \in S_{2i}^*} \frac{1/m_i}{1/M_i} y_{ij} = M_i \sum_{j \in S_{2i}^*} \frac{y_{ij}}{m_i} = M_i \hat{\mu}_i \qquad (11.18)$$

where $\hat{\mu}_i = \sum_{j \in S_{2i}^*} y_{ij} / m_i$ is the mean-per-unit estimator of the mean for primary unit i.

An unbiased estimator of sampling variance of $\hat{T}_i$ for this case is the familiar

$$\hat{V}(\hat{T}_i) = M_i^2 \left(\frac{M_i - m_i}{M_i} \right) \frac{\hat{\sigma}_i^2}{m_i} \tag{11.19}$$

where $\hat{\sigma}_i^2 = \sum_{j \in S_{2i}} (y_{ij} - \hat{\mu}_i)^2 / (m_i - 1)$ is the estimator of σ_i^2 under SRS.

Two-stage estimators

Given Murthy's estimator, $\hat{T}_i$, for the total in primary unit i and the associated estimator of sampling variance, $\hat{V}(\hat{T}_i)$, for the two cases ($S_{2i}^* = U_i$, or $S_{2i}^* = s_{2i}$), the usual two-stage estimators for the overall total and its sampling variance (see Chapter 9) apply for TSCAS. Assuming without replacement selection of primary units, the overall total can be unbiasedly estimated using a two-stage Horvitz–Thompson estimator

$$\hat{T} = \sum_{i \in S_1} \frac{\hat{T}_i}{\pi_i} \tag{11.20}$$

If the first-stage sample size is fixed, then a two-stage version of the Sen–Yates–Grundy estimator [Equation (8.10)] of $V(\hat{T})$ is

$$\hat{V}(\hat{T}) = \frac{1}{2} \sum_{i \in S_1} \sum_{\substack{j \in S_1 \\ j \neq i}} \frac{(\pi_i \pi_j - \pi_{ij})}{\pi_{ij}} \left(\frac{\hat{T}_i}{\pi_i} - \frac{\hat{T}_j}{\pi_j} \right)^2 + \sum_{i \in S_1} \frac{\hat{V}(\hat{T}_i)}{\pi_i} \tag{11.21}$$

If SRS is used at the first stage of sampling, then

$$\hat{T} = \frac{N}{n} \sum_{i \in S_1} \hat{T}_i$$

and

$$\hat{V}(\hat{T}) = N^2 \left(\frac{N-n}{N} \right) \frac{\sum\limits_{i \in S_1} (\hat{T}_i - \sum_i^n \hat{T}_i / n)^2}{n(n-1)} + \frac{N}{n} \sum_{i \in S_1} \hat{V}(\hat{T}_i)$$

Example 11.3. *Sample space illustration for TSCAS.* A full sample space illustration for TSCAS would require construction of all possible first-stage sample selections, S_1, as well as all possible associated second-stage *adaptive* samples, along with associated estimates of primary unit totals, similar to the illustration presented in Section 9.2.3 except that the random second-stage sample selections, S_{2i}, may lead to $S_{2i}^* = s_{2i}$ or $S_{2i}^* = U_i$. Therefore, we restrict our numerical illustration of this approach to estimation of T_i, the total of the target variable within selected primary unit i, using adaptive complete allocation sampling and Murthy's (1957) estimator. Equations (11.16) or (11.18) are used to unbiasedly estimate T_i, and Equations (11.17) or (11.19) are used to estimate $V(\hat{T}_i)$, as appropriate.

Suppose that $M_i = 5$, $\{y_{ij}\} = \{0, 0, 2, 8, 2\}$, $T_i = 12$, $\sigma_i^2 = 10.8$, an SRS initial second-stage sample size of $m_i = 2$, and that adaptive sampling beyond the initially selected subunits will take place whenever $y_{ij} > 3$ (i.e., $c = 3$). Table 11.6 lists for each of the possible second-stage initial samples (s_{2i}) the *realized* adaptive sample subunits (s_{2i}^*) and size (m_i^*), Murthy's estimate of the cluster total [$\hat{T}_i(s_{2i}^*)$], and its associated sampling variance estimate [$\hat{V}(\hat{T}_i)(s_{2i}^*)$]. For this within primary unit adaptive sample space, designated as S_{2i}^*, $E(\hat{T}_i) = \sum_{s_{2i}^* \in S_{2i}^*} \hat{T}_i(s_{2i}^*) p(s_{2i}^*) = (1/10) \cdot 120 = 12 = T_i$, $V(\hat{T}_i) = 78.5$, and $E[\hat{V}(\hat{T}_i)] =$

Table 11.6 Sample space illustration of complete allocation adaptive sampling using Murthy's (1957) estimator within a single primary unit for which $M_i = 5$, $y_{ij} = \{0, 0, 2, 8, 2\}$, $T_i = 12$, $\sigma_i^2 = 10.8$, and an SRS initial second-stage sample size of $m_i = 2$. Complete primary unit sampling takes place whenever $y_{ij} > c, c = 3$, for any one of the s_{2i} subunits, with s_{2i} denoting a particular second-stage selection of subunits, s_{2i}^* denoting the associated adaptive sample of subunits, and m_i^* denoting the realized adaptive sample size (either m_i or M_i).

ID	Sample Units $j, j \in s_{2i}$	Units $j, j \in s_{2i}^*$	m_i^*	$\hat{T}_i(s_{2i}^*)$	$\hat{V}(\hat{T}_i)(s_{2i}^*)$
1	1, 2	1, 2	2	0.0	0.00
2	1, 3	1, 3	2	5.0	15.00
3	1, 4	1, 2, 3, 4, 5	5	22.5	181.25
4	1, 5	1, 5	2	5.0	15.00
5	2, 3	2, 3	2	5.0	15.00
6	2, 4	1, 2, 3, 4, 5	5	22.5	181.25
7	2, 5	2, 5	2	5.0	15.00
8	3, 4	1, 2, 3, 4, 5	5	22.5	181.25
9	3, 5	3, 5	2	10.0	0.00
10	4, 5	1, 2, 3, 4, 5	5	22.5	181.25
Sums:				120.0	785.00

$\sum_{s_{2i}^* \in \mathcal{S}_{2i}^*} \hat{V}(\hat{T}_i)(s_{2i}^*)p(s_{2i}^*) = (1/10) \cdot 785.00 = 78.5 = V(\hat{T}_i)$, numerically verifying for this example that the complete allocation adaptive sampling estimators $\hat{T}_i$ and $\hat{V}(\hat{T}_i)$ are unbiased for T_i and $V(\hat{T}_i)$, respectively.

Note that when complete allocation ($S_{2i}^* = U_i$) takes place, Murthy's estimator does not take on the value T_i, even though all of the cluster subunits have been selected in this case and T_i is therefore known without error. The virtue of using Murthy's (1957) estimator in the TSCAS setting is that it generates unbiased estimates of T_i for each of the primary units in S_1, thereby allowing unbiased estimation of the overall population total, T, which is the target of estimation in the TSCAS survey. If estimates of the primary unit totals, $\hat{T}_i$, $i = 1, 2, \ldots, N$ are of interest in and of themselves, then the usual mean-per-unit estimates (assuming $S_{2i}^* = s_{2i}$) or the known totals (assuming $S_{2i}^* = U_i$) could be used for that purpose. But unbiased estimation of T requires use of Murthy's estimates of T_i, $i = 1, 2, \ldots, N$, even though they may be at odds with the known primary unit totals.

11.3 Chapter comments

Although adaptive sampling has quite clear conceptual attraction, it may prove exceptionally difficult to control the total cost of an adaptive cluster sampling survey because the eventual size of the realized adaptive sample cannot be accurately predicted prior to execution of an adaptive survey. For example, in single-stage adaptive cluster sampling, if c is set too low (e.g., $c = 1$) and $y \geq c$ for most or all units i in N, then successive iterations of adaptive search beyond the initial random sample of size n_1 may lead to a near census of the population. If instead c is set too high, then few or no $y \geq c$ and adaptive sampling may often go no further than the initial sample n_1. There are at least three alternative approaches for addressing this rather serious dilemma: (1) use of **order statistics** (to assist in a reasonable specification of c), (2) use of **stopping rules** (to prevent n^* from potentially spinning out of control), and (3) adoption of TSCAS which provides a well-identified upper limit to the size of the realized two-stage adaptive sample, typically much less than the total number of population units.

When there is little pre-survey information to inform specification of c, Thompson (1996) and Thompson and Seber (1996 Chapter 6) proposed adaptive cluster sampling based on order statistics generated from the initial SRS of size n_1. With this approach, the observed values of y in n_1 are ordered from lowest to highest, implicitly relabeling units from the population to the ordered values of y in the sample (similar to the implicit reordering of $\sum_{i \in s} = \sum_{i=1}^{n}$). The surveyor then adaptively samples in the neighborhood of the $n_1 - r$ units with the largest y values by setting $c = y_{n_1 - r}$ (i.e., r pre-specified instead of c) as the criterion to initiate further adaptive sampling. For example, if $n_1 = 40$, $r = 4$, and the four highest y values in the initial sample are 17, 21, 23, 24, then set $c = 17$. This kind of a procedure has substantial appeal because it should encourage further adaptive sampling (because the distribution of y in the initial sample suggests that about 10% of all y may equal or exceed $c = 17$), and it should also prevent n^* from being "much too large". As Thompson and Seber (1996 p. 164) noted: "If the wrong criterion is used, it may turn out that every unit in the study region, or none of them, satisfies the criterion." Thompson (1996) shows that unbiased adaptive cluster sampling estimators can be formed when order statistics are used for specification of c. The theory behind such estimators is considerably beyond the level of this text as the particular value that c takes on, over the sample space, is conditioned on the particular random sample S_1 that is selected and the particular values y that appear in this sample. Thus, over the sample space, c is a random variable when order statistics are used to specify it.

Another approach to containing the size of the realized adaptive sample is adoption of a stopping rule [e.g., Su and Quinn (2003)] which cuts off further adaptive sampling at the tth iteration of the adaptive search beyond the initial SRS, whether based on a pre-specified c or on order statistics observed in the initial sample, where t is pre-specified. Adoption of a stopping rule will obviously restrict total sampling effort and survey cost, but it also leads to estimator bias.

The two-stage complete allocation adaptive sampling strategy presented in this chapter may be an excellent alternative for reduction of the variability in the total number of subunits that are included in an adaptive two-stage cluster sample. The relative benefits of using adaptive sampling within primary units in a two-stage sampling design, however, may often be relatively modest because adaptive sampling affects only the second-stage component of sampling variance. In typical two-stage settings, the first-stage variance contribution (due to variation in means or totals across primary units) normally dominates total sampling variance.

According to Thompson and Seber (1996), the relative efficiency of adaptive cluster sampling ($\hat{T}_{\text{ad,mpu}}$) as compared to simple random sampling ($\hat{T}_{\text{mpu}}$) will exceed 1 when the following conditions are met: (1) within network variance is a high proportion of the total variance in the population (highly aggregated), (2) the population is *rare*—the number of (spatial) sampling units in the population is large compared to the number of units satisfying the constraint c, (3) the expected final sample size is not much larger than the initial sample size, and (4) the cost of observing units in clusters or in networks is less than the cost of observing the same number of units at random throughout the study region. Of these conditions, we note that in field settings it would seem that statement (4) would frequently be met. The likely lower cost of units selected via adaptive search within clusters as compared to those selected randomly from N has not been accounted for in our assessment of the relative efficiency of $\hat{T}_{ad,mpu}$ as compared to $\hat{T}_{mpu}$ (Section 11.1.4). We also note, however, that it is hard to imagine much reduction in sampling variance when the total realized adaptive sample size, n^*, is "not much larger" than the initial sample size, n_1 [condition (3)], because in this case adaptive sampling has not produced much information beyond the original SRS.

Despite the well-recognized lack of control over realized sample size in adaptive sampling, the approach has seen numerous applications in natural resources and environmental settings, especially for estimation of the abundance of *rare* species. Chapters 5, 6, and 14 of W. L. Thompson's (2004) text on sampling for rare species and Chapter 3 of Manly and Navarro Alberto's (2015) text on ecological sampling are devoted entirely or almost entirely to adaptive sampling designs; the March 2003 issue of Environmental and Ecological Statistics is devoted entirely to adaptive sampling; and Turk and Borkowski (2005) provide a review of developments in adaptive sampling theory from 1990–2003. Seber and Salehi (2013) provide a more recent overview of adaptive sampling strategies. We note, however, that not all applications of adaptive sampling have produced results superior to conventional survey designs [see, e.g., Noon et al. (2006), Mier and Picquelle (2008)]. Given the high uncertainty in realized size of an adaptive sample, considerable thought must be expended before an adaptive sampling strategy can be effectively implemented.

Problems

Problem 11.1. Suppose that the number of buttercups is known for each of three hundred quadrats (100 m^2) in an isolated population, as presented in Salehi and Seber (2017 Figure 1). Ignore the stratification and the shading of cells in this figure and label the individual cells from 1–300, with first (top) row entries (starting from the left) of 1–15, second row entries of 16–30, and so on. Using R, select an initial SRS of size $n = 30$ from these $N = 300$ units using a random seed value based on your birthdate.

```
iseed <- yyyy + dd  # where mm/dd/yyyy is your birthdate
set.seed(iseed)
N <- 300; n <- 30
s1 <- sample(1:N, size=n, replace=FALSE)
```

Given your selected units and assuming an adaptive search criterion of $c = 2$ (i.e., begin an adaptive search if the number of buttercups in an initial SRS cell is at least 2): (A) Find the full adaptive sample that your initial SRS sample generates. (B) Calculate an estimate of the total number of buttercups based on the network means estimator, $\hat{\mu}_{ad,mpu}$ [Equation (11.3)]. (C) Calculate an estimate of sampling variance for your estimate of the total based on Equation (11.5). (D) Calculate the corresponding SRS estimate of the total and associated sampling variance based only on the initial SRS, and compare with the adaptive sampling estimates. Is this a "fair" comparison?

Problem 11.2. Given the realized adaptive sample generated in Problem 11.1: (A) Calculate an estimate of total buttercup abundance based on $\hat{\mu}_{ad,HT}$, Equation (11.7), which ignores the edge units selected from adaptive searching in the neighborhood of the original SRS units. (B) Calculate a corresponding estimate of sampling variance based on Equation (11.13). Note that this will probably require calculation of some second order inclusion probabilities. (C) Compare your estimates $\hat{T}_{ad,HT}$ and $\hat{V}(\hat{T}_{ad,HT})$ with $\hat{T}_{ad,mpu}$ and $\hat{V}(\hat{T}_{ad,mpu})$, respectively, from Problem 11.1.

Problem 11.3. Suppose for a population of $N = 2,000$ individuals that y is a binary (0, 1) variable indicating whether or not a specific genetic trait is carried by a particular individual, and that the trait is quite rare, $\pi = 0.01$. Imagine that you adopt the following sampling strategy known as **inverse sampling**: select individuals with equal probability without replacement until $Q = 3$ individuals possessing the trait of interest have been detected. In this case the number of sampled individuals, n, is random, not fixed. (A) Estimate the proportion of individuals possessing the trait as $\hat{\pi} = (Q - 1)/(n - 1)$, and estimate its associated sampling variance using $\hat{V}(\hat{\pi}) = \hat{\pi}^2 - (Q - 1)(Q - 2)/(n(n - 1))$. (B) Write an R program that simulates 10^5 independently selected samples following this $Q = 3$ *stopping rule*. (C) Numerically verify that $\hat{\pi}$ is an unbiased estimator of π, and that $\hat{V}(\hat{\pi})$ is approximately unbiased for $V(\hat{\pi})$ in inverse sampling. (D) Plot the frequency histogram of realized sample sizes for this sampling scheme. (E) Discuss the possible advantages and disadvantages of this sampling scheme, relating them to adaptive cluster sampling.

CHAPTER 12

Spatially balanced sampling

Fig. 12.1 Spatially representative sampling can be especially useful when a target variable displays substantial spatial variablility, as in this photo of drought-induced tree mortality due to bark beetles on private and public (Sierra National Forest) lands in California, 2016. Photo credit: Steve Dunsky.

12.1 Introduction

All probability sampling designs that have thus far been considered in this text can be regarded as producing *representative* samples in the sense first defined by Neyman (1934). By this definition, a representative sample is one which has been selected by a probability sampling design that allows unbiased or nearly unbiased design-based estimation of a target population parameter. For the most part we have treated populations as simple *lists* of units with associated variable values and there has been little explicit recognition or consideration of the spatial relationships among population units. Implicit recognition of spatial relationships has been acknowledged in at least stratified sampling and cluster sampling, however, where units may be grouped into clusters or strata on the basis of physical proximity (e.g., a cluster is comprised of all households on a city block). And

Sampling Theory: For the Ecological and Natural Resource Sciences. David G. Hankin, Michael S. Mohr, and Ken B. Newman, Oxford University Press (2019). © David G. Hankin, Michael S. Mohr, and Ken B. Newman. DOI: 10.1093/oso/9780198815792.001.0001

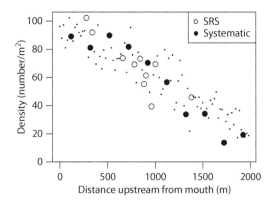

Fig. 12.2 Density of juvenile steelhead (anadromous rainbow trout, *Oncorhynchus mykiss*) within 20 m sections of the lower 2,000 m of a stream (*N* = 100 units). Location of sample units in a particular systematic sample (solid circles) compared to a particular simple random sample (open circles) of the same size (*n* = 10). Note the difference in spatial balance of the systematic sample compared to the SRS.

adaptive sampling relies explicitly on the spatial proximity of units relative to one another in an areal grid frame.

The potential importance of spatial location of units was previously noted in Chapter 4 where we briefly contrasted the spatial locations of a single SRS sample with a single systematic sample of stream units selected on a one-dimensional spatial axis (distance upstream) of length 2,000 m (Figure 12.2, reproduced from Figure 4.2). Many selected units in the SRS are clustered between 600–1,000 m upstream from the mouth of the stream and no selected units are located more than 1,500 m upstream. In contrast, adjacent selected units in the systematic sample are separated by equal stream distances and span nearly the entire 2,000 m length of interest. All possible systematic samples would have similar desirable spatial properties, whereas the spatial properties of an SRS are unpredictable and may be undesirable. By chance, the selected units in an SRS could be spaced like a systematic sample, or they might instead all be clustered from 1,500–2,000 m. As noted in Chapter 4, systematic sampling can be especially efficient (have much lower sampling variance than SRS) when there is a strong linear trend in the target variable value with spatial location (as evident for Figure 12.2). In this kind of setting, individual SRS samples can produce estimates which are seriously off target, whereas that is generally not the case for a systematic sample. For example, if all selected units were clustered at 1,500–2,000 m, then $\hat{T}_{y,\text{mpu}}(s)$ would be much less than T_y; if instead all selected units were clustered at 0–500 m, then $\hat{T}_{y,\text{mpu}}(s)$ would be much greater than T_y. The units in the systematic sample depicted in Figure 12.2 have desirable **spatial balance**. That is, (a) the distances between adjacent selected population units (20 m stream sections) are equal, and (b) the set of selected units spans nearly the entire length of the stream. Such spatial balance is desirable in the stream survey setting and elsewhere because there is often considerable spatial structure underlying the distribution of target variable values. In Figure 12.2, densities of juvenile fish are highest near the stream mouth and trend to lowest densities at 1,500–2,000 m.

In many large-scale natural resources contexts, there may be a high degree of spatial structure underlying the distribution of target variable values over the landscape and therefore it can be desirable to adopt a sampling design which can dependably generate good spatial balance. For example, imagine a nation-wide sample survey of water quality parameters in all continental U.S. lakes exceeding 10 hectares in surface area. Suppose, by chance, that an SRS of 100 lakes resulted in all 100 lakes being selected from the state of Minnesota. Although such a random probability sample would clearly be representative in the Neyman (1934) sense, resource managers would rightly question the utility of such a sample for generating useful inferences concerning the population of all continental U.S. lakes. The statistician's assertion that this method of sample selection, when averaged over all possible SRS selections, leads to unbiased estimation, might not provide much comfort to a resource manager. Instead, the resource manager might quite reasonably question the validity of inference based only on Minnesota lakes and would characterize this as a *bad sample*. The manager would feel much more comfortable if (1) the sampling design ruled out the possibility of selecting a sample consisting of only lakes from Minnesota, and (2) the selected sample contained lakes from many different states and ecozones (i.e., if the sample were spatially balanced). Achieving such spatial balance, while ruling out samples like the set of 100 Minnesota lakes, requires adoption of a *restricted randomization scheme* relative to SRS. Linear systematic sampling is perhaps the most restricted randomization scheme: there are only k ($\approx N/n$) possible samples that can be selected.

Some of the sampling designs already considered in this text can, of course, essentially guarantee selection of a sample of lakes that would have good spatial balance with respect to all continental U.S. lakes. For example, suppose that lakes were stratified by (at least) elevation (high and low), region (East Coast, Mid West, West, West Coast) and latitude (northern and southern), for a total of $2 \times 4 \times 2 = 16$ strata. Then, if a stratified SRS design were adopted with, say, $n_h = 10$ lakes selected from each stratum, then the resulting samples would always include units from every one of these sixteen strata, thereby assuring a substantial degree of spatial balance. An improved degree of spatial balance could be achieved by maximizing the number of strata subject to the variance estimation constraint that all $n_h \geq 2$. The advantage of such stratified schemes is that unbiased variance estimation is possible and stratum-specific estimates can be generated for all strata, in addition to allowing estimates of overall population parameters. As noted in Chapter 5, however, the disadvantage of such highly stratified schemes is that they tend to be most efficient for a single target variable and less useful if a large-scale survey has numerous target variables of different character. For the lakes/water quality target variables, that might not be a serious objection. Other disadvantages of a pre-stratified design include: (a) inability to guarantee spatial balance *within strata* and (b) post-stratification would be ruled out or might become exceptionally complicated.

A similar spatially balanced result might be achieved using systematic sampling. Suppose that the stratification of lakes proposed previously were adopted and that, within each stratum $h = 1, 2, \ldots, 16$, the N_h lakes are listed in order of their latitude (at lake center), and that these lists are then joined end-to-end and the lakes assigned an ID number, i, reflecting their position in this overall list, $i = 1, 2, \ldots, \sum N_h$. Select a sample by drawing a random start, r, on the integers 1 through k, and then select every k^{th} lake thereafter on this list. The resulting sample will have good spatial balance, and the number of lakes selected from the strata used to form this list should be proportional to the numbers of lakes in those strata. (A proportionally allocated stratified sample would have similar expected membership.) A major disadvantage of systematic sampling, of course, is that unbiased estimation of sampling variance is not possible because many $\pi_{ij} = 0$ for a linear systematic design.

Alternatives to the previously described stratified and systematic sampling methods for ensuring selection of spatially balanced samples rely on explicit geographic referencing of the spatial locations of population units. For example, a stream could be referenced by a one-dimensional line with length proportional to total stream length, with particular stream reaches represented as line segment lengths along this line, and with a single spatial coordinate (distance along the line) identifying the midpoint of a given stream reach (line segment). A lake could be represented on a two-dimensional map, having an associated area and a central "location point" with a pair of spatial coordinates (latitude, longitude). These same ideas could be extended to a three-dimensional setting, with locations represented by three spatial coordinates.

In many natural resources and ecological contexts, interest lies in estimation of descriptive parameters of a *study area* rather than of an explicit well-identified finite population of units (McDonald 2012). For example, a resource manager might be interested in the total number of trees of a specific species on a National Forest that have recently died as a direct or indirect result of drought. The National Forest occupies a well-identified geographic area on a map and could be partitioned into population units, but it would seem more natural to be able to select a sample of point locations (latitude, longitude coordinates) in the forest, determine the mean number of dead trees per unit area in a fixed-size field plot around these points, and scale this by the total area of the forest to estimate the total number of dead trees. In this case, the number of points (latitude, longitude coordinates) contained in the forest (study area) is mathematically infinite, and the forest would constitute an *infinite population*. Given the kind of spatial variation in tree mortality across a study area that is implied by Figure 12.1, it would seem desirable for these sample sites to be widely and evenly distributed across the study area, i.e., for the sample points to be spatially balanced. For example, one might randomly superimpose a grid over the National Forest, and then use the intersections of grids lines as sample points. This would, of course, generate a spatially balanced and systematically located set of sample points.

In this chapter, we focus our attention on two recently developed methods for selecting spatially balanced samples: Generalized Random Tessellation Stratified sampling (GRTS, Stevens and Olsen 2004), and Balanced Acceptance Sampling (BAS, Robertson et al. 2013, 2017). These methods can be applied to finite populations, with well-identified population units, but they can also be applied within the context of infinite populations of points within a study area. To simplify our presentation, we restrict our attention to two-dimensional geographic settings, where the location of a population unit or point is defined in terms of (latitude, longitude) coordinates. BAS can also be used in one-dimensional or higher settings. We also restrict our attention to equal probability selection of population units or points. Both GRTS and BAS can accommodate unequal probability selection.

Within this two-dimensional geographic setting, the study population itself may be zero-dimensional (0-D), one-dimensional (1-D), or two-dimensional (2-D), and it is important to keep this distinction in mind when reading this chapter. (We will endeavor to reserve use of the shorthand 0-D, 1-D, and 2-D for referencing the study population itself.) A finite population of units is considered to be 0-D because the number of associated point locations is finite (has no length or area), while an infinite population may be 1-D or 2-D. A 1-D *linear* population consists of a set of lines, possibly connected, within a region. For example, a stream network within a watershed. A 2-D *areal* population consists of a set of areas, possibly connected, within a region. For example, a collection of disjoint private forest holdings, or a single contiguous wilderness area within a county.

We begin with detailed descriptions of the sample selection procedure for GRTS and BAS for finite populations. We follow with estimation of population parameters and sampling variance under equal probability selection. We conclude with an overview of the application of these methods in the infinite population setting.

12.2 Finite populations

In this section, we confine our attention to the selection of spatially balanced samples from finite populations. In this setting, the *study area* consists of the finite population units and their associated spatial reference coordinates. That is, a finite set of *point objects* that are geographically referenced by (latitude, longitude) coordinates. Examples include a set of well-defined stream reaches referenced by their midpoint coordinates, a set of lakes referenced by their centroid coordinates, or a set of centroid coordinates for known areas where a rare species of plant is present. In the finite population setting, the variable(s) of interest are measured with respect to the entire unit, for example, over an entire stream reach or lake rather than at a specific point within the stream reach or lake.

12.2.1 *Generalized random tessellation stratified sampling*

GRTS is a complicated multi-step scheme for selecting spatially balanced samples developed by Stevens and Olsen (2004). A **cover** or **bounding box**, typically a square, is defined such that it completely encloses the study area. That is, all points identifying population units are found within the boundaries of the cover. The first three steps in the GRTS selection procedure solely involve the cover and can be viewed as set-up steps prior to the selection of a sample. In Step 1, the cover (and all points identifying locations of population units within the cover) is rescaled and shifted to a random position within a unit square and associated with a two-dimensional Cartesian coordinate system. In Step 2, the cover is recursively partitioned, generating a grid of cells along with numerical addresses, attached to each of these cells, with addresses indicative of cell location in the two-dimensional geography. In Step 3, these cell addresses are hierarchically randomized, creating a randomly ordered list of the addresses where some of the sequential addresses are spatially proximate in the cover.

The set of randomized cell addresses produced in Steps 1–3 can be viewed as a randomized geographic stratification, or partitioning, of the study area. In Step 4, the cells are then laid out along a line in this randomized order. This can be viewed as a mapping from the two-dimensional geography to a one-dimensional space, a line, where cells that are adjacent on the line tend to be adjacent in the two-dimensional geography. The length of the line is scaled to have length n, where n is the desired sample size. In the equal probability finite population setting, cell length on the line is proportional to the number of point objects (population units) within the cell. A systematic sample of n points is then selected along the line, beginning with a random start on the interval $(0,1)$, similar to the scheme used in PPS systematic sampling (Section 8.5). This procedure leads to the selection of n cells, and then one population unit is selected at random from within each selected cell. In Step 5 (optional), required when nonresponse is anticipated and *oversampling* must be done, the sample units selected in Step 4 are reordered in such a way that the first n response units in this reordered listing should also be spatially balanced. For details on the implementation of the GRTS selection procedure beyond those given in this text, an excellent reference is Olsen et al. (2012).

Procedure

Step 1: Scaling and randomly translating the cover.

The cover can be viewed as simply a square big enough to overlay the locations of all population units (treated as point objects), no matter how irregular the geography of the study area within which the population units are found. That square, along with the N point objects within it, are then re-scaled such that the resulting square, with axes z_1 and z_2, has vertex coordinates $(z_1, z_2) = (0,0), (0,0.5), (0.5,0.5), (0.5,0)$, i.e., the sides of the square are length 0.5. Next, two continuous uniform(0,0.5) random variates, u_1 and u_2, are generated, where u_1 is added to all of the z_1 coordinates, and u_2 is added to all of the z_2 coordinates. Thus, the original square and N point objects are shifted by a random amount to a new location within the *unit square* having vertex coordinates (z_1, z_2) $= (0,0), (0,1), (1,1), (1,0)$, but their orientation is not changed, i.e., no rotation is done. This random translation ensures that there is positive probability that any two distinct point objects in the study area can both be included in the sample (Stevens and Olsen 2004). Thus, all possible second order inclusion probabilities are, theoretically, positive-valued. In particular, point objects that are "right next to each other" can both be randomly selected because the random translation in the unit square ensures that any two point objects have a positive probability of falling into different cells that are formed by the recursive partitioning process described in Step 2.

Step 2: Recursive partitioning and cell addressing.

Recursive partitioning is a method for dividing up the cover into grid cells, and assigning to each cell a spatially referenced address. The subsequent set of addresses can be laid out as a linear sequence of cells, and can be viewed as a mapping of the cells from two-dimensional space to one-dimensional space, i.e., a line.

The number of recursive partitions required to form the cells is determined by the sample size desired and the distribution of point objects over the cells. Generally, the larger the sample size and the greater the between cell variation in the numbers of point objects, the greater the number of required partitioning steps. Recursive partitioning must be continued until the cell sample inclusion probability $\pi_i^* < 1$ for all cells (see Step 4).

For a level-one partitioning, the square cover is partitioned into four equal size sub-squares (Figure 12.3, $r = 1$). For a level-two partitioning, each of the level-one sub-squares is partitioned into four equal size sub-sub-squares (Figure 12.3, $r = 2$). This recursive partitioning (quartering) of the cells from the previous level continues until the $\pi_i^* < 1$ criteria is met, with the cells getting progressively smaller at each level of partitioning. The total number of cells at recursion level r is thus $M = 4^r$, e.g., for $r = 1, 2, 3, 4$, $M = 4, 16, 64, 256$, respectively (Figure 12.3).

Next, a cell addressing system is applied to the resulting cells. We demonstrate this for recursion levels $r = 1, 2, 3, 4$. For $r = 1$, the cell addresses, starting with the top right corner and moving in a clockwise manner[1] are .0, .1, .2, and .3 (Figure 12.3, $r = 1$). For $r = 2$, the addresses for the sub-sub-squares are assigned by appending digits 0, 1, 2, and 3 to the right of the parent sub-square ($r = 1$) address. For example, the addresses within the .0 sub-square are .00, .01, .02, and .03, corresponding to the top right sub-sub-square and again moving in a clockwise manner (Figure 12.3, $r = 2$). Similarly, for $r = 3$, the addresses

[1] Stevens and Olsen (2004) assign the cell addresses in a different order: bottom left, top left, bottom right, top right, but this is inconsequential, particularly given the hierarchical randomization step.

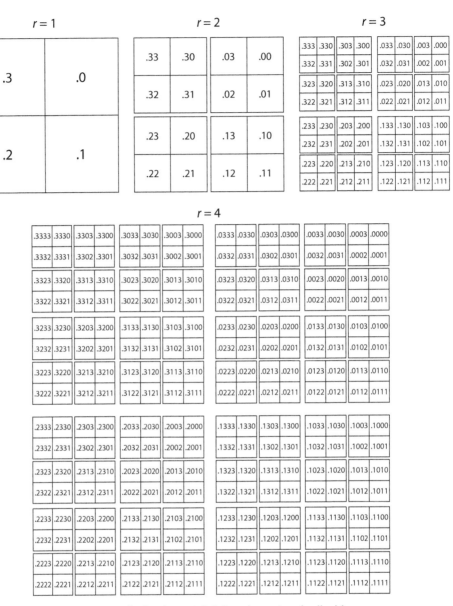

Fig. 12.3 Recursive partitioning for levels $r = 1, 2, 3, 4$, and associated cell addresses.

for the sub-sub-sub-squares are assigned by appending digits 0, 1, 2, and 3 to the right of the parent sub-sub-square ($r = 2$) address. For example in the sub-sub-square labeled .02, the sub-sub-sub-square addresses are .020, .021, .022, and .023 (Figure 12.3, $r = 3$). The addresses for $r = 4$ follow this same scheme (Figure 12.3, $r = 4$), as it does for higher values of r.

The cells can now be ordered, from low to high, by their addresses along a line, with the cells in sub-square .0 appearing first, followed by those in sub-squares .1, .2, and .3. Within each of these sub-squares, the cell ordering again follows this same clockwise pattern, as it does within each recursive partition. For example, in sub-square .0, the cells in sub-sub-square .00 appear first, followed by those in sub-sub-squares .01, .02, and .03. As a result, cells near each other in the two-dimensional geography tend to be near each other in this one-dimensional ordering. For example, referring to Figure 12.3 $r = 2$, the ordered listing of cells would be .00, .01, .02, .03, .10, .11, .12, .13, .20, .21, .22, .23, .30, .31, .32, .33. This one-dimensional spatial ordering keyed by the addressing system is what allows for the drawing of a spatially balanced sample via linear systematic selection.

Step 3: Hierarchical randomization.

Hierarchical randomization is a means of injecting some randomness into the sort order of the one-dimensional list of cell addresses, without compromising the property that cells near each other in the two-dimensional geography will tend to be near each other in the one-dimensional listing. This is done by randomly re-ordering the addresses sequentially from the first digit to the last digit. The effect of this is to randomize the one-dimensional ordering of the sub-squares, and the ordering of the sub-sub-squares within the sub-squares, and the ordering of the sub-sub-sub-squares within the sub-sub-squares, etc. Recalling the $r = 4$ example with 256 cells, the first randomization is of the addresses of the original $r = 1$ four sub-squares, .0, .1, .2, and .3. Suppose the result is the ordering shown in Table 12.1 ($r = 1$). Then, *within* each of these four sub-squares, the digits in the second position of addresses for sub-sub-squares are randomly ordered, an example of which is shown in Table 12.1 ($r = 2$). Within each of the sub-sub-squares, the third digit of the addresses corresponding to the sub-sub-sub-squares are randomly ordered, an example of which is shown in Table 12.1 ($r = 3$). Finally, within each of the sub-sub-sub-squares, the fourth digit of the addresses corresponding to the sub-sub-sub-sub-squares are randomly ordered, an example of which is shown in Table 12.1 ($r = 4$). With this randomization, note that the first (and second, third, and fourth) (1/4) of cells all arise from the same sub-square, and that the first (and second, third, ..., sixteenth) (1/16) of cells all arise from the same sub-sub-square, etc., thus preserving the property that cells near each other in the two-dimensional geography will tend to be near each other with this one-dimensional ordering.

Step 4: Sample selection.

Cell extents and inclusion probabilities. Each of the M cells resulting from the recursive partitioning is assigned an inclusion probability. Cell-specific inclusion probabilities are constructed such that they sum to n, the desired sample size. For a finite population of point objects and equal probability selection from these point objects, cell inclusion probabilities (probabilities that a given cell will be included in the sample) depend on the desired sample size and the number of point objects within a given cell. Cells that do not contain any point objects (population units) are given cell inclusion probabilities of zero.

With M cells and equal probability selection of point objects, the inclusion probability for cell i is $\pi_i^* = n(e_i/E)$, $i = 1, 2, \ldots, M$, where n is the desired sample size or oversample size $n' = \gamma n$ (Step 5), e_i is the *cell extent* (here equal to m_i, the number of point objects in cell i), and E is the *total extent*, here equal to the total number of point objects in the population (Olsen et al. 2012). Thus, for a finite population, $E = \sum_{i=1}^{M} e_i = N$ and $\pi_i^* = n(m_i/N)$. (Note that π_i^* is the probability of selecting *cell i*, but it is *not* equivalent to the probability of a selecting a particular *point object* in the study area. This is discussed further in the Step 4 subsection "Random selection of point object within each selected cell".) Recall that if $\pi_i^* \geq 1$ for one or more of the cells, then another level of recursive partitioning must be performed (i.e., return to Step 2).

Table 12.1 Hierarchical randomization example for partition levels $r = 1, 2, 3, 4$.

$r=1$	$r=2$	$r=3$	$r=4$			
.1	.11	.112	.1120	.1121	.1122	.1123
		.113	.1131	.1132	.1133	.1130
		.111	.1111	.1113	.1112	.1110
		.110	.1102	.1100	.1103	.1101
	.13	.131	.1311	.1313	.1310	.1312
		.130	.1302	.1303	.1301	.1300
		.133	.1330	.1331	.1332	.1333
		.132	.1320	.1323	.1321	.1322
	.12	.122	.1223	.1222	.1220	.1221
		.120	.1201	.1203	.1200	.1202
		.123	.1231	.1230	.1233	.1232
		.121	.1212	.1210	.1213	.1211
	.10	.101	.1012	.1011	.1013	.1010
		.102	.1021	.1020	.1022	.1023
		.103	.1032	.1030	.1033	.1031
		.100	.1000	.1002	.1001	.1003
.0	.02	.020	.0200	.0203	.0201	.0202
		.021	.0210	.0213	.0211	.0212
		.022	.0223	.0222	.0220	.0221
		.023	.0233	.0231	.0230	.0232
	.00	.001	.0010	.0013	.0012	.0011
		.002	.0023	.0022	.0020	.0021
		.003	.0032	.0033	.0031	.0030
		.000	.0003	.0001	.0000	.0002
	.01	.012	.0121	.0122	.0123	.0120
		.011	.0110	.0111	.0113	.0112
		.013	.0131	.0132	.0133	.0130
		.010	.0100	.0102	.0101	.0103
	.03	.030	.0300	.0301	.0303	.0302
		.033	.0331	.0332	.0333	.0330
		.032	.0321	.0322	.0323	.0320
		.031	.0310	.0311	.0313	.0312
.3	.31	.310	.3100	.3103	.3102	.3101
		.313	.3133	.3132	.3130	.3131
		.311	.3112	.3111	.3113	.3110
		.312	.3121	.3123	.3120	.3122
	.33	.330	.3301	.3300	.3303	.3302
		.333	.3333	.3331	.3330	.3332
		.331	.3311	.3312	.3313	.3310
		.332	.3320	.3322	.3321	.3323
	.32	.321	.3211	.3212	.3213	.3210
		.323	.3230	.3233	.3232	.3231
		.322	.3221	.3223	.3222	.3220
		.320	.3201	.3200	.3203	.3202
	.30	.301	.3011	.3012	.3010	.3013
		.303	.3033	.3032	.3030	.3031
		.300	.3003	.3002	.3001	.3000
		.302	.3021	.3020	.3023	.3022

Table 12.1 Continued

r = 1	r = 2	r = 3	r = 4			
.2	.20	.200	.2001	.2000	.2003	.2002
		.202	.2021	.2020	.2022	.2023
		.203	.2032	.2033	.2031	.2030
		.201	.2010	.2013	.2012	.2011
	.21	.212	.2121	.2123	.2120	.2122
		.213	.2132	.2133	.2131	.2130
		.211	.2113	.2111	.2112	.2110
		.210	.2101	.2102	.2100	.2103
	.23	.233	.2332	.2333	.2330	.2331
		.232	.2321	.2323	.2320	.2322
		.231	.2312	.2313	.2311	.2310
		.230	.2301	.2303	.2302	.2300
	.22	.221	.2211	.2212	.2213	.2210
		.223	.2232	.2233	.2230	.2231
		.222	.2221	.2223	.2220	.2222
		.220	.2202	.2200	.2201	.2203

Systematic sample of cells. Once the cell-specific inclusion probability assignments, the π_i^*, have been made for all M cells, these probabilities are "laid out" on the line $(0, n]$ in the order of the hierarchical randomization of cell addresses produced in Step 3. Another way to put this is that the line $(0, n]$ is partitioned into M segments: $(0, \pi_1^*]$, $(\pi_1^*, \pi_1^* + \pi_2^*]$, ..., $(\sum_{i=1}^{M-1} \pi_i^*, n]$, where each segment corresponds to the order of the randomized cell addresses of the previous step, excluding cells with zero inclusion probability. For example, with the Table 12.1 $(r = 4)$ hierarchical randomization, and assuming no cells with zero inclusion probability, the ordered listing of cells would be .1120, .1121, .1122, .1123, .1131, .1132, .1133, .1130, ..., .2221, .2223, .2220, .2222, .2202, .2200, .2201, .2203.

The sample is then selected using a method similar to that described in Section 8.5. Generate a continuous uniform$(0,1)$ random variate, u, and create a list of selected points at locations u, $u+1$, $u+2$, ..., $u+(n-1)$. For example, if $n = 20$ and $u = 0.7$, then the selected points are located at 0.7, 1.7, 2.7, ..., 19.7. In the case of a finite set of point objects, the selected cells are those whose line segments include the selected points.

Random selection of point object within each selected cell. A single point object within each selected cell i is then randomly selected with probability $1/m_i$. For a finite population (finite set of N point objects), the sample selection procedure is thus analogous to a *two-stage* selection procedure which results in a spatially balanced sample for which the first order inclusion probabilities are equal (n/N) for all population units. The first stage of selection involves selection of n cells (primary units) with probabilities $\pi_i^* = n(m_i/N)$, $i = 1, 2, ..., M$. At the second stage of selection, a single point object within selected cell i is selected with probability $1/m_i$. Therefore, the inclusion probability for point object j in cell i is

$$\pi_{j \in \text{cell } i} = \Pr\{\text{cell } i \text{ selected}\} \cdot \Pr\{\text{point object } j \text{ selected} | \text{cell } i \text{ selected}\}$$

$$= \pi_i^*(1/m_i) = n(e_i/E)(1/m_i) = n(m_i/N)(1/m_i) = n/N \tag{12.1}$$

This selection procedure can thus be viewed as a *self-weighting* design, with each point object having the same probability of inclusion in the overall sample, regardless of its cell membership: the first-stage selection probability for cell i is proportional to the number of point objects within cell i, but the second-stage selection probability is inversely proportional to the number of point objects in the cell, resulting in a constant and thus equal overall inclusion probability.

Step 5: Reverse hierarchical ordering and oversampling.

The systematic selection procedure (Step 4) is specifically designed to take advantage of the (restrictive) hierarchical ordering of the cells performed in Step 3, sequentially passing through the sub-squares, the sub-sub-squares, etc., to yield a sample that has good spatial balance. In some applications though, not all of the n selected points or point objects can necessarily be surveyed; some of them may be *nonresponse* units. If there are a substantial number of nonresponse units, then the realized sample size will be substantially less than the desired n. Assuming that the reason for the nonresponse is "pure chance", this possibility can be countered by initially selecting an *oversample* of size $n' = \lambda n$, e.g., $\lambda = 2$, but then limiting the survey to sampling the first n response units (Sections 3.5, 4.6, 8.9). However, given the one-dimensional spatial stratification of cells described previously, doing so may lead to a sample that is heavily concentrated in the first two sub-squares if it turns out that there are few nonresponse units.

To address this possibility, Stevens and Olsen (2004) added a final (optional) step to the GRTS procedure, **reverse hierarchical ordering**, which *reorders* the selected units so that any subsequence of them will also have good spatial balance. Thus, the first n' units, the first n units, or any number of units in between should lead to good spatial balance in the presence of nonresponse. The method consists of secondarily labeling the selected units according to their fractional order of selection $(0/n', 1/n', 2/n', \ldots, (n'-1)/n')$, converting these values from base-10 to base-4, reversing the base-4 digits, and taking the sort order as the new sample order for the selected units. For example, if $n' = 8$, the original fractional order of selection (0/8, 1/8, 2/8, ..., 7/8) in base-4 would be (.00, .02, .10, .12, .20, .22, .30, .32), reversing the digits gives (.00, .20, .01, .21, .02, .22, .03, .23), and taking the sort order (1, 5, 2, 6, 3, 7, 4, 8) as the new sample order for the n' selected units; in this case, every 2nd unit of the original ordering in a circular fashion. Alternative base representations and digit reversal schemes play a central role in BAS as well (Section 12.2.2).

Reverse hierarchical ordering of the selected units is not required if there are few or no nonresponse units in the sample and it is not essential that the realized sample size achieves a pre-specified fixed value. As noted in Section 4.6, if relatively few units in the systematic sample selected in Step 4 are nonresponse units, then the realized sample, though less than the desired size, should still have good spatial balance and, assuming pure chance nonresponse and equal probability selection, then the mean-per-unit estimator would remain unbiased if calculated over the response units (Section 3.5). If an oversample of $n' = \lambda n$ is initially selected to ensure that n response units are surveyed, and it is reasonable to assume that all nonresponse is pure chance, then the first order inclusion probabilities will remain equal to n/N for equal probability selection.

Application: coastal salmonid monitoring

We illustrate equal probability GRTS selection of a sample from a finite population. The illustrative context involves estimation of average spawning escapement (adult spawners/km) of anadromous salmonid fish (steelhead, salmon) along the northern California coast where GRTS selection is applied within the context of the long-term monitoring program for coastal anadromous salmonids (Adams et al. 2011). We show how GRTS can be used to select a spatially balanced sample of $n = 16$ stream reaches from the population of $N = 338$ such reaches along the Mendocino County coast. Most reaches range from about 1.5–3.0 km in length, are designed to have good access points, and allow field personnel to complete surveys in one day of field work. Typical target variables measured include the number of salmon redds (nests constructed by salmon within which eggs are deposited and covered for incubation/hatching) or the number of salmon carcasses (salmon die following spawning) that are observed. Each stream reach is spatially referenced by the (latitude, longitude) coordinates of its midpoint. Thus, the set of stream reaches has a two-dimensional geographic representation as a set of N point objects, each with a unique location.

We simplify our illustration of GRTS by omitting Step 1 (scaling and randomly translating the cover within a unit square). This random translation guarantees that all $\pi_{ij} > 0$, but the essential features of the GRTS selection process remain the same, whether this random translation is made or not. We also do not illustrate oversampling (i.e., $n' = n = 16$), but do illustrate reverse hierarchical ordering (Step 5) of the systematic sample selected in Step 4.

To begin, the study area and stream reach coordinates (in degrees) were projected (from three-dimensional sphere to two-dimensional plane) using the "NAD83 California (Teale) Albers" conical, equal area projection. The projected coordinate values are in meters from the origin $(0,0)$ (near the center of the state). A square cover (bounding box) large enough to enclose the study area was then specified [Figure 12.4(a)].

Proceeding to Step 2, $r = 4$ levels of recursive partitioning were required to ensure that $\pi_i^* < 1$ for all cells given the desired sample size of $n = 16$. Figure 12.4(b) shows the resulting 16×16 grid of 256 cells, along with the midpoints of all $N = 338$ stream reaches. Numbers of stream reach midpoints per cell range from 0–17 and the associated π_i^* range from 0 to $n(m_i/N) = 16(17/338) = 0.80473$. Base-4 addresses were then assigned to each cell as shown in Figure 12.3 $(r = 4)$. Following that, the cells were hierarchically randomized using the example $r = 4$ randomization result listed in Table 12.1.

The hierarchically randomized cell addresses were then laid out on a line of total length $n = 16$ with the cell i line segment length equal to π_i^*. There were 57 cells for which $\pi_i^* > 0$ $(m_i \geq 1)$. Table 12.2 illustrates tabulation of the cumulative sums of the π_i^* for these 57 cell addresses, along with a systematic sample (with random start $u = 0.736331$) which led to the $n = 16$ selected cells indicated in Figure 12.4(b) (dark shade). From each selected cell i, one reach was then selected at random (with probability $1/m_i$). The $n = 16$ selected reach midpoints are shown in Figure 12.4(c) with the order of cell selection noted, and their corresponding reach ID is listed in Table 12.3. (The reverse hierarchical ordering procedure for the selected reaches is demonstrated in Table 12.3, and the result is shown in Figure 12.4(d), which would be needed if oversampling were adopted.)

12.2.2 Balanced acceptance sampling

BAS (Robertson et al. 2013, 2017) is another method for selecting spatially balanced samples. It can be applied in one- or higher-dimensional settings, e.g., a lake where depth is an additional dimension, or the atmosphere where elevation is an additional dimension. In contrast to GRTS, BAS relies explicitly on generating spatially balanced points for a sample. In a finite population setting, the locations of these points are then linked to the selection of specific, spatially-referenced population units.

Computer *random number* generators are not in fact random, they are deterministic. The standard generators produce a sequence of numbers that *appears* to be random (independent draws from the specified distribution) and for this reason are termed *pseudo-random* number generators. Other generators, termed *quasi-random* number generators, produce a sequence of numbers that are more evenly spaced than expected of a pseudo-random

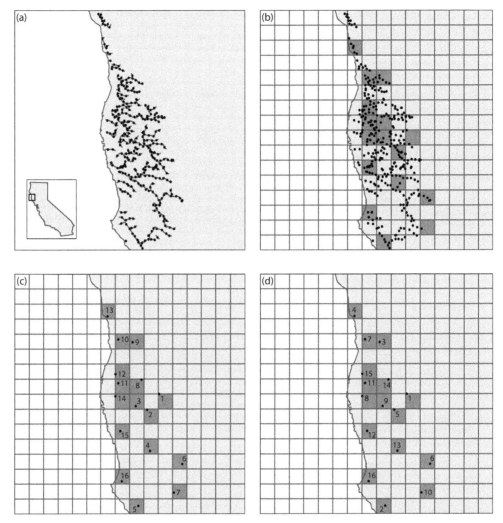

Fig. 12.4 GRTS selection of stream reaches along the Mendocino County coast of California. Panel (a) depicts the population of $N = 338$ stream reaches, with reach midpoints indicated by dots (reach endpoints not shown). Reaches range from 0.1–4.3 km and average 2.5 km. Panel (b) depicts reach midpoints and overlaid partition ($r = 4$), a $16 \times 16 = 256$ cell grid. Number of reach midpoints per cell range from 0–17, and are used to determine the "extent" of each cell for the systematic selection of cells. Selected cells (dark shade) match the selected cells indicated in Table 12.2. Panel (c) shows the randomly selected reach midpoint from each of the selected cells and notes the order in which cells were selected. Panel (d) shows the same selected reach midpoints but notes what their sample order would be under reverse hierarchical ordering (Table 12.3). See text for further details.

sequence. In particular, successive numbers in the sequence are never *very close* to one another, and the sequence achieves good *spatial balance* with respect to the spacing of the points and coverage of the distribution domain.

The BAS procedure for generating spatially balanced points uses a quasi-random number generator, the **Halton sequence** (Halton 1960), which is an extension of the one-dimensional van der Corput sequence (van der Corput 1935) to two or more dimensions. Before describing the details of the van der Corput and Halton sequences (Sections 12.2.2

Table 12.2 GRTS selection of $n = 16$ cells from the 57 Mendocino County coast cells for which $m_i > 0$. Selected cells are those for which the systematically selected points fall within the associated line segments of length $\pi_i^* = n(m_i/N)$. The cell i line segment is $(start_i, end_i]$, with $start_i = \sum_{j=1}^{i-1} \pi_j^*$ and $end_i = \sum_{j=1}^{i} \pi_j^*$. Random start for the systematic selection, generated as a continuous uniform(0,1) variate, was $u = 0.736331$.

	Cell			Line segment		Systematic
ID (i)	address	m_i	π_i^*	$start_i$	end_i	selection
1	.1313	6	0.2840237	0.000000	0.284024	
2	.1310	1	0.0473373	0.284024	0.331361	
3	.1312	4	0.1893491	0.331361	0.520710	
4	.1302	4	0.1893491	0.520710	0.710059	
5	.1303	2	0.0946746	0.710059	0.804734	0.736331
6	.1330	9	0.4260355	0.804734	1.230769	
7	.1331	11	0.5207101	1.230769	1.751479	1.736331
8	.1332	9	0.4260355	1.751479	2.177515	
9	.1333	12	0.5680473	2.177515	2.745562	2.736331
10	.1320	12	0.5680473	2.745562	3.313609	
11	.1323	7	0.3313609	3.313609	3.644970	
12	.1321	7	0.3313609	3.644970	3.976331	3.736331
13	.1322	9	0.4260355	3.976331	4.402367	
14	.1223	4	0.1893491	4.402367	4.591716	
15	.1222	6	0.2840237	4.591716	4.875740	4.736331
16	.1220	3	0.1420118	4.875740	5.017751	
17	.1221	7	0.3313609	5.017751	5.349112	
18	.1203	8	0.3786982	5.349112	5.727811	
19	.1200	4	0.1893491	5.727811	5.917160	5.736331
20	.1202	5	0.2366864	5.917160	6.153846	
21	.1231	2	0.0946746	6.153846	6.248521	
22	.1230	5	0.2366864	6.248521	6.485207	
23	.1233	2	0.0946746	6.485207	6.579882	
24	.1232	2	0.0946746	6.579882	6.674556	
25	.1212	1	0.0473373	6.674556	6.721893	
26	.1210	1	0.0473373	6.721893	6.769231	6.736331
27	.1213	6	0.2840237	6.769231	7.053254	
28	.0212	6	0.2840237	7.053254	7.337278	
29	.0223	4	0.1893491	7.337278	7.526627	
30	.0222	6	0.2840237	7.526627	7.810651	7.736331
31	.0220	9	0.4260355	7.810651	8.236686	
32	.0221	10	0.4733728	8.236686	8.710059	
33	.0233	6	0.2840237	8.710059	8.994083	8.736331
34	.0232	7	0.3313609	8.994083	9.325444	
35	.3100	10	0.4733728	9.325444	9.798817	9.736331
36	.3101	14	0.6627219	9.798817	10.461538	
37	.3112	3	0.1420118	10.461538	10.603550	
38	.3111	17	0.8047337	10.603550	11.408284	10.736331
39	.3113	2	0.0946746	11.408284	11.502959	
40	.3110	15	0.7100592	11.502959	12.213018	11.736331

Table 12.2 Continued

	Cell			Line segment		Systematic
ID (i)	address	m_i	π_i^*	start$_i$	end$_i$	selection
41	.3011	3	0.1420118	12.213018	12.355030	
42	.3012	4	0.1893491	12.355030	12.544379	
43	.3013	5	0.2366864	12.544379	12.781065	12.736331
44	.3030	3	0.1420118	12.781065	12.923077	
45	.3003	6	0.2840237	12.923077	13.207101	
46	.3002	2	0.0946746	13.207101	13.301775	
47	.2001	9	0.4260355	13.301775	13.727811	
48	.2000	12	0.5680473	13.727811	14.295858	13.736331
49	.2003	4	0.1893491	14.295858	14.485207	
50	.2002	2	0.0946746	14.485207	14.579882	
51	.2010	13	0.6153846	14.579882	15.195266	14.736331
52	.2013	2	0.0946746	15.195266	15.289941	
53	.2011	5	0.2366864	15.289941	15.526627	
54	.2111	1	0.0473373	15.526627	15.573964	
55	.2110	3	0.1420118	15.573964	15.715976	
56	.2101	3	0.1420118	15.715976	15.857988	15.736331
57	.2100	3	0.1420118	15.857988	16.000000	

Table 12.3 Reverse hierarchical ordering of the $n = 16$ selected stream reaches for the Mendocino County coast GRTS illustration. rev(base-b) indicates reversal of base b digits, with $b = 4$. See text for further explanation.

Selection	Cell		Reverse hierarchical ordering			Sample
order	address	Reach ID	base-10	base-4	rev(base-4)	order
1	.1303	35	0/16	.00	.00	1
2	.1331	197	1/16	.01	.10	5
3	.1333	178	2/16	.02	.20	9
4	.1321	241	3/16	.03	.30	13
5	.1222	319	4/16	.10	.01	2
6	.1200	245	5/16	.11	.11	6
7	.1210	296	6/16	.12	.21	10
8	.0222	144	7/16	.13	.31	14
9	.0233	94	8/16	.20	.02	3
10	.3100	89	9/16	.21	.12	7
11	.3111	122	10/16	.22	.22	11
12	.3110	116	11/16	.23	.32	15
13	.3013	70	12/16	.30	.03	4
14	.2000	170	13/16	.31	.13	8
15	.2010	239	14/16	.32	.23	12
16	.2101	310	15/16	.33	.33	16

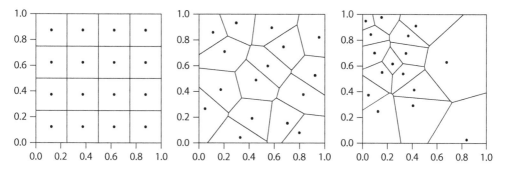

Fig. 12.5 Voronoi polygons (tessellation) constructed around points selected using *perfect* systematic grid (left panel), Halton method (center panel), and Uniform method (right panel). Voronoi polygons and associated areas were determined using functions from the R package `deldir` (Turner 2018). Halton sequence was generated using function `halton()` from the R package `SDraw` (McDonald 2016).

and 12.2.2), and the procedures for selecting BAS samples (Sections 12.2.2 and 12.2.2), in the next section we explore further the notion of spatial balance, and illustrate the ability of a quasi-random Halton sequence to achieve it relative to a pseudo-random uniform sequence.

Spatial balance and Voronoi polygons

In this section we present a simple contrast of the degree of spatial balance that results when points are selected on a two-dimensional unit square, with axes z_1 and z_2, using three very different methods: (a) a deterministically located systematic set of 16 (z_1, z_2) points; and 16 points generated using (b) a random start two-dimensional Halton sequence (see Section 12.2.2 for details); or (c) two independent, continuous uniform(0,1) random sequences for the z_1 and z_2 coordinates, respectively. In this section, we'll refer to (b) and (c) as the "Halton method" and the "Uniform method", respectively.

Figure 12.5 shows the location of a single sample of $n = 16$ points generated by each of these methods. The systematic grid of points is unique in the sense that there is no other set of 16 points that will have this same *perfect* spatial balance. It can thus serve as an "ideal standard for comparison". For the other two methods, of course, the patterns of points reflect just one of the many possible selection outcomes of these methods. Superimposed on the unit squares with plotted points are lines which define the boundaries of *Voronoi polygons* surrounding each point, and the set of these polygons over the unit square is called a *tessellation*. A Voronoi polygon delineates all (z_1, z_2) values that are closer to its identifying point than to any other sample point. It is analogous to a cell tower's *sphere of influence*: if one were to move through the unit square with a cell phone, the cell phone would search for the nearest (all else being equal) cell tower and, after exiting a cell tower's sphere of influence, would switch to the closer cell tower.

The variance in the size (area) of the Voronoi polygons is one possible metric that can be used to compare the degree of spatial balance that is achieved by the three methods for the selection of $n = 16$ points on the unit square. For the perfect systematic grid of 16 points, all of the Voronoi polygons have identical shape (square) and area (1/16 = 0.0625), so there is no variance in the polygon areas. For the $n = 16$ points selected using the Halton and Uniform methods, the average areas of the polygons are again 0.0625, but the variance of the polygon areas for the Halton sequence (0.00009366) is much less than for the random sequence (0.003995).

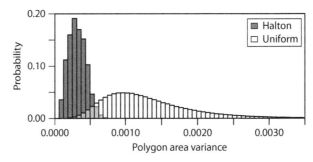

Fig. 12.6 Histogram representations of the sampling distributions of the variance in areas of Voronoi polygons constructed around 16 sample points selected using the Halton and Uniform methods. The number of simulated selections for each method was 10^6.

Of course, this is just one possible selection outcome for the Halton and Uniform methods: the spatial balance of points selected using the Uniform method might sometimes be similar to, or even better than, those produced by the Halton method. To legitimately compare the degree of spatial balance that one might expect to achieve using these two methods, one should simulate selection of a large number of sets of 16 points for each method, and then compare the sampling distributions of variance in polygon areas for each of these two methods. Figure 12.6 demonstrates the dramatically superior performance of the Halton method compared to the Uniform method. The average variance in polygon areas for the Halton method is 0.0003188 as compared to 0.0013249 for the Uniform method. By analogy to our definition of *relative precision* of one sampling strategy compared to another, using the variance of areas metric, we might calculate the *relative balance* of the Halton method compared to the Uniform method as $(1/.0003188)/(1/.0013249) \approx 4.2$. These simple simulation results suggest that selection of points using the Halton method can reliably generate points with good spatial balance, whereas the Uniform method cannot reliably achieve this result. Note, however, that this notion of spatial balance concerns itself only with the locations of the selected points and not with the location of the selected points relative to the spatial distribution of points objects in a finite population. There are several alternative measures of spatial balance in this more relevant finite population framework (Olsen et al. 2012 pp. 141–145).

van der Corput sequence

Suppose we wish to select n points over a one-dimensional continuous [0,1) interval such that: (a) the points are "evenly spread" over the interval, and (b) the points are ordered such that any subsequence of the points would also be relatively evenly spread over the interval. The van der Corput (1935) sequence provides one means to achieve both of these objectives. Its construction involves transforming the sequence of non-negative base-10 integers into some other base-b representation.

For a specified integer base $b \geq 2$, element k, $k = 0, 1, 2, \ldots$, of the **van der Corput sequence** can be obtained as follows. (1) Express k in base-b. Any positive integer k can be expressed as $\sum_{i=0}^{v} a_i (b)^i$ for some integer v with a set of appropriately valued integers $\{a_i\}$, where $a_i \in \{0, 1, \ldots, b - 1\}$, and can thus be written in base-b as $a_v a_{v-1} \cdots a_0$. For example, suppose $k = 42$ and $b = 5$. Expressing 42 as a linear combination of the powers of 5, $42 = 2(5)^0 + 3(5)^1 + 1(5)^2$, and thus 42 can be written in base-5 as 132. (2) Reverse the order of the base-b digits and place a decimal point in front to obtain the new base-b value $.a_0 a_1 \cdots a_v$. For the $k = 42, b = 5$ example, this gives .231. (3) Convert this

Table 12.4 Illustration of van der Corput sequence construction. Black dots mark current value of sequence; grey dots mark previous values of sequence. See text for explanation.

Original sequence			van der Corput sequence	
base-10	base-2	rev(base-2)	base-10	Selections on [0, 1) interval
0	0	.0	0.000	
1	1	.1	0.500	
2	10	.01	0.250	
3	11	.11	0.750	
4	100	.001	0.125	
5	101	.101	0.625	
6	110	.011	0.375	
7	111	.111	0.875	
				0 1/8 2/8 3/8 4/8 5/8 6/8 7/8 1

base-b value to base-10 to yield element k of the sequence, $\sum_{i=0}^{\nu} a_i \left(\frac{1}{b}\right)^{i+1}$. For the $k = 42$, $b = 5$ example, the base-5 representation .231 is equivalent to the base-10 representation $2\left(\frac{1}{5}\right)^1 + 3\left(\frac{1}{5}\right)^2 + 1\left(\frac{1}{5}\right)^3 = 0.528$.

Table 12.4 illustrates the van der Corput sequence construction given a base of $b = 2$ for the first eight integers, $k = 0, 1, \ldots, 7$. All of these integers can be expressed as $a_0(2)^0 + a_1(2)^1 + a_2(2)^2$ for some $a_i \in \{0, 1, 2\}$ and can thus be expressed in base-2 as $a_2 a_1 a_0$ (Table 12.4, column 2). The order of digits is then reversed and a decimal point placed in front, $.a_0 a_1 a_2$ (Table 12.4, column 3). These values are then converted back to base-10 for the sequence value as $a_0 \left(\frac{1}{2}\right)^1 + a_1 \left(\frac{1}{2}\right)^2 + a_2 \left(\frac{1}{2}\right)^3$ (Table 12.4, column 4). The last column of Table 12.4 illustrates the "balanced" nature of the van der Corput sequence construction. The first two sequence values evenly partition the [0,1) interval, and the third and fourth values evenly partition the resulting two subintervals, yielding four equal width subintervals. The next four values evenly partition these four subintervals, but do so by sequentially alternating between the [0,0.5) and [0.5,1) intervals. For larger n, the sequence continues to build in this "halving" manner, alternating between the [0,0.5) and [0.5,1) intervals. This is why the construction achieves relatively even coverage over the [0, 1) interval for any subsequence of points within an overall sequence.

With base integer values of $b > 2$, the sequence results in an initial partitioning of the [0,1) interval into b subintervals, followed by a partitioning of each of those subintervals, in an alternating manner, into b subsubintervals, etc. For example, with a base-3 construction, the first three points of the sequence, 0, 1/3, 2/3, evenly partition the [0,1) interval into three subintervals, and the next six points evenly partition each of these regions into three subsubintervals in an alternating manner, and so on, until the desired number of points are obtained.

Halton sequence

The Halton sequence (Halton 1960) is an extension of the one-dimensional van der Corput sequence (van der Corput 1935) to two or more dimensions. It uses a separate van der Corput sequence for each dimension, each with a different base. Coprime integers are required for the bases (coprime integers are integers which, when taken pairwise, have a greatest common divisor of 1). Common bases used for two-dimensional geographies are 2 and 3.

An important property of a Halton (and van der Corput) sequence is that any sub-sequence shares the same spatial balance properties as the overall sequence. Thus, an element other than $k = 0$ could be used to start the van der Corput sequence in each dimension of the Halton sequence. For the purpose of selecting a random sample of points, this is commonly done by selecting a non-negative integer at random, r_i, $i = 1, 2$, to serve as the starting element for dimension i, and proceeding with the sequence from there, $r_i, r_i + 1, r_i + 2, \ldots$ (as was done for Figure 12.5). This is termed a *random-start* Halton sequence.

Procedure

Selection of an equal probability BAS sample of size n from a finite population of N point objects within a two-dimensional study area is conceptually simple and consists of the following steps (Robertson et al. 2013).

1. Scale and translate the study area and N point object coordinates to a unit square.
2. Partition the unit square into equal size cells, the number of which are large enough to ensure that no more than one population unit is located in a cell. Alternatively, create an equal size, non-overlapping cell around each point object.
3. Use a two-dimensional random-start Halton sequence to generate an initial point. If this point falls into a cell that contains a point object, continue using that sequence to generate a large number of points (much greater than n). If not, discard the entire sequence, choose a new random-start, and return to the beginning of this step (Robertson et al. 2017).
4. Identify the first n points that fall into distinct cells containing a point object.
5. Select for the sample the point objects associated with the cells of these n points.
6. Retain the Halton sequence random-start values in case there is a desire to extend the sample.

First order inclusion probabilities for this procedure may not exactly equal the target value of n/N. Robertson et al. (2013) discuss how the exact values of π_i in a given setting can be directly determined.

There is no need for *oversampling* in BAS. If some selected units are nonresponse, then one may just extend the original Halton sequence until the desired n response units have been encountered. If all nonresponse is due to pure chance, then estimation should remain unbiased.

Procedure with Halton frame

Robertson et al. (2017) introduced some important changes to the original BAS approach designed to enable exact, equiprobable sample selection. The most important changes involve (a) selecting the sample from a *Halton frame* which consists only of cells containing a point object, and (b) *directly* selecting a random subset of these cells for the sample, as an alternative to the generation of Halton points. The cells in this case are termed *Halton boxes* and are defined in a very specific way.

Halton boxes.

For a two-dimensional application and given a Halton sequence with bases $b = (b_1, b_2)$, suppose that each dimension i, $i = 1, 2$, of the unit square is sub-divided into b_i equal length intervals, and that these intervals are used to partition the unit square into $B = b_1 b_2$ boxes (inclusive lower bounds, exclusive upper bounds). Halton (1960) showed that for *any* B successive points in the Halton sequence, each of the points will fall into a different

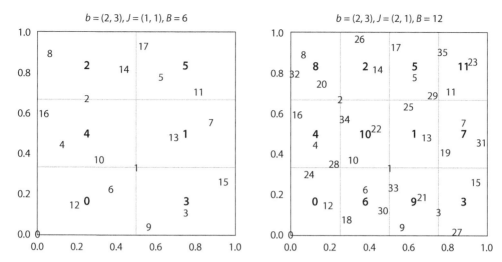

Fig. 12.7 Halton boxes for two-dimensional Halton sequence (no random start). Total number of boxes, B, is determined by the sequence bases $b = (b_1, b_2)$ and positive integers $J = (J_1, J_2)$, where $B = b_1^{J_1} b_2^{J_2}$. Bold numbers at box centers are the box indices; other numbers indicate location of point k of the initial $3B$ points of the sequence.

one of these boxes. More generally, let $J = (J_1, J_2)$ denote any two positive integers. Subdividing each dimension i into $b_i^{J_i}$ equal length intervals and partitioning the unit square into $B = b_1^{J_1} b_2^{J_2}$ boxes leads to the same result: any B successive points in the sequence will each fall into a different one of these boxes. Thus, the entire Halton sequence of points falls into these boxes in a specific order and on a cyclical basis. Therefore, whatever box point k in the sequence falls into, so too will fall the sequence of points $k + B, k + 2B, k + 3B, \dots$. In this sense, the Halton sequence is "quasi-periodic". This reflects the spatial balancing behavior of the Halton sequence, extending to higher dimensions the "interval filling" behavior we observed for the van der Corput sequence in one dimension. The boxes constructed in this manner have been termed **Halton boxes** by Robertson et al. (2017).

Halton boxes are indexed according to the order in which they are occupied by the first B points (no random start) (see Robertson et al. (2018) for an analytic method of how to solve for the index values directly). Figure 12.7 shows the set of Halton boxes resulting from bases $b = (b_1, b_2) = (2, 3)$ matched with two alternative $J = (J_1, J_2)$ values: (1,1) and (2,1), which results in $B = b_1^{J_1} b_2^{J_2} = 6$ and 12, respectively. The box index values are noted at the center of each box, and the first $3B$ Halton points (no random start) are plotted, labeled according to their index in the sequence. Note that the coordinate locations of the first 18 points $(0, 1, 2, \dots, 17)$ are identical for both panels of Figure 12.7, though the number and shape of the Halton boxes differ. Note also the fixed, alternating order in which the points fall into the set of boxes as the sequence proceeds, and that within a box the point index values reflect this, differing by multiples of B. Note also that the sequence of points alternates in a regular way between each of the one-dimensional subintervals as well. Finally, notice the effect of the $J = (J_1, J_2)$ values on the shape of the Halton boxes. $J = (1, 1)$ generates rectangular boxes for which the width is greater than the height, whereas $J = (2, 1)$ generates rectangular boxes for which the height is greater than the width.

Halton frame.

To construct a **Halton frame** (Robertson et al. 2017) for a finite population, the number of Halton boxes B is increased until no box contains more than one point object. The size and shape of the boxes can be adjusted through choice of b and J. The boxes are then indexed, and those boxes that do not contain a point object are discarded. A Halton frame thus consists of N Halton boxes, each containing a single point object.

Sample selection.

Selection of an equal probability BAS with a Halton frame sample of size n selected from a finite population of N point objects within a two-dimensional study area proceeds as follows.

1. Scale and translate the study area and N point object coordinates to a unit square.
2. Construct a two-dimensional Halton frame. Sort (ascending order) the Halton frame boxes by index, and denote them as $\{a_0, a_1, \ldots, a_{N-1}\}$.
3. Generate a random integer j from the set $\{0, 1, 2, \ldots, N-1\}$.
4. Select from the Halton frame the n successive boxes beginning with box a_j (circling back to the top of the sorted frame if necessary). For example, if $n = 4$, $N = 50$, and $j = 7$, select boxes $\{a_7, a_8, a_9, a_{10}\}$; if $j = 48$, select boxes $\{a_{48}, a_{49}, a_0, a_1\}$.
5. Select for the sample the point objects associated with the selected Halton boxes.
6. Retain the random integer j in case there is a desire to extend the sample.

The sample space thus consists of N equally likely samples of n boxes. There are n of these samples that include point object i: those for which j indexes one of the previous $(n-1)$ boxes in the sorted frame, or indexes box i itself. Thus, the target first order inclusion probabilities are achieved exactly, $\pi_i = n/N$. The very restricted sample space also implies that many second order inclusion probabilities equal zero.

Interestingly, with the BAS Halton frame approach the Halton sequence itself is not used to directly generate points in the unit square. Instead, the properties of the sequence in terms of its sequential filling of Halton boxes are exploited to restrict the sample space to N equally likely samples of n boxes each, and then one of these is selected at random.

12.2.3 *Estimation*

Horvitz–Thompson estimation of the population total is recommended for both GRTS and BAS (with or without a Halton frame). With equal probability selection of population units, $\pi_i = n/N$, Horvitz–Thompson estimation is equivalent to mean-per-unit estimation

$$\hat{T}_{y,\text{HT}} = \sum_{i \in S} \frac{y_i}{\pi_i} = N \frac{\sum_{i \in S} y_i}{n} = \hat{T}_{y,\text{mpu}} \tag{12.2}$$

Although both GRTS and BAS produce spatially balanced samples that allow for unbiased estimation of target population parameters, both methods suffer from an absence of unbiased estimators of sampling variance. In this regard, they share some features of systematic sampling. Indeed, all spatially balanced sampling designs by their very nature have second order inclusion probabilities that are close to zero for pairs of units that are close to each other. With GRTS, although it is theoretically true that all $\pi_{ij} > 0$ (because of the random translation of the cover, Step 1), it is also true that $\pi_{ij} \approx 0$ for point objects that are close to one another. If the random translation (Step 1) of GRTS were adopted as the first step in implementation of BAS, then BAS would also theoretically have all $\pi_{ij} > 0$. GRTS and BAS might best be considered *marginally measurable*. In principle, Equation (8.8) or (8.10) could be used for estimation of the sampling variance with GRTS because all $\pi_{ij} > 0$, but no known analytic methods exist to calculate the π_{ij} for GRTS (Anthony R. Olsen, personal communication) and, according to Stevens and Olsen (2003), some π_{ij} are likely to be so small that this would lead to substantial instability in estimates of the sampling variance based on use of Equations (8.8) or (8.10).

In Chapter 4, we explored a number of alternative approaches for variance estimation in systematic sampling and at least one of these might also be used for GRTS or BAS. For example, in BAS the relatively small number of possible samples would lend itself to unbiased estimation of sampling variance via selection of $m \geq 2$ BAS samples selected by SRS from the set of possible samples and application of Equation (4.8). For GRTS, the same general approach could be used but with selection of $m \geq 2$ independent samples and application of Equation (4.7). Another approach from systematic sampling that is somewhat similar in nature to the variance estimation scheme recommended for GRTS and BAS is the following. For systematic sampling (and also for GRTS and BAS), the variance of y within the sample will generally exceed σ_y^2, sometimes by a large amount, due to the high degree of spatial variation or trend in y with unit location. In one-dimensional linear systematic sampling in the presence of a *linear trend*, a mean square successive differences estimator [Equation (4.5)] will generate good estimates of sampling variance; this estimator will be exactly unbiased when the linear correlation between y_i and unit label, i (sometimes a proxy for unit location), is exactly 1.0. This mean square successive differences estimator is based on the average squared difference in y values between successive pairs of units that are adjacent to one another along the line of equidistant unit labels (or locations), i.e., between units that are nearest to one another.

Stevens and Olsen (2003) proposed a **neighborhood variance estimator** for the estimation of sampling variance for GRTS, and Robertson et al. (2013, 2017) recommend this same estimator be used for BAS. The estimator is based on the variation observed within local neighborhoods, which consist of a small number of sample units (usually four) that are spatially close to one another. We describe this estimator following Stevens and Olsen (2003).

Neighborhood variance estimator

Two general properties of the second order inclusion probabilities for a GRTS design with respect to the distance between the locations of population units i and j are: (1) as this distance decreases π_{ij} approaches 0, and (2) as this distance increases π_{ij} approaches $\pi_i \pi_j$, and can equal $\pi_i \pi_j$ beyond some distance. In other words, nearby pairs of units have a relatively small probability of both being included in a GRTS sample, and the probability of selecting a unit relatively far from a previously selected unit is, in effect, independent of the previously selected unit and vice versa. Stevens (1997) provides a geometric explanation for these properties based on a precursor to the GRTS design, the randomized tessellation stratified (RTS) design, and Stevens and Olsen (2003) note that the spatial distribution properties of GRTS and RTS (with the same spatial resolution as GRTS) are very similar. Assuming these properties apply, the pairs of population units that are far from one another make no contribution to $V(\hat{T}_{y,\mathrm{HT}})$ because that contribution includes the multiplicative factor $(\pi_{ij} - \pi_i \pi_j) = 0$ [Equation 8.7]. This means that the Sen–Yates–Grundy estimator for $V(\hat{T}_{y,\mathrm{HT}})$ [Equation 8.10] need only concern itself with pairs of sample units that are in the neighborhood of one another

$$\hat{V}_{\mathrm{SYG}}\left(\hat{T}_{y,\mathrm{HT}}\right) = \sum_{i \in S} \sum_{j \in D_i} \left(\frac{\pi_i \pi_j - \pi_{ij}}{\pi_{ij}}\right)\left(\frac{y_i}{\pi_i} - \frac{y_j}{\pi_j}\right)^2$$

where D_i is the neighborhood of sample units clustered around unit i, and pairs of sample units (i, j) that fall outside of the n neighborhoods in S are excluded from the expression. Stevens and Olsen (2003) note that such neighborhoods can be readily identified for tessellations in an RTS design, and that the expected number of sample points in a neighborhood is four, which they assert is also the case for the GRTS design. The Sen–Yates–Grundy estimator immediately above is still not of practical value, however, as it would require specification of the π_{ij} which are either unknown or sometimes so small as to cause substantial instability in the estimator. Stevens and Olsen (2003) note that $n \geq 30$ usually results in one or more small values of π_{ij}.

Stevens and Olsen (2003) motivate their neighborhood variance estimator by arguing that because of the recursive partitioning and cell addressing (Step 2), and the hierarchical randomization (Step 3), the unit intervals $(0, 1], (1, 2], \ldots, (n-1, n]$ on the line (Step 4) correspond to a random stratification of the study area. They then argue that the systematic selection in GRTS (Step 4), conditional on the realized strata, is equivalent to a stratified random sample with one unit selected per stratum (per unit interval). That is, from the perspective of the Hovitz–Thompson estimator, $\hat{T}_{y,\mathrm{HT}} = \sum_{i \in S} y_i / \pi_i$, unit i is selected from a random stratum with a stratum sample size of one, and the y_i are conditionally independent. Therefore, $E(\hat{T}_{y,\mathrm{HT}}|\mathrm{strata}) = T_y$ and $V[E(\hat{T}_{y,\mathrm{HT}}|\mathrm{strata})] = V(T_y) = 0$ so that, by the law of total variance (Section A.3.7) $V(\hat{T}_{y,\mathrm{HT}}) = E[V(\hat{T}_{y,\mathrm{HT}}|\mathrm{strata})] + V[E(\hat{T}_{y,\mathrm{HT}}|\mathrm{strata})] = E[V(\hat{T}_{y,\mathrm{HT}}|\mathrm{strata})]$. Conditional independence along with the linearity of expectations gives

$$V(\hat{T}_{y,\mathrm{HT}}) = \sum_{i \in S} E[V(\{y_i/\pi_i\}|\mathrm{strata})]$$

For an estimator of $V(\hat{T}_{y,\mathrm{HT}})$, Stevens and Olsen (2003) substitute for $E[V(\{y_i/\pi_i\}|\mathrm{strata})]$ in the equation immediately above a weighted average of the variation observed between y_i/π_i and an average $\{y_k/\pi_k\}$ value (an estimated total) for each of the surrounding neighborhoods

$$\sum_{j \in D_i} w_{ji} \left(\frac{y_i}{\pi_i} - \hat{T}_{D_i} \right)^2, \quad \hat{T}_{D_i} = \sum_{k \in D_i} w_{jk} \frac{y_k}{\pi_k}$$

The D_i neighborhood is defined initially to include its three nearest neighbors, as well as unit i itself. Following this initial specification for all n sample units, if D_i includes unit j but D_j does not include unit i, then D_j is expanded to include unit i. For the weights, $\sum_{i \in S} w_{ij} = \sum_{j \in S} w_{ij} = 1$ which insures that $\sum_{i \in S} \hat{T}_{D_i} = \hat{T}_{y,\mathrm{HT}}$, however the weights in general are not symmetric, i.e., $w_{ij} \neq w_{ji}$. The w_{ij} value is inversely proportional to π_j, decreases as the ranked distance of j from i relative to the other points in the D_i neighborhood increases, and equals zero for j outside the D_i neighborhood (Stevens and Olsen (2003 p. 601) provide further details on the weights). Together, the previous ideas define the neighborhood (NBH) variance estimator

$$\hat{V}_{\mathrm{NBH}}(\hat{T}_{y,\mathrm{HT}}) = \sum_{i \in S} \sum_{j \in D_i} w_{ji} \left(\frac{y_i}{\pi_i} - \hat{T}_{D_i} \right)^2 = \sum_{i \in S} \sum_{j \in D_i} w_{ij} \left(\frac{y_j}{\pi_j} - \hat{T}_{D_i} \right)^2 \qquad (12.3)$$

The second equality in Equation (12.3) follows from the symmetrical definition of the neighborhoods: if $j \in D_i$ then $i \in D_j$. Note that the estimator depends on the π_i, but does not depend on the π_{ij}. With equal probability selection of the population units, $\pi_i = n/N$, the estimator becomes

$$\hat{V}_{\mathrm{NBH}}(\hat{T}_{y,\mathrm{HT}}) = \left(\frac{N}{n} \right)^2 \sum_{i \in S} \sum_{j \in D_i} w_{ij} \left(y_j - \hat{\mu}_{D_i} \right)^2, \quad \hat{\mu}_{D_i} = \sum_{k \in D_i} w_{ik} y_k \qquad (12.4)$$

Stevens and Olsen (2003) report that for the simulation study they conducted $\hat{V}_{\mathrm{NBH}}(\hat{T}_{y,\mathrm{HT}})$ performed relatively well, was approximately unbiased, and appeared to be stable. They also conjectured that the estimator would work well for other spatially balanced sampling designs.

12.3 Infinite populations

GRTS and BAS can both be applied within the context of an **infinite population**, a context that has not thus far been considered in this text. An infinite population is a spatially continuous, bounded *study area* containing an infinite number of *points* (coordinates) rather than a finite set of point objects or units. As noted previously, natural resources and environmental scientists may be interested in descriptive population parameters of a study area such as the mean *per unit area* or total over the study area. Such study areas may not contain distinct, natural units and thus creation of a conventional N unit finite sampling frame may be somewhat arbitrary and it may not be necessary.

In natural resources and ecological research settings, some infinite population variables can be measured at a point (e.g., temperature), but most require measurement *around a point*, i.e., are length-based, area-based, or volume-based measurements (e.g., density). Measurement of the variable value associated with a sample point thus requires specification of a consistent measurement protocol. For example, in vegetation sampling, a circular *field plot* of, say, 10 m radius, centered at point location z might be used to determine the density of a species (number/m^2), y, associated with that point, $y(z)$. In stream sampling, a fixed length of stream channel of, say, 1 m on either side of a point location z might be used to determine the density of an aquatic invertebrate (number/m), y, associated with that point, $y(z)$. There are thus two distinct design elements to consider in the infinite population setting, the sampling design and what Stevens and Urquhart (2000) termed the *response design* or what McDonald (2015) termed the *site design*. The sampling design addresses the selection of sample points, while the response or site design addresses the measurement of the variable (response) at the sample points. In some disciplines, in particular forest mensuration and vegetation analysis, there is a considerable literature on response designs, and guidelines for determining desirable sizes and shapes of field plots depending on the type of variable of interest have been developed. We do not address response design in this text, but Stevens and Urquhart (2000) address this topic in ecological contexts generally, and Gregoire and Valentine (2008 Chapter 7) address it in forestry contexts.

12.3.1 *Generalized random tessellation stratified sampling*

Steps 1, 2, 3, and 5 of the GRTS procedure for a finite population remain the same for an infinite population: scaling and randomly translating the cover (and also the study area including population extent), recursive partitioning and cell addressing, hierarchical randomization, and reverse hierarchical ordering and oversampling, respectively. The differences involve only the sample selection (Step 4) portion of the process: determination of the cell extents and inclusion probabilities, and the random selection of a point within a selected cell, but not the systematic sample selection of the cells themselves. Once the sample points have been selected, reverse hierarchical ordering and oversampling (Step 5) proceeds as in the finite population case.

For an infinite population, the *extent* of cell i, e_i, is the total length (1-D) or total area (2-D) of the population contained within the cell, rather than the total number of units for a finite population. Note that e_i should include only the actual extent of the population within cell i, it should not include non-study area, e.g., for a terrestrial population, any lake, bay, or ocean area. Thus, the extent among cells may not be equal. The total extent of the population over the cover's M cells is $E = \sum_{i=1}^{M} e_i$. For equal probability selection of sample points, the cell inclusion probabilities are, as in the finite population case, proportional to the extent of the population contained within the cells: $\pi_i^* = n(e_i/E)$, $i = 1, 2, \ldots, M$, where n is the desired sample size or oversample size $n' = \gamma n$ (Step 5). As in the finite population case, recursive partitioning (Step 2) must continue until $\pi_i^* < 1$ for all cells to insure that no more than one point can, ultimately, be selected from a cell. Note that while E remains fixed over the recursive process, with each level of recursion the set of cells and their boundaries change requiring recomputation of the e_i (and thus the "clipping" of lines or areas that intersect the cell boundaries). Once the criteria $\pi_i^* < 1$ has been met for all cells, the systematic selection of cells proceeds as in the finite population case.

A single point within the extent of each selected cell i is then randomly selected. If cell i consists of a number of distinct sub-extents, e_{ij}, arising from distinct line-segments or sub-areas within the cell, where $e_i = \sum_j e_{ij}$, this may be accomplished by selecting a line-segment or sub-area at random first, with probabilities e_{ij}/e_i, $j = 1, 2, \ldots$ and then selecting a point at random within the selected line-segment or sub-area. For a line (or line-segment), a continuous uniform$(0, e_i)$ random deviate could be used to determine a random distance along the line to locate the sample point. For an area (or sub-area), two continuous uniform random deviates, one whose domain spans its latitudinal range and one whose domain spans its longitudinal range, could be used to locate the sample point (the first generated deviate pair that falls within the area).

12.3.2 *Balanced acceptance sampling*

BAS can also be used for an equal probability selection of sample points in an infinite population, in a manner that conceptually is quite similar to the two-stage procedure described for GRTS in Section 12.3.1. The study area cover is partitioned into equal size cells, cells are selected with probability proportional to their study area extent, and a point within the extent of each of these cells is selected at random. The means by which BAS does this, of course, is quite different from GRTS. The selection procedure is as follows.

1. Scale and translate the study area and population extent coordinates to a unit square.
2. Partition the unit square into a *large* number of equal size cells. For example, Robertson et al. (2013) recommend a Halton partition with $(b_1, b_2) = (2, 3)$ and $(J_1, J_2) = (8, 5)$, which results in $B = 62{,}208$ approximately square cells. A fine-grained partition such as this helps to insure, ultimately, a well-balanced spatial sample.
3. Measure the *extent* of the population in each of the cells. The extent of cell i, e_i, $i = 1, 2, \ldots, B$ is the total length (1-D) or total area (2-D) of the population contained within the cell. It should not include non-study area, e.g., for a terrestrial population, any lake, bay, or ocean area. Thus, the extent among cells may not be equal. The total extent of the population over the unit square's B cells is $E = \sum_{i=1}^{B} e_i$.

4. Specify the target inclusion probability for each cell i, π_i^* to be proportional to e_i; $\pi_i^* = ne_i/E$, where n is the desired sample size. Check that $\pi_i^* \leq 1$ for all i. (If not, return to Step 2 and use a finer-grained partition of the unit square.)
5. Use a random-start Halton sequence with an extra dimension (for the Bernoulli trial described in the next step), to generate an initial point. If this point falls into a cell with $e > 0$, continue using that sequence to generate a large number of points (vastly greater than n). If not, discard the entire sequence, choose a new random-start, and return to the beginning of this step (Robertson et al. 2017).
6. Starting at the beginning of the sequence, if a point falls into a cell, i, with $e_i > 0$, take the value of the extra dimension in the Halton sequence as the result of a bernoulli(π_i) trial to decide whether or not to accept that point (Robertson et al. 2013). If the value is $\leq \pi_i$ *accept* the point, otherwise *reject* the point. If the point is accepted, select the associated cell (unless that cell has already been selected). Proceed to the next point in the sequence until n distinct cells have been selected.
7. Select for the sample a point at random within the extent of each of the n selected cells in the same way as is done for GRTS in Section 12.3.1.
8. Retain the Halton sequence random-start values in case there is a desire to extend the sample.

12.3.3 *Estimation*

Infinite population target variable parameters are defined somewhat differently than their finite population counterparts. Let z denote a particular point location. Over the study area domain, $\mathcal{R}$, the variable $y(z)$ (Section 12.3), $z \in \mathcal{R}$, has a population total of

$$T_y = \int_{\mathcal{R}} y(z)dz \qquad (12.5)$$

and a population mean of

$$\mu_y = T_y/E \qquad (12.6)$$

where E is the total extent of the population over $\mathcal{R}$. Thus, for an infinite population total, integration over the population domain $\mathcal{R}$ replaces summation over population units N, and for an infinite population mean, the population extent E replaces the number of population units N in the divisor, reflecting the per-length or per-area (rather than per-unit) nature of $y(z)$. In the case of a binary (0, 1) variable (e.g., presence or absence of a species), $\mu_y = \pi_y$ would equal the proportion of the study area for which $y = 1$.

Cordy (1993) extended the Horvitz–Thompson estimator of the finite population total to continuous (infinite) populations

$$\hat{T}_{y,\text{HT}} = \sum_{i \in S} \frac{y(z_i)}{\pi(z_i)} \qquad (12.7)$$

where z_i is the location of sample point i. The estimator has the same general appearance as for a finite population total, but the divisor $\pi(z)$ is now the first order inclusion *density*, rather than *probability*, at location z. Inferences for infinite populations are based on inclusion densities rather than inclusion probabilities. The inclusion density, $\pi(z)$, is sometimes called the *sample intensity function* as it specifies the target number of samples per unit length (1-D) or per unit area (2-D) at location z. It integrates over $\mathcal{R}$ to n, $\int_{\mathcal{R}} \pi(z)dz = n$, just as $\sum_{i=1}^{N} \pi_i = n$ in the finite population case. If $\pi(z)$ is constant for all $z \in \mathcal{R}$, then $\pi(z) = n/E$, analogous to the $\pi_i = n/N$ result for equal probability finite population sampling. In this case,

$$
\hat{T}_{y,\text{HT}} = E \frac{\sum\limits_{i \in S} y(z_i)}{n}
\tag{12.8}
$$

That is, $\hat{T}_{y,\text{HT}}$ is the total study area extent multiplied by the sample mean density.

For the GRTS and BAS sample selection procedures, we can confirm that $\pi(z) = n/E, z \in \mathcal{R}$ as follows. Assume that multiple sub-areas (or line-segments in the 1-D case) $j = 1, 2, \ldots$ may be present in cell i, and let $\mathcal{R}_{ij}$ denote the portion of the study area domain comprised by sub-area j in cell i. Following the logic previously used for Equation (12.1), for $z \in \mathcal{R}_{ij}$

$$
\pi(z) = \Pr\{\text{cell } i \text{ selected}\} \cdot \Pr\{\text{sub-area } j \text{ selected}|\text{cell } i \text{ selected}\}
$$

$$
\cdot \pi(z|\text{cell } i \text{ sub-area } j \text{ selected})
$$

$$
= \pi_i^*(e_{ij}/e_i)(1/e_{ij}) = n(e_i/E)(e_{ij}/e_i)(1/e_{ij}) = n/E
\tag{12.9}
$$

where the conditional inclusion density for cell i sub-area j, assuming it is uniform, is $1/e_{ij}$. Because $\pi(z) = n/E$ does not depend on i or j, this implies that $\pi(z) = n/E$ for all $z \in \mathcal{R}$.

Cordy (1993) and Stevens (1997) also provide expressions for sampling variance and estimation of sampling variance, in terms of first and second order inclusion densities, that are analogous to the Horvitz–Thompson and Sen–Yates–Grundy (fixed n) expressions presented in Chapter 8 for the finite population setting. However, the joint inclusion densities that are required for these expressions suffer from the same issues previously identified for the finite population π_{ij}: lack of specification and some values near zero that lead to estimator instability. The continuous version of Stevens and Olsen's (2003) neighborhood (NBH) variance estimator (Section 12.2.3) does apply here, however, requiring only that the $\pi(z_i)$ be specified

$$
\hat{V}_{\text{NBH}}(\hat{T}_{y,\text{HT}}) = \sum_{i \in S} \sum_{j \in D_i} w_{ij} \left(\frac{y(z_j)}{\pi(z_j)} - \hat{T}_{D_i} \right)^2, \quad \hat{T}_{D_i} = \sum_{k \in D_i} w_{ik} \frac{y(z_k)}{\pi(z_k)}
\tag{12.10}
$$

With constant first order inclusion density, $\pi(z) = n/E$, the estimator becomes

$$
\hat{V}_{\text{NBH}}(\hat{T}_{y,\text{HT}}) = \left(\frac{E}{n} \right)^2 \sum_{i \in S} \sum_{j \in D_i} w_{ij} \left(y(z_j) - \hat{\mu}_{D_i} \right)^2, \quad \hat{\mu}_{D_i} = \sum_{k \in D_i} w_{ik} y(z_k)
\tag{12.11}
$$

12.4 Chapter comments

The fact that GRTS has been used for selection of spatially balanced samples in many large-scale, long-term sample surveys has in part motivated our detailed presentation of this method of generating spatially balanced samples. Large-scale survey applications of GRTS have included at least the U.S. Environmental Protection Agency's National Aquatic Resource Surveys program (National Lakes Assessment, National Coastal Condition Assessment, National Wetland Condition Assessment, and National Rivers & Streams Assessment) (Olsen et al. 2012) and the U.S. National Park Service's Inventory and Monitoring Program (Fancy and Bennetts 2012). Implementation of this method of sample selection typically requires that one is comfortable with the use of GIS software and the R package spsurvey (Kincaid and Olsen 2018) which implements the GRTS algorithms and provides estimates of sampling variance. Successful use of the spsurvey software requires some experience and we provide some illustrations of its use at http://global.oup.com/uk/companion/hankin. The R package SDraw (McDonald 2016) includes functions for drawing equal probability BAS samples from finite populations, and from linear, and areal continuous populations. Presently a Halton frame is not used and thus small departures from equal inclusion probabilities/densities can occur.

Grafström et al. (2012) has proposed an alternative design for producing spatially balanced samples for finite populations called the *local pivotal method*. This method can be applied to either a finite set of N spatially distributed units, or to a study area partitioned into N discrete units. All N inclusion probabilities must be calculated prior to selection of a sample. The inclusion probabilities are calculated sequentially over the units, and have the property that second order inclusion probabilities for units spatially near one another are low compared to those for units distant from one another.

Although both GRTS and BAS can be extended to unequal probability selection of units, we have presented only equal probability implementations in this chapter for two reasons. First, it is easier to develop a basic understanding of the GRTS and BAS selection procedures in the equal probability setting. Second, we believe that generally it would be prudent to select population units with equal probability in large-scale surveys for which spatial balance is judged important. As noted by Overton and Stehman (1996), equal probability selection is a sound practice in long-term, large-scale surveys for several reasons. Imagine a hypothetical national survey of water quality in lakes throughout the U.S. Generally, interest would lie in estimation of water quality parameters not just at the national (population) level but also in various regions (or domains) throughout the study area. Estimation for domains can be easily accomplished via post-stratification when units are selected with equal probability. Normally, multiple target and auxiliary variables would be measured during such a large-scale survey and interest would lie in post-survey model-based analyses of relationships among target variables and/or auxiliary variables. Such model-based analyses are fraught with complications when units are selected with unequal probabilities. Also, such model-based analyses may prove most informative at a regional (post-stratified) level. For example, the well-known sensitivity of Adirondack water bodies to acid rain due to air pollution from SO_2 emissions (Likens and Bailey 2014) revealed a regional level sensitivity. An additional advantage of equal probability selection and post-survey stratification is that ratio and regression estimation may often be used to advantage at a post-stratified regional scale, whereas they may generate no reduction in sampling variance at the population scale. For example, consider a coast-wide survey (California-Washington) based on equal probability GRTS or BAS selection of pre-specified stream segments to establish the average density (number/m) of juvenile steelhead (*Oncorhynchus mykiss*). This species is present throughout the coast, but is present at extremely low levels in southern and south central California as compared to other coastal areas. For the full coast-wide survey data, the correlation between stream segment length and segment-specific abundance might be near zero, but within post-stratified regions (e.g., southern California), the correlation might be quite high and a ratio estimator might work well.

Although GRTS and BAS can be applied in the context of infinite populations, we have emphasized applications to finite populations in this chapter, consistent with our emphasis on finite populations elsewhere in the text. It also reflects our concerns regarding the implementation of field plot sampling around selected points. First, if selected points are located near the boundary of the study area, this can be problematic because the field plot must then either be extended beyond the study area, truncated at the boundary, or "reflected" back into the study area. Such *boundary* or *edge effects* may lead to some degree of bias in estimation (Stevens and Urquhart 2000). Study areas that are relatively narrow and have a high perimeter to area ratio are probably more prone to such problems. Second, it seems unlikely that, in general, the field plot could or would be faithfully located in a rigorous and consistent fashion around selected points. For example, with a stream survey, if the field plot consists of an equal distance of stream above and below the selected points, what is a field crew to do if the selected point falls within a nearly inaccessible, steep and narrow canyon? Could field crews be trusted to consistently execute such a protocol? Might they instead decide to reposition the sample point and associated field plot to some *nearby*, more accessible location?

Although the published literature would appear to suggest that oversampling to achieve a desired fixed n is unique to the GRTS and BAS sampling designs, oversampling can also be successfully applied in simpler sampling design settings whenever it can be reasonably assumed that nonresponse is the result of pure chance (Sections 3.5, 4.6, and 8.9). For example, in an equal or unequal probability spatially stratified design, oversampling could be used to essentially guarantee achievement of desired stratum-specific sample sizes (see Sections 3.5 and 8.9) which could allow a high degree of spatial balance to be achieved in a spatially stratified design with proportional allocation. Estimation would remain unbiased assuming that all nonresponse was pure chance. As noted previously, when nonresponse is not due to pure chance but instead depends on unit- or point-specific y_i, then estimation will no longer be unbiased, no matter what sampling strategy is adopted.

The BAS algorithm for sample selection is considerably simpler than that for GRTS, and that simplicity is a plus for both understanding and explaining the selection procedure. We identify two additional advantages of BAS over GRTS. First, Robertson et al. (2013 Figure 5) present evidence that variance in areas of Voronoi polygons resulting from selection of points using BAS is considerably less than for GRTS, i.e., that BAS delivers better *spatial balance*. Second, in BAS it is unnecessary to adopt a formal oversampling algorithm because one may simply extend the Halton sequence until n response units have been identified. Whether or not one approach yields greater precision for the same sample size appears to depend upon the spatial pattern of the variable of interest, the *spatial field*. Robertson et al. (2013) carried out simulation studies for two different spatial fields partitioned into $N = 400$ units with samples of size $n = 20$, 40, and 60, and compared the performance of BAS and GRTS. Both methods yielded (practically) unbiased estimates of the population mean. For one spatial field, BAS yielded more precise estimates for all three sample sizes, whereas for the other field, GRTS yielded more precise estimates for $n = 20$ and $n = 60$, but BAS was more precise for $n = 40$. Additional research is needed to determine the characteristics of a spatial field, e.g., the degree of spatial autocorrelation, that most affect the relative performance of the two approaches. On "balance", we believe that BAS is a worthy successor to GRTS, and that BAS should be adopted in more studies for which achieving spatial balance is of critical importance.

A very recent development by Robertson et al. (2018) is *Halton iterative partitioning* (HIP), which builds upon the ideas of BAS while overcoming one primary drawback of BAS. For BAS with a Halton frame (Robertson et al. 2017), the number of Halton boxes required for a large-scale survey can become unmanageably large because they must contain no more than one point object (population unit). (Robertson et al. 2017) noted that multiple units could be allowed in Halton boxes if all such units were measured in the context of a single-stage cluster design, but such designs are often inefficient.) HIP overcomes this problem by partitioning the study area into nested boxes and then drawing points from the nested boxes in a particular order so that the sample is spatially balanced. This structure is somewhat analogous to the GRTS procedure of grouping sets of population units (point objects, lines, areas) within cells created during the recursive partitioning step. We conjecture that HIP may become widely adopted in the future, perhaps overshadowing the use of GRTS for selection of spatially balanced samples.

CHAPTER 13

Sampling through time

Fig. 13.1 Adult bald eagle (probably a female), *Haliaeetus leucocephalus*, on its nest with fledgling, Prince William Sound, Alaska. A dual frame sampling strategy (considered in this chapter) has been used to estimate the total number of active bald eagle nest sites on the Kodiak National Wildlife Refuge. Photo credit: Milo Burcham.

Monitoring of natural resources, plants, animals, land, water, and air is a ubiquitous, global, and ever increasing activity. A Google search for "status and trends monitoring" along with key words like wildlife, wetlands, fish, and birds will yield thousands of entries. In many countries, natural resource agencies have been mandated by law to monitor particular environmental and ecological resources.

We define **monitoring** as *repeatedly sampling the same population over some span of time*. The span of time and the sampling frequency determine the number of occasions on which samples are taken and will, of course, vary with the resource and monitoring objectives. The duration of a monitoring program for a redwood forest would be much longer than for monitoring of an algal bloom in an estuary in a given year, for example. In this chapter we consider situations where at a sequence of discrete occasions, $t_1, t_2, \ldots, t_K$, usually equally separated in time, individual samples are taken of the resource. In a

Sampling Theory: For the Ecological and Natural Resource Sciences. David G. Hankin, Michael S. Mohr, and Ken B. Newman, Oxford University Press (2019). © David G. Hankin, Michael S. Mohr, and Ken B. Newman. DOI: 10.1093/oso/9780198815792.001.0001

redwood forest, sample estimates of wood volume might be made every five years for 80 years. An algal bloom in an estuary might be sampled weekly for a period of 12 weeks.

The objectives and duration of monitoring programs are highly variable. For some resources (e.g., salmon populations subject to commercial and recreational fisheries, or endangered or threatened species), indefinite duration **long-term monitoring** may be required by law for harvest management or for assessment of current population status as well as trend in population status. **Status** is generally defined as a population total, either at a particular point in time, or more commonly as an *average total* over a period of time. **Trend** is generally defined as a *change in status* over a period of time. In other contexts, monitoring programs may be of more limited duration. **Effectiveness monitoring** (Reynolds et al. 2016) may be carried out for a limited amount of time prior to and following some management action. For example, an effectiveness monitoring program might be designed to allow determination of whether or not an increase in overwinter water levels in a wildlife refuge has resulted in an increased overwinter abundance of waterfowl. Thus, a change in status may be abrupt, or it may be gradual. The structure and duration of monitoring programs designed to allow assessment of impacts of natural environmental disturbances (e.g., fires) or deliberate environmental interventions (e.g., dam removal) might be similar to those concerning evaluation of the effectiveness of management actions.

Our focus in this chapter is primarily on motivating the conceptual and quantitative basis for monitoring designs. To do that, we begin with the simple two occasion setting and show that it raises the fundamental issues that arise in the design of longer-term monitoring programs. We then extend the two occasion setting to the more general $K > 2$ occasion setting which includes effectiveness and long-term monitoring. Population status and trend measures and their estimation from data collected from monitoring surveys is discussed mostly from a design-based perspective, relying on fundamental properties of linear combinations of random variables (Section A.3). Results are consistently developed in terms of the population total, but can be readily extended to the population mean or proportion. Other approaches to the analysis of sample data collected from long-term monitoring programs are briefly commented on. We also present an overview of dual frame sampling, which may prove useful in a long-term monitoring context.

13.1 Sampling on two occasions

We begin with the simplest possible illustration of sampling through time—sampling on two occasions—and consider two alternative monitoring designs. The first design consists of selecting a single SRS of units (S), and measuring the target variable on these units for each of the two occasions. The second design consists of selecting an independent SRS of units on each occasion (S_1, S_2), and measuring the target variable on the respective units for the two occasions. We later consider a third design which is a combination of the first two designs.

The performance characteristics of these two monitoring designs is illustrated by way of a small example population of $N = 3$ units, with a sample size of $n = 2$ on each occasion. We denote occasion by t with $t = 1, 2$, the target variable by $y_{i,t}, i = 1, 2, \ldots, N$, and the target variable total by $\mathcal{T}_t = \sum_{i=1}^{N} y_{i,t}$. Values of the target variable for the population on the two occasions are

	Occasion 1	Occasion 2
Population unit (i)	$y_{i,1}$	$y_{i,2}$
1	12	10
2	20	20
3	40	32
Sum:	72	62

The population totals for the target variable are $T_1 = 72$ and $T_2 = 62$, resulting in an average total (a measure of status) of $\overline{T} = (T_1 + T_2)/2 = 67$. The difference (change) in the population totals between the two occasions (a measure of trend) is $\Delta_T = T_2 - T_1 = 62 - 72 = -10$. Note from the table that the correlation between unit-specific y values on successive occasions is high and positive $(\mathrm{Cor}(y_1, y_2) = 0.98)$.

We evaluate the following estimators of $\overline{T}$ and Δ_T for each of the two monitoring designs. The mean-per-unit estimator is used on each occasion to estimate the population totals

$$\hat{T}_t = N \sum_{i \in S_t} \frac{y_{i,t}}{n}, \quad t = 1, 2 \tag{13.1}$$

The average total across the two occasions is then estimated as

$$\hat{\overline{T}} = \left(\frac{1}{2}\right)(\hat{T}_1 + \hat{T}_2) \tag{13.2}$$

and the difference between the totals is estimated as

$$\hat{\Delta}_T = \hat{T}_2 - \hat{T}_1 \tag{13.3}$$

The next two sections assess the performance of the two designs applying these estimators to the two occasion population values tabled above, and following that we discuss the expected performance of these two monitoring designs more broadly.

13.1.1 *Design 1: Full retention of units across occasions*

For this design, a single SRS of size $n = 2$ is selected from the $N = 3$ units, and the target variable is measured on these units for each of the occasions. That is, the units selected for occasion 1 are *retained* for occasion 2 (i.e., $S_1 = S_2 = S$). There are thus $\binom{3}{2} = 3$ possible samples over the two occasions. The sample space and corresponding estimates are

Sample units (s)	$\hat{T}_1(s)$	$\hat{T}_2(s)$	$\hat{\overline{T}}(s)$	$\hat{\Delta}_T(s)$
1, 2	48	45	46.5	-3
1, 3	78	63	70.5	-15
2, 3	90	78	84.0	-12

For this design the expected values and sampling variances for $\hat{\overline{T}}$ and $\hat{\Delta}_T$ are

Estimator ($\hat{\theta}$)	$E(\hat{\theta})$	$V(\hat{\theta})$
$\hat{\overline{T}}$	67	240.5
$\hat{\Delta}_T$	-10	26.0

13.1.2 *Design 2: Independent SRS on each occasion*

For this design, *independent* SRS samples of size $n = 2$ are selected from the $N = 3$ units on each of the successive occasions, and the target variable is measured on the selected units. Thus, there may or may not be overlap in the units selected for the two occasions. There are $\binom{3}{2}\binom{3}{2} = 9$ possible sample combinations over the two occasions. The sample space and corresponding estimates are listed in the following table.

Occasion 1		Occasion 2			
Sample units (s_1)	$\hat{T}_1(s_1)$	Sample units (s_2)	$\hat{T}_2(s_2)$	$\hat{\bar{T}}(s_1,s_2)$	$\hat{\Delta}_T(s_1,s_2)$
1, 2	48	1, 2	45	46.5	−3
1, 2	48	1, 3	63	55.5	+15
1, 2	48	2, 3	78	63.0	+30
1, 3	78	1, 2	45	61.5	−33
1, 3	78	1, 3	63	55.5	−15
1, 3	78	2, 3	78	78.0	0
2, 3	90	1, 2	45	67.5	−45
2, 3	90	1, 3	63	76.5	−27
2, 3	90	2, 3	78	84.0	−12

For this design the expected values and sampling variances for $\hat{\bar{T}}$ and $\hat{\Delta}_T$ are

Estimator $(\hat{\theta})$	$E(\hat{\theta})$	$V(\hat{\theta})$
$\hat{\bar{T}}$	67	123.5
$\hat{\Delta}_T$	−10	494.0

13.1.3 *Comparison of full retention and independent SRS designs*

Note first that $\hat{\bar{T}}$ and $\hat{\Delta}_T$ are unbiased estimators regardless of whether sample units are fully retained or independently selected on the two occasions. Unbiasedness ought not to be a surprise since mean-per-unit estimators of the population total (for both designs) are unbiased on each occasion. That is, $E(\hat{T}_t) = T_t, t = 1, 2$, and thus

$$E(\hat{\bar{T}}) = \frac{1}{2}[E(\hat{T}_1) + E(\hat{T}_2)] = \frac{1}{2}(T_1 + T_2) = \bar{T} \tag{13.4}$$

$$E(\hat{\Delta}_T) = E(\hat{T}_2) - E(\hat{T}_1) = T_2 - T_1 = \Delta_T \tag{13.5}$$

The sampling variances of $\hat{\bar{T}}$ and $\hat{\Delta}_T$ are not the same for the two designs, however. When sample units are fully retained across occasions (Design 1), then the sampling variance of $\hat{\Delta}_T$ is much smaller (26.0) than when units are independently selected on each occasion (494.0). In contrast, the sampling variance of $\hat{\bar{T}}$ is greater when sample units are fully retained across occasions (240.5) than when units are independently selected on each occasion (123.5).

The differences between Designs 1 and 2 in the sampling variances of the two estimators are a direct reflection of some fundamental relations concerning variances of linear combinations of random variables (and, therefore, estimators). In particular, for random variables X and Y and constants a and b [Equation (A.56)]

$$V(aX + bY) = a^2 V(X) + b^2 V(Y) + 2ab\text{Cov}(X,Y) \tag{13.6}$$

The sampling variance for the two estimators $\hat{\bar{T}}$ and $\hat{\Delta}_T$ are thus of the general form

$$V(\hat{\bar{T}}) = \left(\frac{1}{2}\right)^2 \left[V(\hat{T}_1) + V(\hat{T}_2) + 2\text{Cov}(\hat{T}_1, \hat{T}_2)\right] \tag{13.7}$$

$$V(\hat{\Delta}_T) = V(\hat{T}_1) + V(\hat{T}_2) - 2\text{Cov}(\hat{T}_1, \hat{T}_2) \tag{13.8}$$

For both designs, the use of SRS implies that $V(\hat{T}_t) = N^2 \left(\frac{N-n}{N}\right) \frac{\sigma_{y_t}^2}{n}$, where $\sigma_{y_t}^2$ is the finite population variance of y on occasion t. Therefore, the difference in $V(\hat{\bar{T}})$ and $V(\hat{\Delta}_T)$ between the two designs observed in this example must arise from the $\text{Cov}(\hat{T}_1, \hat{T}_2)$ term. For Design 2, independent SRS samples on the two occasions implies that $\text{Cov}(\hat{T}_1, \hat{T}_2) = 0$. This is not the case for Design 1, in which a single SRS sample of units is used for both occasions.

For a single SRS sample, with population variables x and y, $\text{Cov}(\hat{\mu}_x, \hat{\mu}_y) = \left(\frac{N-n}{N}\right) \frac{\sigma_{x,y}}{n}$ [Equation (7.11)]. Thus, applying Equation (A.52), $\text{Cov}(\hat{T}_x, \hat{T}_y) = N^2 \left(\frac{N-n}{N}\right) \frac{\sigma_{x,y}}{n}$. This also holds for the Design 1 single SRS setting, where $\hat{T}_1$ takes the place of $\hat{T}_x$, $\hat{T}_2$ (based on these same sample units) takes the place of $\hat{T}_y$, and σ_{y_1,y_2} takes the place of $\sigma_{x,y}$. Therefore,

$$\text{Cov}(\hat{T}_1, \hat{T}_2) = N^2 \left(\frac{N-n}{N}\right) \frac{\sigma_{y_1,y_2}}{n} \tag{13.9}$$

Noting that $\sigma_{y_1,y_2} = \text{Cor}(y_1, y_2)\sigma_{y_1}\sigma_{y_2}$ [Equation (A.119)], with Design 1, if $\text{Cor}(y_1, y_2) > 0$ (as in our example) then the $\text{Cov}(\hat{T}_1, \hat{T}_2)$ term in Equation (13.7) results in an *increase* in $V(\hat{\bar{T}})$ relative to Design 2, and the $\text{Cov}(\hat{T}_1, \hat{T}_2)$ term in Equation (13.8) results in a *decrease* in $V(\hat{\Delta}_T)$ relative to Design 2.

The simple two occasions setting thus frames the essential dilemma of sampling through time. If the primary objective is to estimate or detect a trend (change) in status, then full retention of sample units through time is, theoretically, typically the best monitoring design. If instead the objective is to estimate status (average total), then the best strategy typically involves selection of new sample units on each successive occasion. In resource management settings, however, interest often lies in assessment of both status and trend. In this case, the best strategy may be a *compromise* design consisting of partial retention and partial replacement of sample units, which generally yields *intermediate* results in terms of the sampling variances of $\hat{\bar{T}}$ and $\hat{\Delta}_T$.

13.1.4 *Design 3: Partial retention/partial replacement*

For this design, a single SRS sample S_{ret} of size n_{ret} is selected from the N units, and the target variable is measured on these units on each of the occasions (the *retention* portion of the design). In addition, independent SRS samples $S_{t,\text{rep}}, t = 1, 2$, of size n_{rep} are selected from the N units on each of the occasions, and the target variable is measured on these selected units (the *replacement* portion of the design). On each occasion the overall sample size is $n = n_{\text{ret}} + n_{\text{rep}}$. It is thus possible with this design that a population unit may be selected twice on a given occasion. On each occasion, the retention and replacement samples each allow for unbiased mean-per-unit estimation of T_t

$$\hat{T}_{t,\text{ret}} = N \sum_{i \in S_{\text{ret}}} \frac{y_{i,t}}{n_{\text{ret}}}; \qquad \hat{T}_{t,\text{rep}} = N \sum_{i \in S_{t,\text{rep}}} \frac{y_{i,t}}{n_{\text{rep}}} \tag{13.10}$$

Table 13.1 Simulated expected values and sampling variances of $\hat{\bar{T}}$ and $\hat{\Delta}_T$ for three alternative two occasion sampling designs, with $N = 400$ and $n = 40$ for all designs, with $(y_{i,1}, y_{i,2})$, $i = 1,2,\ldots,N$ generated as described in the text, yielding $\bar{T} = 34{,}260$, $\Delta_T = 20{,}174$, and $\text{Cor}(y_1, y_2) = 0.78$. The three designs are: 1) full retention (Retain), 2) independent SRS (Replace), and 3) partial retention/partial replacement (Partial). Results based on 500,000 independent Monte Carlo simulations of each design.

Estimator ($\hat{\theta}$)	$E(\hat{\theta})$			$V(\hat{\theta})$		
	Retain	Replace	Partial	Retain	Replace	Partial
$\hat{\bar{T}}$	34,261	34,261	34,261	417,240	237,747	346,083
$\hat{\Delta}_T$	20,174	20,175	20,173	230,152	951,378	621,411

and thus a weighted average of these two estimators

$$\hat{T}_t = w\hat{T}_{t,\text{ret}} + (1-w)\hat{T}_{t,\text{rep}}, \quad t = 1,2 \tag{13.11}$$

provides an overall, unbiased estimator for T_t. The most natural weight ($0 \leq w \leq 1$) for this estimator is the fraction of the overall sample size committed to the retention sample, $w = n_{\text{ret}}/n$, but any other weighting would also yield an unbiased estimator, and may be more optimal in terms of reducing the sampling variance of either $\hat{\bar{T}}$ or $\hat{\Delta}_T$. Given the estimates $\hat{T}_t, t = 1,2$, the quantities $\bar{T}$ and Δ_T are then estimated as before for Designs 1 and 2 using Equations (13.2) and (13.3), respectively.

The performance characteristics of this design versus those of Designs 1 and 2 are illustrated with a second example as follows. A population of $N = 400$ units was created with target variable values on occasion 1 generated as $y_{i,1} = 60 + \varepsilon_{i,1}$ with independent $\varepsilon_{i,1} \sim \text{normal}(0, 100)$, $i = 1,2,\ldots,N$, and target variable values on occasion 2 generated as $y_{i,2}|y_{i,1} = y_{i,1} + 50 + \varepsilon_{i,2}$ with independent $\varepsilon_{i,2} \sim \text{normal}(0, 64)$, $i = 1,2,\ldots,N$. Repeated sampling of the population units on the two occasions was then simulated under all three designs with a sample size of $n = 40$ (for Design 3, $n_{\text{ret}} = n_{\text{rep}} = 20$).

Table 13.1 summarizes the simulated expected values and sampling variances of the estimators $\hat{\bar{T}}$ and $\hat{\Delta}_T$ for the three designs. Note first that all three designs resulted in unbiased estimation of $\bar{T}$ and Δ_T. Similar to the results in Sections 13.1.1 and 13.1.2, the independent SRS design is best (has smallest sampling variance) for estimating $\bar{T}$, whereas the full retention design is best for estimating Δ_T. The advantage of the partial retention/partial replacement design is that it produces sampling variances that are intermediate in value between the full retention and independent SRS designs for both $\bar{T}$ and Δ_T. Thus, these simulation results support one's intuition that the partial retention/partial replacement design may be a good compromise design when interest lies in estimation of both status and trend, as is often the case in a resource assessment context.

Sampling variance

The sampling variances of $\hat{\bar{T}}$ and $\hat{\Delta}_T$ for this design have the same general form as for Designs 1 and 2, given by Equations (13.7) and (13.8), but the formulas for $V(\hat{T}_t)$ and $\text{Cov}(\hat{T}_1, \hat{T}_2)$ differ. Application of Equation (13.6) to Equation (13.11) for $\hat{T}_t$ gives

$$V(\hat{T}_t) = w^2 V(\hat{T}_{t,\text{ret}}) + (1-w)^2 V(\hat{T}_{t,\text{rep}}), \quad t = 1,2 \tag{13.12}$$

noting that $\text{Cov}(\hat{T}_{t,\text{ret}}, \hat{T}_{t,\text{rep}}) = 0$, $t = 1,2$, because the retention and replacement SRS samples are selected independently. SRS selection of these samples also implies that $V(\hat{T}_{t,\text{ret}}) = N^2 \left(\frac{N-n_{\text{ret}}}{N}\right) \frac{\sigma_{y_t}^2}{n_{\text{ret}}}$, and $V(\hat{T}_{t,\text{rep}}) = N^2 \left(\frac{N-n_{\text{rep}}}{N}\right) \frac{\sigma_{y_t}^2}{n_{\text{rep}}}$.

For the $\text{Cov}(\hat{T}_1, \hat{T}_2)$ term, substitution of Equation (13.11) for $\hat{T}_t, t = 1,2$ gives

$$\text{Cov}(\hat{T}_1, \hat{T}_2) = \text{Cov}\left(w\hat{T}_{1,\text{ret}} + (1-w)\hat{T}_{1,\text{rep}}, w\hat{T}_{2,\text{ret}} + (1-w)\hat{T}_{2,\text{rep}}\right)$$

which is the covariance of two linear combinations of random variables. In general, for two sets of random variables $\{X_i\}$ and $\{Y_j\}$ and scalar constants $\{a_i\}$ and $\{b_j\}$, the covariance of the resulting two linear combinations is given by [Equation (A.70)]

$$\text{Cov}\left(\sum_i a_i X_i, \sum_j b_j Y_j\right) = \sum_i \sum_j a_i b_j \text{Cov}(X_i, Y_j) \tag{13.13}$$

so that $\text{Cov}(\hat{T}_1, \hat{T}_2)$ is equal to a sum of the scaled covariances between the various pairwise combinations of retention and replacement estimators for occasions 1 and 2. However, except for the retention samples on occasions 1 and 2, all other SRS samples are selected independently of one another, which implies that $\text{Cov}(\hat{T}_{1,\text{ret}}, \hat{T}_{2,\text{rep}}) = 0$, $\text{Cov}(\hat{T}_{1,\text{rep}}, \hat{T}_{2,\text{ret}}) = 0$, and $\text{Cov}(\hat{T}_{1,\text{rep}}, \hat{T}_{2,\text{rep}}) = 0$. Therefore, $\text{Cov}(\hat{T}_1, \hat{T}_2) = w^2 \text{Cov}(\hat{T}_{1,\text{ret}}, \hat{T}_{2,\text{ret}})$, and referring to our earlier result for the covariance of retention-based estimators on the two occasions [Equation (13.9)] leads to

$$\text{Cov}(\hat{T}_1, \hat{T}_2) = w^2 N^2 \left(\frac{N-n_{\text{ret}}}{N}\right) \frac{\sigma_{y_1,y_2}}{n_{\text{ret}}} \tag{13.14}$$

13.2 Monitoring design

The two occasion setting provides a firm foundation from which to consider the more elaborate monitoring designs that are possible for long-term and effectiveness monitoring programs. To describe these elaborations it will be helpful to generalize our terminology. Following McDonald (2003), we define a **monitoring design** as consisting of two interrelated specifications: (a) the sampling designs used to select the units on any given occasion, which are then grouped into one or more sets of units called "panels", and (b) the schedule for surveying these panels. We refer to the sampling design/panel formation component as the **membership design** because it is the mechanism by which the membership of the panels is established, and we refer to the schedule for surveying panels as the **revisit design**. It is important to note that a unit may be a member of more than one panel, but that a panel's membership once established is, in principle, fixed over time. A complete **monitoring strategy**, by analogy with a sampling strategy, consists of combining the monitoring design (which generates observations on randomly selected population units through time) with the estimation methods used broadly for the assessment of status and trend.

For some concrete examples of this terminology we refer back to the monitoring designs considered in the two occasion setting. For the "full retention of units across occasions" design (Design 1), SRS was used to establish a single panel of n units (the membership design), and the panel was surveyed on occasions 1 and 2 (the revisit design). For the "independent SRS on each occasion" design (Design 2), SRS was used independently to establish two panels of n units each (the membership design), with panel 1 surveyed on occasion 1 and panel 2 surveyed on occasion 2 (the revisit design). For the "partial retention/partial replacement" design (Design 3), SRS was used independently to establish three panels: panels 1 and 2 of size n_{rep}, and panel 3 of size n_{ret} (the membership design), with panel 1 surveyed on occasion 1, panel 2 surveyed on occasion 2, and panel 3 surveyed on both occasions (the revisit design). Membership and revist design considerations are discussed in Sections 13.2.1 and 13.2.2, respectively.

13.2.1 *Membership design*

The membership design consists of the sampling designs used to select units that form the panels which are sampled over time. Although a large number of sampling designs have been considered in this text and could in principle be used to form panels for monitoring programs, use of many of these designs could significantly complicate the estimation portion of the overall monitoring strategy (further comments on this to follow), and for this reason most sampling theorists favor relatively simple, equal probability selection methods for membership designs (e.g., McDonald 2003, Overton and Stehman 1996). This means that not only are unequal probability selection methods discouraged, but also stratified sampling methods that typically result in differences in inclusion probabilities for units belonging to different strata. Note, however, that post-stratification can be an effective estimation technique for data collected in monitoring programs (Overton and Stehman 1996).

For long-term monitoring designs, there may be numerous panels formed and sampled over time. A membership design may allow for some random overlap in panel membership, or it may not allow for any overlap in panel membership. For example, if we were to extend the independent SRS selection of units on two successive occasions to the $K > 2$ occasion setting, this would result in a new panel being formed for each occasion (K panels overall), but would allow for some random overlap in panel membership. Alternatively, SRS might be used to select an individual panel's members *conditional* on those units not having already been selected for another panel. In this case, for panel 1 SRS is used to select n_1 units from the N population units, but then for panel 2 SRS is used to select n_2 units from the $N - n_1$ units not selected for panel 1, and so on. This too would result in the formation of K panels, but there would be no overlap in panel membership. An advantage of selecting units from those units not previously selected for other panels is that it ensures, over time, that a larger fraction of population units are examined.

Other membership designs are available that would also prevent overlap in membership between panels. For example, a single SRS of Kn units could be selected from the N population units, with the first n units assigned to panel 1, the next n units assigned to panel 2, and so on. Or, a linear systematic design (assuming N/k is integer-valued) could be used to create k non-overlapping groups of size n and SRS could be used to select $m \geq 2$ of these groups as panels. These panels might have very desirable *spatial coverage* if the sampling units were ordered along some physical dimension or gradient (e.g., stream habitat units listed in an upstream order). Alternatively, a single large GRTS (Chapter 12) oversample could be selected and then reverse hierarchically ordered (Section 12.2.1). As for the SRS approach described in this paragraph, the first n units could be assigned to panel 1, the next n units assigned to panel 2, and so on, to create $m \geq 2$ non-overlapping panels, each of which would be expected to have good spatial coverage. (Non-overlapping sets of spatially balanced units could also be generated as successive sets of n units from a two-dimensional random start Halton sequence using BAS [Chapter 12].) Note that all three of these options for generating non-overlapping panels result in dependent panels, as members of one panel cannot be members of another panel.

Membership designs which produce non-overlapping panels may be more efficient than those based on independent selection of units (i.e., they may result in reduced sampling variances for $\hat{\bar{\mathcal{T}}}$ or $\hat{\Delta}_{\mathcal{T}}$). A disadvantage of such membership designs, however, is that the resulting panels are dependent, thereby complicating the covariance structure of collected survey data and associated formulation of design-based sampling variances and their estimators. We believe that the advantages of these non-overlapping panel selection designs, relative to independent panel selection designs (which allow for some random overlap in panel membership) are realized primarily in long-term monitoring settings rather than in the relatively short-term settings of effectiveness monitoring.

13.2.2 Revisit design

The revisit design consists of the schedule that specifies which panels will be surveyed on which occasions. At one end of the revisit designs spectrum is the **never revisit** (NR) design, in which a different panel is surveyed on every occasion and no panel is ever revisited.[1] At the other end of the spectrum is the **always revisit** (AR) design where only a single panel is formed and it is surveyed on every occasion. The NR and AR revisit designs are thus the $K > 2$ occasion extensions of the revisit schedules for the "independent SRS on each occasion" and "full retention of units across occasions" monitoring designs explored in the two occasion setting, respectively.

As noted in Section 13.1, the AR revisit design is more efficient than the NR revisit design for estimating trend if the correlation between unit-specific y values through time is large and positive, while the NR revisit design is typically more efficient for estimating status. The NR revisit design might also have less of an impact on the population units themselves. With the AR revisit design, repeated sampling of the same unit may lead to what is sometimes called *response burden* (McDonald 2003), where repeated measurements on the same unit affect the value of the variable being measured. For example, certain animals may shy away from the unit over time if it is repeatedly sampled, or repeated sampling may trample or alter the quality of the physical habitat of a unit. With human populations, this type of *sampling fatigue* or *response resistance* (Rao and Graham 1964) due to repeated measurements of the same individual can lead to problematic or untrustworthy measurements of target variable values on successive occasions.

Other problematic issues can arise for an AR revisit design if the physical setting of a population (study area) is highly dynamic or unstable. In such settings, some of the sample units may be destroyed over time, or their variable values may be differentially affected relative to other units. A loss of sample units over time will make the survey less informative at best and unworkable at worst. For example, in some stream systems large storm events and high flows may completely "rearrange" habitat unit sizes and classifications, rendering a stratified AR revisit design based on repeatedly surveying fixed sets of habitat units within strata unworkable. Such large-scale disturbances may cause some pools to become riffles, and so on. Even when habitat unit classification remains stable through time, environmental disturbances may have differential impacts on the variable values of some sample units relative to others, thereby reducing the temporal correlation in AR sample unit values and lessening the advantage of an AR revisit design for estimating $\Delta_{\mathcal{T}}$. For example, intense but spatially limited lightning-caused fires may dramatically alter the physical nature and vegetation structure of some sample forest plots

---

[1] Whether or not a particular unit is ever revisited, however, depends on the membership design employed (i.e., whether the sampling design allows for a particular unit to be a member of more than one panel).

in an AR sample, but may have no influence on others, thereby altering the temporal trend in the variable values (e.g., production or biomass) for some units.

The NR and AR revisit designs typically consist of one panel of units being surveyed per occasion. More elaborate, multi-panel revisit designs allow for a large number of variations in the revisit pattern over time, with some panels being retained for sequences of occasions then dropped permanently or revisited again on a later occasion, new panels of units being added permanently or temporarily, and perhaps one or more panels being fully retained through time. These variations, as can be imagined, share some of the advantages and disadvantages of the NR and AR revisit designs. Two broad categories of such multi-panel revisit designs are the so-called rotating panel and split panel revisit designs.

With a **rotating panel** (RP) revisit design, all panels are surveyed for an equal number of occasions in a row, then not surveyed for an equal number of occasions in a row, and this cycle is repeated indefinitely. The name for this design comes from a panel being *rotated in* to the survey for some number of occasions, and then being *rotated out* of the survey for some number of occasions. In order for all occasions to be sampled, multiple panels are created and their individual revisit schedules are staggered over time. For example, suppose five panels are created, and all are surveyed for two occasions in a row, then not surveyed for three occasions in a row, and that this cycle is repeated indefinitely. Suppose further that panel 1 is surveyed on occasions 1 and 2, and then rotated out. Panel 2 is not surveyed on occasion 1, and then rotated in for occasions 2 and 3, and then rotated out. Panel 3 is not surveyed on occasions 1 and 2, and then rotated in for occasions 3 and 4, and then rotated out, etc. This would result in two panels being surveyed on every occasion (except for the first occasion, as outlined here). The rotating nature of revisits for the RP design takes advantage of the probable positive correlation of the unit-specific $y_{i,t}$ on successive occasions when panels are rotated in, and reduces the potential for response burden by rotating panels out for successive occasions as well.

A **split panel** (SP) revisit design consists of two or more sets of panels that have different revisit design configurations (e.g., not all "rotate in for 2 occasions, rotate out for 3 occasions"). A simple example of an SP revisit design is the $K > 2$ occasion generalization of the revisit schedule for the "partial retention/partial replacement" monitoring design (Design 3) considered in the two occasion setting. That is, a combination of the NR and AR revisit designs. A more complicated SP revisit design might consist of a combination of the NR, AR, and RP designs! The RP and SP revisit designs can provide relatively broad coverage of the sampling frame through time, and with some panel being revisited for at least two successive occasions provides good information about patterns over time in the variable or parameter of interest. Thus, the RP and SP revisit designs can deliver good estimation performance for both status and trend, but the estimation is more complicated than for the simple NR and AR revisit designs.

Shorthand notation

McDonald (2003) developed a shorthand notation that concisely communicates the specifics of a given revisit design, and also facilitates determination of the number of panels required to implement a specific revisit design. We describe this notation and work through a few examples.

The notation represents a revisit design as one or more (comma separated) even numbered, hypenated sequences of numbers and/or letters enclosed in square brackets. Examples of the notation for some simple AR, NR, RP, and SP revisit designs are listed in the following table and depicted graphically in Table 13.2.

Revisit design	Acronym	Notation (example)
always revisit	AR	[1-0]
never revisit	NR	[1-n]
rotating panel	RP	[2-3]
split panel	SP	[1-0,1-n]

For each sequence, values in odd-numbered positions (i.e., positions $1, 3, 5, \ldots$) indicate the number of consecutive occasions that a panel is surveyed before it is rotated out (either temporarily or permanently), whereas values in even-numbered positions (i.e., positions $2, 4, 6, \ldots$) indicate the number of consecutive occasions that a panel is not surveyed before it is rotated in. For an even-numbered position, a value of 0 means the panel is never rotated out (always surveyed), and a value of "n" means the panel is never surveyed again. Thus, for the examples listed immediately above, AR [1-0] translates to "a panel is surveyed once, but then never rotated out"; NR [1-n] translates to "a panel is surveyed once, and never surveyed again"; RP [2-3] translates to "a panel is surveyed on two consecutive occasions before being rotated out, and then not surveyed on three consecutive occasions before being rotated in (i.e., "2 on, 3 off")"; and SP [1-0,1-n] translates to a combined revisit design: AR [1-0] + NR [1-n].

Table 13.2 Visual depictions of the always revisit, never revisit, rotating panel, and split panel revisit design examples reviewed in the text. The symbol $\times$ denotes surveying of a particular panel on a particular occasion.

Always revisit [1-0]

Panel	Occasion 1	2	3	4	5	$\cdots$
1	$\times$	$\times$	$\times$	$\times$	$\times$	

Never revisit [1-n]

Panel	Occasion 1	2	3	4	5	$\cdots$
1	$\times$	$\cdot$	$\cdot$	$\cdot$	$\cdot$	
2	$\cdot$	$\times$	$\cdot$	$\cdot$	$\cdot$	
3	$\cdot$	$\cdot$	$\times$	$\cdot$	$\cdot$	
4	$\cdot$	$\cdot$	$\cdot$	$\times$	$\cdot$	
5	$\cdot$	$\cdot$	$\cdot$	$\cdot$	$\times$	
$\vdots$						$\ddots$

Rotating panel [2-3]

Panel	Occasion 1	2	3	4	5	$\cdots$
1	$\times$	$\cdot$	$\cdot$	$\cdot$	$\times$	
2	$\times$	$\times$	$\cdot$	$\cdot$	$\cdot$	
3	$\cdot$	$\times$	$\times$	$\cdot$	$\cdot$	
4	$\cdot$	$\cdot$	$\times$	$\times$	$\cdot$	
5	$\cdot$	$\cdot$	$\cdot$	$\times$	$\times$	

Split panel [1-0,1-n]

Panel	Occasion 1	2	3	4	5	$\cdots$
1	$\times$	$\times$	$\times$	$\times$	$\times$	
2	$\times$	$\cdot$	$\cdot$	$\cdot$	$\cdot$	
3	$\cdot$	$\times$	$\cdot$	$\cdot$	$\cdot$	
4	$\cdot$	$\cdot$	$\times$	$\cdot$	$\cdot$	
5	$\cdot$	$\cdot$	$\cdot$	$\times$	$\cdot$	
6	$\cdot$	$\cdot$	$\cdot$	$\cdot$	$\times$	
$\vdots$						$\ddots$

Table 13.3 Two startup options for a rotating panel [3-2] revisit design. Five panels are required for this particular design, with a rotation survey schedule of "3 on, 2 off" for each panel, resulting in three of the five panels being surveyed on each occasion. The symbol × denotes surveying of a particular panel on a particular occasion.

Rotating panel [3-2]

Panel	Occasion 1	2	3	4	5	...
1	.	.	×	×	×	
2	×	.	.	×	×	
3	×	×	.	.	×	
4	×	×	×	.	.	
5	.	×	×	×	.	

Rotating panel [3-2]

Panel	Occasion 1	2	3	4	5	...
1	×	×	.	.	×	
2	×	×	×	.	.	
3	.	×	×	×	.	
4	.	.	×	×	×	
5	×	.	.	×	×	

For a string of values that does not end with an "n", the total number of panels required to implement a given revisit design equals the sum of the first odd–even position pair of numbers in which the odd position number is non-zero. Thus, for the AR [1-0] example, a total of $1 + 0 = 1$ panel is required (it is surveyed on all occasions). For the RP [2-3] example, a total of $2 + 3 = 5$ panels are required.

For the NR [1-n] example, the string of values ends in "n", which means a new panel is surveyed on every occasion, and thus K panels are required. For the SP [1-0,1-n] example, one panel is required for the AR [1-0] portion, and K panels are required for the NR [1-n] portion, so that a total of $K + 1$ panels are required.

The revisit design notation does not specify how the individual panel revisit schedules are to be staggered through time. Thus, there may be more than one *startup option* for initiating a given revisit design. For example, two startup options for an RP [3-2] revisit design are shown in Table 13.3. Both options specify a "3 on, 2 off" survey schedule for each panel, but assign the five distinct revisit schedules to different panels.

Example 13.1. A rotating panel revisit design was developed by Adams et al. (2011) for estimation of status and trends of adult coastal salmonids (salmon, steelhead) in California (Table 13.4). The sampling frame consists of well-identified stream reaches (units) that are intended to allow reasonably convenient access, and to allow field personnel to complete surveys of individual selected units in one day of field work. The membership design consists of a single GRTS sample (Chapter 12) allocated to 46 panels. One panel is visited every year (Panel 1); three panels are visited once every three years (Panels 2–4); 12 panels are visited once every 12 years (Panels 5–16); and 30 panels are visited once every 30 years (Panels 17–46)—an RP [1-0,1-2,1-11,1-29] revisit design. Panel 1 contains approximately 40% of the reaches to be surveyed annually. The remaining panel sets each contain approximately 20% of the reaches to be surveyed annually. Thus, each year the sampled units will have a variety of revisit histories, with approximately 40% of them surveyed every year. This is expected to provide a reasonable compromise between status and trend estimation. The 3-, 12-, and 30-year panel revisit periods are designed to facilitate tracking of individual cohorts of coho salmon (*Oncorhynchus kisutch*) which mature at age three, as well as Chinook salmon (*Oncorhynchus tshawytscha*) and steelhead trout (*Oncorhynchus mykiss*) which mature primarily at age three or four.

Table 13.4 Rotating panel revisit design proposed for the California coastal salmonid monitoring program. The symbol × denotes surveying of a particular panel on a particular occasion. Redrafted from Adams et al. (2011).

Rotating panel [1-0,1-2,1-11,1-29]

Panel	Occasion 1	2	3	4	5	6	7	8	9	10	11	12	13	14	15	16	17	18	19	20	21	22	23	24	25	26	27	28	29	30
1	×	×	×	×	×	×	×	×	×	×	×	×	×	×	×	×	×	×	×	×	×	×	×	×	×	×	×	×	×	×
2	×			×			×			×			×			×			×			×			×			×		
3		×			×			×			×			×			×			×			×			×			×	
4			×			×			×			×			×			×			×			×			×			×
5	×												×												×					
6		×												×												×				
7			×												×												×			
8				×												×												×		
9					×												×												×	
10						×												×												×
11							×												×											
12								×												×										
13									×												×									
14										×												×								
15											×												×							
16												×												×						
17	×																													
18		×																												
19			×																											
20				×																										
21					×																									
22						×																								
23							×																							
24								×																						
25									×																					
26										×																				
27											×																			
28												×																		
29													×																	
30														×																
31															×															
32																×														
33																	×													
34																		×												
35																			×											
36																				×										
37																					×									
38																						×								
39																							×							
40																								×						
41																									×					
42																										×				
43																											×			
44																												×		
45																													×	
46																														×

13.3 Estimation of status and trend

Status is generally defined as a population total, either at a particular point in time, or more commonly as an average total over a period of time. But trend, defined as a change in status over a period of time, may be measured in various ways depending on the context. For effectiveness monitoring, it might be defined as the difference between the population totals at two neighboring points in time or, more generally, as the difference between the average population totals over two neighboring periods of time. For long-term monitoring, trend might be defined as the *average* change in status over some period of time, where change is measured as a difference in neighboring totals, or as a ratio of neighboring totals. Alternatively, trend might be left undefined a priori, and measured as a smoothed series of population totals over the monitoring period, for example as a *k*-year moving average of the population totals (assuming that sampling occasions are one year apart).

All of these quantities can be estimated in a straightforward way using the design-based estimates of the occasion-specific population totals, assuming for some that the sampling occasions are equally spaced in time. For status, the average total can be estimated simply as the average of the occasion-specific totals. For the trend measures, the difference between the average totals over two periods of time can be estimated as the difference between the similarly estimated average totals for the two periods. With respect to the average change in status, the average difference in neighboring population totals can be estimated as $\sum_{t=1}^{K-1}(\hat{\mathcal{T}}_{t+1} - \hat{\mathcal{T}}_t)/(K-1) = (\hat{\mathcal{T}}_K - \hat{\mathcal{T}}_1)/(K-1)$, while the average ratio of neighboring population totals can be estimated by the geometric mean of the ratios of estimated totals $[\prod_{t=1}^{K-1}(\hat{\mathcal{T}}_{t+1}/\hat{\mathcal{T}}_t)]^{1/(K-1)} = [\hat{\mathcal{T}}_K/\hat{\mathcal{T}}_1]^{1/(K-1)}$. Finally, each of the estimated *k*-year average totals can be used to provide an estimate of the *k*-year moving average of the population totals (assuming that sampling occasions are one year apart).

Of course, other measures and estimators are possible, as well as alternative, model-based approaches to the analysis of status and trends data, particularly for long-term trends. With a model-based approach, the $\{y_{i,t}\}$ would be regarded as the realized outcomes of random variables $\{Y_t\}$, and a model would be used to relate the $\{Y_t\}$ to covariates, which may or may not include t. Because the $\{Y_t\}$ are random variables, so too are their totals $\{\mathcal{T}_t\}$. The status and trend measures of interest may be model parameters to be estimated, or they might be estimated outside of the model.

Model-based approaches may, or may not, attempt to account for the probability structure of the monitoring design. For example, one model-based option would be to fit a regression model to the survey-estimated population totals as a function of time, but to do this in a way that accounts for the estimated sampling variances of, and possible covariances between, the estimated population totals (e.g., Lewis and Linzer 2005). Another option would be to fit a regression model directly to the observed $\{y_{i,t}\}$, but account for the probability structure of the monitoring design through the use of weights based on the first order sample inclusion probabilities, $\{\pi_{i,t}\}$ (e.g., Nathan 1988, Lumley and Scott 2017). Among the model-based approaches that ignore the probability structure of the monitoring design, is the approach presented in Section 7.4.2. With this approach, a regression model would be fit to the observed $\{y_{i,t}\}$ and the fitted model would be used to predict the y values of the non-sampled units, which would then be summed together with the observed y values for the sampled units to predict the $\{\mathcal{T}_t\}$. The resulting $\{\hat{\mathcal{T}}_t\}$ could then be used in the status and trend estimators presented previously, for example. Another option would be to use one of the many model-based approaches suggested by Bart and Beyer (2012) for the estimation of trend (e.g., estimating the slope of a linear or log-linear model). For the remainder of this section we focus on a design-based approach, as consistent with the emphasis of this text.

13.3.1 *Design-based estimation*

All of the design-based estimators for measures of status and trend described previously, except for the geometric mean of the ratios of neighboring estimated totals, are simple functions of an estimated average total, or of a difference between two estimated average totals. We thus limit ourselves to finding the expectation and sampling variance of these two basic estimators, from which the sampling properties of the various other estimators of status and trend measures can be readily derived.[2] The derived formulas for the expectation and sampling variance of these estimators are general, applicable to any of the monitoring designs contemplated in this chapter.

For estimation of the difference between the average totals over two periods of time, Periods I and II, let $t = 1, 2, \ldots, c$ denote the Period I occasions, and let $t = c + 1, c + 2, \ldots, K$ denote the Period II occasions. We generalize the two-occasion setting definitions of $\overline{\mathcal{T}}$ and $\Delta_{\overline{\mathcal{T}}}$ as

$$\overline{\mathcal{T}} = \frac{1}{K} \sum_{t=1}^{K} \mathcal{T}_t \tag{13.15}$$

$$\Delta_{\overline{\mathcal{T}}} = \overline{\mathcal{T}}_{\mathrm{II}} - \overline{\mathcal{T}}_{\mathrm{I}}, \quad \text{with } \overline{\mathcal{T}}_{\mathrm{I}} = \frac{1}{c} \sum_{t=1}^{c} \mathcal{T}_t, \text{ and } \overline{\mathcal{T}}_{\mathrm{II}} = \frac{1}{K-c} \sum_{t=c+1}^{K} \mathcal{T}_t \tag{13.16}$$

Were there to be an interruption of sampling within a period, then the average total would be defined over the sampled occasions rather than over the full period.

Given occasion-specific estimators of the population total, $\hat{\mathcal{T}}_t, t = 1, 2, \ldots, K$, estimators for $\overline{\mathcal{T}}$ and $\Delta_{\overline{\mathcal{T}}}$ can be obtained by substituting $\hat{\mathcal{T}}_t$ for $\mathcal{T}_t$ in Equations (13.15) and (13.16)

$$\hat{\overline{\mathcal{T}}} = \frac{1}{K} \sum_{t=1}^{K} \hat{\mathcal{T}}_t \tag{13.17}$$

$$\hat{\Delta}_{\overline{\mathcal{T}}} = \hat{\overline{\mathcal{T}}}_{\mathrm{II}} - \hat{\overline{\mathcal{T}}}_{\mathrm{I}}, \quad \text{with } \hat{\overline{\mathcal{T}}}_{\mathrm{I}} = \frac{1}{c} \sum_{t=1}^{c} \hat{\mathcal{T}}_t, \text{ and } \hat{\overline{\mathcal{T}}}_{\mathrm{II}} = \frac{1}{K-c} \sum_{t=c+1}^{K} \hat{\mathcal{T}}_t \tag{13.18}$$

It should be clear from the form of these estimators (linear combinations of the $\{\hat{\mathcal{T}}_t\}$) that if $\hat{\mathcal{T}}_t$ is unbiased for $\mathcal{T}_t$, then $\hat{\overline{\mathcal{T}}}$ and $\hat{\Delta}_{\overline{\mathcal{T}}}$ will be unbiased for $\overline{\mathcal{T}}$ and $\Delta_{\overline{\mathcal{T}}}$, respectively, regardless of the monitoring design employed

$$E(\hat{\overline{\mathcal{T}}}) = \frac{1}{K} \sum_{t=1}^{K} E(\hat{\mathcal{T}}_t) = \frac{1}{K} \sum_{t=1}^{K} \mathcal{T}_t = \overline{\mathcal{T}} \tag{13.19}$$

$$E(\hat{\Delta}_{\overline{\mathcal{T}}}) = E(\hat{\overline{\mathcal{T}}}_{\mathrm{II}}) - E(\hat{\overline{\mathcal{T}}}_{\mathrm{I}}) = \frac{1}{K-c} \sum_{t=c+1}^{K} E(\hat{\mathcal{T}}_t) - \frac{1}{c} \sum_{t=1}^{c} E(\hat{\mathcal{T}}_t) \tag{13.20}$$

$$= \frac{1}{K-c} \sum_{t=c+1}^{K} \hat{\mathcal{T}}_t - \frac{1}{c} \sum_{t=1}^{c} \hat{\mathcal{T}}_t = \overline{\mathcal{T}}_{\mathrm{II}} - \overline{\mathcal{T}}_{\mathrm{I}} = \Delta_{\overline{\mathcal{T}}} \tag{13.21}$$

---

[2] For the geometric mean of the ratios of neighboring estimated totals, the Delta method (Section A.8) could be used to derive an approximate expression for its sampling variance.

The sampling variance of $\hat{\bar{\mathcal{T}}}$ can be expressed as the sampling variance of a linear combination of random variables. In general, for a set of random variables $\{X_i\}$ and constants $\{c_i\}$ [Equation (A.56)]

$$V\left(\sum_i c_i X_i\right) = \sum_i c_i^2 V(X_i) + \sum_i \sum_{i\neq j} c_i c_j \text{Cov}(X_i, X_j) \tag{13.22}$$

Thus, the sampling variance of $\hat{\bar{\mathcal{T}}}$ is given by

$$V(\hat{\bar{\mathcal{T}}}) = \left(\frac{1}{K}\right)^2 \left[\sum_{t=1}^{K} V(\hat{\mathcal{T}}_t) + 2\sum_{t=1}^{K-1} \sum_{t'=t+1}^{K} \text{Cov}(\hat{\mathcal{T}}_t, \hat{\mathcal{T}}_{t'})\right] \tag{13.23}$$

which is a generalization of Equation (13.7) which was derived for the two occasion setting.

For the sampling variance of $\hat{\Delta}_{\bar{\mathcal{T}}} = \hat{\bar{\mathcal{T}}}_{\text{II}} - \hat{\bar{\mathcal{T}}}_{\text{I}}$, application of Equation (13.22) provides $V(\hat{\Delta}_{\bar{\mathcal{T}}}) = V(\hat{\bar{\mathcal{T}}}_{\text{I}}) + V(\hat{\bar{\mathcal{T}}}_{\text{II}}) - 2\text{Cov}(\hat{\bar{\mathcal{T}}}_{\text{I}}, \hat{\bar{\mathcal{T}}}_{\text{II}})$. Because $\hat{\bar{\mathcal{T}}}_{\text{I}}$ and $\hat{\bar{\mathcal{T}}}_{\text{II}}$ are also linear combinations of random variables, their sampling variances have the same form as that for $V(\hat{\bar{\mathcal{T}}})$, and their covariance is given by Equation (13.13). Substitution of these results thus gives

$$V(\hat{\Delta}_{\bar{\mathcal{T}}}) = \left(\frac{1}{c}\right)^2 \left[\sum_{t=1}^{c} V(\hat{\mathcal{T}}_t) + 2\sum_{t=1}^{c-1} \sum_{t'=t+1}^{c} \text{Cov}(\hat{\mathcal{T}}_t, \hat{\mathcal{T}}_{t'})\right]$$
$$+ \left(\frac{1}{K-c}\right)^2 \left[\sum_{t=c+1}^{K} V(\hat{\mathcal{T}}_t) + 2\sum_{t=c+1}^{K-1} \sum_{t'=t+1}^{K} \text{Cov}(\hat{\mathcal{T}}_t, \hat{\mathcal{T}}_{t'})\right]$$
$$- 2\left(\frac{1}{c}\right)\left(\frac{1}{K-c}\right)\left[\sum_{t=1}^{c} \sum_{t'=c+1}^{K} \text{Cov}(\hat{\mathcal{T}}_t, \hat{\mathcal{T}}_{t'})\right] \tag{13.24}$$

which is a generalization of Equation (13.8) which was derived for the two occasion setting.

The linearity of Equations (13.23) and (13.24) implies that unbiased estimators of $V(\hat{\bar{\mathcal{T}}})$ and $V(\hat{\Delta}_{\bar{\mathcal{T}}})$ can be obtained by substituting unbiased estimators of $V(\hat{\mathcal{T}}_t)$ and $\text{Cov}(\hat{\mathcal{T}}_t, \hat{\mathcal{T}}_{t'})$ into Equations (13.23) and (13.24). Given such estimates, a confidence interval could be built around $\hat{\bar{\mathcal{T}}}$ or, in an effectiveness monitoring context, around $\hat{\Delta}_{\bar{\mathcal{T}}}$ to test if there was statistical evidence of a change in $\bar{\mathcal{T}}$ following a disturbance or management action.

Note that $V(\hat{\bar{\mathcal{T}}})$ and $V(\hat{\Delta}_{\bar{\mathcal{T}}})$ do not depend on the interannual variation in the actual population totals. It is also important to note that all of the results presented in this section apply regardless of the monitoring design employed. What will differ depending on the monitoring design employed are the $\{\hat{\mathcal{T}}_t\}$ estimators and the $\{V(\hat{\mathcal{T}}_t)\}$ and $\{\text{Cov}(\hat{\mathcal{T}}_t, \hat{\mathcal{T}}_{t'})\}$ terms. We provide some specific examples of this in the next section.

13.3.2 *Estimators for some specific designs*

Next we develop estimators of $\bar{\mathcal{T}}$ and $\Delta_{\bar{\mathcal{T}}}$ and their sampling variances for the three monitoring designs that were explored in Section 13.1. The membership designs all use independent SRS selection of units to form panels, and the three revisit designs are AR [1-0], NR [1-n], and SP [1-0,1-n] (Table 13.2). The SP [1-0,1-n] revisit design is a combination of the AR [1-0] and NR [1-n] revisit designs. In the remainder of this section, we drop the square bracket specification and refer to these three revisit design configurations simply as AR, NR, and SP. A single panel per occasion is monitored with these AR and NR revisit designs, and two panels per occasion are monitored with this SP revisit design (an AR panel component, and an NR panel component). Designs 1, 2, and 3 in Section 13.1 correspond, respectively, to the AR, NR, and SP revisit designs evaluated here (when coupled with the independent SRS membership design).

We assume for convenience that the number of population units is fixed, as is the sample size over time, so that $N_t = N$ and $n_t = n$ for $t = 1, 2, \ldots, K$. The T_t estimators assumed here are the same as those assumed for the two occasion case. For the single panel per occasion AR and NR revisit designs, the occasion-specific mean-per-unit estimator is

$$\hat{T}_t = N \sum_{i \in S_t} \frac{y_{i,t}}{n}, \quad t = 1, 2, \ldots, K \tag{13.25}$$

where for the AR design $S_1 = S_2 = \cdots = S$. For the two panel per occasion SP revisit design, T_t is estimated by a weighted average of the panel-specific, mean-per-unit based estimators

$$\hat{T}_t = w \hat{T}_{t,\mathrm{AR}} + (1 - w) \hat{T}_{t,\mathrm{NR}}, \quad t = 1, 2, \ldots, K \tag{13.26}$$

where

$$\hat{T}_{t,\mathrm{AR}} = N \sum_{i \in S_{\mathrm{AR}}} \frac{y_{i,t}}{n_{\mathrm{AR}}}; \qquad \hat{T}_{t,\mathrm{NR}} = N \sum_{i \in S_{t,\mathrm{NR}}} \frac{y_{i,t}}{n_{\mathrm{NR}}} \tag{13.27}$$

with the AR and NR subscripts refering here to the two panel components of the SP design.

Use of independent SRS selection to form the panels coupled with use of the mean-per-unit estimator for the total implies that $\hat{T}_t$ is unbiased for T_t for all three monitoring designs, including for the SP revisit design where a weighted average of the panel-specific unbiased estimators is also an unbiased estimator. Therefore, as demonstrated in the previous section, this implies that $\hat{\bar{T}}$ and $\hat{\Delta}_{\bar{T}}$ are also unbiased estimators of $\bar{T}$ and $\Delta_{\bar{T}}$ for all three monitoring designs.

The sampling variances $V(\hat{\bar{T}})$ and $V(\hat{\Delta}_{\bar{T}})$ depend on the individual components $\{V(\hat{T}_t)\}$ and pairwise components $\{\mathrm{Cov}(\hat{T}_t, \hat{T}_{t'})\}$, and thus our results for these monitoring designs in the Section 13.1 two occasion setting apply here as well. In particular, the independent SRS membership design implies

$$V(\hat{T}_t) = \begin{cases} N^2 \left(\dfrac{N-n}{N} \right) \dfrac{\sigma_{y_t}^2}{n}, & \text{for AR, NR designs} \\[3mm] w^2 N^2 \left(\dfrac{N-n_{\mathrm{AR}}}{N} \right) \dfrac{\sigma_{y_t}^2}{n_{\mathrm{AR}}} + (1-w)^2 N^2 \left(\dfrac{N-n_{\mathrm{NR}}}{N} \right) \dfrac{\sigma_{y_t}^2}{n_{\mathrm{NR}}}, & \text{for SP design} \end{cases} \tag{13.28}$$

and

$$\mathrm{Cov}(\hat{T}_t, \hat{T}_{t'}) = \begin{cases} 0, & \text{for NR design} \\[3mm] N^2 \left(\dfrac{N-n}{N} \right) \dfrac{\sigma_{y_t, y_{t'}}}{n}, & \text{for AR design} \\[3mm] w^2 N^2 \left(\dfrac{N-n_{\mathrm{AR}}}{N} \right) \dfrac{\sigma_{y_t, y_{t'}}}{n_{\mathrm{AR}}}, & \text{for SP design} \end{cases} \tag{13.29}$$

for $t \neq t'$, where the AR and NR subscripts refer to the two panel components of the SP design.

Estimators for $V(\hat{\bar{T}})$ and $V(\hat{\Delta}_{\bar{T}})$ can be obtained by substituting estimators for $\sigma_{y_t}^2$ and $\sigma_{y_t, y_{t'}}$ into Equations (13.28) and (13.29), and then substituting the resulting $V(\hat{T}_t)$ and

$\mathrm{Cov}(\hat{T}_t, \hat{T}_{t'})$ estimators into Equations (13.23) and (13.24), respectively. For the AR and NR designs the finite population variances, and for the AR design the finite population covariances, can be unbiasedly estimated in the usual way, with $\hat{\mu}_t = \hat{T}_t/N$, as

$$\hat{\sigma}^2_{y_t} = \frac{\sum_{i \in S_t}(y_{i,t} - \hat{\mu}_t)^2}{n-1} \qquad \hat{\sigma}_{y_t,y_{t'}} = \frac{\sum_{i \in S}(y_{i,t} - \hat{\mu}_t)(y_{i,t'} - \hat{\mu}_{t'})}{n-1} \qquad (13.30)$$

where for the AR design $S_1 = S_2 = \cdots = S$. For the SP design, panel-specific estimates can be used for $\hat{\sigma}^2_{y_t}$, or the panel data could be pooled. For $\hat{\sigma}_{y_t,y_{t'}}$, the data are restricted to the AR panel component.

13.4 Sample size determination

In this section we demonstrate how one can use the sampling variance formulas of the previous sections to determine what values of n and K are necessary to achieve a desired level of estimator precision. To simplify matters we assume a fixed annual budget, and that $N_t = N$ and $n_t = n$ are fixed constants for $t = 1, 2, \ldots, K$. We also assume that the most straightforward monitoring design (in terms of sampling variance) of the examples previously considered will be used: an independent SRS membership design coupled with a NR [1-0] revisit design.

We first consider estimation of $\overline{T}$, and for planning purposes assume that $\sigma^2_{y_t} = \sigma^2_y$ is constant for $t = 1, 2, \ldots, K$. In this case, Equation (13.23) becomes

$$V(\hat{\overline{T}}) = \left(\frac{1}{K}\right)^2 \sum_{t=1}^{K} V(\hat{T}_t) = \left(\frac{1}{K}\right)^2 \sum_{t=1}^{K} N^2 \left(\frac{N-n}{N}\right)\frac{\sigma^2_y}{n} = N^2 \left(\frac{N-n}{N}\right)\frac{\sigma^2_y}{Kn} \qquad (13.31)$$

This sampling variance is analogous to that for stratified random sampling (each year being a stratum, the number of strata equal to K) with independent SRS within strata [Equation (5.11)], except that Equation (5.11) concerns estimation of the total across strata rather than the average stratum total. From the perspective of stratified random sampling, if N_t, $\sigma^2_{y_t}$, and the cost per unit sampled are constant across strata, then optimal allocation would call for a constant value of n_t across strata, as we have assumed here. Note also that for large N (i.e., $(N-n)/N \approx 1$), the $V(\hat{\overline{T}})$ formula [Equation (13.31)] is equivalent to that for a single SRS sample of Kn units. Thus, for large N, the value of K and n have an equivalent affect on $V(\hat{\overline{T}})$, e.g., doubling K for a given n would have the same effect on $V(\hat{\overline{T}})$ as doubling n for a given K.

Let C denote the annual budget and assume a cost function of $C = c_0 + c_1 n$, where c_0 is the survey annual overhead cost, and c_1 is the cost per unit sampled. In this case, $n = (C - c_0)/c_1$. Solving Equation (13.31) for K gives

$$K = \frac{N^2}{V(\hat{\overline{T}})} \left(\frac{N-n}{N}\right)\frac{\sigma^2_y}{n}, \quad \text{with } n = \frac{C - c_0}{c_1} \qquad (13.32)$$

Thus, given values for C, c_0, c_1, N, σ^2_y, and a desired $V(\hat{\overline{T}})$, the necessary number of survey years K and annual sample size n can be determined. As stated elsewhere in the text, if the desired precision is expressed in terms of the coefficient of variation of the estimator, $\mathrm{CV}(\hat{\overline{T}})$, or the large sample bounds on the error of estimation, $\mathrm{B}(\hat{\overline{T}})$, then $(\mathrm{CV}(\hat{\overline{T}}) \cdot \overline{T})^2$ or $\mathrm{B}(\hat{\overline{T}})^2/4$ could be substituted for $V(\hat{\overline{T}})$ in Equation (13.32) as appropriate, assuming in the case of $\mathrm{CV}(\hat{\overline{T}})$ that a guess can be made for the value of $\overline{T}$.

Alternatively, if the survey's total budget over all years, C^*, was fixed, then the cost equation $C^* = K(c_0 + c_1 n)$ and Equation (13.31) could be solved simultaneously to determine the necessary K and n (two equations, two unknowns). Or, if the number of survey years were fixed at K rather than the annual budget C or C^*, Equation (13.31) could be rearranged to give n as a function of $V(\hat{\mathcal{T}})$ and K, resulting in a total cost of $K(c_0 + c_1 n)$.

Continuing with this monitoring design example, suppose now that interest is on estimating $\Delta_{\overline{\mathcal{T}}} = \overline{\mathcal{T}}_{II} - \overline{\mathcal{T}}_I$. Under the same assumptions made for Equation (13.31), Equation (13.24) becomes

$$V(\hat{\Delta}_{\overline{\mathcal{T}}}) = \left(\frac{1}{c}\right)^2 \sum_{t=1}^{c} V(\hat{\mathcal{T}}_t) + \left(\frac{1}{K-c}\right)^2 \sum_{t=c+1}^{K} V(\hat{\mathcal{T}}_t) = \frac{K}{c(K-c)} N^2 \left(\frac{N-n}{N}\right) \frac{\sigma_y^2}{n} \qquad (13.33)$$

Assuming an equal number of survey years for each of the two periods (i.e., $c = K/2$), solving $V(\hat{\Delta}_{\overline{\mathcal{T}}})$ for K gives

$$K = \frac{4N^2}{V(\hat{\Delta}_{\overline{\mathcal{T}}})} \left(\frac{N-n}{N}\right) \frac{\sigma_y^2}{n}, \quad \text{with } n = \frac{C - c_0}{c_1} \qquad (13.34)$$

Thus, given values for C, c_0, c_1, N, σ_y^2, and a desired $V(\hat{\Delta}_{\overline{\mathcal{T}}})$, the necessary number of survey years K and annual sample size n can be determined. Alternative scenarios along the lines described previously would be possible here as well.

For effectiveness monitoring, one might also be interested in knowing the number of monitoring years necessary to detect, with probability p and Type I error rate α, a difference $\Delta_{\overline{\mathcal{T}}}$ that is greater than or equal to some specified amount, δ. This is a standard question posed in the context of statistical tests for the comparison of two sample means (e.g., Zar 2010), the results for which we adapt here. With the same basic assumptions made for Equation (13.31), and large n, to satisfy these conditions the precision would need to be

$$V(\hat{\Delta}_{\overline{\mathcal{T}}}) = \frac{\delta^2}{(z_p + z_{1-\alpha})^2} \qquad (13.35)$$

where z_p is the p quantile of the standard normal distribution (e.g., $z_{0.975} = 1.96$). If one was interested in detection of $|\Delta_{\overline{\mathcal{T}}}| \geq \delta$, $z_{1-\alpha}$ would be replaced by $z_{1-\alpha/2}$. Equation (13.35) for $V(\hat{\Delta}_{\overline{\mathcal{T}}})$ can then be substituted into Equation (13.34) to give

$$K = \frac{4N^2 (z_p + z_{1-\alpha})^2}{\delta^2} \left(\frac{N-n}{N}\right) \frac{\sigma_y^2}{n}, \quad \text{with } n = \frac{C - c_0}{c_1} \qquad (13.36)$$

Thus, the number of years required increases with larger σ_y^2, N, and p, and decreases with larger δ, n, and α. This equation can also be rearranged to determine how small of a difference will be detectable, with probability p and Type I error rate α, for a given number of survey years

$$\delta = (z_p + z_{1-\alpha})\sqrt{\frac{4N^2}{K}\left(\frac{N-n}{N}\right)\frac{\sigma_y^2}{n}} \quad \text{with } n = \frac{C - c_0}{c_1} \qquad (13.37)$$

Practical application of Equations (13.35) and (13.36) in an effectiveness monitoring setting requires that one specify an *anticipated* value for δ, or alternatively for the mean total for the post-disturbance/intervention period assuming that an estimate of the mean total for the pre-disturbance/intervention period is available. Such anticipated values will be uncertain as the future is, of course, unknown.

13.5 Dual frame sampling

Dual frame sampling is a special case of *multiple frame* sampling (Lohr 2009) in which samples are selected from more than one sampling frame for the population of interest, some of which may be incomplete, and the resulting estimates are combined to estimate an overall population parameter. While a frame may be incomplete, it can lead to increased sampling efficiency, especially when sampling for a rare item, and thereby result in improved estimation performance when appropriately combined with complete frame information. Long-term monitoring programs that include annual updating and sampling from a *list frame* as well as from a fixed *area frame*, such as certain bird breeding surveys, provide an example of this.

We illustrate the utility of this approach for estimating the number of active (adults active on nest site) or successful (young produced) nest sites for territorial birds that rear young in just a single nest in a given breeding season and that often return to the same nest sites in following breeding seasons (Haines and Pollock 1998). Based on the most recent past breeding season of nest use surveys, biologists have a list frame that identifies the locations of all active (or successful) nests that have been found in the most recent breeding season(s). This list frame can be very helpful for locating nests in an upcoming breeding season, but it must be recognized as *incomplete* for two reasons: (1) not all active (or successful) nest sites may have been located in the previous breeding season, and (2) new nest sites may be used in the current breeding season. Thus, exclusive reliance on sampling from the list frame would lead to negative bias in estimates of the number of active (or successful) nests in the current year. With dual frame sampling, however, we can augment sampling from the list frame with sampling from a conventional area frame (or grid) in such a way as to allow for unbiased estimation of the total number of active (or successful) nests. Figure 13.2 provides a visual illustration of the distinction between the list and (stratified) area frames that could be used to estimate active bald eagle (*Haliaeetus leucocephalus*) nests on the Kodiak National Wildlife Refuge in Alaska (Reynolds and Shelly 2010).

We describe dual frame survey estimation for this setting of bird breeding surveys. Let N_A be the number of areal units in an area frame that includes the entire extent of the region over which an estimate of the number of active nests is desired, and let N_L be the number of previously identified active nest locations within this region (the list frame). For the list frame, let $I_i, i = 1, 2, \ldots, N_L$, be an indicator variable for active nest use currently, with $I_i = 1$ if nest i is currently active, and $I_i = 0$ otherwise. The proportion and total number of list frame nests that are currently active is thus $\pi_L = \sum_{i=1}^{N_L} I_i / N_L$ and $T_L = N_L \pi_L$, respectively. For the area frame, denote by $y_i, i = 1, 2, \ldots, N_A$, the number of currently active nests within area i that are *not included in the list frame*. The mean number per area and total number of these currently active, non-list frame nests is thus $\mu_A = \sum_{i=1}^{N_A} y_i / N_A$ and $T_A = N_A \mu_A$, respectively. The total number of active nests, $T = T_L + T_A$, is the target of estimation.

For the survey, suppose that independent SRS samples are taken from the list and area frames of size n_L and n_A resulting in samples S_L and S_A, respectively. Assuming that the sample values $\{I_i; i \in S_L\}$ and $\{y_i; i \in S_A\}$ can be determined without error[3], T can then be estimated as the sum of the estimated totals for the two frames

$$\hat{T} = \hat{T}_L + \hat{T}_A = N_L \hat{\pi}_L + N_A \hat{\mu}_A \tag{13.38}$$

where $\hat{\pi}_L = \sum_{i \in S_L} I_i / n_L$, and $\hat{\mu}_A = \sum_{i \in S_A} y_i / n_A$.

---

[3] Shyvers et al. (2018) used a dual frame survey approach, in combination with occupancy analysis, to adjust estimates for imperfect detection (e.g., not all active nest sites are detected at sampled locations).

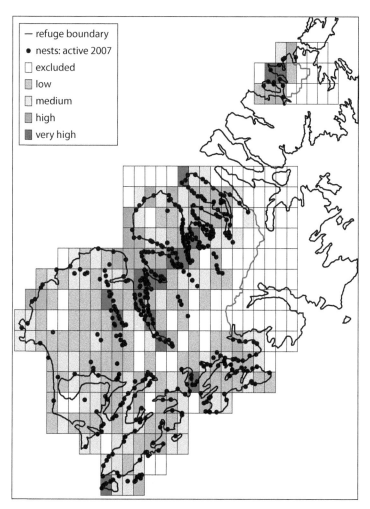

Fig. 13.2 Bald eagle nesting sites on the Kodiak National Wildlife Refuge. Visual illustration of list frame (individual nests active in 2007 identified by black dots) and area frame (rectangular grid cells) for estimating active nest sites in 2008 on the refuge. Color coding of grid cells indicates stratification into use levels based on historic and list frame data and would allow more efficient estimation of the number of active nest sites not included in the list frame. White grid cells are excluded from survey (cell is 100% water, cell area is <15% refuge, or >80% of cell area is above 2000 foot elevation with no known nesting or food source). Redrawn from Reynolds and Shelly (2010).

For the sampling variance of $\hat{T}$, independent SRS samples for the list and area frames imply that (Chapter 3)

$$V(\hat{T}) = V(\hat{T}_L) + V(\hat{T}_A) = N_L^2 V(\hat{\pi}_L) + N_A^2 V(\hat{\mu}_A)$$

$$= N_L^2 \left(\frac{N_L - n_L}{N_L} \right) \frac{\sigma_L^2}{n_L} + N_A^2 \left(\frac{N_A - n_A}{N_A} \right) \frac{\sigma_A^2}{n_A} \qquad (13.39)$$

where $\sigma_L^2 = \pi_L(1 - \pi_L)N_L/(N_L - 1)$ and $\sigma_A^2 = \sum_{i=1}^{N_A} (y_i - \mu_A)^2/(N_A - 1)$.

As a result, $V(\hat{T})$ can be unbiasedly estimated as

$$\hat{V}(\hat{T}) = N_L^2 \left(\frac{N_L - n_L}{N_L}\right)\frac{\hat{\sigma}_L^2}{n_L} + N_A^2 \left(\frac{N_A - n_A}{N_A}\right)\frac{\hat{\sigma}_A^2}{n_A} \tag{13.40}$$

where $\hat{\sigma}_L^2 = \hat{\pi}_L(1 - \hat{\pi}_L)n_L/(n_L - 1)$ and $\hat{\sigma}_A^2 = \sum_{i \in S_A}(y_i - \hat{\mu}_A)^2/(n_A - 1)$ are unbiased estimators of σ_L^2 and σ_A^2, respectively.

Given the selection of independent samples for the two frames, it is clear that allocation of sampling effort to the list frame as opposed to the area frame can have substantial effect on $V(\hat{T})$. To find the optimal allocation of effort, assume that the total cost of the dual frame survey is $C = c_0 + c_L n_L + c_A n_A$, where c_0 is the overhead cost, c_L is the per unit cost for surveying a list frame unit, and c_A is the per unit cost for surveying an area frame unit. To minimize $V(\hat{T})$ with respect to choice of n_L and n_A subject to the cost constraint C, form the Lagrange function (Section A.9) $\mathcal{L}(n_L, n_A, \lambda) = V(\hat{T}) + \lambda(C - c_0 - c_L n_L - c_A n_A)$, and partially differentiate $\mathcal{L}$ with respect to n_L, n_A, and λ

$$\frac{\partial \mathcal{L}(n_L, n_A, \lambda)}{\partial n_L} = -\frac{N_L^2 \sigma_L^2}{n_L^2} - \lambda c_L$$

$$\frac{\partial \mathcal{L}(n_L, n_A, \lambda)}{\partial n_A} = -\frac{N_A^2 \sigma_A^2}{n_A^2} - \lambda c_A$$

$$\frac{\partial \mathcal{L}(n_L, n_A, \lambda)}{\partial \lambda} = C - c_0 - c_L n_L - c_A n_A$$

Setting these partial derivatives equal to zero and solving for n_L, n_A, and λ gives the optimal solution [minimum $V(\hat{T})$] for sample size allocation to the list and area frames

$$n_{L,\text{opt}} = \frac{C - c_0}{c_L + \dfrac{N_A \sigma_A \sqrt{c_L c_A}}{N_L \sigma_L}}, \quad n_{A,\text{opt}} = \frac{C - c_0}{c_A + \dfrac{N_L \sigma_L \sqrt{c_L c_A}}{N_A \sigma_A}} \tag{13.41}$$

Example 13.2. *Bald eagle active nests.* Suppose that an area frame that covers the entire region of interest has $N_A = 250$ units within which habitat is suitable for bald eagle nesting. A pre-existing list frame identifies the locations of $N_L = 250$ known, previously active bald eagle nests within this region from the most recent year's surveys. From past research it is hypothesized that $\pi_L = 0.7$ (i.e., 70% of these nests will be active in the current breeding season) so that $\sigma_L^2 = 0.2108$, and it is guessed that $\sigma_A^2 = 7.49$ based on previous years' research. The budget available for the overall survey is $C = \$200,000$ and it is estimated that $c_0 = \$25,000$ will be required for overhead. The estimated cost for establishing nest use within the list frame is $c_L = \$250$ per nest, whereas the estimated cost for determining y within the area frame is $c_A = \$2,000$ per unit. Substituting these guessed or estimated values into Equation (13.41) yields $n_{L,\text{opt}} = 39.2$ and $n_{A,\text{opt}} = 82.6$. Thus, the optimal allocation would call for surveying about 16% of the units from the list frame and about 33% of the units from the area frame. If an optimally allocated stratified sampling design were instead used to estimate $\mathcal{T}_A$, then allocation of sampling effort to the area frame could probably be substantially reduced so that more effort could be allocated to the list frame.

13.6 Chapter comments

A complete strategy for sampling through time has three components: a membership design, a revisit design, and procedures for estimation and inference using the collected data. In this Chapter we reviewed a number of membership designs that can be used to select panels or groups of sample units; we presented a range of revisit designs which specify when these panels will or will not be sampled; and we derived generic, design-based methods for (a) estimating average population totals (a measure of status) over a period of time, and (b) estimating differences in average population totals between two periods of time (a measure of trend). Derived sampling variance formulas for these estimators are, in principle, applicable to any of the monitoring designs discussed in this chapter, and were used to develop sampling variance formulas specific to the AR [1-0], NR [1-n], and SP [1-0,1-n] revisit designs assuming that panel membership is formed using independent SRS selection. We believe that these design-based methods for estimating measures of status and trend may be useful in many ecological and natural resources contexts, particularly for effectiveness monitoring (before/after comparisons) for which the number of years of before and after sampling may be relatively small (say, 3-6 years for each period) and an SP [1-0,1-n] revisit design with independent SRS formation of panels might be worth considering due to its simplicity and performance.

Smith (1978) refers to analyses based on design-based estimates alone as a *secondary analysis*, under an assumption that the only data available to the analyst are these design-based estimates but not also the observed $\{y_{i,t}\}$ that underly them. For example, a linear regression of $\hat{T}(t)$ against t, used to estimate a trend parameter under an assumption of a linear model, would be regarded as a secondary analysis. The design-based expressions for sampling variance presented in Section 13.3.1, however, rely directly on the underlying $\{y_{i,t}\}$ for estimation of covariances of estimated totals across time intervals, and so it may be inappropriate to view these as secondary analyses in the sense used by Smith (1978).

Nevertheless, as noted in the last paragraph of Section 13.3, the design-based analyses presented in Section 13.3.1 are distinctly different from model-based analyses which rely directly on observed $\{y_{i,t}\}$ and which regard the observed $\{y_{i,t}\}$ as realizations of random variables $\{Y_{i,t}\}$ associated with time and population unit or site. Smith (1978) refers to such model-based analyses of data collected from repeated surveys as *primary analyses* and provides sketches of possible time series methods that might be used for analysis of change through time. These time series methods explicitly incorporate the correlation between $\{y_{i,t}\}$ and $\{y_{i,t-1}\}$ across population units that are retained across two adjacent time periods, but they do not address unequal inclusion probabilities.

Such model-based primary analyses often have more ambitious objectives (i.e., establish/identify relationships among survey variables) than the design-based analyses that we provide. Many long-term ecological and natural resources monitoring programs may result in collection of data from numerous variables at selected individual sample units or "sites" (e.g., mean temperature, elevation, vegetation status, soil moisture, abundance of amphibian species) and it would be natural to wish to explore the relationships among these variables through time. Long-term monitoring designs with membership designs that rely on stratification for creation of panels and that adopt complex rotating panel revisit designs may generate unequal inclusion probabilities as well as complex covariance structures. Such complications are by not limited to ecological and natural resources surveys, of course. They are also generated in many complex long-term public health, agricultural, and social science or economic surveys. Theoretically, the complexity of long-term sampling designs should be considered/addressed in model-based analyses of long-term monitoring data collected in a wide variety of contexts. The $64,000 question is "How?"

Many of the issues raised in model-based analyses of data from complex repeated surveys (like long-term monitoring programs) are raised in complex sample surveys generally (i.e., in *any* surveys that have stratification, clustering, unequal probability selection, etc.) and the literature on this topic is therefore broad and extensive. Readers seriously interested should therefore expand their horizons well beyond ecological and natural resource contexts. For example, Korn and Graubard (1999) is a full text devoted to analysis of data collected from complex, often repeated, health surveys. Pfeffermann (1993) and Pfeffermann (1996) review the use of *sample weights* (also called *design weights*, usually the inverse of first order inclusion probabilities) in model-based analyses of survey data. (Lohr (2010 Chapters 10, 11) provides numerous simple illustrations of the importance of using design weights in graphical portrayal of survey data and in simple model-based analyses, illustrating that biased inferences may often result if analyses do not incorporate weights.) Steel and McLaren (2009) summarize two broad classes of analysis approaches for surveys repeated over time: a classical approach where the population parameters are considered fixed quantities and a time series approach where the population parameters are considered random quantities generated by a stochastic process.

In the long-term ecological and natural resource monitoring setting, the specific problem of trend detection has been a primary focus. Urquhart and Kincaid (1999) and Urquhart (2012) consider how the revisit design affects ability to detect trend, using a linear model and components of variance approach, and stress the advantage of rotating panel designs. Components of variance reflect variation in y across sites and across years as well as residual variation. Skalski (2012) and Gray (2012) provide additional illustrations of components of variance analyses, including for multi-stage designs.

A particularly relevant publication concerning assessment of trends from long-term monitoriong data is the simulation study reported by Starcevich et al. (2018). This study explored the degree to which incorporation of design weights was or was not critical for three different approaches to assessment of trend of lake water quality indicators. Equal probability generalized random tesselation stratified sampling (GRTS, Chapter 12) was used for selection of panel membership and three alternative revisit designs were explored ([1-0], [1-0,1-3], [1-3]) The three trend assessment approaches used were a linear mixed model that did not incorporate design weights (i.e., a "naive" analysis that ignored the sample structure); a linear regression approach applied to design-based estimates of finite population means, unweighted or weighted by the reciprocal of estimated sampling variance (a *secondary analysis*); and a probability-weighted (inverse of panel inclusion probabilities for site i in year j) iterative generalized least-squares (PWIGLS) approach applied to the observed $\{y_{ij}\}$ that explicitly recognized the sample structure.

Results of their Starcevich et al.'s (2018) simulations are qualitatively summarized in Starcevich et al. (2018 Figure 6) and suggest a very complex answer to the $64,000 question. First, the "naive" linear model analysis, which ignores the underlying sample structure, provided the most reliable assessment of trend in many of the simulated situations, including all settings for which there was no evidence of *subpopulation trends*, i.e., no evidence for different sets of population units having different trends over time. The naive approach also had the best performance in many settings for which subpopulation trends were evident. For "short" (12 year) monitoring periods, weighted or unweighted linear regressions of design-based estimates outperformed other assessment approaches when the year-to-year component of variation was *large*. Finally, the PWIGLS approach outperformed other methods only for the long (24 year) monitoring period with a low between year variance component and unequal probability selection. We note that these complex simulation results are not inconsistent with the far simpler setting considered in Chapter 8. Theoretically, if a survey had unequal inclusion probabilities, it would seem a priori necessary to incorporate these inclusion probabilities if sample data were used for unbiased or nearly unbiased inference. However, the generalized HT estimator of finite population variance, which explicitly incorporates these inclusion probabilities, is superior to the "naive" unbiased estimator of finite population variance under equal probability sampling only when the correlation between x and y is "high" ($\text{Cor}(x,y) > 0.4$) (Courbois and Urquhart 2004).

APPENDIX A

Mathematical foundations

Fig. A.1 When sampling theorists land a good trout, they engage in deep numerical thought. Illustration credit: Alex Nabaum.

As noted in the Introduction to this text, some statisticians (e.g., Royall 1970, Valliant et al. 2000) have objected to design-based sampling theory on the grounds that, unlike almost all other areas of statistics, posited statistical models or relationships between target and auxiliary variables are not central to the theory. In this appendix we present a number of critical mathematical relationships and fundamental principles of probability theory that provide the theoretical foundation for much of model-based statistics. These same mathematical relationships and probability theory principles also provide the theoretical foundation for design-based sampling theory. Because knowledge of the results presented in this appendix is required to fully appreciate sample selection methods, estimator properties, and estimator derivations, we reference pertinent sections of this appendix throughout the text. We encourage readers to study pertinent sections of this appendix as they are referenced.

An expanded version of this appendix, including proofs of all important results and some additional results and concepts, can be found at www.oup.co.uk/companion/hankin. There are a vast number of books on probability and statistics that you may wish to also consult for additional coverage and depth beyond this appendix. A couple of recent texts include Carlton and Devore (2017), which is written at a similar mathematical level as this appendix and also covers continuous probability distributions and random variables, and Rohatgi and Ehsanes Saleh (2015) which is written at a more advanced mathematical level.

A.1 Counting techniques

In this section we describe techniques for determining the number of different arrangements or combinations of items that are possible under various scenarios (e.g., sample selection methods). Knowledge of these techniques is required to understand the nature of discrete probability distributions and to determine the statistical properties of probability sampling designs.

The **multiplication principle** will be used frequently in our development of these techniques, and we define it here as follows. Suppose that a sequence of tasks $1, 2, \ldots, m$ are to be completed and there are $n_1, n_2, \ldots, n_m$ ways of completing these respective tasks. The total number of ways of completing the overall sequence of tasks is then $n_1 n_2 \cdots n_m$. For example, if there are three different travel routes from point A to point B, and two different travel routes from point B to point C, the total number of distinct travel routes from point A to B to C is $3 \cdot 2 = 6$.

A.1.1 *Permutations*

A permutation of a set of n items is one of the possible orderings of these items. A **permutation without repetition** is one of the possible ordered selections of n items from an original set of n items. Thus, two such permutations of a set of items are distinct if the items are arranged in a different order. For example, with a set of two items, $\{a_1, a_2\}$, the permutations (a_1, a_2) and (a_2, a_1) are distinct. By the multiplication principle, the total number of distinct permutations without repetition of a set of n items, $P(n)$, is

$$P(n) = n(n-1)(n-2) \cdots 2 \cdot 1 = n! \tag{A.1}$$

For example, for the set of two items $\{a_1, a_2\}$, the $P(2) = 2! = 2$ possible permutations without repetition are (a_1, a_2) and (a_2, a_1).

A **permutation with repetition** of a set of n items is one of the possible ordered selections with replacement of n items from the original set of n items, but with repetition of items allowed. Thus, two such permutations are distinct if the selected items include the same items a different number of times, or include the same items the same number of times but arranged in a different order. For example, with a set of two items, $\{a_1, a_2\}$, the permutations (a_1, a_1), (a_2, a_1), (a_1, a_2), and (a_2, a_2) are all distinct. By the multiplication principle, the total number of distinct permutations with repetition of a set of n items, $P'(n)$, is

$$P'(n) = n \cdot n \cdots n \ (n \text{ times}) = n^n \tag{A.2}$$

For example, for the set of two items $\{a_1, a_2\}$, the $P'(2) = 2^2 = 4$ possible permutations with repetition are $(a_1, a_1), (a_2, a_1), (a_1, a_2)$ and (a_2, a_2).

A.1.2 *k-Permutations*

A *k*-**permutation without repetition** is one of the possible ordered selections without replacement of k items taken from n items (repetition of items not allowed). Thus, two *k*-permutations without repetition are distinct if the selected items (a) do not include the

Table A.1 All possible k-permutations and combinations of $k = 2$ items taken from $n = 3$ items, $\{a_1, a_2, a_3\}$, with repetition (WR) and without repetition (WOR) as marked by $\times$. The k-permutations and combinations also represent all possible ordered and unordered samples, respectively, of $n = 2$ items selected from $N = 3$ items $\{a_1, a_2, a_3\}$ with replacement (WR) and without replacment (WOR).

	k-Permutations		Combinations	
Items	WR	WOR	WR	WOR
(a_1, a_1)	$\times$		$\times$	
(a_1, a_2)	$\times$	$\times$	$\times$	$\times$
(a_1, a_3)	$\times$	$\times$	$\times$	$\times$
(a_2, a_1)	$\times$	$\times$		
(a_2, a_2)	$\times$		$\times$	
(a_2, a_3)	$\times$	$\times$	$\times$	$\times$
(a_3, a_1)	$\times$	$\times$		
(a_3, a_2)	$\times$	$\times$		
(a_3, a_3)	$\times$		$\times$	

same items, or (b) include the same items arranged in a different order. For example, with a set of three items, $\{a_1, a_2, a_3\}$ and $k = 2$, the k-permutations (a_1, a_2), (a_1, a_3), and (a_3, a_1) are all distinct. By the multiplication principle, the total number of distinct k-permutations without repetition of a set of n items, $P(n, k)$, is

$$P(n, k) = n(n-1)(n-2)\cdots(n-k+1) = \frac{n!}{(n-k)!}, \quad k = 0, 1, 2, \ldots, n \quad \text{(A.3)}$$

For example, for the set of three items $\{a_1, a_2, a_3\}$ and $k = 2$, there are $P(3, 2) = 3!/(3-2)! = 6$ possible k-permutations without repetition (Table A.1).

Example A.1. *Number of possible ordered samples selected without replacement.* Suppose there are a total of N items of which n are selected without replacement. The total number of distinct, ordered samples is $P(N, n) = \frac{N!}{(N-n)!}$.

A **k-permutation with repetition** is one of the possible ordered selections with replacement of k items taken from n items (repetition of items allowed). Thus, two k-permutations with repetition are distinct if the selected items (a) do not include the same items, (b) include the same items but a different number of times, or (c) include the same items the same number of times but arranged in a different order. For example, with a set of four items, $\{a_1, a_2, a_3, a_4\}$ and $k = 3$, the k-permutations (a_1, a_2, a_3), (a_1, a_1, a_3), (a_1, a_3, a_3), and (a_3, a_1, a_3) are all distinct. By the multiplication principle, the total number of distinct k-permutations with repetition of a set of n items, $P'(n, k)$, is

$$P'(n, k) = n \cdot n \cdots n \ (k \text{ times}) = n^k, \quad k = 0, 1, 2, \ldots, n \quad \text{(A.4)}$$

For example, with a set of three items $\{a_1, a_2, a_3\}$ and $k = 2$, there are $P'(3, 2) = 3^2 = 9$ possible k-permutations with repetition. Table A.1 contrasts the k-permutations that are possible with and without repetition.

Example A.2. *Number of possible ordered samples selected with replacement.* Suppose there are a total of N items of which n are selected with replacement. The total number of distinct, ordered samples is $P'(N, n) = N^n$.

A.1.3 *Combinations*

A **combination without repetition** is one of the possible *unordered* without replacement selections of k items taken from n items. Two combinations without repetition are distinct if the selected items are not identical. For example, the combinations (a_1, a_2) and (a_1, a_3) are distinct, but (a_1, a_3) and (a_3, a_1) are not distinct. The total number of distinct combinations of k items without repetition from a set of n items, $C(n, k)$, is

$$C(n,k) = \frac{P(n,k)}{P(k)} = \frac{n!}{k!(n-k)!} = \binom{n}{k}, \quad k = 0, 1, 2, \ldots, n \tag{A.5}$$

For example, with a set of three items $\{a_1, a_2, a_3\}$ and $k = 2$, there are $C(3, 2) = \binom{3}{2} = 3$ possible combinations without repetition (Table A.1).

The $C(n, k)$ numerator, $P(n, k)$, gives the number of possible ordered selections without replacement of k items from n items, while the divisor, $P(k)$, accounts for the number of these selections that are reorderings of the same k items. The term $\binom{n}{k}$ is often referred to as "n choose k" and is also known as the *binomial coefficient* because it is the coefficient of the polynomial terms in the binomial theorem: $(x+y)^n = \sum_{k=0}^{n} \binom{n}{k} x^k y^{n-k}$.

Example A.3. *Number of possible unordered samples selected without replacement.* Suppose there are a total of N items of which n are selected without replacement. The total number of distinct, unordered samples is $C(N, n) = \binom{N}{n} = \frac{N!}{n!(N-n)!}$.

A **combination with repetition** is one of the possible *unordered* with replacement selections of k items taken from n items. Two combinations with repetition are distinct if the selected items (a) do not include the same items, or (b) include the same items a different number of times. For example, with a set of four items, $\{a_1, a_2, a_3, a_4\}$ and $k = 3$, the combinations (a_1, a_2, a_3), (a_1, a_1, a_3), and (a_1, a_3, a_3) are all distinct. The total number of distinct combinations of k items with repetition from a set of n items, $C'(n, k)$, is

$$C'(n,k) = \frac{(k+n-1)!}{k!(n-1)!} = \binom{k+n-1}{k}, \quad k = 0, 1, 2, \ldots, n \tag{A.6}$$

For example, with a set of three items $\{a_1, a_2, a_3\}$ and $k = 2$, there are $C'(3, 2) = \binom{2+3-1}{2} = 6$ possible combinations with repetition. Table A.1 contrasts the k-permutations and combinations that are possible with and without repetition.

Example A.4. *Number of possible unordered samples selected with replacement.* Suppose there are a total of N items of which n are selected with replacement. The total number of distinct, unordered samples is $C'(N, n) = \binom{n+N-1}{n} = \frac{(n+N-1)!}{n!(N-1)!}$.

A.1.4 *Partitions*

A **partition** is a subdivision of n items into m groups (unordered, without repetition) consisting of k_i items, $i = 1, 2, \ldots, m$, so that $\sum_{i=1}^{m} k_i = n$. Thus, two partitions are distinct if the grouped item sets are not identical. For example, two distinct partitions of the $n = 5$ items $\{a_1, a_2, a_3, a_4, a_5\}$ into $m = 3$ groups of sizes $k_1 = 2, k_2 = 1, k_3 = 2$ are $[\{a_1, a_2\}, \{a_3\}, \{a_4, a_5\}]$ and $[\{a_1, a_3\}, \{a_2\}, \{a_4, a_5\}]$. The total number of distinct partitions of n items into m groups of sizes $k_1, k_2, \ldots, k_m$, $C^*(n, k_1, k_2, \ldots, k_m)$, is

$$C^*(n, k_1, k_2, \ldots, k_m) = \frac{n!}{k_1! k_2! \cdots k_m!} = \binom{n}{k_1, k_2, \ldots, k_m}, \quad \sum_{i=1}^{m} k_i = n \tag{A.7}$$

Table A.2 Number of possible selections with repetition (WR) and without repetition (WOR) by arrangement type. In a sampling context, WR and WOR refer to with and without replacement, respectively, and the number of possible ordered and unordered samples are equivalent to the number of possible k-permutations and combinations, respectively.

		Number of possible selections	
Arrangement	Type	WOR	WR
ordered	permutation	$P(n) = n!$	$P'(n) = n^n$
	k-permutation	$P(n,k) = n!/(n-k)!$	$P'(n,k) = n^k$
unordered	combination	$C(n,k) = \binom{n}{k}$	$C'(n,k) = \binom{k+n-1}{k}$
	partition	$C^*(n,k_1,k_2,\ldots,k_m) = \binom{n}{k_1,k_2,\ldots,k_m}$	

For example, for the set of $n = 5$ items $\{a_1, a_2, a_3, a_4, a_5\}$ and $m = 3$ groups of sizes $k_1 = 2, k_2 = 2, k_3 = 2$, there are $C^*(5, 2, 1, 2) = \binom{5}{2,1,2} = 30$ possible partitions.

The number of distinct partitions can be derived as follows. The number of ways of forming the first group of k_1 items from the n items overall is equal to the total number of combinations without repetition, $C(n, k_1)$. Given a particular combination of items for group 1, the number of ways of forming the second group of k_2 items from the remaining $n - k_1$ items is $C(n - k_1, k_2)$, and this procedure is continued for the remaining groups. By the multiplication principle, the total number of distinct partitions is equal to the product of these step-wise combinations

$$\binom{n}{k_1}\binom{n-k_1}{k_2}\binom{n-k_1-k_2}{k_3}\cdots\binom{n-k_1-k_2-\cdots-k_{m-1}}{k_m} = \frac{n!}{k_1!k_2!\cdots k_m!} \quad (A.8)$$

The term $\binom{n}{k_1,k_2,\ldots,k_m}$ is sometimes referred to as the *multinomial coefficient* because it is the coefficient of the polynomial terms in the multinomial theorem: $(x_1 + x_2 + \cdots + x_m)^n = \sum \binom{n}{k_1,k_2,\ldots,k_m} x_1^{k_1} x_2^{k_2} \cdots x_m^{k_m}$, where the sum is over all possible non-negative integer sets $\{k_i\}$ such that $\sum_{i=1}^{m} k_i = n$. Note that for $m = 2$ with $k_1 + k_2 = n$, the multinomial coefficient $\binom{n}{k_1,k_2} = \frac{n!}{k_1!k_2!}$ reduces to the binomial coefficient $\binom{n}{k_1} = \frac{n!}{k_1!(n-k_1)!}$, and thus $C^*(n,k_1,k_2) = C(n,k_1)$.

Table A.2 summarizes the number of possible selections with and without repetition by arrangement type as presented in Section A.1.

A.2 Basic principles of probability theory

In this section we provide a brief introduction to the basic principles of probability theory, including random experiment, sample space, outcome probability, event, event relations, probability relations, and conditional probability. Presentation is limited to finite sample spaces. Probability sampling theory rests on these principles, and understanding them will allow for a deeper understanding of the theory and methods of probability sampling.

A.2.1 *Random experiment*

We define a **random experiment** as an experiment, procedure or process having a well-defined set of possible **outcomes**, the actual outcome of which is subject to chance and thus cannot be known with certainty prior to the experiment taking place. It is assumed that the experiment is, at least conceptually, infinitely repeatable under identical conditions so that the likelihood of the respective outcomes can be characterized in probabilistic terms.

Example A.5. *Random sample.* Consider selection of a random sample of n units from a finite population of N units using a well-defined, random selection procedure. The set of all possible samples using this procedure is well-defined, and the sample actually selected is subject to chance, thus meeting the definition of a random experiment.

A.2.2 Sample space

It is assumed that the possible outcomes of a random experiment are mutually exclusive, i.e., that the experiment can yield only one outcome at a time. The set of all possible outcomes for the experiment, $\mathcal{S}$, is termed the **sample space**, with an individual outcome or element of $\mathcal{S}$ being denoted by s. Performing the experiment results in a random outcome, S, which is equal to one of these possible outcomes, $S = s$, where $s \in \mathcal{S}$. Characterization of the sample space is essential for the definition and interpretation of the outcome probabilities in an experiment.

Example A.6. *Random sample (continuation).* The set of all possible samples of n units from a finite population of N units using a random selection procedure constitutes a sample space, $\mathcal{S}$. An outcome, $s \in \mathcal{S}$, is one of the possible samples of units.

A.2.3 Outcome probability

We define the **probability** of a random experiment outcome, $\Pr\{S = s\}, s \in \mathcal{S}$, in the frequentist sense as the limit of the long-term relative frequency of outcome s were the experiment to be repeated ad infinitum. Denoting the number of repeated experiments as r, and the number of those experiments resulting in outcome s as r_s,

$$\Pr\{S = s\} = \lim_{r \to \infty} \frac{r_s}{r}, \quad s \in \mathcal{S} \tag{A.9}$$

This definition of probability is consistent with the **law of large numbers** which states that a sample mean (here, the relative frequency of outcome s) taken over r repeated independent experiments converges to its expected value (here, the probability of outcome s) as $r \to \infty$.

To simplify notation, we introduce the probability function, $p(s)$, to denote the probability that $S = s$ occurs

$$p(s) = \Pr\{S = s\} \quad \text{for } s \in \mathcal{S} \tag{A.10}$$

From Equation (A.9), it is clear that the outcome probabilities are bounded by 0 and 1, and that their sum over the sample space is equal to 1, meaning that if the experiment is performed one of the outcomes in the sample space is certain to occur. Thus, for each value s, $p(s)$ specifies the probability that $S = s$ occurs, and considered over all possible s values, it defines the **probability distribution** of S, with

$$0 \le p(s) \le 1, \ s \in \mathcal{S} \quad \text{and} \quad \sum_{s \in \mathcal{S}} p(s) = 1 \tag{A.11}$$

For some experiments, all of the possible outcomes are, by construction, equally likely. In this case, $p(s) = 1/T, s \in S$, where T is the number of possible outcomes. For some other experiments, the probabilities of the possible outcomes are known, but they may be unequal. In some cases, the probabilities of the outcomes may be unknown.

Example A.7. *Random sample (continuation).* Suppose that the sample of n units is selected without replacement from the population of N units, and that every unit has an equal chance of being selected. The number of possible unordered samples without replacement is $C(N,n) = \binom{N}{n}$ (Section A.1.3). Each outcome (selected sample) is equally likely with probability

$$p(s) = \frac{1}{\binom{N}{n}}, \quad s \in S \tag{A.12}$$

For example, if $n = 5$ and $N = 10$, $C(10, 5) = \binom{10}{5} = 252$ and $p(s) = 1/252$.

A.2.4 *Event*

In addition to identifying the possible individual outcomes of a random experiment and their associated probabilities, it is useful to be able to refer collectively to various subsets of the possible outcomes and the associated probabilities of these subsets. Any defined subset of the sample space outcomes is termed an **event**. If an event consists of just a single outcome, it is termed an *elementary event*. Defined events are usually denoted by capital letters.

If the observed outcome of an experiment, $S = s$, is one of the outcomes of the event subset, then the event is said to occur, otherwise it is said not to occur. For example, if $S = \{s_1, s_2, s_3, s_4, s_5\}$, event $A = \{s_2, s_4\}$, and an experiment outcome is $S = s_2$ or $S = s_4$ then event A has occurred; if the outcome is $S = s_1$ or $S = s_3$ or $S = s_5$, then event A has not occurred.

Because the outcomes in the sample space are **disjoint** (non-overlapping), the probability of an event occurring is equal to the sum of the probabilities of the outcomes that together define the event. Thus, for an event A

$$\Pr\{A\} = \sum_{s \in A} p(s) \tag{A.13}$$

For the example given in the preceding paragraph, $\Pr\{A\} = p(s_2) + p(s_4)$.

A.2.5 *Event relations*

Outcomes may correspond to more than one defined event. Therefore, not all defined events are **mutually exclusive**. Different events may include some of the same outcomes, and if one of those outcomes occurs then different events may co-occur. To determine the outcomes associated with a new event that is defined by the union or intersection of two events, we can appeal to the following known algebraic relationships that exist between sets. The **union** (merging) of two events A and B, which we will express as "A or B" (in set algebraic notation, $A \cup B$), is the set of distinct outcomes among A and B. Thus, if the event "A or B" occurs, this implies that either A occurs or B occurs or A and B co-occur. The **intersection** of two events A and B, which we express as "A, B" (in set algebraic notation, $A \cap B$), includes only the outcomes that A and B have in common. Thus, if the event "A, B" occurs, this implies that A and B co-occur.

A.2.6 *Union and intersection*

The probability of the union of two events A and B follows directly from the probabilities of the individual events, taking care not to count the probability of their intersection twice

$$\Pr\{A \text{ or } B\} = \begin{cases} \Pr\{A\} + \Pr\{B\} - \Pr\{A, B\}, & \text{always} \\ \Pr\{A\} + \Pr\{B\}, & \text{if } A, B \text{ are mutually exclusive} \end{cases} \tag{A.14}$$

For the latter case, if A, B are mutually exclusive events (no outcomes in common, cannot co-occur), then $\Pr\{A, B\} = 0$ and $\Pr\{A \text{ or } B\}$ reduces to $\Pr\{A\} + \Pr\{B\}$.

The probability of the intersection of two events is given by

$$\Pr\{A, B\} = \begin{cases} \Pr\{A\} \Pr\{B|A\} = \Pr\{B\} \Pr\{A|B\}, & \text{always} \\ \Pr\{A\} \Pr\{B\}, & \text{if } A, B \text{ are independent} \end{cases} \tag{A.15}$$

where $\Pr\{B|A\}$ is the conditional probability (Section A.2.7) of B given that A occurred, and $\Pr\{A|B\}$ is similarly defined. If A, B are **independent**, then knowing that A occurred does not change the probability of whether B also occurred, and vice-versa, so that $\Pr\{B|A\} = \Pr\{B\}$ and $\Pr\{A|B\} = \Pr\{A\}$. Thus, for the latter case of Equation (A.15), if A, B are independent $\Pr\{A, B\}$ reduces to $\Pr\{A\} \Pr\{B\}$.

A.2.7 *Conditional probability*

Conditional probability refers to the probability of an event given that another event occurred. The conditional probability of event B given that event A occurred is denoted as $\Pr\{B|A\}$. It can be determined by first restricting the sample space to the subset of outcomes defining event A (since event A occurred), and then evaluating within that restricted space the probability of event B occurring (i.e., events A and B co-occurring). This process is sometimes referred to as **conditioning** on event A. Thus

$$\Pr\{B|A\} = \frac{\Pr\{A, B\}}{\Pr\{A\}} \tag{A.16}$$

is the probability of B normalized to this A-restricted sample subspace, assuming that $\Pr\{A\} \neq 0$. Note that, in general, $\Pr\{B|A\} \neq \Pr\{A|B\}$.

The conditional probability of B given A can also be thought of as the limit of the long-term relative frequency of event B given that event A occurred, were the experiment to be repeated ad infinitum. Denoting the number of repeated experiments as r, the number of those experiments in which event A occurs as r_A, and the number of those experiments in which events A and B co-occur as $r_{A,B}$, then

$$\Pr\{B|A\} = \lim_{r \to \infty} \frac{r_{A,B}}{r_A} = \frac{\lim_{r \to \infty} r_{A,B}/r}{\lim_{r \to \infty} r_A/r} = \frac{\Pr\{A, B\}}{\Pr\{A\}} \tag{A.17}$$

A.2.8 *Other probability relations*

There are a number of other probability relations that prove useful for developing the material presented in this text.

Law of total probability. Suppose that the sample space can be fully partitioned into a set of mutually exclusive events, denoted as $\{A_i\}$, such that $\sum_i \Pr\{A_i\} = 1$. The probability of an event B can then be expressed as

$$\Pr\{B\} = \sum_i \Pr\{A_i, B\} = \sum_i \Pr\{A_i\} \Pr\{B|A_i\} \tag{A.18}$$

This can prove useful in cases when determining the conditional probabilities $\Pr\{B|A_i\}$ and $\Pr\{A_i\}$ are more straightforward than determining $\Pr\{B\}$ directly.

Bayes' theorem. With conditional probabilities it is sometimes easier to determine $\Pr\{A|B\}$ than $\Pr\{B|A\}$, and as long as the associated probabilities $\Pr\{A\}$ and $\Pr\{B\}$ can be specified, then $\Pr\{B|A\}$ can be formulated in terms of $\Pr\{A|B\}$. Note that $\Pr\{A\}\Pr\{B|A\} = \Pr\{A, B\} = \Pr\{B\}\Pr\{A|B\}$, and therefore

$$\Pr\{B|A\} = \frac{\Pr\{B\}\Pr\{A|B\}}{\Pr\{A\}} \tag{A.19}$$

Thus, $\Pr\{B|A\}$ can be found from $\Pr\{A|B\}$, $\Pr\{A\}$, and $\Pr\{B\}$.

A.3 Discrete random variables

Discrete random variables are central to probability sampling, and in this section we discuss their definition, probability distributions, probability relations, and summary measures over a finite sample space. Many of the relations described in Section A.3 apply also to continuous random variables, but their proofs rely on integration rather than summation.

A.3.1 *Definition*

A real-valued **random variable** is a function that associates a numeric value to each of the possible outcomes of a random experiment, which may or may not be numeric themselves. It is thus a *mapping* from the sample space to the real number line, $\mathcal{R}$. More generally, a vector random variable, or **multivariate random variable**, of length d is a mapping from the sample space to d-dimensional real space, $\mathcal{R}^d$. Each time the experiment is performed, an outcome occurs and the random variable takes on the numeric value associated with that outcome. It is "variable" in the sense that its numeric value varies over the sample space, and it is "random" in the sense that its value depends on the outcome of a random experiment. For each outcome the random variable has a single value, but these values are not necessarily unique over the sample space. Several random variables of interest may be defined for a random experiment.

We use capital letters toward the end of the alphabet to denote random variables, and their corresponding lower case letter to symbolize a particular, but unspecified, value that the random variable takes on. Thus, $X(s) = x$ states "random variable X for realized outcome $S = s$ is equal to x", where x is some particular number (e.g., $x = 3$ or $x = 7.1$). We will also be interested in various events based on the value of X. For example, the event $X = x$ occurs for that subset of outcomes for which $X(s) = x$, $\{s \in \mathcal{S} : X(s) = x\}$.

Example A.8. *Coin flipping.* Suppose that the random experiment is to toss a balanced (fair) coin twice to observe which side lands face up on each occassion, Head (H) or Tail (T). The sample space consists of all possible (joint) outcomes: $\mathcal{S} = \{(H,H), (H,T), (T,H), (T,T)\}$. Let X equal the number of times H lands face up, a random variable. The possible values of $X(s), s \in \mathcal{S}$, are 0, 1, or 2, and there are two outcomes in $\mathcal{S}$ for which $X(s) = 1$: $s = (H,T)$ and $s = (T,H)$.

A.3.2 *Probability distributions*

In this section we provide the general definitions and notation for probability distributions of *discrete* univariate, multivariate, conditional, and marginal random variables. Discrete random variables have a finite number of possible values, as opposed to continuous random variables, which can take on any value over a continuum of values which may or may not be finite in their extent. For specific discrete random variables, various distributional properties such as the expectation and variance can be determined as described in the sections following. Certain key discrete probability distributions and their properties are reviewed in Section A.4.

For a discrete random variable X, the outcomes for which $X = x$ constitute a subset of the sample space. Therefore, by definition, $X = x$ is an event and its probability of occurrence is

$$\Pr\{X = x\} = \sum_{\substack{s \in \mathcal{S} \\ X(s) = x}} p(s) \tag{A.20}$$

where the summation is over $s \in \mathcal{S}$ for which $X(s) = x$. To simplify notation, we introduce the probability function, $p(x)$, to give the probability that $X = x$ occurs

$$p(x) = \Pr\{X = x\} \quad \text{for all } x \tag{A.21}$$

and refer to $p(x)$ as the **probability mass function (pmf)** of X. Because events are subsets of the sample space, it is clear that their probabilities are bounded by 0 and 1, and because they form a partition of the sample space their probabilities sum to 1, meaning that if the experiment is performed some value of X is certain to occur. Thus, for each value x, $p(x)$ specifies the probability that $X = x$ occurs, and considered over all possible x values, it defines the **probability distribution** of X, with

$$0 \le p(x) \le 1, \text{ for all } x \quad \text{and} \quad \sum_x p(x) = 1 \tag{A.22}$$

Example A.9. *Coin flipping (continuation).* Referring to the previous example of two tosses of a fair coin, where the four possible outcomes are equally likely and X is the number times a Head lands face up: $p(0) = 1/4$, $p(1) = 1/2$, $p(2) = 1/4$, and $\sum_{x=0}^{2} p(x) = 1$.

For a **multivariate discrete random variable** (vector), $\mathbf{X} = (X_1, X_2, \dots, X_m)$, the pmf is the joint probability function

$$\Pr\{X_1 = x_1, X_2 = x_2, \dots, X_m = x_m\} = p(x_1, x_2, \dots, x_m) \tag{A.23}$$

where, as for $p(x)$, for all $\mathbf{x} = (x_1, x_2, \dots, x_m)$

$$0 \le p(x_1, x_2, \dots, x_m) \le 1 \quad \text{and} \quad \sum_{x_1} \sum_{x_2} \cdots \sum_{x_m} p(x_1, x_2, \dots, x_m) = 1 \tag{A.24}$$

For a **conditional discrete random variable**, for example, $X_1 | X_2, X_3, \dots, X_m$

$$\Pr\{X_1 = x_1 | X_2 = x_2, X_3 = x_3, \dots, X_m = x_m\} = p(x_1 | x_2, x_3, \dots, x_m) \tag{A.25}$$

The pmf is defined, over all possible values of x_1, as the joint probability of all the variables divided by the joint probability of the "conditioned on" variables

$$p(x_1 | x_2, x_3, \dots, x_m) = \frac{p(x_1, x_2, x_3, \dots, x_m)}{p(x_2, x_3, \dots, x_m)} \tag{A.26}$$

The pmf of a **marginal discrete random variable** is defined as the sum of the joint probability of all the variables over all possible values of the *other* variables. For example, the marginal pmf for X_1 is defined as

$$p(x_1) = \sum_{x_2} \sum_{x_3} \cdots \sum_{x_m} p(x_1, x_2, x_3, \dots, x_m) \tag{A.27}$$

for all possible values of x_1. As for all probability distributions of discrete random variables, the event probabilities of conditional and marginal random variables are bounded by 0 and 1, and the sum of the probabilities over all events is equal to 1.

Example A.10. *Bivariate discrete random variable.* For a bivariate discrete random variable (X, Y) with joint pmf $p(x, y)$

$$0 \le p(x, y) \le 1 \quad \text{and} \quad \sum_x \sum_y p(x, y) = 1$$

the conditional pmfs are

$$p(y|x) = p(x, y)/p(x) \quad \text{and} \quad p(x|y) = p(x, y)/p(y)$$

and the marginal pmfs are

$$p(y) = \sum_x p(x, y) \quad \text{and} \quad p(x) = \sum_y p(x, y)$$

A.3.3 *Probability relations*

The probability relations previously outlined for events in Sections A.2.4–A.2.8 necessarily apply to discrete random variable events as well, including their pmfs. Thus, letting $p(x \text{ or } y)$ denote $\Pr\{X = x \text{ or } Y = y\}$

$$p(x \text{ or } y) = \begin{cases} p(x) + p(y) - p(x, y) \\ p(x) + p(y), & \text{if } X, Y \text{ independent} \end{cases} \tag{A.28}$$

$$p(x, y) = \begin{cases} p(x)p(y|x) = p(y)p(x|y) \\ p(x)p(y), & \text{if } X, Y \text{ independent} \end{cases} \tag{A.29}$$

$$p(y) = \sum_x p(x)p(y|x) \tag{A.30}$$

$$p(y|x) = \frac{p(x, y)}{p(x)} = \frac{p(y)p(x|y)}{p(x)} = \frac{p(y)p(x|y)}{\sum_y p(y)p(x|y)} \tag{A.31}$$

$$p(x_1, x_2, \ldots, x_m) = \begin{cases} p(x_1)p(x_2|x_1)p(x_3|x_1, x_2) \cdots p(x_m|x_1, x_2, \ldots, x_{m-1}) \\ p(x_1)p(x_2) \cdots p(x_m), & \text{if } \{X_i\} \text{ mutually independent} \end{cases} \tag{A.32}$$

A.3.4 *Expectation*

In this and following sections, we provide proofs for some of the most fundamental results when proofs are brief, and otherwise present results without proof. In Example A.11 we provide a small sample space illustration of direct calculation of expectation and variance for a binomial random variable.

The **expectation** (**expected value**) of a discrete random variable X, denoted as $E(X)$, is defined as the sum of the possible x values that X takes on, each weighted by its probability of occurrence $p(x)$

$$E(X) = \sum_x x p(x) \tag{A.33}$$

The expectation is thus a measure of central tendency (a weighted average) of X. Alternatively, the expectation may be directly determined over the sample space as the sum of the individual $X(s)$ values weighted by their associated outcome probabilities $p(s)$.

The sample space formulation of expectation is often used in the development of sampling theory.

$$E(X) = \sum_{s \in \mathcal{S}} X(s)p(s) \tag{A.34}$$

Proof. Partition the sample space into outcomes resulting in the same value $X = x$.

$$E(X) = \sum_{x} xp(x) = \sum_{x} x \sum_{\substack{s \in \mathcal{S} \\ X(s) = x}} p(s) = \sum_{x} \sum_{\substack{s \in \mathcal{S} \\ X(s) = x}} X(s)p(s) = \sum_{s \in \mathcal{S}} X(s)p(s)$$

■

Similar direct sample space analogs are possible for all of the expectation formulas presented below.

For two discrete random variables X, Y, the **conditional expectation** of $Y|X = x$ is defined as

$$E(Y|X = x) = \sum_{y} yp(y|x) \tag{A.35}$$

and the expectation of a function of a discrete random variable, $g(X)$, is defined as

$$E[g(X)] = \sum_{x} g(x)p(x) \tag{A.36}$$

For a constant c, if $g(X) = c$ or $g(X) = cX$, then

$$E(c) = c \tag{A.37}$$

$$E(cX) = cE(X) \tag{A.38}$$

For two discrete random variables X, Y

$$E(X + Y) = E(X) + E(Y) \tag{A.39}$$

Proof.

$$E(X + Y) = \sum_{x} \sum_{y} (x + y)p(x,y) = \sum_{x} \sum_{y} xp(x,y) + \sum_{y} \sum_{x} yp(x,y)$$

$$= \sum_{x} x \sum_{y} p(x,y) + \sum_{y} y \sum_{x} p(x,y) = \sum_{x} xp(x) + \sum_{y} yp(y) = E(X) + E(Y)$$

■

More generally, for a set of constants $\{c_i\}$ and discrete random variables $\{X_i\}$

$$E\left(\sum_{i} c_i X_i\right) = \sum_{i} c_i E(X_i) \tag{A.40}$$

This property is known as the **linearity of expectations**; that the expectation of a linear combination of discrete random variables is the linear combination of the random variable expectations.

For a product of two discrete random variables, XY, the definition of covariance, $\text{Cov}(X, Y) = E(XY) - E(X)E(Y)$ (Section A.3.6), can be rearranged to give

$$E(XY) = \begin{cases} E(X)E(Y) + \text{Cov}(X, Y) \\ E(X)E(Y), & \text{if } X, Y \text{ independent} \end{cases} \tag{A.41}$$

Proof. The latter result can be proven directly as follows. If X and Y are independent, $p(x,y) = p(x)p(y)$ [Equation (A.29)] so that

$$E(XY) = \sum_x \sum_y xyp(x,y) = \sum_x \sum_y xyp(x)p(y) = \left[\sum_x xp(x)\right]\left[\sum_y yp(y)\right] = E(X)E(Y)$$

∎

More generally,

$$E\left(\prod_i X_i\right) = \prod_i E(X_i), \quad \text{if } \{X_i\} \text{ mutually independent} \tag{A.42}$$

A.3.5 *Variance and coefficient of variation*

The **variance** of a discrete random variable X, denoted as $V(X)$, is defined as the expected value of the squared deviation of X from $E(X)$. That is, the sum of the possible $[x - E(X)]^2$ values weighted by their probabilities of occurrence, $p(x)$

$$V(X) = E\left([X - E(X)]^2\right) = \sum_x [x - E(X)]^2 p(x) \tag{A.43}$$

The variance is thus a measure of the average variability in X about its expectation. Because the deviations are squared, $V(X) \geq 0$. Alternatively, the variance may be directly determined over the sample space as the sum of the individual $[X(s) - E(X)]^2$ values weighted by the corresponding outcome probabilities $p(s)$

$$V(X) = \sum_{s \in S} [X(s) - E(X)]^2 p(s) \tag{A.44}$$

Equation (A.44) sees use throughout sampling theory and similar direct sample space analogs are possible for all of the variance formulas presented below.

The variance of a discrete random variable X can also be expressed as

$$V(X) = E(X^2) - [E(X)]^2 \tag{A.45}$$

Proof.

$$V(X) = E\left([X - E(X)]^2\right) = E\left(X^2 - 2XE(X) + [E(X)]^2\right)$$
$$= E(X^2) - 2[E(X)]^2 + [E(X)]^2 = E(X^2) - [E(X)]^2$$

∎

For two discrete random variables X, Y, the **conditional variance** of $Y|X = x$ is defined as

$$V(Y|X = x) = E\left([Y - E(Y|X = x)]^2|X = x\right) = \sum_y [y - E(Y|X = x)]^2 p(y|x) \tag{A.46}$$

which can also be expressed as

$$V(Y|X = x) = E(Y^2|X = x) - [E(Y|X = x)]^2 \tag{A.47}$$

The variance of a function of a discrete random variable, $g(X)$, is defined as

$$V[g(X)] = E\left[(g(X) - E[g(X)])^2\right] = \sum_x (g(x) - E[g(X)])^2 p(x) \tag{A.48}$$

which can also be expressed as

$$V[g(X)] = E\left([g(X)]^2\right) - (E[g(X)])^2 \tag{A.49}$$

For a constant c, if $g(X) = c$, $g(X) = c + X$, or $g(X) = cX$, then

$$V(c) = 0 \tag{A.50}$$
$$V(c + X) = V(X) \tag{A.51}$$
$$V(cX) = c^2 V(X) \tag{A.52}$$

For two discrete random variables X, Y

$$V(X + Y) = V(X) + V(Y) + 2\text{Cov}(X, Y) \tag{A.53}$$
$$V(X - Y) = V(X) + V(Y) - 2\text{Cov}(X, Y) \tag{A.54}$$

Proof. Proof for $V(X + Y)$ (proof for $V(X - Y)$ is directly analogous):

$$V(X + Y) = E\left[(X + Y)^2\right] - [E(X + Y)]^2 = E\left[X^2 + 2XY + Y^2\right] - [E(X) + E(Y)]^2$$

$$= \left[E(X^2) + 2E(XY) + E(Y^2)\right] - \left[[E(X)]^2 + 2E(X)E(Y) + [E(Y)]^2\right]$$

$$= \left[E(X^2) - [E(X)]^2\right] + \left[E(Y^2) - [E(Y)]^2\right] + 2\left[E(XY) - E(X)E(Y)\right]$$

$$= V(X) + V(Y) + 2\text{Cov}(X, Y)$$

∎

More generally, for a set of constants $\{c_i\}$ and discrete random variables $\{X_i\}$

$$V\left(\sum_i c_i X_i\right) = \sum_i \sum_j c_i c_j \text{Cov}(X_i, X_j) \tag{A.55}$$

$$= \begin{cases} \sum_i c_i^2 V(X_i) + \sum_i \sum_{j \neq i} c_i c_j \text{Cov}(X_i, X_j) \\ \sum_i c_i^2 V(X_i), & \text{if } \{X_i\} \text{ mutually independent} \end{cases} \tag{A.56}$$

since if $\{X_i\}$ are mutually independent, $\text{Cov}(X_i, X_j) = 0$ for $i \neq j$. In particular,

$$V(X + Y) = V(X - Y) = V(X) + V(Y), \quad \text{if } X, Y \text{ independent} \tag{A.57}$$

For a product of two discrete random variables, XY

$$V(XY) = \begin{cases} (V(X) + [E(X)]^2)(V(Y) + [E(Y)]^2) + \text{Cov}(X^2, Y^2) - [\text{Cov}(X, Y) + E(X)E(Y)]^2 \\ V(X)V(Y) + [E(Y)]^2 V(X) + [E(X)]^2 V(Y), & \text{if } X, Y \text{ independent} \end{cases} \tag{A.58}$$

More generally, for independent discrete random variables

$$V\left(\prod_i X_i\right) = \prod_i (V(X_i) + [E(X_i)]^2) - \prod_i [E(X_i)]^2, \quad \text{if } \{X_i\} \text{ mutually independent} \tag{A.59}$$

The **coefficient of variation** of a discrete random variable X, denoted by $\text{CV}(X)$, is defined as

$$\text{CV}(X) = \frac{\sqrt{V(X)}}{E(X)} \tag{A.60}$$

and is often reported as a percentage. Note that CV(X) is not defined in the special case where $E(X) = 0$, nor is it commonly used in the case where $E(X) < 0$. The numerator, $\sqrt{V(X)}$, termed the **standard deviation** of X, has the same units as X. The coefficient of variation can be a more meaningful quantity than the variance in that it is a dimensionless measure of the variation in X *relative to* the expectation of X. For example, suppose $V(X) = 625$ and $E(X) = 100$. By itself, $V(X) = 625$ may not mean much, whereas CV(X) = $\sqrt{625}/100 = 25\%$ means that one standard deviation of X is equal to 25% of its expectation, whereas if $E(X) = 25$ then $V(X) = 625$ would equate to a CV(X) of 100%; quite a different level of variation relative to the expectation.

A.3.6 *Covariance and correlation*

The **covariance** of two discrete random variables X and Y, denoted by Cov(X, Y), is defined as the expected value of the product of deviations $[X - E(X)][Y - E(Y)]$. That is, the sum of the possible $[x - E(X)][y - E(Y)]$ values weighted by their corresponding probabilities of occurrence, $p(x, y)$

$$\text{Cov}(X, Y) = E\big([X - E(X)][Y - E(Y)]\big) = \sum_x \sum_y [x - E(X)][y - E(Y)]p(x, y) \tag{A.61}$$

For a given (x, y) value, if both deviations are positive (or negative) their product is positive, whereas if the deviations differ in sign their product is negative. Covariance is a measure of the degree to which X and Y linearly covary, and may be positive valued, negative valued, or equal to zero. Alternatively, the covariance may be directly determined over the sample space as the sum of the $[X(s) - E(X)][Y(s) - E(Y)]$ values weighted by the outcome probabilities $p(s)$

$$\text{Cov(X,Y)} = \sum_{s \in \mathcal{S}} [X(s) - E(X)][Y(s) - E(Y)]p(s) \tag{A.62}$$

Similar direct sample space analogs are possible for all of the covariance formulas presented in this section.

The covariance of two discrete random variables X and Y can also be expressed as

$$\text{Cov}(X, Y) = E(XY) - E(X)E(Y) \tag{A.63}$$

Proof.

$$\text{Cov}(X, Y) = E\big([X - E(X)][Y - E(Y)]\big) = E\,[XY - XE(Y) - E(X)Y + E(X)E(Y)]$$
$$= E(XY) - 2E(X)E(Y) + E(X)E(Y) = E(XY) - E(X)E(Y)$$

∎

It is clear from Equations (A.63), (A.45), and (A.41) that

$$\text{Cov}(X, X) = V(X) \tag{A.64}$$
$$\text{Cov}(X, Y) = \text{Cov}(Y, X) \tag{A.65}$$
$$\text{Cov}(X, Y) = 0, \quad \text{if } X, Y \text{ are independent} \tag{A.66}$$

For constants c and d

$$\text{Cov}(c, X) = 0 \tag{A.67}$$
$$\text{Cov}(c + X, d + Y) = \text{Cov}(X, Y) \tag{A.68}$$
$$\text{Cov}(cX, dY) = cd\text{Cov}(X, Y) \tag{A.69}$$

In general, for two linear combinations of random variables $\sum_i a_i X_i$ and $\sum_j b_j Y_j$

$$\text{Cov}\left(\sum_i a_i X_i, \sum_j b_j Y_j\right) = \sum_i \sum_j a_i b_j \text{Cov}(X_i, Y_j) \tag{A.70}$$

For the covariance of products of random variables see Bohrnstedt and Goldberger (1969) and Gray (1999).

The **correlation** of two random variables X and Y, denoted by $\text{Cor}(X, Y)$, is defined as

$$\text{Cor}(X, Y) = \frac{\text{Cov}(X, Y)}{\sqrt{V(X)V(Y)}}, \quad -1 \leq \text{Cor}(X, Y) \leq 1 \tag{A.71}$$

Correlation scales the covariance to a dimensionless quantity bounded by -1 and +1, and for this reason is more readily interpretable than the covariance itself. For example, if X and Y are the total weight and the dressed weight, respectively, of harvested yellowfin tuna, then their covariance when the unit of weight is pounds will differ from their covariance when the unit of weight is kilograms.[1] Their correlation will be identical in both cases, however. Correlation measures the *linear* association between two random variables, and the closer it is to -1 (negative association) or 1 (positive association) the more highly correlated the variables are. For independent random variables $\text{Cor}(X, Y) = 0$, but the converse is not true in general: if $\text{Cor}(X, Y) = 0$ the random variables are not necessarily independent. When random variables are perfectly correlated ($|\text{Cor}(X, Y)| = 1$), knowledge of a value of X allows perfect prediction of the value of Y and vice versa.

A.3.7 *Total expectation and variance*

When deriving the expectation or variance of a discrete random variable, it is sometimes easier to do so by first conditioning on another random event, and then using the law of total probability [Equation (A.18)] to expand this over the set of possible events. This technique sees effective use in multi-stage sampling (Chapter 9).

Total expectation

Suppose that the sample space can be fully partitioned into a set of mutually exclusive events, $\{A_i\}$, such that $\sum_i \text{Pr}\{A_i\} = 1$. Let A denote the corresponding random event, and let Y denote the discrete random variable of interest. The **law of total expectation** provides that

$$E(Y) = E_A[E(Y|A)] \tag{A.72}$$

where $E(Y|A)$ is the conditional expectation of Y given A, and $E_A(\cdot)$ denotes expectation with respect to $\text{Pr}\{A\}$. Note that $E(Y|A)$ is a random variable as its value depends on the random event A, and that it therefore has an expectation. In contrast, $E(Y|A = A_i)$ can be viewed as a function of A_i, but given $A = A_i$ it is a single numeric value and is not a random variable. Thus, the overall expectation of Y is equal to the weighted (by $\text{Pr}\{A\}$) average of its conditional expectations.

---

[1] 1 kg ≈ 2.2 lb. Therefore, by Equation (A.69), $\text{Cov}(X_{\text{lb}}, Y_{\text{lb}}) \approx (2.2)^2 \cdot \text{Cov}(X_{\text{kg}}, Y_{\text{kg}})$.

Proof.

$$E(Y) = \sum_y y \cdot p(y) \qquad\qquad \text{definition of expectation}$$

$$= \sum_y y \left[\sum_i p(y|A_i) \Pr\{A = A_i\} \right] \qquad \text{law of total probability}$$

$$= \sum_y \sum_i y \cdot p(y|A_i) \Pr\{A = A_i\} \qquad \text{reposition } y$$

$$= \sum_i \left[\sum_y y \cdot p(y|A_i) \right] \Pr\{A = A_i\} \qquad \text{switch summation order, group}$$

$$= \sum_i E(Y|A = A_i) \Pr\{A = A_i\} \qquad \text{definition of conditional expectation}$$

$$= E_A[E(Y|A)] \qquad\qquad \text{definition of expectation}$$

∎

Often the random event A is a random variable, say X, in which case

$$E(Y) = E_X[E(Y|X)] \tag{A.73}$$

Total variance

The law of total variance expresses the variance of a random variable in terms of its conditional expectations and variances given another random event. Suppose that the sample space can be fully partitioned into a set of mutually exclusive events, $\{A_i\}$, such that $\sum_i \Pr\{A_i\} = 1$. Let A denote the corresponding random event, and let Y denote the discrete random variable of interest. The **law of total variance** provides that

$$V(Y) = E_A[V(Y|A)] + V_A[E(Y|A)] \tag{A.74}$$

where $E(Y|A)$ and $V(Y|A)$ are the conditional expectation and variance of Y given A, respectfully, and $E_A(\cdot)$ and $V_A(\cdot)$ are the expectation and variance with respect to $\Pr\{A\}$, respectfully. Note that $V(Y|A)$ and $E(Y|A)$ are random variables as their values depend on the random variable A, and that they therefore have an expectation and variance. Given $A = A_i$, however, $V(Y|A = A_i)$ and $E(Y|A = A_i)$ have single numerical values and are not random variables. Thus, the overall variance of Y is equal to a weighted (by $\Pr\{A\}$) average of its conditional variances plus the variance of its conditional expectations. This is analogous to ANOVA, where the overall variation can be decomposed into within-group variation plus the variation of the between-group means.

If the random event A is a discrete random variable, say X, then

$$V(Y) = E_X[V(Y|X)] + V_X[E(Y|X)] \tag{A.75}$$

The concepts of total expectation and total variance see powerful applications in multistage sampling. Further details concerning these principles are presented in Chapter 9.

A.3.8 *Indicator variables*

Indicator variables often offer a convenient means for finding the expectation and variances of other, more complicated discrete random variables (as demonstrated particularly in the expanded version of Section A.4 which can be found at www.oup.co.uk/

companion/hankin). An **indicator variable** is a discrete random variable that indicates whether or not an event occurred, and is an example of a Bernoulli random variable (Section A.4.2). The indicator variable for an event A is defined as

$$I_A(s) = \begin{cases} 1, & s \in A \\ 0, & \text{otherwise} \end{cases} \tag{A.76}$$

This simple $(0, 1)$ binary structure implies that the following relationships exist between indicator variables for events A and B

$$I_A^2 = I_A \tag{A.77}$$

$$I_{A,B} = I_A I_B \tag{A.78}$$

$$I_{A \text{ or } B} = I_A + I_B - I_A I_B \tag{A.79}$$

The probability distribution for an indicator random variable is also simple. For an event A

$$\Pr\{I_A = x\} = \begin{cases} \sum_{s \in A} p(s) = \Pr\{A\}, & x = 1 \\ \sum_{s \notin A} p(s) = 1 - \Pr\{A\}, & x = 0 \end{cases} \tag{A.80}$$

As a consequence, the expectation, variance, and covariance of indicator random variables are simple functions of the event probabilities. For events A and B

$$E(I_A) = \Pr\{A\} \tag{A.81}$$

$$V(I_A) = \Pr\{A\}(1 - \Pr\{A\}) \tag{A.82}$$

$$\text{Cov}(I_A, I_B) = \Pr\{A, B\} - \Pr\{A\}\Pr\{B\} \tag{A.83}$$

Proof.

$$E(I_A) = \sum_{x=0}^{1} x \Pr\{I_A = x\} = 0 \cdot \Pr\{I_A = 0\} + 1 \cdot \Pr\{I_A = 1\} = \Pr\{A\}$$

$$V(I_A) = E(I_A^2) - [E(I_A)]^2 = E(I_A) - [E(I_A)]^2 = \Pr\{A\}(1 - \Pr\{A\})$$

$$\text{Cov}(I_A, I_B) = E(I_A I_B) - E(I_A)E(I_B) = E(I_{A,B}) - E(I_A)E(I_B) = \Pr\{A, B\} - \Pr\{A\}\Pr\{B\}$$

■

Equations (A.81), (A.82), and (A.83) apply equally to events involving discrete random variables. For example, for the event $X = x$, the indicator variable can be written as $I_{X=x}$, where $I_{X=x} = 1$ occurs with probability $p(x)$, and $I_{X=x} = 0$ occurs with probability $1 - p(x)$. As a result, $E(I_{X=x}) = p(x)$ and $V(I_{X=x}) = p(x)[1 - p(x)]$.

A.4 Key discrete probability distributions

Probability distributions for key discrete random variables are presented in this section, along with their respective expectation, variance, and covariance, as appropriate. The notation "$X \sim$" used below is short for "X is distributed as".

A.4.1 *Uniform*

A **uniform distribution** is the probability distribution of a random variable whose possible values are all equally likely. The random variable, X, is often defined to take on the consecutive integer values from a to b (a total of $b - a + 1$ values). In this case the pmf of X is

$$p(x) = 1/(b - a + 1), \quad x = a, a+1, \ldots, b-1, b \tag{A.84}$$

This specification is often written as $X \sim \text{uniform}(a,b)$, and X is said to be a *uniform random variable* or *uniformly distributed*. The expectation and variance of X are given by

$$E(X) = (a+b)/2 \tag{A.85}$$

$$V(X) = ([b-a+1]^2 - 1)/12 \tag{A.86}$$

There is a similarly defined uniform distribution for a continuous random variable, which is also typically written as $X \sim \text{uniform}(a,b)$ for distribution endpoints a and b. For a continuous uniform random variable

$$E(X) = (a+b)/2 \tag{A.87}$$

$$V(X) = (b-a)^2/12 \tag{A.88}$$

A.4.2 Bernoulli

A **Bernoulli distribution** is the probability distribution of an indicator random variable (Section A.3.8). The discrete random variable, X, takes on only two possible values: $X=1$ if the event of interest occurs, and $X=0$ otherwise. Letting $\Pr\{X=1\}=q$ and $\Pr\{X=0\}=1-q$, the pmf of X can be expressed as

$$p(x) = \begin{cases} q, & \text{if } x=1 \\ 1-q, & \text{if } x=0 \end{cases} \tag{A.89}$$

or as

$$p(x) = q^x(1-q)^{1-x}, \quad x=0,1 \tag{A.90}$$

This specification is often written as $X \sim \text{bernoulli}(q)$, and X is said to be a *Bernoulli random variable* or *Bernoulli distributed*, and the corresponding experiment is sometimes referred to as a (single) *Bernoulli trial*. Because X is an indicator random variable, the expectation and variance of X are given by (Section A.3.8)

$$E(X) = q \tag{A.91}$$

$$V(X) = q(1-q) \tag{A.92}$$

A.4.3 Multinoulli

A **multinoulli distribution** is a generalization of the Bernoulli distribution to characterize multiple events of interest. It is the probability distribution of a multivariate indicator random variable. This discrete random variable is vector-valued, $\mathbf{X} = (X_1, X_2, \ldots, X_m)$, where $X_i=1$ if event i occurs and $X_i=0$ otherwise, and the m events are mutually exclusive and fully partition the sample space. Thus, $\mathbf{X}$ is a vector of 0's except for a 1 in the position corresponding to the event that occurs. Letting $\Pr\{X_i=1\}=q_i, i=1,2,\ldots,m$ with $\sum_{i=1}^m q_i = 1$, the pmf of $\mathbf{X}$ is

$$p(\mathbf{x}) = p(x_1, x_2, \ldots, x_m) = q_1^{x_1} q_2^{x_2}, \cdots q_m^{x_m}, \quad x_i=0,1; \quad \sum_{i=1}^m x_i = 1 \tag{A.93}$$

This specification is often written as $\mathbf{X} \sim \text{multinoulli}(q_1, q_2, \ldots, q_m)$, and $\mathbf{X}$ is said to be a *multinoulli random variable*, and the corresponding experiment is sometimes referred to as a (single) *multinoulli trial*. Because X_i is also an indicator random variable for event i, the expectation and variance of X_i, and the covariance of X_i and X_j for $i \neq j$ are given by

$$E(X_i) = q_i \tag{A.94}$$

$$V(X_i) = q_i(1-q_i) \tag{A.95}$$

$$\text{Cov}(X_i, X_j) = -q_i q_j, \quad i \neq j \tag{A.96}$$

Note that X_i and X_j cannot both be equal to 1 for $i \neq j$, so that $\text{Pr}\{X_i = 1, X_j = 1\} = 0$ for $i \neq j$ (Section A.3.8).

A.4.4 Binomial

A **binomial distribution** is the probability distribution of a discrete random variable, X, that is equal to the number of times the event of interest occurs in a sequence of n *independent* bernoulli(q) trials. Thus, for independent $Y_k \sim$ bernoulli(q), $k = 1, 2, \ldots, n$, the random variable $X = \sum_{k=1}^{n} Y_k$ takes on possible values $0, 1, \ldots, n$, and the pmf of X is

$$p(x) = \binom{n}{x} q^x (1-q)^{n-x}, \quad x = 0, 1, \ldots, n \tag{A.97}$$

Proof. $q^x (1-q)^{n-x}$ is the probability of n Bernoulli trials resulting in a *particular* sequence of outcomes consisting of x events of interest (e.g., event of interest occurs in first x trials, and does not occur in following $n - x$ trials). There are $\binom{n}{x}$ such distinct sequences (combinations) possible that generate x events of interest (Section A.1.3) and each of these distinct sequences has the same probability $q^x (1-q)^{n-x}$. Note that the sum of the pmf probabilities equals 1, as required for a proper random variable: $\sum_{x=0}^{n} p(x) = \sum_{x=0}^{n} \binom{n}{x} q^x (1-q)^{n-x} = [q + (1-q)]^n = 1$ (binomial theorem; see Section A.1.3).   ∎

This specification is often written as $X \sim$ binomial($n; q$), and X is said to be a *binomial random variable* or *binomially distributed*, and the corresponding overall experiment is sometimes referred to as a *binomial trial*. The expectation and variance of X are given by

$$E(X) = nq \tag{A.98}$$

$$V(X) = nq(1-q) \tag{A.99}$$

Example A.11. *Coin flipping (continuation): numerical demonstration of equivalence of expectation and variance calculation results over sample space and random variable pmf.* Suppose that a sequence of $n = 3$ independent Bernoulli trials are performed with $Y_k \sim$ bernoulli(0.5), $k = 1, 2, 3$, and let $X = \sum_{k=1}^{3} Y_k$ where $Y_k = 1$ if the event of interest occurs and $Y_k = 0$ otherwise. This is an extension of the previous coin flipping example to three tosses, with the event of interest being a Head. The experiment random outcome is $S = (Y_1, Y_2, Y_3)$, and the sample space, S, consists of all possible outcomes of which there are $P'(2, 3) = 2^3 = 8$. Because $\text{Pr}\{Y_k = 0\} = \text{Pr}\{Y_k = 1\} = 0.5$, the probability of any given outcome $s \in S$ is $p(s) = 0.5 \cdot 0.5 \cdot 0.5 = 0.125$. The sample space outcomes, outcome probabilities, and associated values of the random variable X are tabled below.

Outcome			
ID	s	$p(s)$	$X(s)$
1	(0, 0, 0)	0.125	0
2	(1, 0, 0)	0.125	1
3	(0, 1, 0)	0.125	1
4	(0, 0, 1)	0.125	1
5	(1, 1, 0)	0.125	2
6	(1, 0, 1)	0.125	2
7	(0, 1, 1)	0.125	2
8	(1, 1, 1)	0.125	3
Sum:		1.000	

Direct calculation of $E(X)$ and $V(X)$ over the sample space gives

$$E(X) = \sum_{s \in S} X(s)p(s) = 0.125[0+1+1+1+2+2+2+3] = 1.5$$

$$V(X) = \sum_{s \in S}[X(s) - E(X)]^2 p(s) = 0.125[(-1.5)^2 + (-0.5)^2 + (-0.5)^2 + (-0.5)^2$$
$$+ (0.5)^2 + (0.5)^2 + (0.5)^2 + (1.5)^2] = 0.75$$

$E(X)$ and $V(X)$ can also be calculated over the pmf of the random variable X. From the Example A.11 table, the pmf of X, $p(x)$, is given by

x	$p(x)$
0	0.125
1	0.375
2	0.375
3	0.125
Sum:	1.000

Calculation of $E(X)$ and $V(X)$ over the pmf of X gives

$$E(X) = \sum_x xp(x) = (0 \cdot 0.125) + (1 \cdot 0.375) + (2 \cdot 0.375) + (3 \cdot 0.125) = 1.5$$

$$V(X) = \sum_x [x - E(X)]^2 p(x) = (-1.5^2 \cdot 0.125) + (-0.5^2 \cdot 0.375)$$
$$+ (0.5^2 \cdot 0.375) + (1.5^2 \cdot 0.125) = 0.75$$

in agreement with the values of $E(X)$ and $V(X)$ calculated directly over the sample space.

Of course, given the construction of this example, $X \sim$ binomial$(3, 0.5)$, with the tabled $p(x)$ values matching those provided by Equation (A.97), and the calculated values of $E(X)$ and $V(X)$ previously found matching results generated by Equations (A.98) and (A.99): $E(X) = nq = 3 \cdot 0.5 = 1.5$, and $V(X) = nq(1-q) = 3 \cdot 0.5 \cdot 0.5 = 0.75$.

Example A.12. *SWR sample.* For an equal probability with replacement (SWR) sample of size n, if X is equal to the number of defined "type-1" items obtained from a population of size N that consists of N_1 type-1 items, then $X \sim$ binomial$(n; N_1/N)$.

In Example A.11, if instead of a sequence of three Bernoulli trials there was a population consisting of $N = 2$ items of which $N_1 = 1$ was of type-1, then an equal probability with replacement sample of size $n = 3$ would result in the same ordered sample space and binomial pmf for X with $q = N_1/N = 0.5$. For $n = 3$ and $N > 2$ such that $N_1/N = 0.5$, the sample space would be larger than for the previous example, but the pmf for X would be identical.

A.4.5 *Multinomial*

A **multinomial distribution** is a generalization of the binomial distribution to characterize multiple events of interest. It is the probability distribution of a **multivariate discrete random variable** (vector), $\mathbf{X} = (X_1, X_2, \ldots, X_m)$, that is equal to the number of times each of the m events of interest occur in a sequence of n *independent* multinoulli$(q_1, q_2, \ldots, q_m)$ trials. For independent $\mathbf{Y}_k \sim$ multinoulli$(q_1, q_2, \ldots, q_m)$, $k = 1, 2, \ldots, n$, the discrete random variable $\mathbf{X} = \sum_{k=1}^{n} \mathbf{Y}_k$ is vector-valued and has pmf

$$p(\mathbf{x}) = p(x_1, x_2, \ldots, x_m) = \binom{n}{x_1, x_2, \ldots, x_m} q_1^{x_1} q_2^{x_2} \cdots q_m^{x_m}, \quad x_i = 0, 1, \ldots, n; \sum_{i=1}^{m} x_i = n \quad \text{(A.100)}$$

Proof. $q_1^{x_1} q_2^{x_2} \cdots q_m^{x_m}$ is the probability of n multinoulli trials resulting in a *particular* sequence of outcomes consisting of $(x_1, x_2, \ldots, x_m)$ events (e.g., event 1 occurs in first x_1 trials, event 2 occurs in next x_2 trials, ..., and event m occurs in last x_m trials). There are $\binom{n}{x_1, x_2, \ldots, x_m}$ such distinct sequences (partitions) possible (Section A.1.4). Note that the sum of the pmf probabilities equals 1, as required for a proper random variable: $\sum p(\mathbf{x}) = \sum \binom{n}{x_1, x_2, \ldots, x_m} q_1^{x_1} q_2^{x_2} \cdots q_m^{x_m} = (q_1 + q_2 + \cdots + q_m)^n = 1$ (multinomial theorem; see Section A.1.4) where the sum is over all possible $\{x_i \geq 0\}$ such that $\sum_i x_i = n$. ∎

This specification is often written as $\mathbf{X} \sim \text{multinomial}(n; q_1, q_2, \ldots, q_m)$, and $\mathbf{X}$ is said to be a *multinomial random variable* or *multinomially distributed*, and the corresponding overall experiment is sometimes referred to as a *multinomial trial*. The expectation and variance of X_i, and the covariance of X_i and X_j for $i \neq j$ are given by

$$E(X_i) = nq_i \tag{A.101}$$

$$V(X_i) = nq_i(1 - q_i) \tag{A.102}$$

$$\text{Cov}(X_i, X_j) = -nq_i q_j, \quad i \neq j \tag{A.103}$$

Example A.13. *SWR sample (continuation).* For an equal probability with replacement (SWR) sample of size n, if $\mathbf{X} = (X_1, X_2, \ldots, X_m)$ is equal to the number of defined "type-i" items, $i = 1, 2, \ldots, m$, obtained from a population of size N that consists of $(N_1, N_2, \ldots, N_m)$ items of each type, with $N = \sum_{i=1}^{m} N_i$, then $\mathbf{X} \sim \text{multinomial}(n; N_1/N, N_2/N, \ldots, N_m/N)$.

A.4.6 Hypergeometric

A **hypergeometric distribution** is the probability distribution of a discrete random variable, X, that is equal to the number of defined "type-1" items obtained in a simple random sample (SRS) of size n (Section 3.1) selected from a population of size N that consists of N_1 type-1 items and $N - N_1$ other items. The discrete random variable X takes on values ranging over the non-negative integers $0, 1, \ldots, n$, subject to being no less than $n - (N - N_1)$ and no more than N_1. The pmf of X is

$$p(x) = \frac{\binom{N_1}{x} \binom{N - N_1}{n - x}}{\binom{N}{n}}, \quad x = \max\{0, n - (N - N_1)\}, \ldots, \min\{n, N_1\}. \tag{A.104}$$

Proof. There are a total of $\binom{N}{n}$ possible without replacement samples (combinations of n from N items), all of which are equally likely. The total number of unordered samples that include x type-1 items and $n - x$ other items is, by the multiplication principle: $\binom{N_1}{x}$ possible combinations of x from N_1 type-1 items multiplied by $\binom{N - N_1}{n - x}$ possible combinations of $n - x$ from $N - N_1$ other items. The ratio of these two quantities is $p(x)$. ∎

This specification is often written as $X \sim \text{hypergeometric}(n; N, N_1)$, and X is said to be a *hypergeometric random variable*. The expectation and variance of X are given by

$$E(X) = n\frac{N_1}{N} \tag{A.105}$$

$$V(X) = n\frac{N_1}{N}\left(1 - \frac{N_1}{N}\right)\left(\frac{N-n}{N-1}\right) \tag{A.106}$$

Example A.14. *SRS estimation of a proportion.* The population proportion of type-1 items is $\pi = N_1/N$. With simple random sampling, an estimator for π is the sample proportion of type-1 items, $\hat{\pi} = X/n$, which therefore has pmf $p_{\hat{\pi}}(x/n) = p(x)$. From Equations (A.105) and (A.106), $E(\hat{\pi}) = E(X)/n = \pi$ and $V(\hat{\pi}) = V(X)/n^2 = [\pi(1-\pi)/n][(N-n)/(N-1)]$, in agreement with the results derived for $\hat{\pi}$ in Section 3.12 [Equation (3.18)].

A.4.7 *Multivariate hypergeometric*

A **multivariate hypergeometric distribution** is a generalization of the hypergeometric distribution to characterize multiple item types. It is the probability distribution of a multivariate discrete random variable (vector), $\mathbf{X} = (X_1, X_2, \ldots, X_m)$, that is equal to the number of defined "type-i" items, $i = 1, 2, \ldots, m$, obtained in an SRS of size n selected from a population of size N that consists of $(N_1, N_2, \ldots, N_m)$ items of each type, respectively $(N = \sum_{i=1}^{m} N_i)$. The random variable X_i takes on values ranging over the non-negative integers $0, 1, \ldots, n$, subject to $X_i \le N_i$ and $\sum_{i=1}^{m} X_i = n$. The pmf of $\mathbf{X}$ is

$$p(\mathbf{x}) = p(x_1, x_2, \ldots, x_m) = \frac{\binom{N_1}{x_1}\binom{N_2}{x_2}\cdots\binom{N_m}{x_m}}{\binom{N}{n}}, \quad x_i \le N_i; \sum_{i=1}^{m} x_i = n$$

Proof. There are a total of $\binom{N}{n}$ possible without replacement samples (combinations of n from N items), all of which are equally likely. Similarly, there are $\binom{N_i}{x_i}$ possible combinations of x_i from N_i items, so the multiplication principle gives $\prod_{i=1}^{m}\binom{N_i}{x_i}$ as the total number of combinations resulting in $\{x_i\}$, all equally likely. The ratio of these two quantities is $p(\mathbf{x})$. ∎

This specification is often written as $\mathbf{X} \sim$ multivariate hypergeometric$(n; N_1, N_2, \ldots, N_m)$, and $\mathbf{X}$ is said to be a *multivariate hypergeometric random variable*. Because the sampling is without replacement, the expectation and variance of X_i, and the covariance of X_i and X_j are given by

$$E(X_i) = n\frac{N_i}{N} \tag{A.107}$$

$$V(X_i) = n\frac{N_i}{N}\left(1 - \frac{N_i}{N}\right)\left(\frac{N-n}{N-1}\right) \tag{A.108}$$

$$\text{Cov}(X_i, X_j) = -n\frac{N_i}{N}\frac{N_j}{N}\left(\frac{N-n}{N-1}\right), \quad i \ne j \tag{A.109}$$

A.5 Population variables

A.5.1 *Definition*

For a finite population of units, there may be many unit-level variables of interest. For example, for a population of N mallard ducks on a refuge, the variables of weight, length, age, and gender may be of interest for an assessment of body condition. Typically, one or more of these **population variables** may be of primary interest. These we refer to as *target variables*. Other *auxiliary variables* may be associated with the same population units (here, ducks). In the context of probability sampling, we regard the unit variable values as *fixed* rather than *random* (e.g., the age or weight of an individual mallard duck at a point in time is a fixed value, not a random value).

Lower case letters toward the end of the alphabet are typically used to denote these population variables, which are then indexed over the population units. For example, for a population of N units, we denote a particular target variable by y with unit-specific values $y_1, y_2, \ldots, y_N$.

A.5.2 *Population parameters*

A population parameter is a function of the target population variable values $y_1, y_2, \ldots, y_N$, and in some cases the auxiliary variable values $x_1, x_2, \ldots, x_N$ as well. The primary population parameters of interest are defined in this section. These definitions are consistent with the definitions for the expectation, variance, and covariance of random variables provided in Sections A.3.4–A.3.6 if the pmfs $p(y)$ and $p(x, y)$ are instead interpreted as relative frequency functions, with $p(y_i) = p(x_i, y_i) = 1/N, i = 1, 2, \ldots, N$.

For a single population variable, y, the **population mean** is defined as

$$\mu_y = \frac{\sum_{i=1}^{N} y_i}{N} \tag{A.110}$$

In the case of a binary categorical variable, where the two possible states (e.g., male and female) are scored as 0 and 1, respectively, the population mean is equivalent to the **population proportion** of units scored as 1

$$\pi_y = \frac{\sum_{i=1}^{N} y_i}{N} \quad \text{where} \quad y_i = \begin{cases} 1, & \text{if category of interest} \\ 0, & \text{otherwise} \end{cases} \tag{A.111}$$

Another parameter often of interest in finite population surveys is the **population total**

$$T_y = \sum_{i=1}^{N} y_i = N\mu_y \tag{A.112}$$

The **population variance** is defined as

$$V(y) = \frac{\sum_{i=1}^{N} (y_i - \mu_y)^2}{N} \tag{A.113}$$

We also define the closely related population parameter, σ_y^2, which we refer to as the **finite population variance**, to distinguish it from the population variance, though we recognize that both parameters refer to a finite population.

$$\sigma_y^2 = \frac{\sum_{i=1}^{N} (y_i - \mu_y)^2}{N-1} = \frac{N}{N-1} V(y) \tag{A.114}$$

The **population coefficient of variation** is a measure of variation relative to the mean, and is defined as

$$CV(y) = \frac{\sqrt{V(y)}}{\mu_y} \tag{A.115}$$

and often expressed as a percentage. Note that $CV(y)$ is not defined in the special case where $\mu_y = 0$, nor is it commonly used in the case where $\mu_y < 0$.

For two population variables, x and y, the **population ratio** is defined as

$$R = \frac{\sum_{i=1}^{N} y_i}{\sum_{i=1}^{N} x_i} = \frac{T_y}{T_x} = \frac{\mu_y}{\mu_x} \tag{A.116}$$

The **population covariance** is defined as

$$\text{Cov}(x,y) = \frac{\sum_{i=1}^{N}(x_i - \mu_x)(y_i - \mu_y)}{N} \tag{A.117}$$

We also define the closely related population parameter, $\sigma_{x,y}$, which we refer to as the **finite population covariance**, to distinguish it from the population covariance, again recognizing that both parameters refer to a finite population.

$$\sigma_{x,y} = \frac{\sum_{i=1}^{N}(x_i - \mu_x)(y_i - \mu_y)}{N-1} = \frac{N}{N-1}\text{Cov}(x,y) \tag{A.118}$$

Finally, the **population correlation** and **finite population correlation** are equivalent, and defined as

$$\text{Cor}(x,y) = \frac{\text{Cov}(x,y)}{\sqrt{V(x)V(y)}} = \frac{\sigma_{x,y}}{\sigma_x \sigma_y} \tag{A.119}$$

Note that some sampling theorists define σ_y^2 as equal to $V(y)$, and also $\sigma_{x,y}$ as equal to $\text{Cov}(x,y)$. We choose to define σ_y^2 and $\sigma_{x,y}$ as in Equations (A.114) and (A.118), respectively, while retaining the closely related definitions of $V(y)$ and $\text{Cov}(x,y)$, because these definitions allow for simpler formulations of finite population sampling theory results. Note also that in the sampling theory literature, the parameters $(\mu_y, \pi_y, \mathcal{T}_y, \sigma_y^2, \sigma_{x,y})$ are often denoted as $(\bar{Y}, P, Y, S^2, S_{yx})$ or variants thereof.

A.6 Probability sampling

A.6.1 *Sampling experiment*

In **probability sampling**, all population units have a positive probability (greater than zero) of being included in a sample, what is referred to in Section A.6.3 as an inclusion probability. Because every unit has a chance of being selected, and because the inclusion of a unit in a sample depends on a random mechanism (is subject to chance), selection of a probability sample is thus a type of random experiment (Section A.2.1), in this case a **sampling experiment**.

The finite population of N units are labeled $i = 1, 2, \ldots, N$. An outcome of the "sampling experiment", a sample s, is a combination of n units (labels) selected from these N population units (labels). Note that s may include repetitions of individual unit labels when units are selected with replacement. The set of all possible samples under the selection method used constitutes the sample space, $\mathcal{S}$, and often differs between selection methods. The random outcome of the sampling experiment, S, is equal to one of these possible samples, $S = s$, where $s \in \mathcal{S}$. Thus, S is a *set-valued* random outcome. The sampling method used, along with the sampling frame (Section 2.2) to which it is applied, are together referred to as the **sampling design**. (In this appendix, we assume that a *simple frame* is used, a frame equivalent to the listing of population units, $i = 1, 2, \ldots, N$.) The specific sampling design used determines the probability distribution of S, $\{p(s), s \in \mathcal{S}\}$, a consequence of the experiment's random mechanism. We refer to selection of a sample according to $\{p(s), s \in \mathcal{S}\}$, $\sum_{s \in \mathcal{S}} p(s) = 1$, as *probability sampling*.

Example A.15. *Sampling with and without replacement.* The size of the sample space differs for with and without replacement sampling. For sampling without replacement, the number of possible combinations *without repetition* (number of distinct samples) is $C(N, n) = \binom{N}{n}$. For sampling with replacement, the number of possible combinations *with repetition* is $C'(N, n) = \binom{n+N-1}{n}$.

A.6.2 *Sampling designs*

There are many different probability sampling designs—the details of which are considered throughout this book—that may be advantageous to use in one or more application settings, but there are certain basic characteristics that they all share when sampling from a finite population. First, the set of all possible samples, $\mathcal{S}$, that could be selected using a particular design is well defined and can be specified *a priori*. Second, the probability that a particular sample s is the one actually selected, $p(s)$, may vary among $s \in \mathcal{S}$ but it is specifiable for all $s \in \mathcal{S}$ and conforms to the definition of a pmf with

$$0 < p(s) < 1, \ s \in \mathcal{S} \quad \text{and} \quad \sum_{s \in \mathcal{S}} p(s) = 1 \tag{A.120}$$

where $p(s)$ is determined by the sampling design. Third, each population unit (or sampling unit in a *complex frame*, Section 2.2) is included in at least one of the possible samples. That is, each population unit has a chance of being selected and included in the sample, i.e., its inclusion probability (Section A.6.3) is greater than zero.

A.6.3 *Inclusion probabilities*

A sample **inclusion probability** refers to the probability that a particular unit, or a particular pair of units, is included in a probability sample selected according to a particular sampling design with associated sample selection probabilities, $\{p(s), s \in \mathcal{S}\}$. For a given sampling design, the **first order inclusion probability** for unit i, denoted as $\pi_i, \ i = 1, 2, \ldots, N$, is defined as the probability of the event that unit i is included in the random sample S. This inclusion probability is equal to the sum of the sample selection probabilities, $p(s)$, of all of those samples in $\mathcal{S}$ that include unit i

$$\pi_i = \Pr\{\text{unit } i \text{ in random sample } S\} = \sum_{\substack{s \in \mathcal{S} \\ i \in s}} p(s) \tag{A.121}$$

where the summation is over all the particular samples s in $\mathcal{S}$ that include unit i.

The **second order inclusion probability** for units i and j, denoted as $\pi_{ij}, \ i = 1, 2, \ldots, N$, $j = 1, 2, \ldots, N$, is defined as the probability of the event that units i and j are both included in the random sample S. This inclusion probability is equal to the sum of the sample selection probabilities, $p(s)$, of all of those samples in $\mathcal{S}$ that include both unit i and unit j

$$\pi_{ij} = \Pr\{\text{unit } i \text{ and unit } j \text{ in random sample } S\} = \sum_{\substack{s \in \mathcal{S} \\ i,j \in s}} p(s) \tag{A.122}$$

where the summation is over all the particular samples s in $\mathcal{S}$ that include unit i and unit j. Note that $\pi_{ii} = \pi_i$. The sampling design thus determines through its set of possible samples and their associated probabilities, $\{p(s), s \in \mathcal{S}\}$, the sets of first and second order inclusion probabilities $\{\pi_i\}$ and $\{\pi_{ij}\}$, respectively.

A.6.4 *Inclusion indicator variables*

When deriving the properties of estimators, it is often convenient to do so using indicator variables to denote whether or not a unit, or pair of units, is included in a sample. In particular, the expectation and variance of estimators can often be reduced to a linear function of the expectation, variance, and covariance of the sample inclusion indicator variables, which in turn depend on the associated inclusion probabilities.

The sample **inclusion indicator variable** for unit i, indicates whether or not unit i is included in the random sample S. It is thus a random variable (because S is random) and for completeness would be denoted $I_{i\in S}$ (Section A.3.8), but we will use the simpler notation I_i unless it is ambiguous.

$$I_i = \begin{cases} 1 & \text{if } i \in S \\ 0 & \text{otherwise} \end{cases} \tag{A.123}$$

For a particular sample $s \in \mathcal{S}$, we'll write $I_i(s)$ with $I_i(s) = 1$ if $i \in s$ and $I_i(s) = 0$ otherwise. An inclusion indicator variable, $I_{i,j}$, can also be used to identify whether or not both units i and j appear in a sample (Section A.3.8).

These sample inclusion indicator variables are linked to the sample inclusion probabilities as follows

$$\Pr\{I_i = 1\} = \pi_i \tag{A.124}$$

$$\Pr\{I_{i,j} = 1\} = \Pr\{I_i = 1, I_j = 1\} = \pi_{ij} \tag{A.125}$$

with the first equality of the latter equation following from Equation (A.78). Therefore, from Section A.3.8, $I_i \sim \text{bernoulli}(\pi_i)$, $i = 1, 2, \ldots, N$ and

$$E(I_i) = \pi_i \tag{A.126}$$

$$V(I_i) = \pi_i(1 - \pi_i) \tag{A.127}$$

$$\text{Cov}(I_i, I_j) = \pi_{ij} - \pi_i \pi_j \tag{A.128}$$

The sample inclusion indicator variables and probabilities are linked in turn to the sample size. The sum of the sample inclusion indicator variables over the population units, by definition, is equal to the sample size, $\sum_{i=1}^{N} I_i = n$, which may be a random variable depending on the sampling design. Therefore

$$\sum_{i=1}^{N} \pi_i = \sum_{i=1}^{N} E(I_i) = E\left(\sum_{i=1}^{N} I_i\right) = E(n) \tag{A.129}$$

For fixed sample size designs, n is not a random variable and this leads to the following relationships between the inclusion probabilities and n that can be useful for establishing estimator properties.

$$\sum_{i=1}^{N} \pi_i = n, \quad \sum_{i=1}^{N}\sum_{j\neq i}^{N} \pi_{ij} = n(n-1), \quad \sum_{\substack{i=1 \\ i\neq j}}^{N} \pi_{ij} = (n-1)\pi_j \tag{A.130}$$

Thus, for example, with fixed sample size designs

$$\sum_{i=1}^{N} \text{Cov}(I_i, I_j) = \sum_{i=1}^{N}\left[\pi_{ij} - \pi_i\pi_j\right] = \left[\sum_{i=1}^{N}\pi_{ij}\right] - \left[\pi_j\sum_{i=1}^{N}\pi_i\right] = \left[(n-1)\pi_j + \pi_j\right] - \left[n\pi_j\right] = 0 \tag{A.131}$$

A.7 Estimators

A.7.1 *Definition*

A **statistic** is a real-valued function of sample data, and in sampling theory an **estimator** is a statistic that is used to estimate a population parameter. In general, we denote an estimator of an arbitrary population parameter θ as $\hat{\theta}$. Once a specific sample of units is selected $(S = s)$, it is assumed that the target variable value (y) for the selected units can then be measured and that auxiliary information (x) in the sampling frame may also be available, so that an estimator (formula) can then be applied to the sample data to generate a numerical **estimate**, $\hat{\theta}(s)$. For example, suppose $\theta = \mu$. The mean-per-unit (mpu) estimator for θ is $\hat{\theta} = \sum_{i \in S} y_i / n$, which depends on S. Given a realized sample selection $S = s$, the mpu estimate of θ would be calculated as $\hat{\theta}(s) = \sum_{i \in s} y_i / n$ using the sample s unit values of y. Although the y values are considered to be fixed values for the population units, an estimator is a random variable because of its dependence on the random set of units selected, S. Being a random variable, $\hat{\theta}$ has a probability distribution—the distribution of $\hat{\theta}$ over all possible samples that could be selected—which is referred to as its sampling distribution. The sampling distribution of $\hat{\theta}$ is key to establishing the statistical performance of $\hat{\theta}$ as an estimator of θ.

A.7.2 *Sampling distribution*

In design-based sampling theory, the pmf of an estimator is termed its **sampling distribution** because the distribution of the estimator's possible values is induced through the random (probability) sampling mechanism (the sampling design). An estimator's sampling distribution depends on, at a minimum, the frequency distribution of the population variable, the sampling design including the sample size, and the form of the estimator. As noted in Chapter 2, we refer to the sampling design and estimator jointly as the **sampling strategy**.

Formally, the sampling distribution of $\hat{\theta}$ consists of the unique values of $\{\hat{\theta}(s), s \in S\}$ and their associated probabilities under the sampling design (the $p(s)$ summed over the $s \in S$ that result in the same value of $\hat{\theta}$). Letting $\hat{\Theta} = \{\hat{\theta}_1, \hat{\theta}_2, \ldots, \hat{\theta}_K\}$ denote the set of unique-valued $\hat{\theta}$ over the sample space, the sampling distribution of an estimator $\hat{\theta}$ is given by

$$p(\hat{\theta}_k) = \Pr\{\hat{\theta} = \hat{\theta}_k\} = \sum_{\substack{s \in S \\ \hat{\theta}(s) = \hat{\theta}_k}} p(s) \quad \text{for } \hat{\theta}_k \in \hat{\Theta} \tag{A.132}$$

where the summation is over those samples s in S for which $\hat{\theta}(s) = \hat{\theta}_k$.

Specification of the sampling distribution of an estimator requires that the frequency distribution of the population target variable (y) be known in order to determine $\{\hat{\theta}(s), s \in S\}$, and thus can be empirically established only in a hypothetical setting. Indeed, throughout this text, we provide illustrative small sample space examples $(y, S, \{p(s)\}$ and $\{\hat{\theta}(s)\}$ for small N and n) to illustrate the performance of alternative sampling strategies. However, in most real-life settings, this can be computationally impractical due to the very large number of random samples possible under most sampling designs. For example, for an SRS of $n = 10$ units selected from a population of $N = 100$ units, the number of possible samples is on the order of 10^{13}! In such settings, **Monte Carlo simulation** can instead be used to simulate the sampling distribution of an estimator as follows (assuming a simple frame and no reliance on auxiliary variables).

1. Specify the population units and target variable values $(y_1, y_2, \ldots, y_N)$.
2. Following a specified sample selection method, draw a random sample of units $S = s$, and record the associated variable values.
3. Calculate $\hat{\theta}(s)$.
4. Repeat steps 2 and 3 a large number of times (independent trials).

By the law of large numbers (Section A.2.3), the distribution of the resulting $\{\hat{\theta}(s_t), t = 1, 2, \ldots, \#\text{trials}\}$ approximates the sampling distribution of $\hat{\theta}$. The approximation tends to improve as the number of trials is increased, but it is often adequate for many purposes with as few as 10^4 or 10^5 trials.

A.7.3 *Statistical properties*

In this section we define the statistical properties of sampling distributions that are of greatest importance in assessing the expected performance of an estimator $\hat{\theta}$. Two fundamental metrics relating to the "location" and "spread" of the sampling distribution are, respectively, the expectation, $E(\hat{\theta})$, and sampling variance, $V(\hat{\theta})$. These metrics can be defined over the pmf, as for any other discrete random variable (Sections A.3.4 and A.3.5), as

$$E(\hat{\theta}) = \sum_{\hat{\theta}_k \in \hat{\Theta}} \hat{\theta}_k p(\hat{\theta}_k) \tag{A.133}$$

$$V(\hat{\theta}) = E\left([\hat{\theta} - E(\hat{\theta})]^2\right) = \sum_{\hat{\theta}_k \in \hat{\Theta}} [\hat{\theta}_k - E(\hat{\theta})]^2 p(\hat{\theta}_k) \tag{A.134}$$

where $\hat{\Theta} = \{\hat{\theta}_1, \hat{\theta}_2, \ldots, \hat{\theta}_K\}$ denotes the set of unique-valued $\hat{\theta}$ over the sample space. Alternatively, expectation and sampling variance can be defined directly over the sample space (Sections A.3.4 and A.3.5) as

$$E(\hat{\theta}) = \sum_{s \in \mathcal{S}} \hat{\theta}(s) p(s) \tag{A.135}$$

$$V(\hat{\theta}) = \sum_{s \in \mathcal{S}} [\hat{\theta}(s) - E(\hat{\theta})]^2 p(s) \tag{A.136}$$

The sample space formulation of expectation and variance is more often used in the development of sampling theory. Expectation is a measure of the average value of the estimator, and the sampling variance is a measure that is inversely related to estimator **precision**.

Two other metrics of fundamental importance for an estimator are its bias and mean square error. The **bias**, $B(\hat{\theta})$, is defined as

$$B(\hat{\theta}) = E(\hat{\theta}) - \theta \tag{A.137}$$

Bias measures the extent to which the expected value of $\hat{\theta}$ differs from the target of estimation. An estimator can be positively biased ($B(\hat{\theta}) > 0$), negatively biased ($B(\hat{\theta}) < 0$), or unbiased ($B(\hat{\theta}) = 0$). Note well that estimator bias refers to its expected value over all possible samples—individual estimates are neither "biased" nor "unbiased", although they can be overestimates or underestimates, and typically are. It is desirable that an estimator be unbiased regardless of the sample size, or at least have little bias if the sample size is large, otherwise it will not be possible to construct valid confidence intervals associated with individual estimates (Section A.7.4).

The **mean square error**, $\text{MSE}(\hat{\theta})$, is defined as

$$\text{MSE}(\hat{\theta}) = E\left([\hat{\theta} - \theta]^2\right) = \sum_{\hat{\theta}_k \in \hat{\Theta}} [\hat{\theta}_k - \theta]^2 p(\hat{\theta}_k) = \sum_{s \in \mathcal{S}} [\hat{\theta}(s) - \theta]^2 p(s) \tag{A.138}$$

Mean square error measures the average squared deviation of the possible estimates, not from their expected value, but from θ itself, and is thus a measure that is inversely related to estimator *accuracy*. If an estimator is unbiased then its expected value is equal to θ, in which case $V(\hat{\theta})$ is equivalent to $\text{MSE}(\hat{\theta})$. In general, however

$$\mathrm{MSE}(\hat{\theta}) = V(\hat{\theta}) + B(\hat{\theta})^2 \qquad \text{(A.139)}$$

Proof.

$$\mathrm{MSE}(\hat{\theta}) = E\left([\hat{\theta} - \theta]^2\right) = E(\hat{\theta}^2) - 2\theta E(\hat{\theta}) + \theta^2$$

$$= E(\hat{\theta}^2) - [E(\hat{\theta})]^2 + [E(\hat{\theta})]^2 - 2\theta E(\hat{\theta}) + \theta^2$$

$$= E(\hat{\theta}^2) - [E(\hat{\theta})]^2 + [E(\hat{\theta}) - \theta]^2 = V(\hat{\theta}) + B(\hat{\theta})^2$$

∎

Two other related estimator metrics are the **standard error**, $\mathrm{SE}(\hat{\theta})$, and the **coefficient of variation**, $\mathrm{CV}(\hat{\theta})$, which are defined as

$$\mathrm{SE}(\hat{\theta}) = \sqrt{V(\hat{\theta})} \qquad \text{(A.140)}$$

$$\mathrm{CV}(\hat{\theta}) = \mathrm{SE}(\hat{\theta})/E(\hat{\theta}) \qquad \text{(A.141)}$$

with $\mathrm{SE}(\hat{\theta})$ a measure inversely related to precision on the $\hat{\theta}$ scale, and $\mathrm{CV}(\hat{\theta})$ this same measure of precision but relative to the expected value of $\hat{\theta}$ (and often expressed as a percentage). Note that $\mathrm{CV}(\hat{\theta})$ is not defined in the special case where $E(\hat{\theta}) = 0$, nor is it commonly used in the case where $E(\hat{\theta}) < 0$. Confidence intervals (Section A.7.4) are typically constructed based on estimates of $\mathrm{SE}(\hat{\theta})$. Sample size is often set so as to achieve a specified $\mathrm{CV}(\hat{\theta})$ target.

A.7.4 *Confidence intervals*

The **central limit theorem** (**CLT**) is a fundamental result in the theory of statistics, and its importance for sampling theory cannot be overstated. It states that, in general, the distribution of the sample mean converges to a normal (Gaussian) distribution as $n \to \infty$ *regardless of the shape of the sampled distribution*. Specifically, in the CLT's most basic sampling setting, if a random sample $Y_1, Y_2, \ldots, Y_n$ is taken from an arbitrary distribution in such a way that the $\{Y_i\}$ are independent random variables having a common mean, μ, and finite variance, σ^2, then the estimator $\hat{\mu} = \sum_{i=1}^{n} Y_i/n$ has expectation μ and variance $V(\hat{\mu}) = \sigma^2/n$, and the CLT provides that as $n \to \infty$ the standardized estimator converges to the standard normal distribution

$$\frac{\hat{\mu} - \mu}{\sqrt{V(\hat{\mu})}} \xrightarrow{D} \mathrm{normal}(0, 1) \qquad \text{(A.142)}$$

where the symbol $\xrightarrow{D}$ means "converges in distribution to", and $\mathrm{normal}(0, 1)$ denotes a normal distribution with mean 0 and variance 1. The speed of convergence to the $\mathrm{normal}(0, 1)$ distribution with increasing n depends on the shape of the Y distribution. When Y is highly skewed, for example, the distribution of $\hat{\mu}$ will converge more slowly than when Y is symmetrically distributed.

This setting, however, does not quite match most survey sampling settings because the sampled population is finite and there is usually also a lack of independence between the selected units (unless unit selection is with-replacement). Unequal probability selection methods and complex sampling designs further complicate matters. In these settings, one approach used to investigate the asymptotic distribution of an estimator is to mathematically construct a sequence of finite populations of increasing size with $N \to \infty$, and to assume that both $n \to \infty$ and $N - n \to \infty$ (Lehmann 1998). Using this and other approaches, suitably modified versions of the CLT have been shown to apply in a number of finite population sampling settings (e.g., Hájek 1960, Madow 1948, Prášková and

Sen 2009, Scott and Wu 1981), and together with other Monte Carlo simulation study experience, there is now general comfort in assuming with finite population sampling and an estimator $\hat{\theta}$ of $\theta = \mu, \pi,$ or T that asymptotically

$$\frac{\hat{\theta} - \theta}{\sqrt{V(\hat{\theta})}} \xrightarrow{D} \text{normal}(0, 1) \tag{A.143}$$

Typically, for $n \geq 30$ the distribution of $\hat{\theta}$ is fairly well approximated by the normal distribution, and sometimes for values of n much less than this (see Figure 3.3).

Moreover, so long as a **consistent estimator**, $\hat{V}(\hat{\theta})$, of $V(\hat{\theta})$ is available, then it is also true (Slutzky's theorem) that asymptotically

$$\frac{\hat{\theta} - \theta}{\sqrt{\hat{V}(\hat{\theta})}} \xrightarrow{D} \text{normal}(0, 1) \tag{A.144}$$

A consistent estimator is one that converges in probability to the quantity being estimated as $n \to \infty$. That is, for any particular value of $c > 0$, $\Pr\{|\hat{V}(\hat{\theta}) - V(\hat{\theta})| < c\} \to 1$ as $n \to \infty$, which states that no matter how small c is, it is virtually certain that $\hat{V}(\hat{\theta})$ differs from $V(\hat{\theta})$ by less than c if n is large enough. A sufficient condition for this to be true is if $\text{MSE}[\hat{V}(\hat{\theta})] \to 0$ as $n \to \infty$.

Assuming that n is large, these results enable construction of a confidence interval for θ from a sample in the usual way as

$$\hat{\theta} \pm z_{1-\alpha/2}\sqrt{\hat{V}(\hat{\theta})} \tag{A.145}$$

where z_v is the quantile for a normal$(0, 1)$ random variable Z such that $P(Z \leq z_v) = v$, and $1 - \alpha$ is the associated confidence level. For example, for a 95% confidence interval, $\alpha = 0.05$ and $z_{1-\alpha/2} = z_{0.975} = 1.96$. The confidence level is, of course, approximate due to the use of the normal approximation. The interpretation of a confidence interval in the context of probability sampling designs is that the confidence level is equal to the probability of having selected one of the samples for which the associated confidence interval contains θ. Section 3.4.2 presents methods for construction of 95% confidence intervals, based on the t-statistic, when $n < 30$, again assuming that the sampling distribution of the estimator is approximately normal in shape. If the adequacy of the normal approximation is in doubt, one can simulate the actual sampling distribution of $\hat{\theta}$ for a given sampling design, assuming a specific population variable distribution, as discussed in Section A.7.2.

In sampling from a finite population of size N, an estimator $\hat{\theta}$ of a population parameter θ is **finite population consistent** if $\hat{\theta} = \theta$ for $n = N$. For example, with simple random sampling, the mean-per-unit estimator of the population mean, $\hat{\mu}_y = \sum_{i \in s} y_i/n$, is finite population consistent because for $n = N$, $\hat{\mu}_y = \sum_{i=1}^{N} y_i/N = \mu_y$. Similarly, $\hat{\sigma}_y^2 = \sum_{i \in s}(y_i - \hat{\mu}_y)^2/(n-1)$ is finite population consistent for σ_y^2 which has a divisor of $N - 1$, but not for $V(y)$ which has a divisor of N.

A.8 Delta method

In many contexts, it is difficult or impossible to obtain an exact expression for the variance of a function of one or more random variables and instead an approximation is made. One very useful method for doing this is commonly called the **delta method**, and is based on a Taylor series first order linear approximation of the function. As long as the function is smooth (differentiable) and the approximation is adequate over the range of the component random variables, the delta method should provide an adequate approximation of the variance of the function, and also provides an estimator for this variance.

We illustrate the method by first applying it to a function of a single random variable, $g(X)$. The Taylor series first order approximation of $g(X)$ at the point $X = \mu_X$ is

$$g(X) \approx g(\mu_X) + \left(\frac{dg(X)}{dX}\right)(X - \mu_X) \tag{A.146}$$

where $dg(X)/dX$ is the *derivative* of $g(X)$ with respect to X, and is evaluated at $X = \mu_X$. Note that this equation is linear in X. The variance of $g(X)$ can then be approximated as the variance of this linear function

$$V[g(X)] \approx V\left[g(\mu_X) + \left(\frac{dg(X)}{dX}\right)(X - \mu_X)\right] = \left(\frac{dg(X)}{dX}\right)^2 V(X) \tag{A.147}$$

which follows from application of Equations (A.51) and (A.52) with μ_X, $g(\mu_X)$, and $dg(X)/dX$ evaluated at $X = \mu_X$, all being constants.

Similarly, for a function of two random variables, $g(X,Y)$, the Taylor series linear approximation about the point $(X,Y) = (\mu_X, \mu_Y)$, and its variance, are given by

$$g(X,Y) \approx g(\mu_X, \mu_Y) + \left(\frac{\partial g(X,Y)}{\partial X}\right)(X - \mu_X) + \left(\frac{\partial g(X,Y)}{\partial Y}\right)(Y - \mu_Y) \tag{A.148}$$

$$V[g(X,Y)] \approx \left(\frac{\partial g(X,Y)}{\partial X}\right)^2 V(X) + \left(\frac{\partial g(X,Y)}{\partial Y}\right)^2 V(Y)$$
$$+ 2\left(\frac{\partial g(X,Y)}{\partial X}\right)\left(\frac{\partial g(X,Y)}{\partial Y}\right)\text{Cov}(X,Y) \tag{A.149}$$

where $\partial g(X,Y)/\partial X$ and $\partial g(X,Y)/\partial Y$ are the *partial derivatives* of $g(X,Y)$ with respect to X and Y, and both are evaluated at $(X,Y) = (\mu_X, \mu_Y)$, with the variance following from application of Equation (A.56).

The method can be extended to a function of any number of random variables, including estimators. Let $g(\hat{\boldsymbol{\theta}})$ be the function of interest with vector-valued $\hat{\boldsymbol{\theta}} = (\hat{\theta}_1, \hat{\theta}_2, \ldots, \hat{\theta}_k)$. Application of the delta method gives

$$V[g(\hat{\boldsymbol{\theta}})] \approx \sum_{i=1}^{k}\left(\frac{\partial g(\hat{\boldsymbol{\theta}})}{\partial \hat{\theta}_i}\right)^2 V(\hat{\theta}_i) + \sum_{i=1}^{k}\sum_{j\neq i}^{k}\left(\frac{\partial g(\hat{\boldsymbol{\theta}})}{\partial \hat{\theta}_i}\right)\left(\frac{\partial g(\hat{\boldsymbol{\theta}})}{\partial \hat{\theta}_j}\right)\text{Cov}(\hat{\theta}_i, \hat{\theta}_j) \tag{A.150}$$

where $\partial g(\hat{\boldsymbol{\theta}})/\partial \hat{\theta}_i$ is the *partial derivative* of $g(\hat{\boldsymbol{\theta}})$ with respect to $\hat{\theta}_i$, and the partial derivatives are evaluated at the expected values of the component estimators.

The covariance of two functions of estimators, $g(\hat{\boldsymbol{\theta}})$ and $h(\hat{\boldsymbol{\theta}})$, that have one or more component estimators in common can be approximated as

$$\text{Cov}[g(\hat{\boldsymbol{\theta}}), h(\hat{\boldsymbol{\theta}})] \approx \sum_{i=1}^{k}\left(\frac{\partial g(\hat{\boldsymbol{\theta}})}{\partial \hat{\theta}_i}\right)\left(\frac{\partial h(\hat{\boldsymbol{\theta}})}{\partial \hat{\theta}_i}\right)V(\hat{\theta}_i) + \sum_{i=1}^{k}\sum_{j\neq i}^{k}\left(\frac{\partial g(\hat{\boldsymbol{\theta}})}{\partial \hat{\theta}_i}\right)\left(\frac{\partial h(\hat{\boldsymbol{\theta}})}{\partial \hat{\theta}_j}\right)\text{Cov}(\hat{\theta}_i, \hat{\theta}_j) \tag{A.151}$$

For estimators of $V[g(\hat{\boldsymbol{\theta}})]$ and $\text{Cov}[g(\hat{\boldsymbol{\theta}}), h(\hat{\boldsymbol{\theta}})]$, substitute estimators of the component variances and covariances into Equations (A.150) and (A.151) and evaluate the partial derivatives at the sample estimates $\hat{\boldsymbol{\theta}}(s) = [\hat{\theta}_1(s), \hat{\theta}_2(s), \ldots, \hat{\theta}_k(s)]$

$$\hat{V}[g(\hat{\boldsymbol{\theta}})] = \sum_{i=1}^{k}\left(\frac{\partial g(\hat{\boldsymbol{\theta}})}{\partial \hat{\theta}_i}\right)^2 \hat{V}(\hat{\theta}_i) + \sum_{i=1}^{k}\sum_{j\neq i}^{k}\left(\frac{\partial g(\hat{\boldsymbol{\theta}})}{\partial \hat{\theta}_i}\right)\left(\frac{\partial h(\hat{\boldsymbol{\theta}})}{\partial \hat{\theta}_j}\right)\widehat{\text{Cov}}(\hat{\theta}_i, \hat{\theta}_j) \tag{A.152}$$

$$\widehat{\mathrm{Cov}}[g(\hat{\boldsymbol{\theta}}), h(\hat{\boldsymbol{\theta}})] = \sum_{i=1}^{k} \left(\frac{\partial g(\hat{\boldsymbol{\theta}})}{\partial \hat{\theta}_i} \right) \left(\frac{\partial h(\hat{\boldsymbol{\theta}})}{\partial \hat{\theta}_i} \right) \hat{V}(\hat{\theta}_i) + \sum_{i=1}^{k} \sum_{j \neq i}^{k} \left(\frac{\partial g(\hat{\boldsymbol{\theta}})}{\partial \hat{\theta}_i} \right) \left(\frac{\partial h(\hat{\boldsymbol{\theta}})}{\partial \hat{\theta}_j} \right) \widehat{\mathrm{Cov}}(\hat{\theta}_i, \hat{\theta}_j) \quad \text{(A.153)}$$

Example A.16. *Approximate sampling variance of a product estimator.* Suppose that two estimators, $\hat{\mu}_y$ and $\hat{\mu}_x$, are based on independent samples, and consider the sampling variance of their product $\hat{\mu}_y \hat{\mu}_x$. Using the general notation, $\hat{\boldsymbol{\theta}} = (\hat{\theta}_1, \hat{\theta}_2) = (\hat{\mu}_y, \hat{\mu}_x)$, $g(\hat{\boldsymbol{\theta}}) = \hat{\theta}_1 \hat{\theta}_2$, and we want $V[g(\hat{\boldsymbol{\theta}})]$. Appealing to Equation (A.150), $\mathrm{Cov}(\hat{\theta}_1, \hat{\theta}_2) = 0$ given the independence of the samples, and $\partial g(\hat{\boldsymbol{\theta}})/\partial \hat{\theta}_1 = \hat{\theta}_2$, $\partial g(\hat{\boldsymbol{\theta}})/\partial \hat{\theta}_2 = \hat{\theta}_1$. Evaluating the partial derivatives at the expected values of the component estimators and applying Equation (A.150) gives

$$V(\hat{\mu}_y \hat{\mu}_x) \approx [E(\hat{\mu}_x)]^2 V(\hat{\mu}_y) + [E(\hat{\mu}_y)]^2 V(\hat{\mu}_x) \quad \text{(A.154)}$$

Note that this approximate result is less than the exact variance of a product of independent random variables [Equation (A.58)] in that it omits the term $V(\hat{\mu}_y) V(\hat{\mu}_x)$ from the sum. Goodman (1960) showed that the accuracy of this approximate variance formula for a product in independent random variables depends on the coefficients of variation of the component estimators.

A.9 Lagrange multipliers

The method of **Lagrange multipliers** is designed to find the critical values leading to the extrema (minima/maxima) of a differentiable function of several variables subject to constraints, or side conditions. That is, among the set of variable values satisfying the constraints, which yield the minima and maxima of the function? We outline the method for problems involving a single equality constraint.

We consider a function of several variables, $f(x_1, x_2, \ldots, x_m)$, subject to an equality constraint $g(x_1, x_2, \ldots, x_m) = 0$. Ignoring the constraint for a moment, recall that unconstrained function minimization/maximization problems are solved by partially differentiating the function f with respect to each of the component variables, and then setting each partial derivative equal to zero

$$\frac{\partial f(x_1, x_2, \ldots, x_m)}{\partial x_1} = 0, \quad \frac{\partial f(x_1, x_2, \ldots, x_m)}{\partial x_2} = 0, \quad \ldots, \quad \frac{\partial f(x_1, x_2, \ldots, x_m)}{\partial x_m} = 0 \quad \text{(A.155)}$$

Solving this system of m equations for the m unknowns $x_1, x_2, \ldots, x_m$ yields the critical values. Now, if the constraint $g(x_1, x_2, \ldots, x_m) = 0$ is also included, this results in a system of $m + 1$ equations involving m unknowns.

The method of Lagrange multipliers was developed specifically for this type of problem. The first step is to define the Lagrange function, $\mathcal{L}$, as

$$\mathcal{L}(x_1, x_2, \ldots, x_m, \lambda) = f(x_1, x_2, \ldots, x_m) + \lambda g(x_1, x_2, \ldots, x_m) \quad \text{(A.156)}$$

where λ denotes the Lagrange multiplier. The second step is to partially differentiate the Lagrange function with respect to $x_1, x_2, \ldots, x_m$ and λ, and set each result equal to zero, yielding a system of $m + 1$ equations involving $m + 1$ unknowns. The last of these equations, $\partial \mathcal{L}(x_1, x_2, \ldots, x_m, \lambda)/\partial \lambda = g(x_1, x_2, \ldots, x_m) = 0$, recovers the constraint. The third step is to solve the system of equations for the unknowns $(x_1, x_2, \ldots, x_m, \lambda)$, which gives the critical values. The last step is to evaluate whether a critical value corresponds to a minimum or maximum, and whether it is a global minimum or maximum. The most direct way of doing this is to simply evaluate $f(x_1, x_2, \ldots, x_m)$ over the set of critical values.

Example A.17. *Minimization of a paraboloid subject to a planar constraint.* Consider the function $f(x_1, x_2) = 2x_1^2 + x_2^2$, an elliptic paraboloid. Find the (x_1, x_2) values that minimize f subject to the constraint that $x_1 + x_2 = 10$.

To solve this using the method of Lagrange multipliers, specify the constraint as $g(x_1, x_2) = x_1 + x_2 - 10 = 0$, and the Lagrange function as

$$\mathcal{L}(x_1, x_2, \lambda) = 2x_1^2 + x_2^2 + \lambda(x_1 + x_2 - 10)$$

Find the partial derivatives of the function $\mathcal{L}$ with respect to x_1, x_2, and λ, and set these derivatives equal to zero

$$\frac{\partial \mathcal{L}(x_1, x_2, \lambda)}{\partial x_1} = 4x_1 + \lambda = 0$$

$$\frac{\partial \mathcal{L}(x_1, x_2, \lambda)}{\partial x_2} = 2x_2 + \lambda = 0$$

$$\frac{\partial \mathcal{L}(x_1, x_2, \lambda)}{\partial \lambda} = x_1 + x_2 - 10 = 0$$

The first two equations imply $x_2 = 2x_1$. Substituting $2x_1$ for x_2 in the third equation gives $x_1 = 10/3$, and thus $x_2 = 20/3$. By inspection of the form of f, it is clear that $f(x_1, x_2) \geq 0$, with an unconstrained minimum value of zero at $(x_1, x_2) = (0, 0)$, and is unbounded above. Therefore, the critical value $(10/3, 20/3)$ yields the constrained minimum value, $f(10/3, 20/3) = 600/9 = 200/3$.

References

Adams, P.B., Boydstun, L.B., Gallagher, S.P., Lacy, M.K., McDonald, T., and Shaffer, K.E. 2011. "California coastal salmonid population monitoring: Strategy, design, and methods". California Department of Fish and Game, *Fish Bulletin* 180.

Amstrup, S.C., McDonald, T.L., and Manly, B.F.J. 2005. *Handbook of Capture-Recapture Analysis*. Princeton University Press, Princeton.

Aronow, P.M. and Samii, C. 2013. "Conservative variance estimation for sampling designs with zero pairwise inclusion probabilities". *Survey Methodology* **39**(1): 231–241.

Bart, J. and Beyer, H.L. 2012. "Analysis options for estimating status and trends in long-term monitoring". In *Design and Analysis of Long-term Ecological Monitoring Studies*, edited by R.A. Gitzen, J.J. Millspaugh, A.B. Cooper, and D.S. Licht, Cambridge University Press, New York, pp. 253–278.

Basu, D. 1958. "On sampling with and without replacement". *Sankhya* **20**(3/4): 287–294.

Bellhouse, D.R. 1988. "A brief history of random sampling methods". In *Handbook of Statistics*, volume 6, edited by P.R. Krishnaiah and C.R. Rao, Elsevier, pp. 1–14.

Berger, Y.G. 1998. "Rate of convergence to normal distribution for the Horvitz–Thompson estimator". *Journal of Statistical Planning and Inference* **67**(2): 209–226.

Berger, Y.G. 2005. "Variance estimation with Chao's sampling scheme". *Journal of Statistical Planning and Inference* **127**(1–2): 253–277.

Berger, Y.G. and Tillé, Y. 2009. "Sampling with unequal probabilities". In *Handbook of Statistics*, volume 29A, edited by D. Pfeffermann and C.R. Rao, Elsevier, Netherlands, pp. 39–54.

Blackwell, D. 1947. "Conditional expectation and unbiased sequential estimation". *The Annals of Mathematical Statistics* **18**(1): 105–110.

Bohrnstedt, G.W. and Goldberger, A.S. 1969. "On the exact covariance of products of random variables". *Journal of the American Statistical Association* **64**(328): 1439–1442.

Bowley, A.L. 1926. "Measurement of the precision attained in sampling". *Bulletin of the International Statistical Institute* **22**(Part 3): 6–62.

Brewer, K. and Gregoire, T.G. 2009. "Introduction to survey sampling". In *Handbook of Statistics*, volume 29A, edited by D. Pfeffermann and C.R. Rao, Elsevier, pp. 9–37.

Brewer, K.R.W. 1963. "Ratio estimation and finite populations: some results deducible from the assumption of an underlying stochastic process". *The Australian Journal of Statistics* **5**(3): 93–105.

Brewer, K.R.W. 2002. *Combined Survey Sampling Inference: Weighing Basu's Elephants*. Oxford University Press, New York.

Brewer, K.R.W. and Hanif, M. 1983. "Sampling with Unequal Probabilities", *Lecture Notes in Statistics*, volume 15. Springer-Verlag, New York.

Carlton, M.A. and Devore, J.L. 2017. "Probability with applications in engineering, science, and technology". *Springer Texts in Statistics*. Springer International Publishing, second edition.

Cassel, C.M., Särndal, C.E., and Wretman, J.H. 1977. *Foundations of Inference in Survey Sampling*. Wiley, New York.

Chambers, R.L. and Clark, R.G. 2012. "An Introduction to Model-Based Survey Sampling with Applications", *Oxford Statistical Science Series*, volume 37. Oxford University Press, New York.

Chao, M.T. 1982. "A general purpose unequal probability sampling plan". *Biometrika* **69**(3): 653–656.

Chaudhary, M.A. and Sen, P.K. 2002. "Reconciliation of asymptotics for unequal probability sampling without replacement". *Journal of Statistical Planning and Inference* **102**(1): 71–81.

Chaudhuri, A. and Stenger, H. 2005. *Survey Sampling: Theory and Methods*. Chapman & Hall/CRC, Boca Raton, second edition.

Cochran, W.G. 1946. "Relative accuracy of systematic and stratified random samples for a certain class of populations". *The Annals of Mathematical Statistics* **17**(2): 164–177.

Cochran, W.G. 1953. *Sampling Techniques*. Wiley, New York, first edition.

Cochran, W.G. 1977. *Sampling Techniques*. Wiley, New York, third edition.

Cordy, C.B. 1993. "An extension of the Horvitz–Thompson theorem to point sampling from a continuous universe". *Statistics & Probability Letters* **18**(5): 353–362.

van der Corput, J.C. 1935. "Verteilungsfunktionen. (erste mitteilung)". Proceedings of the Royal Academy of Science in Amsterdam **38**(8): 813–821.

Courbois, J.Y.P. and Urquhart, N.S. 2004. "Comparison of survey estimates of the finite population variance". *Journal of Agricultural, Biological and Environmental Statistics* **9**(2): 236–251.

Deming, W.E. 1950. *Some Theory of Sampling*. Wiley, New York.

Deville, J.C. and Tillé, Y. 2004. "Efficient balanced sampling: The cube method". *Biometrika* **91**(4): 893–912.

Everhart, W.H. and Youngs, W.D. 1975. *Principles of Fishery Science*. Cornell University Press, Ithaca, first edition.

Fancy, S.G. and Bennetts, R.E. 2012. "Institutionalizing an effective long-term monitoring program in the US National Park Service". In *Design and Analysis of Long-term Ecological Monitoring Studies*, edited by R.A. Gitzen, J.J. Millspaugh, A.B. Cooper, and D.S. Licht, Cambridge University Press, New York, pp. 481–497.

Fisher, R.A. 1959. *Statistical Methods and Scientic Inference*. Hafner, New York, second edition.

Godambe, V.P. 1955. "A unified theory of sampling from finite populations". *Journal of the Royal Statistical Society*. Series B (Methodological) **17**(2): 269–278.

Goodman, L.A. 1960. "On the exact variance of products". *Journal of the American Statistical Association* **55**(292): 708–713.

Grafström, A., Lundström, N.L.P., and Schelin, L. 2012. "Spatially balanced sampling through the pivotal method". *Biometrics* **68**(2): 514–520.

Gray, B.R. 2012. "Variance components estimation for continuous and discrete data, with emphasis on cross-classified sampling designs". In *Design and Analysis of Long-term Ecological Monitoring Studies*, edited by R.A. Gitzen, J.J. Millspaugh, A.B. Cooper, and D.S. Licht, Cambridge University Press, New York, pp. 200–227.

Gray, G. 1999. "Covariances in multiplicative estimates". *Transactions of the American Fisheries Society* **128**(3): 475–482.

Gregoire, T.G. and Valentine, H.T. 2008. *Sampling Strategies for Natural Resources and the Environment*. Chapman & Hall/CRC, Boca Raton.

Haines, D.E. and Pollock, K.H. 1998. "Estimating the number of active and successful bald eagle nests: an application of the dual frame method". *Environmental and Ecological Statistics* **5**(3): 245–256.

Hájek, J. 1960. "Limiting distributions in simple random sampling from a finite population". *Publication of the Mathematical Institute of the Hungarian Academy of Sciences* **5**: 361–374.

Halton, J.H. 1960. "On the efficiency of certain quasi-random sequences of points in evaluating multi-dimensional integrals". *Numerische Mathematik* **2**(1): 84–90.

Hanif, M. and Brewer, K.R.W. 1980. "Sampling with unequal probabilities without replacement: A review". *International Statistical Review* **48**(3): 317–335.

Hankin, D.G. 1980. "A multistage recruitment process in laboratory fish populations: Implications for models of fish population dynamics". *Fishery Bulletin* **78**(3): 555–578.

Hankin, D.G. 1984. "Multistage sampling designs in fisheries research: Applications in small streams". *Canadian Journal of Fisheries and Aquatic Sciences* **41**: 1575–1591.

Hankin, D.G., Mohr, M.S., and Voight, H. 2009. "Estimating the proportions of closely related species: Performance of the two-phase ratio estimator". *Journal of Agricultural, Biological and Environmental Statistics* **14**(1): 15–32.

Hansen, M.H. 1987. "Some history and reminiscences on survey sampling". *Statistical Science* **2**(2): 180–190.

Hansen, M.H., Dalenius, T., and Tepping, B.J. 1985. "The development of sample surveys of finite populations". In *A Celebration of Statistics: The ISI Centenary Volume*, edited by A.C. Atkinson and S.E. Fienberg, Springer-Verlag, New York, pp. 327–354.

Hansen, M.H. and Hurwitz, W.N. 1943. "On the theory of sampling from finite populations". *The Annals of Mathematical Statistics* **14**(4): 333–362.

Hansen, M.H., Hurwitz, W.N., and Madow, W.G. 1953a. *Sample Survey Methods and Theory*. Volume I: *Methods and Applications*. Wiley, New York.

Hansen, M.H., Hurwitz, W.N., and Madow, W.G. 1953b. *Sample Survey Methods and Theory*. Volume II: *Theory*. Wiley, New York.

Hansen, M.H., Madow, W.G., and Tepping, B.J. 1983a. "An evaluation of model-dependent and probability-sampling inferences in sample surveys". *Journal of the American Statistical Association* **78**(384): 776–793.

Hansen, M.H., Madow, W.G., and Tepping, B.J. 1983b. "An evaluation of model-dependent and probability-sampling inferences in sample surveys: Rejoinder". *Journal of the American Statistical Association* **78**(384): 805–807.

Hansen, M.M., Nielsen, E.E., and Mensberg, K.L.D. 1997. "The problem of sampling families rather than populations: relatedness among individuals ins samples of juvenile brown trout *Salmo trutta* L". *Molecular Ecology* **6**(5): 469–474.

Hartley, H.O. and Rao, J.N.K. 1962. "Sampling with unequal probailities and without replacement". *The Annals of Mathematical Statistics* **33**(2): 350–374.

Hartley, H.O. and Ross, A. 1954. "Unbiased ratio estimators". *Nature* **174**(4423): 270–271.

Hassan, Y., Shahbaz, M.Q., and Hanif, M. 2009. "Empirical comparison of some approximate variance formulae of Horvitz-Thompson estimator". *World Applied Statistics Journal* **7**(5): 597–599.

Haziza, D., Mecatti, F., and Rao, J.N.K. 2008. "Evaluation of some approximate variance estimators under the Rao-Sampford unequal probability sampling design". *Metrons–International Journal of Statistics* **LXVI**(1): 91–108.

Hedayat, A.S. and Sinha, B.K. 1991. *Deisgn and Inference in Finite Population Sampling*. Wiley, New York.

Henderson, T. 2006. "Estimating the variance of the Horvitz–Thompson estimator". Thesis, Bachelor of Commerce with Honours in Statistics, School of Finance and Applied Statistics, The Australian National University.

Holt, D. and Smith, T.M.F. 1979. "Post stratification". *Journal of the Royal Statistical Society*. Series A (General) **142**(1): 33–46.

Horvitz, D.G. and Thompson, D.J. 1952. "A generalization of sampling without replacement from a finite universe". *Journal of the American Statistical Association* **47**(260): 663–685.

Iachan, R. 1982. "Systematic sampling: A critical review". *International Statistical Review* **50**(3): 293–303.

Jacobs, S.E. and Cooney, C.X. 1997. "Oregon coastal salmon spawning surveys, 1994 and 1995". Information Report 97–5, Oregon Department of Fish and Wildlife.

Jacobs, S.E. and Nickelson, T.E. 1998. "Use of stratified random sampling to estimate the abundance of Oregon coastal coho salmon". Final Report, Fish Research Project F-145-R-09, Oregon Department of Fish and Wildlife.

Jessen, R.J. 1978. *Statistical Survey Techniques*. Wiley, New York.

Kendall, M.G. and Stuart, A. 1977. *The Advanced Theory of Statistics*. Volume 1: *Distribution Theory*. Macmillan, New York, fourth edition.

Kendall, M.G. and Stuart, A. 1979. *The Advanced Theory of Statistics*. Volume 2: *Inference and Relationship*. Macmillan, New York, fourth edition.

Kendall, M.G., Stuart, A., and Ord, J.K. 1983. *The Advanced Theory of Statistics*. Volume 3: *Design and Analysis, and Time-Series*. Macmillan, New York, fourth edition.

Kiaer, A.N. 1897. "The representative method of statistical surveys" (1976 English translation of the original Norwegian). Central Bureau of Statistics of Norway, Oslo.

Kincaid, T.M. and Olsen, A.R. 2018. spsurvey: "Spatial survey design and analysis". R package version 3.4. https://CRAN.R-project.org/package=spsurvey/.

Korn, E.L. and Graubard, B.I. 1999. *Analysis of Health Surveys*. Wiley, New York.

Krishnaiah, P.R. and Rao, C.R., editors 1988. "Sampling", *Handbook of Statistics*, volume 6. Elsevier, Netherlands.

Kruskal, W. and Mosteller, F. 1979a. "Representative sampling, i: Non-scientific literature". *International Statistical Review* **47**(1): 13–24.

Kruskal, W. and Mosteller, F. 1979b. "Representative sampling, ii: Scientific literature, excluding statistics". *International Statistical Review* **47**(2): 111–127.

Kruskal, W. and Mosteller, F. 1979c. "Representative sampling, iii: The current statistical literature". *International Statistical Review* **47**(3): 245–265.

Kruskal, W. and Mosteller, F. 1980. "Representative sampling, iv: The history of the concept in statistics, 1895–1939". *International Statistical Review* **48**(2): 169–195.

Kuhn, T.S. 1970. *The Structure of Scientific Revolutions*. University of Chicago Press, Chicago, second edition.

Lahiri, D.B. 1951. "A method of sample selection providing unbiased ratio estimates". *Bulletin of the International Statistical Institute* **33**(2): 133–140.

Lai, R. 2018. iterpc: "Efficient iterator for permutations and combinations". R package version 0.4.1. https://CRAN.R-project.org/package=iterpc/.

Lehmann, E.L. 1998. *Elements of Large-Sample Theory*. Springer-Verlag, New York.

Leu, C.H. and Tsui, K.W. 1996. "New partially systematic sampling". *Statistica Sinica* **6**(3): 617–630.

Lewis, J.B. and Linzer, D.A. 2005. "Estimating regression models in which the dependent variable is based on estimates". *Political Analysis* **13**(4): 345–364.

Likens, G.E. and Bailey, S.W. 2014. "The discovery of acid rain at the Hubbard Brook Experimental Forest: A story of collaboration and long-term research". In *USDA Forest Service Experimental Forests and Ranges: Research for the Long Term*, edited by D.C. Hayes, S.L. Stout, R.H. Crawford, and A.P. Hoover, Springer, pp. 463–482.

Lohr, S.L. 2009. "Multiple-frame surveys". In *Handbook of Statistics*, volume 29A, edited by D. Pfeffermann and C.R. Rao, Elsevier, Netherlands, pp. 71–88.

Lohr, S.L. 2010. *Sampling: Design and Analysis*. Brooks/Cole, Boston, second edition.

Lumley, T. and Scott, A. 2017. "Fitting regression models to survey data". *Statistical Science* **32**(2): 265–278.

Madow, W.G. 1948. "On the limiting distributions of estimates based on samples from finite universes". *The Annals of Mathematical Statistics* **19**(4): 535–545.

Madow, W.G. and Madow, L.H. 1944. "On the theory of systematic sampling, i". *The Annals of Mathematical Statistics* **15**(1): 1–24.

Mahalanobis, P.C. 1940. "A sample survey of the acreage under jute in Bengal". *Sankhya* **4**(4): 511–530.

Manly, B.F.J. and Navarro Alberto, J.A., editors 2015. *Introduction to Ecological Sampling*. CRC Press, Taylor & Francis Group, Boca Raton.

McDonald, T. 2012. "Spatial sampling designs for long-term ecological monitoring". In *Design and Analysis of Long-term Ecological Monitoring Studies*, edited by R.A. Gitzen, J.J. Millspaugh, A.B. Cooper, and D.S. Licht, Cambridge University Press, New York, pp. 101–125.

McDonald, T. 2015. "Sampling designs for environmental monitoring". In *Introduction to Ecological Sampling*, edited by B.F.J. Manly and J.A. Navarro Alberto, CRC Press, Taylor & Francis Group, Boca Raton, pp. 145–165.

McDonald, T. 2016. "SDraw: spatially balanced sample draws for spatial objects". R package version 2.1.3. https://CRAN.R-project.org/package=SDraw/.

McDonald, T.L. 2003. "Review of environmental monitoring methods: Survey designs". *Environmental Monitoring and Assessment* **85**(3): 277–292.

Midzuno, H. 1952. "On the sampling system with probability proportional to sum of sizes". *Annals of the Institute of Statistical Mathematics* **3**(2): 99–107.

Mier, K.L. and Picquelle, S.J. 2008. "Estimating abundance of spatially aggregated populations: comparing adaptive sampling with other survey designs". *Canadian Journal of Fisheries and Aquatic Sciences* **65**(2): 176–197.

Mostafa, S.A. and Ahmad, I.A. 2018. "Recent developments in systematic sampling: A review". *Journal of Statistical Theory and Practice* **12**(2): 290–310.

Mueller-Dombois, D. and Ellenberg, H. 1974. *Aims and Methods of Vegetation Ecology*. John Wiley & Sons, New York.

Murthy, M.N. 1957. "Ordered and unordered estimators in sampling without replacement". *Sankhyā: The Indian Journal of Statistics* **18**(3/4): 379–390.

Murthy, M.N. 1967. *Sampling Theory and Methods*. Statistical Publishing Society, Calcutta, India.

Nathan, G. 1988. "Inference based on data from complex sample designs". In *Handbook of Statistics*, volume 6, edited by P.R. Krishnaiah and C.R. Rao, Elsevier, Netherlands, pp. 247–266.

Nelson, G.A. 2014. "Cluster sampling: A pervasive, yet little recognized survey design in fisheries research". *Transactions of the American Fisheries Society* **143**(4): 926–938.

Newman, K.B., Buckland, S.T., Morgan, B.J.T., King, R., Borchers, D.L. Cole, D.J., Besbeas, P., Gimenez, O., and Thomas, L. 2014. *Modelling Population Dynamics: Model Formulation, Fitting and Assessment Using State-Space Methods*. Springer, New York.

Neyman, J. 1934. "On the two different aspects of the representative method: The method of stratified sampling and the method of purposive selection". *Journal of the Royal Statistical Society* **97**(4): 558–625.

Neyman, J. 1938. "Contribution to the theory of sampling human populations". *Journal of the American Statistical Association* **33**(201): 101–116.

Noon, B.R., Ishwar, N.M., and Vasudevan, K. 2006. "Efficiency of adaptive cluster and random sampling in detecting terrestrial herpetofauna in a tropical rainforest". *Wildlife Society Bulletin* **34**(1): 59–68.

Olsen, A.R., Kincaid, T.M., and Payton, Q. 2012. "Spatially balanced survey designs for natural resources". In *Design and Analysis of Long-term Ecological Monitoring Studies*, edited by R.A. Gitzen, J.J. Millspaugh, A.B. Cooper, and D.S. Licht, Cambridge University Press, New York, pp. 126–150.

Overton, W.S. and Stehman, S.V. 1995. "The Horvitz–Thompson theorem as a unifying perspective for probability sampling: With examples from natural resource sampling". *The American Statistician* **49**(3): 261–268.

Overton, W.S. and Stehman, S.V. 1996. "Desirable design characteristics for long-term monitoring of ecological variables". *Environmental and Ecological Statistics* **3**(4): 349–361.

Pasek, J. 2018. weights: weighting and weighted statistics. R package version 1.0. https://CRAN.R-project.org/package=weights/.

Pfeffermann, D. 1993. "The role of sampling weights when modeling survey data". *International Statistical Review* **61**(2): 317–337.

Pfeffermann, D. 1996. "The use of sampling weights for survey data analysis". *Statistical Methods in Medical Research* **5**(3): 239–261.

Pfeffermann, D. and Rao, C.R., editors 2009a. "Sample Surveys: Design, Methods and Applications", *Handbook of Statistics*, volume 29A. Elsevier, Netherlands.

Pfeffermann, D. and Rao, C.R., editors 2009b. *Sample surveys: Inference and analysis, Handbook of Statistics*, volume 29B. Elsevier, Netherlands.

Pradhan, B.K. 2013. "Three phase stratified sampling with ratio method of estimation". *Statistica* **73**(2): 235–251.

Prášková, Z. and Sen, P.K. 2009. "Asymptotics in finite population sampling". In *Handbook of Statistics*, volume 29B, edited by D. Pfeffermann and C.R. Rao, Elsevier, Netherlands, pp. 489–522.

R Core Team 2018. R: *A language and environment for statistical computing*. R Foundation for Statistical Computing, Vienna, Austria. https://www.R-project.org/.

Raj, D. 1968. *Sampling Theory*. McGraw-Hill, New York.

Rao, C.R. 1945. "Information and accuracy attainable in the estimation of statistical parameters". *Bulletin of the Calcutta Mathematical Society* **37**(3): 81–91.

Rao, J.N.K. 1973. "On double sampling for stratification and analytical surveys". *Biometrika* **60**(1): 125–133.

Rao, J.N.K. and Graham, J.E. 1964. "Rotation designs for sampling on repeated occasions". *Journal of the American Statistical Association* **59**(306): 492–509.

Rao, T.J. 1966. "On the variance of the ratio estimator for Midzuno-Sen sampling scheme". *Metrika* **10**(1): 89–91.

Reynolds, J.H., Knutson, M.G., Newman, K.B., Silverman, E.D., and Thompson, W.L. 2016. "A road map for designing and implementing a biological monitoring program". *Environmental Monitoring and Assessment* **188**(7): 399.

Reynolds, J.H. and Shelly, A. 2010. "Study design assessment for surveys of bald eagle nesting and productivity on Kodiak NWR". doi:10.13140/RG.2.2.26698.88009. Alaska Refuges Report Series no. 10-001. U.S. Fish and Wildlife Service, National Wildlife Refuge System, Anchorage, Alaska.

Robertson, B., McDonald, T., Price, C., and Brown, J. 2018. "Halton iterative partitioning: Spatially balanced sampling via partitioning". *Environmental and Ecological Statistics* **25**(3): 305–323.

Robertson, B.L., Brown, J.A., McDonald, T., and Jaksons, P. 2013. "Bas: Balanced acceptance sampling of natural resources". *Biometrics* **69**(3): 776–784.

Robertson, B.L., McDonald, T., Price, C.J., and Brown, J.A. 2017. "A modification of balanced acceptance sampling". *Statistics and Probability Letters* **129**: 107–112.

Rohatgi, V.K. and Ehsanes Saleh, A.K.M. 2015. *An Introduction to Probability and Statistics*. John Wiley and Sons, New Jersey, third edition.

Royall, R.M. 1970. "On finite population sampling theory under certain linear regression models". *Biometrika* **57**(2): 377–387.

Royall, R.M. 1983. "An evaluation of model-dependent and probability-sampling inferences in sample surveys: Comment". *Journal of the American Statistical Association* **78**(384): 794–796.

Royall, R.M. 1992. "The model based (prediction) approach to finite population sampling theory". In *Current Issues in Statistical Inference: Essays in Honor of D. Basu*, Lecture Notes-Monograph Series, volume 17, edited by M. Ghosh and P.K. Pathak, Institute of Mathematical Statistics, pp. 225–240.

Royall, R.M. 1994. "[Sample surveys 1975–1990; an age of reconciliation?]: Discussion". *International Statistical Review* **62**(1): 19–21.

Royall, R.M. and Herson, J. 1973. "Robust estimation in finite populations i". *Journal of the American Statistical Association* **68**(344): 880–889.

Salehi, M. and Brown, J.A. 2010. "Complete allocation sampling: An efficient and easily implemented adaptive sampling design". *Population Ecology* **52**(3): 451–456.

Salehi, M.M. 2003. "Comparison between Hansen–Hurwitz and Horvitz–Thompson estimators for adaptive cluster sampling". *Environmental and Ecological Statistics* **10**(1): 115–127.

Salehi, M.M. 2017. "Erratum: Two-stage complete allocation sampling". *Environmetrics* **28**(7). doi: 10.1002/env.2461.

Salehi, M.M. and Seber, G.A.F. 2017. "Two-stage complete allocation sampling". *Environmetrics* **28**(3). doi:10.1002/env.2441.

Särndal, C.E. 2010. "Models in survey sampling". In *Official Statistics—Methodology and Applications in Honour of Daniel Thorburn*, edited by M. Carlson, II. Nyquist, and M. Villani, Department of Statistics, Stockhom University, Stockholm, Sweden, pp. 15–27. Available at http://officialstatistics.wordpress.com.

Särndal, C.E. and Swensson, B. 1987. "A general view of estimation for two phases of selection with applications to two-phase sampling and nonresponse". *International Statistical Review* **55**(3): 279–294.

Särndal, C.E., Swensson, B., and Wretman, J. 1992. *Model Assisted Survey Sampling*. Springer-Verlag, New York.

Särndal, C.E. and Lundström, S. 2005. *Estimation in Surveys with Nonresponse*. Wiley, Hoboken, NJ.

Satterthwaite, F.E. 1946. "An approximate distribution of estimates of variance components". *Biometrics Bulletin* **2**(6): 110–114.

Schreuder, H.T., Gregoire, T.G., and Wood, G.B. 1993. *Sampling Methods for Multiresource Forest Inventory*. John Wiley and Sons, New York, first edition.

Scott, A. and Wu, C.F. 1981. "On the asymptotic distribution of ratio and regression estimators". *Journal of the American Statistical Association* **76**(373): 98–102.

Seber, G.A.F. 1982. *The Estimation of Animal Abundance and Related Parameters*. Macmillan Publishing Company, New York, second edition.

Seber, G.A.F. and Salehi, M.M. 2013. *Adaptive Sampling Designs: Inference for Sparse and Clustered Populations*. Springer, New York.

Sen, A.R. 1952. "Present status of probability sampling and its use in the estimation of farm characteristics [abstract]". *Econometrica* **20**(1): 103.

Sen, A.R. 1953. "On the estimate of variance in sampling with varying probabilities". *Journal of the Indian Society of Agricultural Statistics* **5**(2): 119–127.

Sengupta, S. 1989. "On Chao's unequal probability sampling plan". *Biometrika* **76**(1): 192–196.

Shyvers, J.E., Walker, B.L., and Noon, B.R. 2018. "Dual-frame lek surveys for estimating greater sage-grouse populations". *The Journal of Wildlife Management* **82**(8): 1689–1700.

Skalski, J.R. 2012. "Estimating variance components and related parameters when planning long-term monitoring programs". In *Design and Analysis of Long-term Ecological Monitoring Studies*, edited by R.A. Gitzen, J.J. Millspaugh, A.B. Cooper, and D.S. Licht, Cambridge University Press, New York, pp. 174–199.

Smith, T.M.F. 1976. "The foundations of survey sampling: A review (with discussion)". *Journal of the Royal Statistical Society.* Series A (General) **139**(2): 183–204.

Smith, T.M.F. 1978. "Principles and problems in the analysis of repeated surveys". In *Survey Sampling and Measurement*, edited by N. Krishnan Namboodiri, Academic Press, London, pp. 201–216.

Smith, T.M.F. 1991. "Post-stratification". *Journal of the Royal Statistical Society.* Series D (The Statistician) **40**(3): 315–323.

Smith, T.M.F. 1994. "Sample surveys 1975–1990; an age of reconciliation"? *International Statistical Review* **62**(1): 5–19.

Springer, M.D. 1979. *The Algebra of Random Variables.* John Wiley and Sons, New York.

Starcevich, L.A.H., McDonald, T., Chung-MacCoubrey, A., Heard, A., Nesmith, J., and Philippi, T. 2018. "Trend estimation for complex survey designs of water chemistry indicators from Sierra Nevada Lakes". *Environmental Monitoring and Assessment* **190**(596). doi:10.1007/s10661-018-6963-1.

Steel, D. and McLaren, C. 2009. "Design and analysis of surveys repeated over time". In *Handbook of Statistics*, volume 29B, edited by D. Pfeffermann and C.R. Rao, Elsevier, Netherlands, pp. 289–313.

Stehman, S.V. and Overton, W.S. 1994. "Comparison of variance estimators of the Horvitz–Thompson estimator for randomized variable probability systematic sampling". *Journal of the American Statistical Association* **89**(425): 30–43.

Stevens, Jr., D.L. 1997. "Variable density grid-based sampling designs for continuous spatial populations". *Environmetrics* **8**(3): 167–195.

Stevens, Jr., D.L. and Olsen, A.R. 2003. "Variance estimation for spatially balanced samples of environmental resources". *Environmetrics* **14**(6): 593–610.

Stevens, Jr., D.L. and Olsen, A.R. 2004. "Spatially balanced sampling of natural resources". *Journal of the American Statistical Association* **99**(465): 262–278.

Stevens, Jr., D.L. and Urquhart, N.S. 2000. "Response designs and support regions in sampling continuous domains". *Environmetrics* **11**(1): 13–41.

Stuart, A.S. 1984. *The Ideas of Sampling.* Oxford University Press, New York, third edition.

Su, Z. and Quinn, II, T.J. 2003. "Estimator bias and efficiency for adaptive cluster sampling with order statistics and a stopping rule". *Environmental and Ecological Statistics* **10**(1): 17–41.

Sunter, A. 1986. "Solutions to the problem of unequal probability sampling without replacement". *International Statistical Review* **54**(1): 33–50.

Sunter, A.B. 1977. "List sequential sampling with equal or unequal probabilities without replacement". *Journal of the Royal Statistical Society.* Series C (Applied Statistics) **26**(3): 261–268.

Thompson, S.K. 1990. "Adaptive cluster sampling". *Journal of the American Statistical Association* **85**(412): 1050–1059.

Thompson, S.K. 1991. "Adaptive cluster sampling: Designs with primary and secondary units". *Biometrics* **47**(3): 1103–1115.

Thompson, S.K. 1992. *Sampling.* John Wiley & Sons, Inc., New York, first edition.

Thompson, S.K. 1996. "Adaptive cluster sampling based on order statistics". Environmetrics 7(2): 123–133.

Thompson, S.K. 2002. *Sampling.* John Wiley & Sons, Inc., New York, second edition.

Thompson, S.K. 2012. *Sampling.* John Wiley & Sons, Inc., Hoboken, New Jersey, third edition.

Thompson, S.K. and Seber, G.A.F. 1996. *Adaptive Sampling.* John Wiley & Sons, Inc., New York.

Thompson, W.L., editor 2004. *Sampling Rare or Elusive Species: Concepts, Designs, and Techniques for Estimating Populations Parameters.* Island Press, Washington, D.C.

Tillé, Y. 1996. "Some remarks on unequal probability sampling designs without replacement". *Annales d'Economie et de Statistique* **44**: 177–189.

Tillé, Y. 2011. "Ten years of balanced sampling with the cube method: An appraisal". *Survey Methodology* **37**(2): 215–226.

Tout, J. 2009. "An analysis of the adaptive cluster sampling design with rare plant point distributions". Thesis, Master of Arts in Biological Studies, Humboldt State University, Arcata, CA, USA.

Turk, P. and Borkowski, J.J. 2005. "A review of adaptive cluster sampling: 1990–2003". *Environmental and Ecological Statistics* **12**(1): 55–94.

Turner, R. 2018. "deldir: Delaunay triangulation and Dirichlet (Voronoi) tessellation". R package version 0.1-15 https://CRAN.R-project.org/package=deldir/.

Urquhart, N.S. 2012. "The role of monitoring design in detecting trend". In *Design and Analysis of Long-term Ecological Monitoring Studies*, edited by R.A. Gitzen, J.J. Millspaugh, A.B. Cooper, and D.S. Licht, Cambridge University Press, New York, pp. 151–173.

Urquhart, N.S. and Kincaid, T.M. 1999. "Designs for detecting trend from repeated surveys of ecological resources". *Journal of Agricultural, Biological and Environmental Statistics* **4**(4): 404–414.

Valliant, R., Dorfman, A.H., and Royall, R.M. 2000. *Finite Population Sampling and Inference: A Prediction Approach*. John Wiley & Sons, Inc., New York.

Wakimoto, K. 1970. "On unbiased estimation of the population variance based on the stratified random sample". *Annals of the Institute of Statistical Mathematics* **22**(1): 15–26.

Welsh, Jr., H.H., Ollivier, L.M., and Hankin, D.G. 1997. "A habitat-based design for sampling and monitoring stream amphibians with an illustration from Redwood National Park". *Northwestern Naturalist* **78**(1): 1–16.

Williams, L.R., Warren, Jr., M.L., Adams, S.B., Arvai, J.L., and Taylor, C.M. 2004. "Basin Visual Estimation Technique (BVET) and representative reach approaches to wadeable stream surveys: methodological limitations and future directions". *Fisheries* **29**(8): 12–22.

Wolter, K.M. 2007. *Introduction to Variance Estimation*. Springer, New York, second edition.

Wright, T. 2012. "The equivalence of Neyman optimum allocation for sampling and equal proportions for apportioning the U.S. House of Representatives". *The American Statistician* **66**(4): 217–224.

Wright, T. 2014. "A simple method of exact optimal sample allocation under stratification with any mixed constraint patterns". U.S. Census Bureau, Center for Statistical Research and Methodology, Research Report Series, Statistics #2014-07.

Wright, T. 2016. "Two optimal exact sample allocation algorithms: Sampling variance decomposition is key". U.S. Census Bureau, Center for Statistical Research and Methodology, Research Report Series, Statistics #2016-03.

Yates, F. 1960. *Sampling Methods for Censuses and Surveys*. Hafner, New York, third edition.

Yates, F. and Grundy, P.M. 1953. "Selection without replacement from within strata with probability proportional to size". *Journal of the Royal Statistical Society*. Series B (Methodological) **15**(2): 253–261.

Zar, J.H. 2010. *Biostatistical Analysis*. Pearson Prentice Hall, Upper Saddle River, New Jersey, fifth edition.

Index